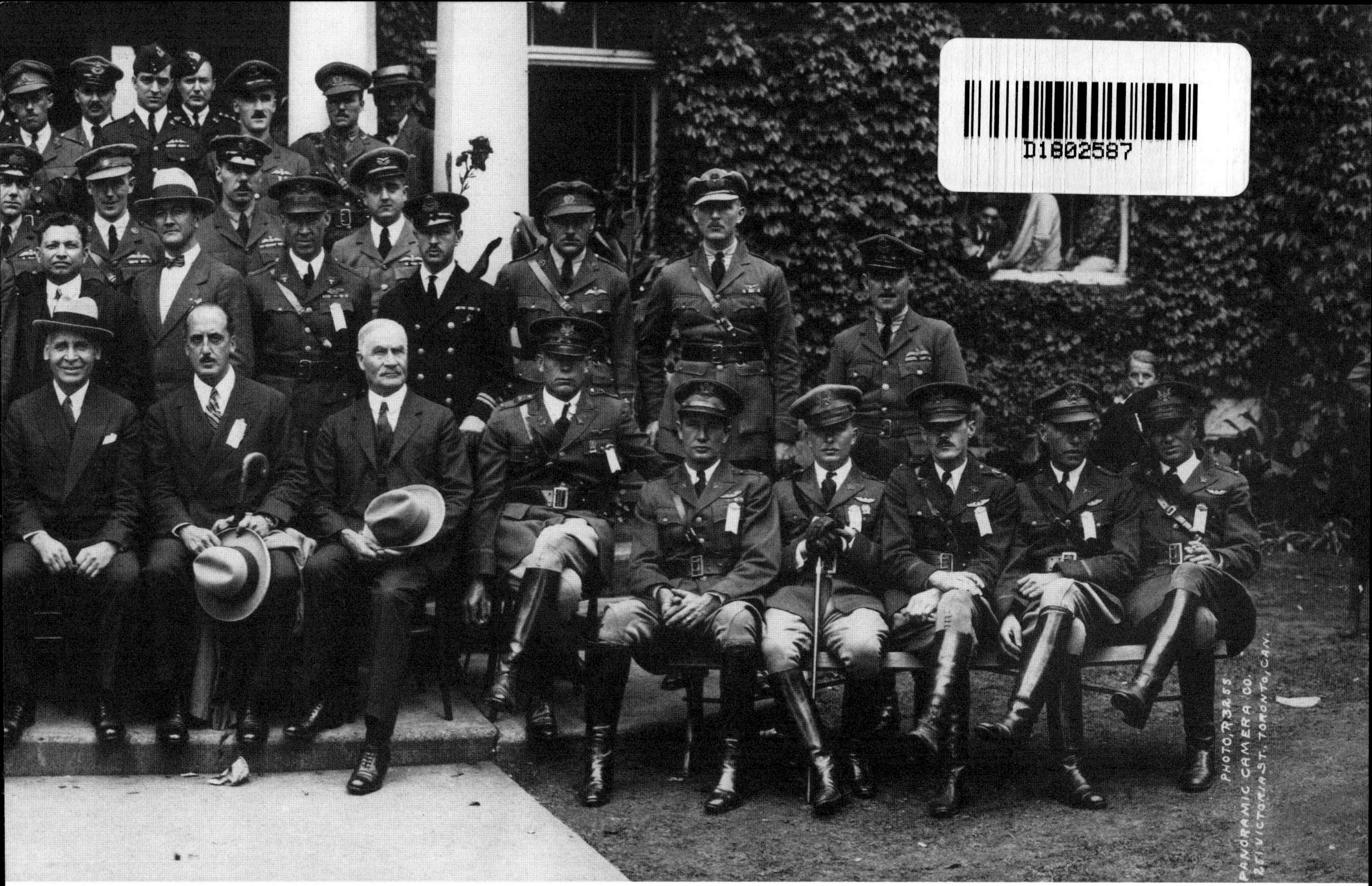
PHOTO, R3R55
PANORAMIC CAMERA CO.
281 VICTORIA ST., TORONTO, CAN.

G-CAHE
G

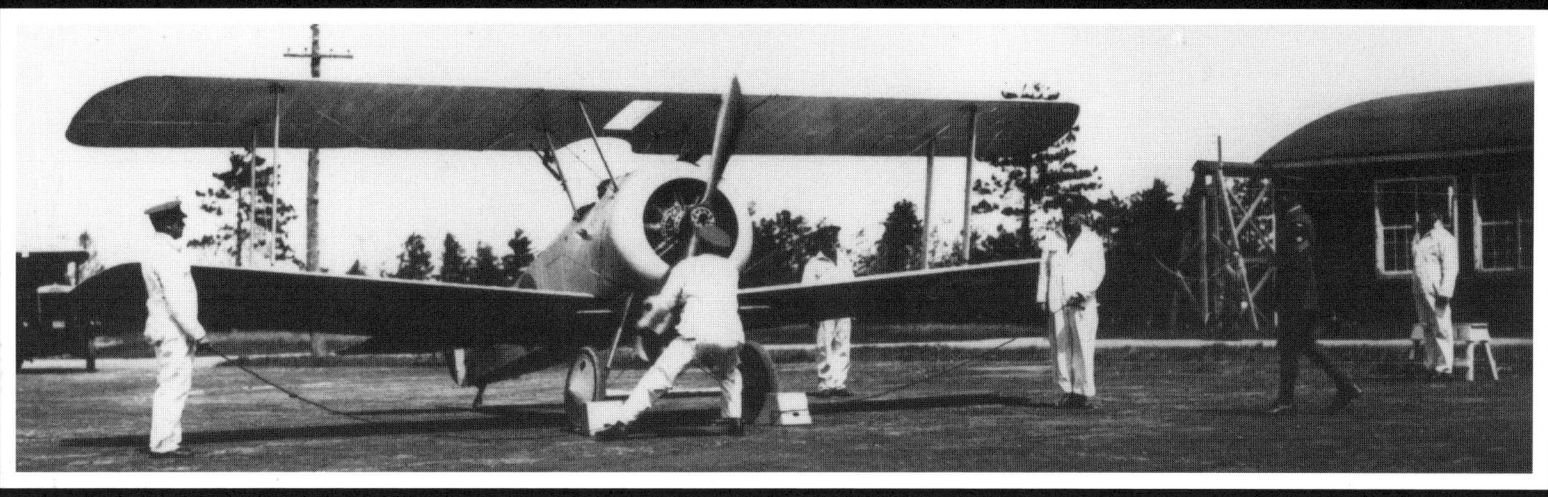

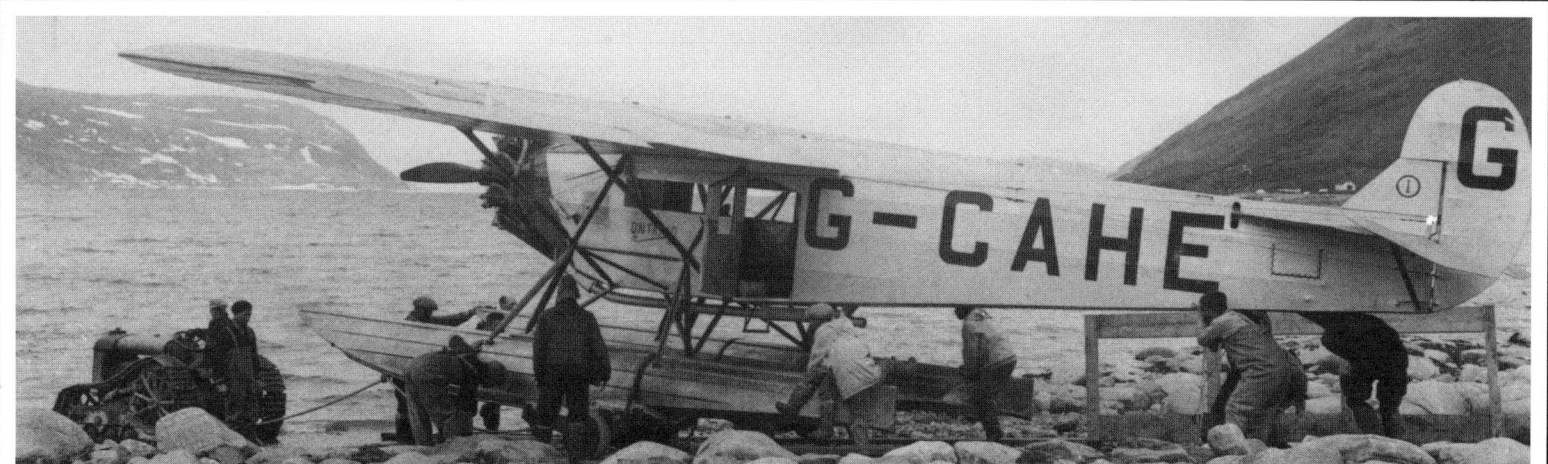

# CANADA'S AIR FORCE
## AT WAR AND PEACE

*Larry Milberry*

VOLUME ONE

# CANADA'S AIR FORCE
## AT WAR AND PEACE
### VOLUME ONE

LARRY MILBERRY

CANAV Books

*Front endpaper:* September 5, 1928 was the Canadian National Exhibition's first Aviation Day. The highlight was a good will visit from south of the border. Here everyone gathers at the CNE, the crowd including CNE and civic officials, RCAF and a US Army contingent. The US pilots had flown in with 20 Hawk fighters from Selfridge, Michigan, landing at the old Leaside airfield.

A famous scene at Jericho Beach in early RCAF days. Curtiss HS-2L G-CYGA goes up the ramp after a day's flying, perhaps on training, or on fisheries patrol, photo survey, or after a day working forest fires. 'GA survived into 1928, by when most HS-2Ls were worn out. (NAC PA133584)

Swinging the prop on a Camel at Camp Borden in CAF days. Ten Camels were on CAF/RCAF strength, the last into 1929. They formed no unit and flew mainly when some senior instructor wanted to fire his wartime nostalgia. (CF RE17958)

The first RCAF Arctic flying took place in 1927-28, when the air force supported the Hudson Strait Expedition. Here one of the Fokker Universals involved is launched in June 1928 for some reconnaissance from Wakeham Bay, Ungava. (NAC PA123505)

*Back endpaper:* Prewar Hurricanes of No.1 Squadron ready at Sea Island in 1939. No.311 later went overseas with No.1. (NMST 5269)

Murray Castator, a 118 Squadron rigger, poses with a damaged Kittyhawk during his tour at Annette Island, Alaska in 1942. The Kittyhawk served well in the defence of the northwest. Later, 118 went overseas to become 438 Squadron on Typhoons. (Murray Castator Col.)

When No.110 Squadron went overseas in February 1940, it began on Lysanders in RAF Army Co-operation Command at Old Sarum. In March 1940 it re-formed as 400 Squadron on Tomahawks. Later came tactical recce Mustangs, then PR Mosquitos and Spitfires. These 110 "Lizzies" were at Odiham in 1940. (CF)

First flown in 1938, the Westland Whirlwind was an advanced RAF fighter of the early war. Powered by 885-hp Peregrine engines, it could speed along at 300+ mph. Whirlwinds equipped Nos.137 and 263 Squadrons, where RCAF pilots had the pleasure of flying them, mainly on anti-shipping and cross-channel Rhubarbs. Successful though it was, the Whirlwind was not mass-produced. After a brief career it gave way to the Typhoon. This beautifully-posed photo shows Whirlwind P6969 of 263 Squadron. The last seen of it was on February 8, 1941, as it tangled with a Luftwaffe Ar.196 a few miles off the Devon coast. (IWM CH4998)

Many Canadians flew Kittyhawks in North Africa. This sturdy fighter performed well in ground attack, but also in air-to-air combat. Here ET611 of 112 Squadron cruises over coastal dunes. (CF PL18097)

A Nova Scotian, F/L G.W.A. "Gordy" Troke flew Kittyhawks with 250 Squadron. This photo dates to April 1943, when 250 moved base 3 times in Libya and Tunisia, while pushing the Germans from North Africa. Troke later flew Spitfires on 443 Squadron. On September 29, 1944 he shot down two Me.109s near Nijmegen. Flying Sabre 19265 of 430 Squadron on June 24, 1952, he had engine failure and crashed fatally near Callendar, Ontario. (CF PL10240)

Fitter LAC Al Mitchell of 439 Squadron guides a Typhoon pilot as he taxis at Eindhoven. Riding the wing was standard procedure, since it was difficult for a pilot to see over the Typhoon's nose. Note the ordnance load—900-pound cluster bombs, each full of anti-personnel bomblets. (CF PL42813)

Mustang KH495/9G-S in camouflage, invasion stripes and full 441 Squadron lettering. The scheme is a mystery. KH495 was not delivered to the RAF till the fall of 1944, by which time invasion stripes were becoming passé. Then, 441 did not accept its first Mustang until May 1945. One wonders if KH495 might have been specially painted for film making or another special occasion, perhaps some media event. KH495 was SOC in November 1946. (David Thompson Col.)

*Half title page:* The RCAF's first fighter was the Armstrong Whitworth Siskin, two of which were acquired for trials in 1926. These were followed by 10 more advanced Siskins. The RCAF Siskin aerobatic flight became famous coast to coast in the early 1930s. J7758 was lost in a crash at High River in June 1927. (H.S. and Gordon Diller Col.)

*Title page:* The Supermarine Stranraer, built under licence by Canadian Vickers in Montreal, played a vital role in the home defence of Canada. From November 1938 there were 40 on RCAF strength. Most lasted till late 1944, by when they were replaced by Cansos. Many "Strannies" then had useful commercial careers, mainly on the British Columbia coast and in the Caribbean. Stranraer 957, shown in a classic BC setting, was delivered to the RCAF in November 1941. (NAC PA115764)

Eastern Air Command Hudson Mk.I No.771 while on 11 (BR) Squadron. This type proved its worth in the U-Boat. (David Thomspon Col.)

Copyright © Larry Milberry, 2000.
All rights reserved. No part of this book may be reproduced in any form or by any means without prior written permission of the publisher.

### Canadian Cataloguing in Publication Data

Milberry, Larry, 1943-
  Canada's Air Force at war and peace

Includes bibliographical references and index.
ISBN 0-921022-11-5 (v.1)

1. Canada. Royal Canadian Air Force – History. 2. World War, 1914-1918 – Aerial operations, Canadian. 3. World War, 1939-1945 – Aerial operations, Canadian. 4. Aeronautics, Military – Canada – History. I. Title.

UG635.C2M533 2000       358.4'00971       C00-930841-5

Design: James Jones and David O'Malley
  Aerographics Creative Services, Ottawa
Photo retouching: Stephen Ng/ML Studio, Toronto and
  James Jones, Aerographics Creative Services, Ottawa
Proofreading: Ralph Clint, Toronto and Lambert Huneault, Windsor, Ontario
Printed and bound in Canada by Friesen Printers Ltd., Altona, Manitoba

Published by
CANAV Books
Larry Milberry, publisher
51 Balsam Avenue
Toronto, Ontario M4E 3B6
Canada

# Contents

Preface .......................... 7

**Chapter One**
Canadians in the First Great Air War ........... 11

**Chapter Two**
Between Two Wars ........................ 49

**Chapter Three**
The War on the Homefront ........................ 115

**Chapter Four**
The Day Fighter Game ............................ 189

Glossary ............................ 279

Bibliography................................. 281

Index.......................................... 283

Short of transports at home, the RCAF settled for a hodgepodge of obsolete types in 1939-40, including the Boeing 247D. Eight were taken on strength, starting with 7635. Once more suitable transports like the Lodestar began arriving, the "247s" were sold, 7635 becoming CF-BVZ with CPA. (David Thompson Col.)

The BCATP at a glance. First, a typical scene at RCAF Station Trenton showing Central Flying School Harvards, Cornells, Cranes and Fleets. Some of these later had postwar civilian careers. Harvard 3275, for example, flew for many years as CF-MGI, while 3222 became CF-MKA. Then, an aerial view of No.3 SFTS near Calgary, a BCATP training station, showing the typical triangular runway arrangement Beyond are the Elbow River and Glenmore Dam. Finally, a look at the Link trainer room at Camp Borden in July 1940; and a great PR photo of young BCATP airmen from far and wide: R.W.N. Bennetts (Australia), T.S. Knapman (New Zealand), C. Jackson (England), L.R. Young (Newfoundland), W. McElwee, Jr., (RCAF from Philadelphia, USA) and W.J.D. MacLaren (RCAF). (CF PL11645, '3752, '8155)

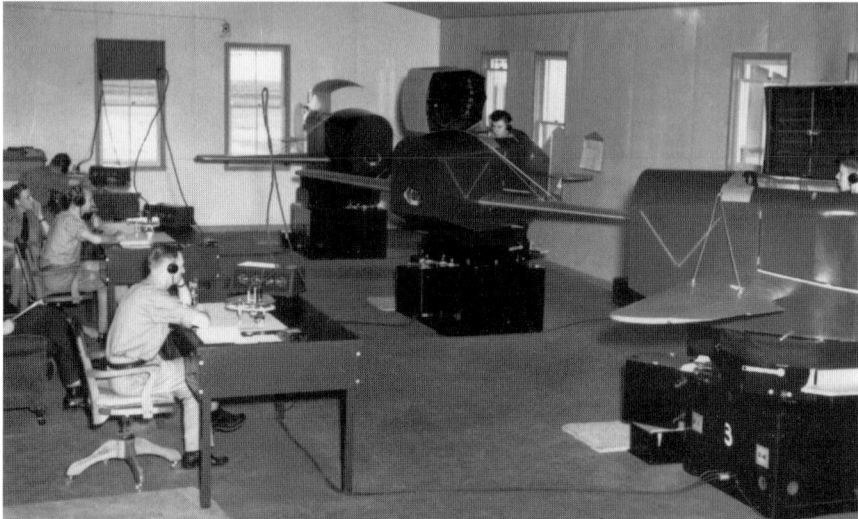

# ❦ Preface ❦

*Canada's Air Force at War and Peace* is an all-encompassing three part series. The only previous book with such broad coverage in text and photos is *Sixty Years: The RCAF and CF Air Command 1924-1984*, published by CANAV. There have been many other books, but none ever attempted such vivid coverage. The first great RCAF work appeared in 1944-49: *The RCAF Overseas* in three volumes from the RCAF Historical Section. Later, the Canadian War Museum published John Griffin's *Canadian Military Aircraft: Serials and Photographs 1920-1968*, and Griffin collaborated with Sam Kostenuk in *RCAF Squadrons and Aircraft 1924-1968*. Between 1980-94 the DND Directorate of History put out a superb series: *Canadian Airmen and the First World War, The Creation of a National Air Force*, and *The Crucible of War*. Anyone wishing to discuss RCAF history intelligently must be familiar with all such books.

CAFWP begins in 1914, when the Canadian "air force" comprised a lone Burgess-Dunne biplane. This experiment failed, making room for something truly impressive—thousands of Canadians fighting in Britain's WWI air arms: the RFC, RNAS and RAF. Naturally, a few scowlers will jump up, yelling, "Hey, you can't include those people! Canada didn't have an air force till late 1918." Well, in typical CANAV style, this book will not be enslaved by primitive, artificial constraints. The intelligent reader will understand our point and enjoy every page.

Astounding exploits are recounted in Chapters 1 and 2. Hugh A. Halliday did the main body of research here, his aim being to avoid anything stereotypical or mythical, and to find new material. He succeeded on both scores. Little of what he put on the table has appeared anywhere, except in scholarly journals. Hugh's sources included such treasures as the Fred Hitchins Papers in the Directorate of History, the Harry Creagen Papers in the National Aviation Museum, many relevant files from the National Archives of Canada, and his own data base, which includes much from the Public Records Office in London, England. Airmen's log books and diaries also were consulted. In one case I returned to the diary of WWI bomber pilot Christian Bergener, excerpts from which first were used in my original book, *Aviation in Canada*. Such rich sources embody all that 1914-18 meant to Canada's first aerial warriors.

Tactical recce Mustang Is of 414 Squadron at Dunsford in October 1942. The scene suggests some aircraft just returned from operations. The despatch riders would be waiting to rush the film to the photo lab for processing. Then, F/O Lew May, a 414 tactical recce pilot. His tour was typically hair-raising for anyone in this dangerous trade. In late October 1943 he was in on a rare kill, sharing a Luftwaffe Yale with F/O R.C.J. Brown. Note the recce camera mounted behind the cockpit. (CF PL10951, '22854)

Like other CANAV titles (notably *Air Transport in Canada*) CAFWP demands a fair level of involvement. The reader should be in an "aviation state of mind" before getting started.

Pilots of 401 "Ram" Squadron in early postwar days at B.152 Fassberg. On the wing are Art Wood, Bill Morris, R.C. "Gudge" Gudgeon, F.E. "Freddy" Thayer, Jr. Beteau, J.P. "Red" Francis, D.B. "Jughead" Dack, Mel Gummer, J.A. Ballantine, J.H. Ashton, "Saint" St. Arneault and Max Atkens. Standing are Don Jarvis, Jack Marson (Adj), L.W. "Red" Woods, A.K. "Woody" Woodill, V.E. Cottrell, Stan Knight, S/L E.A. "Bud" Ker (CO), L.H. "Len" Watt, Steve Stevens, R.H. "Dick" Cull, Chiefy Gillis, F/O Kay and A.E. Sawyer. This was an illustrious unit, having begun in the Battle of Britain as No.1 Squadron. Even this 1945 group could boast dozens of kills. On April 20, 1945 alone the "Rams" tallied 18 destroyed. Dick Cull had several victories. Woods got an Me.109 near Arnhem on January 17, 1945 (he was shot down on April 25, but survived). Len Watt had shot down an Me.262 on March 12, 1945. Francis got an Me.109 of April 17, 1945. Woodill had shared in an Me.109 and an Me.262, and had destroyed an He.111 on the ground at Schonberg. S/L Ker previously had served in the RAF, including a Spitfire tour in North Africa, Malta and Sicily with 145 Squadron. (via Bill Stowe)

You may feel a bit overwhelmed by all the data, and wonder how to find anything in such giant volumes. Get a start with the table of contents. Then, use the index, which has almost any element listed, often cross-indexed. If you need more or different material, refer to the selected bibliography. If you don't have a particular book, you can try finding it through your public library system or local bookstore. For further details about a story or individual, use micro-film resources available through in the public library, the internet, etc. Don't expect all the "good gen" to be in one book, even in a dozen. As to place names, hundreds occur, so have your atlases handy.

Many aviation terms and acronyms appear in *CAFWP*. For unfamiliar ones, refer to the glossary in each volume. CANAV appreciates all terminology—life is too short to argue whether "aeroplane", "airplane" or "plane" is best. Nor will we waste good time fighting over "WWI" versus "First World War", etc. Life must have come to a screeching halt for those with nothing better to do than debate the merits of one phrase compared to another. Similarly, what's the point of jumping up and down if someone calls a C-47 a DC-3, or a DC-3 a Dakota? I can only suggest that you don't read any further if you have weird hang-ups in this regard. Then, in the 21st Century it is strange how publishers still suffer from metric-imperial obsessive compulsiveness. For most, every reference in one system must be accompanied by the equivalent of the other. With CANAV either might be used, but equivalents are not slavishly provided—we take for granted that today's educated reader knows both systems and can convert if desired.

No matter how much is in a book, any author expects to be chastised for not including more, as if the complaining reader has any right to set the agenda, or is paying the publishing bills. Even though this is the grandest Canadian air force history, some will moan that so-and-so isn't mentioned. Even goofier are the letters of outrage complaining that another person was mentioned! Other soreheads will complain that such-and-such gets too much or too little coverage. Perhaps they should try writing their own books!

The commonest question about CANAV Books is "How do you do it?" People everywhere have wondered about a small publisher turning out such glitsy books. All are amazed that none of these has had anything to do with a handout from the Canada Council or other grant giver. From the beginning it was CANAV's goal to be entrepreneurial. As Canadian as maple syrup though it may be in publishing, taking the tax-payer's hard-earned dollars to support a private venture is two-

Dance night at Torbay, Newfoundland for 160 Squadron, an important EAC Canso unit. From whence all the girls? Some were WDs, others were from local communities. (via Harry Mosher)

Another interesting scene from Torbay: 160 Squadron was getting set up for a grand squadron photo. The photographer had his work cut out, getting the mob organized atop and in front of the Canso. (via Harry Mosher)

bit. If a book is worth publishing, readers will take note and buy a copy. Rather than grovelling in the arts council troughs, any writer can organize financing without picking his neighbour's pocket.

As with all CANAV books there are many people and organizations to thank. Of course, I have tried to list all, but apologize in advance, if you have been missed. To begin, Hugh A. Halliday was my chief collaborator and supporter. Not only did he provide most of the raw material for Chapters 1 and 2 of Volume 1, but handled many additional queries. I had him bouncing from National Archives to Directorate of History, National Aviation Museum, and Canadian Forces Photographic Unit, sometimes to more than one in the same day.

Lambert Huneault and Ralph Clint were reliable proof readers of manuscript and galleys. Ralph was an important researcher, double-checking tricky facts and patiently putting up with my incessant phone calls. Dave Thompson and Bob Finlayson also were indispensable, Dave providing photos, technical gen and reference books; Bob digging up photos and spending long days making prints in his dark room. Meanwhile, Stephen Ng sat through sessions retouching bent, scratched and smudged photos. Andrew Cline devoted months updating the RCAF/CF chronology from Sixty Years (see Volume 3), and Tony Cassanova was an important source of obscure data and photos. Also indispensable, especially for his encouragement, was Mike Valenti.

My many other helpers and sources during this project include: Howard Anderson Col., Bill and Nan Baggs, M.G. "Mal" Beverly, John A. Biehler, Vernon L. "Bill" Bowman, Robert Bracken, Robin Brass, Louise Brazeau, J.D. "Danny" Browne, John Buzza, Tony Cassanova, Murray Castator, Roy Clarke, William H. "Butch" Cleaver, W.C. "Bill" Clifford, Andrew Cline, Mike Comar Col., Les and Norm Corness, James P. Coyne, Donald and Georgina Crumb, Chuck Darrow, David Davies, Tom Dietrich, M.F. "Mike" Doyle, H.S. and Gordon Diller Col., Bill Dunphy, Eaker Col., Vince Elmer, Donald Evans, Robert W. Fackrell, J.S.T. Fall Col., Rae R. Farrell, Mike Filey, Arthur Fleming Col., C.W. "Charlie" Fox, W.E. "Ted" Freeman Col., David Frost, Don Graham, John Griffin, Bob Halford, Murray Hallford, Pat Hare, Fred W. Hotson, Norman C. Howe, A.T. "Tony" Jarvis, Arthur B. Jewett, Vic Johnson, John Joseph Kelly, Gordon Kemp, Janet Lacroix, Ross Lennox, J.A. Omer Levesque, E.C.R. "Ed" Likeness, Donald Macfie, Douglas MacKellar, H.M. "Hal" MacLeod, Bruce E. and Pauline Macpherson, Norm Malayney, Stanley Malouf, Stafford D. Marlatt, Walter J. McCarthy, Wesley H., McIntosh, Edmund J. McKay, Jack McNulty Col., Doug McPhail, John McQuarrie, William R. McRae, William H. "Bill" and Coreta Meaden, Carl Mills, K.M. Molson

Col., Stephen Mouncy, Brian Musson, Jack Myles, A.R. Deane Nesbitt, Ian C. Ormston, J.A. "Joe" Ouellette, Alf Pedley, Penhold Col., Frank Phripp, Rae Reid, R.L. "Dick" Reeves, Jack Ritch, Doug Rose, Herbert J. Russell, John Scammell, Rayne D. "Joe" Schultz, Rob Schweyer, J.T. "Jack" Seaman, Larry Seeth, Eric G. Smith, Ken Smith, T.R. "Ted" Smith, Steve St. Martin, George Stewart, Lloyd A. Stewart, C.H. "Smokey" Stover, W.N. "Bill" Stowe, Kenneth I. Swartz, Mel Swift, H.T.C. "Cro" Taylor, Stewart Taylor, George Thompson, A.W. "Tony" Tinmouth, Mike Vacheresse, Mike Valenti, Bertal T. "Bert" Walsh, Douglas "Duke" Warren, E.J. Watkins, John Wegg, Christopher Weicht, William J. Wheeler, Jack Whorwood, Gordon S. Williams, Vernon A. Williams, Rosalie Woodland. These organizations also provided assistance: Air Crew Association of Canada, Alberta Provincial Archives, Canadair Ltd., Canadian Aviation Historical Society, Canadian Forces Photographic Unit, Canadian Harvard Association, City of Toronto Archives (James Col.), de Havilland Canada, Glenbow-Alberta Institute, Imperial War Museum, Museum of the Highwood, National Archives of Canada, National Aviation Museum, National Film Board, National Museum of Science and Technology, Western Canada Aviation Museum.

*Larry Milberry*

The proudest moment for a young airmen was receiving his wings. That day he exchanged the LAC's propeller for sergeant's stripes, and put away the white flash on his wedgy. Here LAC Edgar Leroy Clary of Milwaukee is congratulated at his 2 SFTS wings parade by Air Marshall W.A. Bishop. Clary soon was overseas, flying Beaufighters with 47 Squadron. On November 13, 1943 he and his RAF navigator, FSgt W.E. Finbow, were killed in action off Greece. Thousands of Canadian airmen also would receive decorations. Shown is F/O Horrace B. Hillcoat, AFM receiving the AFC from Governor General, The Earl of Athlone. On January 23, 1944 Hillcoat had piloted his 168 Squadron Fortress to a safe landing at Predannack, Wales, following a mid-air collision with a Wellington. On December 15, 1944 he and his crew went missing while on Mailcan operations between Morocco and the Azores. (CF PL2270)

The Royal Aircraft Factory produced the B.E.-series single- and two-seat fighters and bombers. Many Canadians in the RFC flew these on operations, one of the best known being Lt. L.P. Watkins of Toronto. On the night of June 16/17, 1917 he shot down Zeppelin L.48 while flying B.E.12 No.6610. The standard B.E.12 had a wing span of 37' and was 27'3" long. Gross weight was 2352 pounds. Time to 12,000 feet was more than 47 minutes, top speed at 10,000 feet—91 mph. Usual armament was two .303 machine guns. (CANAV Col.)

Sopwith Camel pilots Raymond Collishaw of Nanaimo, BC, and Arthur T. Whealy of Toronto of Naval 3, a famed RNAS fighter squadron that counted many Canadians. As a boy Collishaw crewed aboard a fisheries enforcement vessel. He enlisted in the RNAS in early 1916, becoming a superb pilot, especially on the Sopwith Triplane. By war's end he was credited with 59 enemy aircraft. Whealy got 19. (NAC PA2789)

# Canadians in the First Great Air War

## The Legacy

Canada owes much to the first RCAF historians, especially W/C Fred H. Hitchins (1904-1972), the Air Historian from 1946 until 1961. Quiet and scholarly he was in Britain as a Historical Officer from 1940-45. At first he had little to do overseas, since the RCAF component was small until the autumn of 1941. Then a flying officer, Hitchins came to his job with an intense curiosity about Canadian airmen in the First World War. In the Public Records Office (Britain's equivalent of Canada's National Archives) he mined thousands of Royal Flying Corps, Royal Naval Air Service and Royal Air Force documents, compiling notes and lists which would not be fully exploited until the 1960s. In 1973 the Canadian War Museum published Hitchins' work *Air Board, CAF, RCAF, 1919-1939*. This long remained the most important document concerning RCAF interwar history. In a word, Hitchins was a giant in Canadian historiography. Anyone following was exploring territory he had pioneered.

Also noteworthy was Sgt Harry Creagen (1922-97). As a member of the Royal Canadian Army Corps of Electrical and Mechanical Engineers, he was an unlikely contributor to Canadian aviation records. He was largely self-taught, but made up for this in enthusiasm, coupled with a fine sense of organization. Through travel and correspondence he collected many clippings, made interview notes and transcribed private documents, all of which enhances Hitchins' work. Creagen also shared his findings generously, whether in articles to the Canadian Aviation Historical Society *Journal*, or through letters. He maintained a lively correspondence with scores of people, until overtaken by Alzheimer's Disease.

Ron Dodds (1913-1994), a journalist and public relations specialist, came to the subject late as a member of the Air Historian's office (1962, retiring in 1973). His output of articles published in *Roundel, Cross and Cockade* and the CAHS *Journal* was impressive. He authored two books: *The Brave Young Wings* and *Air Command*. The former is probably the best popular history of Canadians in the first great air war. The second, supposedly the autobiography of Raymond Collishaw was, for all practical purposes, ghost-written by Dodds.

The RCAF dragged its feet in writing an official history. Its commanders had little interest in the subject beyond recruiting and training literature. Even in the 1960s, when RCAF appropriations used half the Department of National Defence budget, the office of Air Historian was so small (never more than five staff) that it could do little beyond answering questions from the public and contributing articles to *Roundel*. This contrasted with the Army which, over 20 years, had maintained a larger establishment with a solid record of major and minor publications. Successive Air Members for Personnel (to whom the Air Historian reported) patronized W/Cs Fred Hitchins and Ralph Manning, but secured no significant resources for their work.

The integration of the Canadian Armed Forces may have had undesirable consequences but, from the standpoint of aviation history, it was helpful. The first head of the new Directorate of History was Charles P. Stacey, Canada's most esteemed military historian. He recognized the need for a multi-volume air force history, mobilized the manpower for the task, and launched a program of research into primary documents, making up for years of neglect. This continued under S.F. Wise, who also authored the initial volume of the official history—*Canadian Airmen and the First World War*. People who believed the book should have been merely an expanded version of George Drew's reverential, but shallow, *Canada's Fighting Airmen*, would be disappointed. Wise's book broke new ground in countless ways; henceforth writers (whether serious authors or plagiarists) could ignore it only at their peril.

As the RCAF's leading historian W/C F.H. "Fred" Hitchins lay the foundation upon which researchers and writers have counted for decades. (CF PL112642)

W/C R. Manning, DFC, CD, psc, B.A., Director of Air Force History, 1960-1965 (CF PL104989)

The dust jacket of Ron Dodds history of Canadian airmen in the First World War. (Canada's Wings)

Although Alexander Graham Bell financed aeronautical research and development in Canada before WWI, what little flying took place was mainly at air meets and fairs, where huge crowds paid admission to watch the proceedings. The first to fly a powered airplane in Canada was one of Bell's protégés, John A. D. McCurdy, seen at the controls of his machine during the August 1911 Toronto Air Meet. (City of Toronto Archives, James Col.)

J.A.D. McCurdy had made Canada's first powered airplane flight on February 23, 1909, going aloft at Baddeck, Nova Scotia in this airplane—the Silver Dart. In 1909 he and his cohort, F.W. Baldwin, demonstrated two of their designs to the Canadian Militia at Camp Petawawa. Few in the audience cared about aviation, so many would have chuckled when both machines piled up. Following the demonstration, Militia Colonel E. Fiset gave his opinion to the Toronto Star: "I don't think it [the airplane] can ever be an offensive weapon… You cannot expect a young country like Canada to strike out and adopt an airship policy… We must wait a great many years … before the true use of these machines can be demonstrated." (National Geographic Society)

## Military Aviation in Canada 1909-1919

Indifference is the best word to summarize the attitude of the Canadian government towards military aviation from 1909 to 1919. From 1909 Alexander Graham Bell, the most prestigious scientist resident in Canada (albeit part-time) urged that the Militia sponsor efforts by the Aerial Experiment Association and its successor, the Canadian Aerodrome Company. Bell was particularly anxious to advance the careers of his protégés, John A.D. McCurdy and F.W. "Casey" Baldwin. The Militia consented to a demonstration at Petawawa in August 1909, when the two young pilots made a few hops, but crashed their machines "Silver Dart" and "Baddeck No.1". Individual officers subsequently visited Bell and the CAC facilities. On March 9, 1910 Maj G.S. Maunsell went up with McCurdy in "Baddeck No.2", thus becoming the first Canadian soldier to fly.

Before the First World War, at least four Canadian soldiers—Capt P.S. Benoit, Lts B.M. Hay and R.H. Irwin, and Lance-Corporal F.S. Brown applied for pilot training. All were denied, although Hay finally joined the RFC in 1916. In September 1913 Acting Sub-Lt Charles T. Beard, RCN, attending courses in Britain, unsuccessfully applied for appointment to the RFC Airship Section. Also in 1913 the 6th Field Company, Royal Canadian Engineers (North Vancouver), suggested that an aviation section be formed. The Department of Militia did nothing.

This contrasted sharply with other nations. Britain, France, Germany and the United States formed air services. Nations engaged in the Balkan Wars (Serbia, Bulgaria, Turkey) used aircraft. Canada, with no threat to its territory and no overseas ambitions, should be compared with other countries in comparable situations. When one draws the parallel, its record remains dismal. Several South American republics had invested in military aviation before the First World War. In the same period that McCurdy and Baldwin were attempting to impress the Canadian Militia, Australia was offering £5,000 for a "flying machine... for military purposes". None was bought, but in January 1911 Australia's Military Board seriously considered establishing a flying unit, hesitating chiefly because not even Britain had taken such a step. In August 1912 Australia hired a civilian pilot, H.A. Petre; the following month a minuscule Flying Corps was authorized. Shortly before the outbreak of war, a Central Flying School was established near Melbourne with one British and one Australian flying instructor, and five aircraft. Training commenced in August 1914, four officers taking a three-month course. Teachers and pupils became the nucleus of the Mesopotamian Half Flight, organized as a self-contained unit for overseas service in early 1915.

Meanwhile, in 1912 the Union of South Africa had provided for an Aviation Corps in its Defence Act. The government purchased a bankrupt flying school and, in May 1913, advertised for pupils. Hundreds applied, 10 were selected, and in April 1914 six were commissioned as probationary lieutenants for posting to Britain for advanced training. Five served with the RFC in France during the first months of the war. When South African troops occupied Germany's colony in Southwest Africa (now Namibia), they were accompanied by Henry Farman scouts piloted by British and South African airmen.

## Tactical Air Power

Initially, air power was strictly tactical. Tethered balloons that scouted and directed artillery fire on behalf of their armies—those of Revolutionary France or the Union forces in the American Civil War—were directly assisting front line troops. Balloons made good observation posts—better than church steeples, but no different in concept. Although some novel ballooning was tried, such as delivering explosives (Austria against Vienna, 1848) or transporting leaders and messages from a besieged city (Paris, 1870-71), the only practi-

**Trenches on Vimy Ridge photographed from a kite balloon in November 1917. The value of such intelligence can be appreciated at a glance. (NAC PA2366)**

**The gas-filled kite balloon was valuable from earliest days in WWI. Floated aloft, then tethered at a pre-selected height, a balloon allowed one or two skilled observers in its suspended gondola car, to observe, sketch or photograph enemy activities. Balloons, however, were vulnerable—hundreds were shot down, often taking their crews to their deaths. (NAC PA958)**

cal use of air power during the age of lighter-than-air flight was tactical.

Direct support of ground forces remained air power's principal role after the appearance of airplanes, but their mobility (compared to fixed balloons) enhanced their significance. In spite of early attempts at bombing in such theatres as Libya (1912) and the Balkans (1912-13), it was clear that aircraft mainly would be scouts for conventional armies. The first major success of air power came in late August 1914, when reports by RFC pilots indicated changes in the direction of advancing German forces. This opened a way for counter-attacks on the exposed enemy flank. The Battle of the Marne was a direct consequence of tactical air power.

The First World War saw aircraft used in every role (save one), as they were wielded in later wars. The exception was troop transport (although aerial resupply of isolated or advanced units was practiced, and individual spies were delivered by airplane). However, the popular imagination (and much of the attention of subsequent writers) was captured by two aspects of the air war—fighter operations (with their accompanying "Knights of the Air" mythology) and the Zeppelin bombing raids (spectacular in concept and technology, ineffective in execution). The mass of writing on these themes has obscured the fact that air power in the First World War was most important in its tactical application; as such it revolutionized the nature of land warfare.

The most significant aircraft on either side were the Corps machines that mapped opposing trenches, scouted behind lines and directed artillery. Of increasing importance as the war progressed were aircraft that strafed and bombed enemy defences and maintained contact with advancing troops, whose own communications systems (runners, field telephones, pigeons) were inadequate in battle. The role of fighters (called "scouts", a term that misrepresented their tasks)

was to protect their own army co-operation aircraft and destroy the enemy's tactical machines. Dogfights involving fighters were not so important as the tasks of blinding the enemy's aerial eyes, protecting troops from aerial observation, and ensuring safe passage for tactical airplanes.

Since the invention of smokeless powder, modern wars had increasingly become wars of concealment. Observers of the Russo-Japanese War (1904-05) remarked that the battlefield had been like a desert, with men and weapons elaborately hidden. Moreover, it was a desert swept by artillery and machine gun fire that rendered all movement extremely hazardous. Cavalry, the traditional tool of reconnaissance, was obsolete even before 1914. Meantime, armies were developing more deadly ordnance. Tactical air power arrived at a particularly opportune time; in combination with the new fire power, it helped create and maintain the stalemate which was the Western Front. Aircraft spotted coming offensives, guaranteeing that most would fail. Although aircraft seldom slew directly (compared to other weapons), they were the handmaidens to the most deadly killing machines invented to date—the great guns.

The Zeppelin campaign against Britain was front-page news, especially when a "Zep" went down. Torontonian 2Lt L.P. "Don" Watkins, flying a B.E.12, and Capt R.H.M.S. Saundby, in a D.H.2, destroyed Zeppelin L.48. The attached news item appeared on the front page of the Toronto Globe on September 17, 1917. Watkins was killed in July 1918, while on F.E.2b night operations. (CAHS Col.)

## TORONTO AIRMAN DESTROYS A ZEPP

### Loudon Pierce Watkins is Promoted and Given a Military Cross

A Canadian Associated Press cable last night announced that Lieut. Loudon Pierce Watkins of the Royal Flying Corps, and son of Mr. E. J. Watkins of the W. R. Johnston Company living at 95 Breadalbane street, had been promoted to be Captain and given the Military Cross for destroying a Zeppelin.

He was a student at Upper Canada College, and with his brother, E. J., took his aviation course at Toronto Island. They went overseas in December, 1915, direct to France, and for a year were daily over the German lines.

Early this year Lieut. Loudon Pierce Watkins was recalled to England to join a special Zeppelin lookout. His chance came on June 16, for it is recorded that he destroyed the L48 at East Anglia. His parents had received word that he had been promoted to be Captain and that he had received the Military Cross.

A third brother, John Francis Watkins, is with the Royal Divisional Signal Corps, and a brother-in-law, Lieut. R. G. Lewis, is serving in France with a Forestry unit.

Gunners have a saying: "Artillery lends dignity to what would otherwise would be a vulgar brawl". The First World War was far more than a "vulgar brawl"; it was industrialized slaughter, with artillery at its heart. It was the first war when more people died from wounds than disease. Estimates vary, but roughly 60% of all battle deaths were due to artillery fire. Machine guns, rifles, gas and bombs were far behind. Every exploding shell flung deadly shrapnel into trenches or across lines of advancing troops. At the Battle of Messines in June 1917 Britain's Second Army artillery fired 3,258,000 rounds in eight days; in a single 24 hours (September 28-29, 1918) British guns hurled 943,847 shells at the Hindenburg Line. The awful effects were counted by surgeons and burial parties. Those shells were not fired randomly, but directed at targets previously mapped or located by aircraft, under the guidance of near-anonymous men in the cockpits of hundreds of reconnaissance machines that daily criss-crossed the skies.

turn to his training and experience), and the willingness of officers to believe those reports and take the initiative. Air power could influence a battle, but not determine an outcome. Of July 1, 1916, S.F. Wise wrote: "Two conclusions are evident. The first is that the RFC had seized and maintained air supremacy over the battlefield; the second is that even had RFC airmen been able to provide intelligence of the situation on the ground with flawless accuracy, it would have made virtually no difference to the outcome of an attack doomed to failure through faulty conception."

With time and experience air power improved. Thus, the Canadian victory at Vimy Ridge in April 1917 was related directly to the accuracy and scale of the preliminary bombardment (achieved through previous aerial mapping of targets before the first shells were fired), followed by accurate spotting for the guns. At the Battle of Amiens (August 1918) the RAF helped the army achieve surprise by frustrating German reconnaissance, then supporting advancing troops with

Canadian artillery in action during an advance east of Arras in September 1918. Artillery was the most destructive weapon of WWI. Gunners received their best target intelligence from observers in kite balloons and roving reconnaissance aircraft. (NAC PA3133)

Since tactical air power contributed so heavily to the Western Front, it must also be credited (or blamed) for the various attempts to by-pass the trench system. These culminated in the tank (intended to crash through opposing defences) and the strategic bomber (for striking at the enemy's industries and war workers). Both had significant trial runs after 1914, were discussed and developed afterwards, and attained terrible maturity during the Second World War.

The effectiveness of tactical air power varied with time and circumstances. Airmen's reports were not always accurate and army officers were not always correct in assessing reports. Canadian Airmen and the First World War describes how, upon launching the Battle of the Somme (July 1, 1916), British commanders gave the RFC unprecedented responsibilities. An accurate report by No.9 Squadron, that German resistance was weak in one sector, was ignored and an opportunity was lost to gain at least a local victory. An inaccurate report by No.4 Squadron of weak German counter-attacks was believed; the resulting action was a British defeat. Much depended on an observer's accuteness (related in

some of the deadliest ground strafing of the war. Camel and SE.5 "trench fighters" made their most significant contributions by attacking enemy soldiers, rather than battling German aircraft.

Reconnaissance and artillery spotting aside, the aerial task most relevant to troops was the "contact patrol", conducted at low level, reporting the positions of forward troops to their rear-echelon commanders, then dropping fresh intelligence, ammunition and rations to the most isolated units engaged. Such patrols often had to locate enemy troops in the midst of a battle, the best technique being to fly low, then see who fired at one's machine. A contact sortie report from Lt C.M.G. Farrell of No.24 Squadron was typical: (August 8, 1918, SE.5a D6937) "Five of our tanks going northeast from Harbonnieres towards Foucaucourt; one train on fire north of Rosieres. Flew along road from Caix to Vrely at 500 feet to locate machine guns alleged to be holding up cavalry, but was not fired on at all."

Another Canadian in No.24 Squadron, Lt George B. Foster, described a patrol which was representative of the work done for the army, if laconic regarding the casual ruthlessness applied

The nimble Sopwith Camel (only 18' 9" long), first flown in late 1916, was one of the most successful British fighters. Powered by a 110- or 130-hp rotary, air-cooled engine, it could hit 115 mph. According to Peter Lewis in *The British Fighter since 1912*, Camels accounted for 1294 enemy aircraft destroyed. Many Canadians flew this type. In the cockpit of this 203 Squadron Camel (B6378) is 2Lt F.T.S. Sehl of Victoria, BC. (Stewart Taylor Col.)

Near Allonville, France in July 1918: Lt. Arthur Whealy watches his 203 Squadron Camel being armed with light bombs. The Camel, whose chief armament was two .303 Vickers machine guns, proved effective against motor transport, troop columns, trenches and gun emplacements. (NAC PA2796)

The vulnerability of low-flying strafers like the Camel led to purpose-built "trench-fighting" machines, with armour-protected cockpits, fuel tanks and engines. The hefty Junkers J.I was one of the best such fighters. In 1919 this war trophy J. I was displayed at Toronto's CNE. That year there were numerous German aircraft in Canada, especially Fokker D VIIs, some of which were flown privately. Of these historic machines, only the J.I and a D VII survive, the former in the National Aviation Museum, the latter in the town museum in Knowlton, Quebec. (City of Toronto Archives, James Col.)

in doing the job: (August 10, 1918, SE.5a D6918) "50 infantry and three machine guns holding up our troops from a pit in 66D/A.8.d [map reference]; one train going southeast from Chaulnes. Nine of our big tanks in action followed by 4-500 infantry in 66D/A.7.b and d. Five enemy running east across road in 66D/A.2.c without kit or helmets—two shot down." By war's end various nations were developing and producing special aircraft for tactical warfare. Late in 1917 the Germans introduced the armoured Junkers J.I, the ancestor of such army support machines as the Junkers Ju.87 Stuka and the Republic A-10 Wart Hog. The J.I's strength and reliability ensured survival in the fierce, low-level world of tactical aviation. From November 1918 the RAF began delivering Sopwith Salamanders to operational units. This was an armoured, single-seater for trench strafing. With 650 pounds of armour plate it was built to survive small arms fire. Although armament was routine—two .30 calibre machine guns—it carried 2,000 rounds of ammunition, more than double the normal fighter. An experimental Salamander had eight downward-firing machine guns for raking trenches.

Lt Victor Henry McElroy of Richmond, Ontario was a fighter pilot whose ground support exploits were far more significant than anything he achieved dogfighting. Formerly with the Royal Canadian Engineers, he joined the RFC in August 1917, and was posted to No.3 Squadron (Morane "P") in France in February 1918. He was hospitalized, but on April 24 rejoined No.3 (Camel). RAF communiques subsequently mention him twice—on August 27, 1918 (for having destroyed an enemy aircraft in company with a Lt D.J. Hughes) and on September 20, 1918 (awarded the Distinguished Flying Cross). The DFC was formally announced in the London Gazette of December 2, 1918: "This officer has been conspicuous for his courage and determination in attacking enemy troops, transport, huts, etc... Carrying out this service at low altitudes... he has inflicted heavy casualties on the enemy, his machine being frequently badly shot about..." The account of McElroy's actions pales in comparison with another report. On September 1, 1918 his commanding officer wrote to the Headquarters, No.13 Wing, RAF, asking that McElroy be considered for "such award as you may think fit". He then went on:

*On the 1st August 1918 this officer took a prominent part in the daylight raid on Epinoy aerodrome, which he attacked from a height of 200 feet downwards, doing a large amount of damage with his bombs to the sheds and personnel, and attacking an enemy machine on the ground...*
*On the 21st August 1918 at 2.50 p.m., when employed on low flying, this officer attacked enemy troops at Favreuil with bombs and machine gun fire from very low altitude with good effect... He also engaged an enemy kite balloon south of Thilloy under heavy machine gun fire from the ground and drove it down... on the evening of the same day he... attacked a*

*dump at Sailly-Saillisel, dropping four bombs and obtaining two direct hits which caused large explosions in the dump. On the way home he again engaged enemy troops from a low height... On the 22nd August ... he attacked troops and transport on the Bapaume-Biefvillers road with bombs and machine gun fire, obtaining a direct hit on one wagon and causing many casualties.*

*On the 27th August, Lieutenant McElroy ... fired 700 rounds on transport on the Les Boeufs-Le Transloy road, killing six or seven horses and at least ten men. On 28th August this officer was again employed on low flying ... with bombs and machine gun fire, obtaining two hits on the road and knocking out one complete four-horse team... On the 26th August, when in company with Lieutenant Hughes, this officer saw an enemy aircraft two seater at 600 feet southwest of Bapaume. He attacked it... enemy aircraft dived towards the ground. Lieutenant McElroy followed it to 200 feet, and saw it crash north of Beaumetz. During the recent operations, in the absence of the Flight Commander, this officer has been in charge of the Flight, and has led them continually on low bombing with great dash and determination, showing a splendid example of courage and leadership to all pilots, and on many occasions returning with his machine badly shot about. Lieutenant McElroy has been with the squadron since 18th February 1918, and during the German offensive of 21st March to April 1918 took an active part in a large number of low bombing patrols...* (The Commanding Officer's recommendation was timely—Lt McElroy was killed in action on September 2, 1918, while attacking ground targets.)

Lt Richard Duncan was old by RFC standards. Born in 1888, he was in Prince Albert, Saskatchewan, when war broke out. Enlisting in the 32nd Battalion, Canadian Expeditionary Force, he went overseas, but in mid-1916 transferred to the RFC. He first attended No.1 School of Military Aeronautics, then took flying training. He reported to No.5 Squadron (B.E.2d) on December 27, 1916, remaining until November 23, 1917. RFC reports noted him attacking an enemy aircraft on March 11, 1917. On July 18 he flew reconnaissance for XII Corps in bad weather. Operating between 600 and 1000 feet, he obtained useful information about enemy dispositions and dropped messages at divisional and Corps headquarters. A reconnaissance in poor weather on November 19 also was cited in communiques. During one sortie he located German artillery in the face of brisk ground fire.

Even before completing his tour, Duncan had received a Military Cross (*London Gazette*, September 17, 1917). He was posted to instructional duties in Britain, but returned (as a captain) to 106 Squadron in April 1918. This work brought him a Mention in Despatches. A postwar summary of his career noted that he had flown Caudron, Curtiss, Armstrong-Whitworth, B.E.2c, B.E.2e, B.E.12, and R.E.8 types, adding that he had "knowledge of every branch of Corps work and night bombing".

Records of Lt Herbert Lee Holland of Toronto describe an industrious Corps pilot. He arrived in Britain in May 1916 with a cyclist battalion, but was seconded to the RFC in July 1917. After pilot training he was posted to Italy in January 1918—the RFC was then providing three squadrons to stiffen Allied resistance, following a disastrous defeat at Caporetto the previous year. Holland reported to No.34 Squadron (R.E.8) on January 23, 1918, serving there until October 7. On September 18, 1918 he was awarded the Military Cross with the following citation: "For conspicuous gallantry and devotion to duty when working with artillery in carrying out six successful shoots whereby many enemy gun pits were destroyed and explosions caused. In one case he descended to 100 feet and found all pits totally destroyed. He carried out a good low reconnaissance of two suspected hostile batteries and also obtained other very useful information." The citation does only the barest justice to his record. Hitchins' notes for 1918 list numerous "shoots" on enemy positions. This may not seem comparable to a fighter pilot's combat report. Yet, the sobering fact is that, on any given day, the guns under direction of pilots like Holland and Duncan could kill more Germans or Austrians than any of the great aces in their entire careers. Of course it wasn't "chivalrous"—it was war waged on an industrial scale:

May 9 - *Shoot on Battery C.91; one explosion; four pits destroyed; two damaged; descended to 100 feet and found pits caved in and filled with earth. Dropped two bombs on enemy transport.*

June 1 - *Sent General Force call on ten motor transport parked at M.135.965 (Mezzaselva); nine were set on fire and completely burnt. Numerous explosions.*

July 23 - *with Lieutenant Walker; 317 Siege Battery; 400 rounds; destroyed one pit; damaged other. Huge explosion during the shoot. [later] With Lieutenant Ward; 316 Siege Battery; parts of guns could be seen lying about position after the shoot.*

August 13 - *with Lieutenant H.W. Minish [Canadian, Gilbert Plains, Manitoba] in cloudy weather, completed the whole Eastern reconnaissance, obtaining many excellent photographs and a great deal of information.*

Many Canadians served on the R.E.8, an important observation type. But it was vulnerable to enemy fighters. With its gas tank adjacent to the engine, it could become a flying coffin. Trevor Henshaw, author of *The Sky Their Battlefield*, determined that 661 R.E.8s were lost in combat on the Western Front. An action of May 11, 1917 involved Lt T.W. McConkey of Bradford, Ontario (59 Squadron, observer, R.E.8 A3472). His report suggests the mayhem that reigned as men fought for their lives, yet shows the survivor's matter-of-fact outlook: "... while photographing about 9000 yards into Hunland... attacked by 5 Albatros scouts. Between us we shot down 2 ... and drove another down ... My pilot, Captain [F.D.] Pemberton from [Victoria] BC, manoeuvred ... in a most excellent fashion ... giving me every opportunity to bring my Lewis gun into play ... He received a spent bullet in the back, necessitating his spending a week in the casualty clearing station. I came off less fortunately with 4 bullet wounds in the right thigh, 1 in shoulder, 1 in face ..." Pemberton was shot down and killed in an R.E.8 on August 11, 1917. (David Thompson Col.)

Roy Edward Dodds (pilot, from St. Thomas, Ontario) and John Bernard Russell (observer, from Ottawa) often teamed in army co-operation work. They came to No.103 Squadron (D.H.9) by different routes. Dodds enlisted early in the RFC (Canada), arrived overseas in January 1918, and reported to his unit on May 26. Russell had gone overseas with the Canadian Engineers, switched to the RFC in August 1917, and reported to No.103 in April 1918. Russell flew his first sortie on May 19; his fourth (to Douai) on May 25 was also the longest of his tour (4:15 hours). On June 16 he flew his 19th sortie—it was Dodds' first— and thereafter they operated together almost constantly until about August 8, when Dodds began flying with another Canadian observer, 2Lt Irving B. Corey of Barnston, Quebec. Russell's new pilot was Capt J.A. Sparks, although he occasionally flew with Capt J.S. Stubbs. July 4, 1918 was not a typical mission. Two D.H.9s were on a dusk recce at 10,000 feet; No.6150 was piloted by Capt Stubbs (2Lt C.C. Dance, gunner); Dodds and Russell were in 2877. The subsequent report tells the story:

*Crossed lines at La Bassee and were attacked by 15 enemy aircraft at about 8.30 a.m. The fight proceeded towards Estames and Captain Stubbs fired two red lights to attract eight SE.5s which joined the D.H.9s but retired again towards the north without engaging the enemy aircraft. The fight continued until about 9:20 p.m... Captain Stubbs (pilot) shot down one enemy aircraft out of control and 2/Lt. J.B. Russell (observer) shot down another ... pilots saw both these machines go down but did not see them crash, owing to presence of other enemy aircraft. Having exhausted ammunition the D.H.9s made for the lines and enemy aircraft left them at 9.30 p.m. Captain Stubbs then returned east of the lines to attempt to continue the reconnaissance at 4000 feet, but darkness prevented observation.*

Russell would fly 83 sorties (169 hours) that included level bombing, at least 3 low bombing raids and 16 reconnaissance or photo missions. In a typical week (August 11-17, 1918) he flew 13:20 hours, crossed the lines 7 times, fired 100 rounds, saw 17 enemy aircraft and was in aerial combat once. He dropped six 112-pound bombs and took 13 photographs. Apart from his victory of July 4 he was credited with enemy aircraft shot down out of control on July 31 and September 18. He was wounded by anti-aircraft fire on September 29 and awarded a DFC on November 2, 1918.

Dodds' career totalled 64 sorties (130:30 hours). In a typical week (September 3-9) he flew 8:25 hours, crossed the lines 4 times, fired 150 rounds, had a combat with an enemy aircraft, dropped six 112-pound bombs and took 21 photographs. His last sortie was a reconnaissance (1:50 hours) on November 9, 1918. Several of Dodds' flights incorporated adventure normally associated with fighters. On August 11 he was doing late afternoon photography at 16,000 feet, when 7 white-tailed Pfalz scouts attacked. Dodds manoeuvred to enable his gunner (Corey) to fire into one, which went down out of control, vanishing into cloud. He flew directly at 3 more, scored hits on one, and saw it spin away. He resumed photographing, in spite of the remaining Pfalz, which hung about until driven away by 3 Dolphins. October 23 Dodds lead a bombing formation, which was attacked by 10 Fokkers. The de Havillands kept perfect station, claiming an e/a shot down in flames, 2 out of control. On October 30, at the head of 6 D.H.9s, he first bombed his target, then attacked 9 Fokkers and a 2-seater; 3 of the enemy went down, one at Dodds' hands. When recommended for a DFC (awarded in February 1919), his commanding officer wrote of his "very gallant conduct" and concluded by stating: "This officer's vigorous gallantry has been an inspiration and a splendid example not only to his own Flight, but also to the whole squadron".

Charles Ley King of Sault Ste.Marie joined the RFC in April 1917, going to No.34 Squadron (R.E.8) as a pilot on September 21, 1917. He served there to war's end, with time out (missing in action) between October 27 and November 6, 1918. His two gallantry awards only hint at the exacting nature of his work: Military Cross, announced in September 1918—"For conspicuous gallantry and devotion to duty when working with artillery in carrying out thirteen successful shoots, by which numerous enemy gun pits were destroyed and fires and explosions caused. He also carried out two very long reconnaissances, taking excellent photographs and obtaining valuable information." Distinguished Flying Cross, announced in November 1918— "This officer has done excellent work both on reconnaissance duty and in co-operation with our artillery. In the latter service he shows remarkable skill and keen observation. In carrying out a shoot on the 31st August, 848 rounds were fired in five and a half hours, and four pits were destroyed— a fine performance, reflecting great credit on this officer's capability."

Hitchins records some flights in greater detail than others. On May 4, 1918, for example, King was co-operating with the 6-inch howitzers of 293 Siege Battery, engaging a German artillery concentration. Some 400 rounds were fired, as King observed and reported the fall of shot. Three gun pits were destroyed and one damaged. The shoot was punctuated by heavy explosions among the German guns and concluded with a raging fire. This was accomplished in 2:22 hours. King subsequently conducted two shoots on September 14, when the guns he was directing themselves came under artillery fire. He broke away, located the enemy guns, and thus allowed swift counter-battery action. King remained in the postwar RAF and in 1929 was Mentioned in Despatches for services during a campaign against Iraqi tribesmen. In January 1942 he received the Air Force Cross following service with an air gunnery school.

Capt Alfred Cross of Regina joined No.15 Squadron (R.E.8s) in November 1917. Eventually, he was awarded the DFC for daring low level

Westland-built D.H.9A F1612 of No.2 Squadron, Canadian Air Force. Developed from the well-liked D.H.4, the D.H.9 bomber at first was a flop—it could not perform at the desired 15,000 feet, but had to cruise lower, where enemy fighters enjoyed easy pickings. Re-engined with the American "Liberty", the D.H.9A became a success, surviving into the postwar era. (Jack McNulty Col.)

A German artillery emplacement at Pozières, France destroyed by counter barrage in November 1916. Usually, targeting was aided, if not fully directed, by aerial observers. (NAC PA923)

John Deremo's crew with R.E.8 B6661 at 15 Squadron, Vert Galland, France in June 1918: Skinner (engine mechanic), Graham (rigger), Deremo (pilot), Atkinson (observer) and an unknown photo section type. Deremo and observer Wreford almost didn't make it through the war. At the last moment (November 9, 1918) they were shot down and injured. Notice the observer's Lewis gun, and the pilot's side-mounted Vickers. (CF PMR72-378)

reconnaissance. On four successive days in August 1918, while engaged in contact patrols, he drew fire in order to locate German positions. His most spectacular flight, however, was on October 8, 1918, flying D4848. Historians R. Vann and R.C. Bowyer described the event in *Cross and Cockade*, Vol.4, No.2, Summer 1973:

*Another example of the dangers of low flying took place on October 8 when Captain R.A. Cross, the squadron expert on low level oblique photography, and his observer, S. Adler, went out at 0730 for a counter-attack sortie. Near Walincourt... German troops in and around several orchards were seen and machine gun fire started from the ground. The R.E.8 crew flew at 500 feet, replying with front and rear guns. As Cross returned over the British lines, a direct shell hit tore out several interplane struts and the "Harry Tate" came down heavily in an orchard at Ardissart Farm, on the Lesdain-Villers road. Cross was slightly injured. Both officers ran to a shelter in the farm buildings and took refuge in a cellar from a heavy artillery bombardment then in progress. They had hardly entered the cellar when it received a direct hit, killing six other occupants. Both men escaped further injury, however, and within two hours were back at their unit.*

The same source describes a sortie by 2Lt John C. Deremo, DFC of Toronto. He had served in No.15 Squadron (R.E.8) since November 20, 1917 and taken upwards of 1200 photographs of enemy positions. In this instance he was flying C2796:

*On November 9, Lts. J.C. Deremo (pilot) and his observer, W.J. Wreford were on counter-attack patrol just after 0800, flying at less than 500 feet, when they ran straight into the crossfire of some ten machine guns from the ground. The rudder control was shot away, both petrol and oil tanks ruptured and the R.E.8 became uncontrollable. It finally made a heavy landing not far from the village of Berelles and ran into a wire fence. Withdrawing to a nearby wood, they buried all documents except one map and eventually, by various stages, reached safety.*

Deremo and Wreford's story would have been merely one of evading capture, but a British report subsequently stated that they brought back "a most valuable report as to the enemy's movements and intentions, part of which information was gleaned from civilians in the vicinity of his hiding place."

Lt Jonathan Martin Brown, born in Woodstock, Ontario, was living in Saginaw, Michigan when he enlisted in the RFC in Toronto in August 1917. He trained in Canada and sailed for Britain in January 1918. He served briefly in No.8 Squadron (F.K.8), then joined No.35 Squadron (F.K.8) on April 13. The unit was engaged in efforts (ultimately successful) to halt the German spring offensive. On August 9 Brown survived a damaging attack by an enemy fighter. On September 15 he flew a series of remarkable sorties. Photographs were needed of the Hindenburg Line and the St. Quentin Canal. He was up five times, and although heavily machine gunned, obtained oblique pictures that proved invaluable to the troops. To protect himself, yet attain the best results, he sought cloud cover, waiting for favourable light and then dashing out to reconnoitre.

Three days later a special request came from a divisional commander to photograph an objective 2000 yards behind the fortified German line. Brown descended to 1200 feet and got the coverage. Back at his airfield, the films were developed; 200 copies, dropped to British troops within 3 hours of the mission, were of immense value during an attack the next day. Before they had reached their goal, Brown was overhead, photographing the front 1000 yards beyond the objective. On November 2, 1918 the London Gazette announced that he had been awarded a DFC with the following citation (Brown never read this—he was shot down and killed by anti-aircraft fire on October 3): "This officer has shown exceptional skill and courage in obtaining oblique photographs of enemy positions… These photographs were of vital importance in carrying out our attack; realizing this, Lieutenant Brown, despite most adverse weather conditions, succeeded in taking them…" (Brown's story is related by Don Neate and Mick Davis in the Winter 1996 edition of *Cross and Cockade International*.)

That Brown had been downed by ack-ack was significant. A survey of No.35 Squadron's losses in 1918 show that enemy aircraft and ground fire were roughly equal in claiming F.K.8s. A more thorough study of another Corps squadron— No.9—concluded that its losses, accidents aside, had been attributable to ground fire (37%), enemy aircraft (52%) and unknown causes (11%). Nevertheless, from March 1918 onwards anti-aircraft fire was more significant, while German fighters dwindled in efficiency, reflecting declining enemy training standards. Thus, in August-September 1917 No.9 Squadron had 60 air combats resulting in 13 R.E.8s lost, or roughly 1 per 4 encounters, while 1 enemy aircraft was destroyed; for August-September 1918, the squadron had 23 aerial combats and lost 3 R.E.8s (1 per 8 clashes), while 3 enemy fighters were destroyed. March to November 1918 saw 4 R.E.8s destroyed by ack-ack, 5 by fighters, 1 by friendly fire and 1 by unknown causes. No.9's pilots included Capt James Eric Croden of London, Ontario, who survived 6 combats with enemy aircraft between August 29 and October 30, 1918. He was awarded a DFC in November 1918 for reconnaissance work, as well as low-level infantry strafing.

## The Anti-Submarine War

Aircraft were deployed with varying success against submarines during the First World War. Although there were few sinkings (fewer than claimed at the time), the effect of air patrols on submarine movements was marked. U-boats dived more often and were more cautious approaching convoys, when patrolling airplanes were about.

Harold Morrison Gonyon was born in Chatham, Ontario in May 1896 and raised in nearby Wallaceburg. He was on the way to becoming an electrical engineer, when he decided to join the RNAS. He first obtained an Aero Club of America pilot's certificate at the Curtiss School in Newport News, Virginia. On June 30, 1916 he was commissioned as a Temporary Probationary Flight Sub-Lieutenant and sailed for Britain. His log and a few papers, held by the Canadian War Museum, describe a varied career. He began training September 10, 1916 on FBA flying boat (No.9606) at Lake Windemere; after seven flights totalling 139 minutes, he soloed. Next he went to Calshot for more advanced instruction on FBAs, then Short 225s. On December 28 he first reported practice bomb dropping, air firing, spotting and photography. He now was posted to Dover, commencing operations on February 22, 1917 on the Short 830.

Gonyon's first flights were local familiarization hops. His first patrol was on March 11 (aircraft 1340) to the North Foreland and mid-Channel (86 minutes). Many reconnaissance and patrol missions followed. Most were without incident, but a few log entries are intriguing for the conditions they illustrate:

April 4, 1917, Short 830 #1344,
130 minutes at 2,500 feet.
*Patrol to Dungeness. Engine failed. High tension leads repaired and patrol carried on.*

April 8, 1917, Short 830 #1346,
180 minutes at 2,000 feet.
*Calais-Gravelines-North Foreland patrol. Forced landing at St. Margarets Bay.*

April 28, 1917, Short 184 #8102,
60 minutes at 3,000 feet.
*Hostile aircraft patrol. Short scrap over Goodwins [Goodwin Sands]; Boche beat it.*

May 14, 1917, Short 830 # 1344,
135 minutes at 1,000 feet.
*Patrol in search of machine from Manston. Two enemy mines sighted on the Buoy Line.*

June 16, 1917, Short 830 #1337,
50 minutes at 1,500 feet.
*Gris Nez-Calais and mid-Channel; landed 1½ miles off Calais. Engine overheated and copper jacket loosened from steel cylinder allowing water to escape. Towed into Calais harbour.*

June 17, 1917, Short 830 #1337.
*Towed from Calais to Dover by M [Motor Launch] 239, two hours.*

On August 7, 1917 Gonyon was introduced to the Curtiss H.12 during a flight with Flight Sub-Lieutenant Norman A. Magor (Montreal). From October 1917 he was an H.12 captain. He made his first flight on a D.H.4 on January 2, 1918: "Fair take off and landing; control stick a trifle strange and machine found very sensitive on rudder". However, by February 3 he had warmed to the new type; "Very nice machine and you

A Short seaplane launches from an RN ship. While U-Boats initially operated with impunity, as WWI progressed, they grew wary as RNAS and RFC patrols increased. (Jack Whorwood Col.)

The D.H.4 played a wide role during the war, including in coastal and anti-submarine work in which Canadians like Harold Gonyon took part. This type carried three 112- or 12 25-pound bombs. (E.J. Watkins Col.)

know you're flying". He was soon transferred to the other side of the Channel for anti-submarine patrols; his arrival in France was not auspicious:

February 27, D.H.4 A7773,
35 minutes at 2,000 feet. Dover to Dunkirk.
*Landed well but through not knowing the aerodrome did not slow up as soon as expected and rolled slowly into BHP 6403 damaging chassis of A7773 and tail of 6403.*

February 28, D.H.4 A7867,
30 minutes at 2,000 feet.
*Test flight with two 230-pound bombs and machine took off well in a wind of 20-25 mph. Bombs would not release so had to land with them. Landing good.*

After several days of low-level bombing practice he resumed operations on March 12. Patrols usually comprised formations of 2 or 3 D.H.4s. Flight Sub-Lieutenant Read flew as his regular gunner. Gonyon's log is not entirely clear—some terminology is not explained, but the sense of adventure comes through:

March 13, D.H.4 A7867,
85 minutes at 2,000 feet.
*Patrol for submarine up to Dutch coast. Three destroyers ¼ miles off the Mole at Zeebrugge. Prepared to attack but they were too close in. Flight Sub-Lieutenant Rutter (starting) could not pick up his formation. Failed to pick up listening trawlers off Ostende after searching the whole area. Priceless look at Mole, Ostend and Blankenbughe.*

March 16, DH.4 A7867,
70 minutes at 1,500 feet.
*Dutch coast patrol with machines 7878 and 7870. Trawler bombed off Ostend on return and damaged; one bomb failed to release.*

March 17, D.H.4 A7870,
75 minutes at 2,000 feet.
*Dutch coast patrol. Three enemy destroyers attacked two miles off Zeebrugge Mole and 2-230 pound bombs dropped. One ... seen to explode on the deck ... just aft the rear funnel. The other bomb missed by about 30 feet...*

March 18, D.H.4 A7863,
90 minutes at 2,000 feet.
*Dutch coast patrol. Two trays of Lewis fired at three enemy seaplanes off Zeebrugge. Four hostile seaplanes sighted off Blankenbughe. They were about 300 feet up and made no attempt to attack.*

March 21, D.H.4 A7867,
55 minutes at 4,000 feet.
*Special patrol to bomb 9 enemy destroyers 3 miles north of Ostend. Bombs released from 2,500 feet... All bombs missed from 20 to 50 feet. Heavy gunfire before and after attack.*

March 24, D.H.4 A7867,
100 minutes at 15,000 feet.
*Dutch coast patrol with four machines. Blankenbughe EMB and CMB base bombed from 15,000. Attacked twice by enemy aircraft... Heavy anti-aircraft fire from Ostende and Zeebrugge. Three machines hit.*

March 30, D.H.4 A7867,
  80 minutes at 2,000 feet.
  *West Hinder-Thornton Ridge and Belgian coast. Ran into rain clouds and had to turn back. Attacked a trawler off Ostende piers. Six 230 pound bombs. Probably damaged hull of ship badly.*

April 3, 1918, D.H.4 serial A7867,
  25 minutes at 1,000 feet.
  *Patrol to look for reported submarine. Bombed submarine 8 miles out from Dunkirk, course 20 degrees. First bomb direct hit just ahead of conning tower.*

Gonyon's action of April 3 (D.H.4 A7867, 55 minutes at 500 feet) was the subject of a special report by No.5 Group:

*Anti-Submarine Patrol by No.217 Squadron, D.H.4s. Captain Gonyon and Lieutenant Brown on special mission observed at 16:10 enemy submarine with conning tower awash at position 10 miles N. of Dunkirque. Captain Gonyon released one 230-lb bomb from 700 feet which hit just ahead of the conning tower. Submarine disappeared and quantities of air bubbles and oil were observed. Same pilot dropped another 230-lb bomb, and Lieutenant Brown 230-lb bombs on same position. Having re-bombed their machines, pilots returned and reaching the position where bombs were dropped they observed two large patches of oil and a floating spar. Four 230-lb bombs were dropped on the position by other machines. There appears to be no doubt that enemy submarine was destroyed. Reconnaissance of No.213 Squadron reports having observed a very large patch of oil in the above position.*

S.F. Wise, in *Canadian Airmen and the First World War*, cites German records as indicating no submarines sunk on this date. Nevertheless, Gonyon and his superiors were entitled to believe at the time that they had grievously wounded the enemy. The action was undoubtedly a factor in his receipt of a DFC (June 3, 1918). Although his anti-submarine and anti-shipping patrols continued, Gonyon was increasingly involved in bombing attacks on Zeebrugge with its adjacent defences, sometimes flying two sorties per day:

April 12, 1918, D.H.4 A7867,
  105 minutes, 16,000 feet.
  *Bomb raid on Zeebrugge Mole and government buildings. Direct hits on Mole power house observed.*

April 20, D.H.4 A7867,
  81 minutes, 1500 feet.
  *Bombed what appeared to be periscope of a submarine in the northeast channel; nothing unusual observed after.*

April 27, D.H.4 A7867,
  55 minutes, 2000 feet.
  *Raid on Zeebrugge lock gates. Visibility very bad and continual rainstorm. Forced to turn back near Westende so bombed battery there making direct hits. Captain Lusk got into side slip in cloud and came out 100 feet over battery... badly shot up by machine gun.*

May 1, D.H.4 A7867,
  85 minutes at 3000 feet.
  *Bombed trenches east of Nieuport and blew up ammunition dump. Visibility was too bad for lock gates.*

May 3, D.H.4 A8065,
  105 minutes at 14,500 feet.
  *Bomb raid on Zeebrugge lock gates. Fair shooting but poor formation keeping (6 machines); 400 rounds into trenches.*

May 9, D.H.4 A8065,
  95 minutes at 12,000 feet.
  *Bomb raid on Zeebrugge. Met 7 Huns over Blankenbughe so turned and bombed Ostende as there were only 3 of us.*

May 29, D.H.4 A8089,
  90 minutes at 13,500 feet.
  *Bomb raid on Zeebrugge. Engine failed off Blankenbughe. Managed to scrape over Nieuport piers and land. Machine shelled on beach and set on fire. I was hit and spent fortnight at Petit Cynthe (hospital).*

Gonyon returned to flying on June 25 (test flight), followed by renewed combat, his observer being Gunlayer Day. Gonyon flew his last patrol on June 28, 1918, although he seems to have remained with No.217 Squadron until July 21, when posted to Dover to command an airfield repair detachment. His most interesting entry for this period was faintly amusing: November 11, 1918 (Avro 504 E9370, 40 minutes at 1,000 feet)—"Joyride after armistice had been signed. (Told off by Colonel Halahan)." It may have been that Gonyon had celebrated the armistice with some aerobatics or low flying (celebratory stunts which killed some young men), but Col F.L. Halahan forgave him, providing a letter of reference, describing the Canadian as having shown "very great zeal and officer-like qualities and a fine command of men". Privately, Halahan wrote, "You are just the type of fellow we can ill afford to lose". Gonyon remained at Dover doing occasional test flying and instructing until May 19, 1919, although his last flight had been on February 24, when the final D.H.9s were ferried away. He returned to Canada upon demobilization, but afterwards sought employment in Britain as manager of a sugar factory. During the Second World War he served in the Home Guard. Ill-health forced his retirement to Canada in 1945.

**In the RNAS the Franco-British Aviation flying boat was used mainly in gunnery spotting and scouting. It was the first RNAS flying boat, having entered service in 1912. Top speed was 60 mph. This Italian-built FBA is seen at Otranto, Italy (at the bottom of the Adriatic Sea) in 1916-17, F.C. Henderson of Montreal occupying the right seat. (Finlayson Col.)**

**Large Felixstowe flying boats served the RNAS on shipping, submarine and Zeppelin patrols. Some versions could cruise at nearly 100 mph at 11,000 feet for 5-6 hours, carrying 4 x 100- or 2 x 230-lb bombs. F/S/L Claude C. Purdy of Prince Rupert, BC flew this type. On February 15, 1918 his and another F-boat crew shot it out with three German Friedrichshafen seaplanes. Purdy's machine went into the sea aflame. Three men were seen clinging to the wreckage, but all were lost in rough seas. Here three F-boats are seen cast ashore in rough weather at Killingholme (Humber River estuary) near Grimsby. (Stewart Taylor Col.)**

One of Britain's grand Handley Page O/100 night bombers. Design began in 1915, the Director of the Air Department, Commodore Murray F. Sueter, asking industry for a "bloody paralyser" of a bomber. The O/100 flew at Hendon on December 17, 1915 with pilot LCDR John Babington and engineer LCDR E.W. Stedman (later to become the father of RCAF R&D). Along with Lt Hain, they also flew the O/100 into battle. Operating from Ochey, France with No.3 (N) Wing, they raided the railroad station at Moulin-les-Metz on March 16-17, 1917, putting 11 of their 12 100-lb bombs on target. The O/100 carried 16 x 100-lb bombs, 350 gal. fuel, 5 Lewis guns, 4 crew (all-up weight: 14,000 lb), using two 250-hp Rolls-Royce Eagle engines. It flew at 75-85 mph. H.P. bomber production totalled 636: O/100—46, O/400—549, and V/1500—41. Operational losses were 34. Shown at Ochey is O/100 No.1459 "Le Tigre". (Stewart Taylor Col.)

## Heavy Bombers

The First World War saw heavy bombers deployed in strategic and tactical roles; technology was not sufficiently advanced to envisage area bombing. Even so, the similarities from one world war to the next are striking. Nothing illustrates this better than a poem, published in the RAF first yearbook (1919), written in the heat of the Great War, yet eerily appropriate for 1939-45:

*THE NIGHT BOMBERS*
*by Captain Paul Bewsher, DSC*

*Dusk is our dawn, and midnight is our noon;*
*And for the sun we have the silver moon;*
*We love the darkness, and we hate the light;*
*For we are wedded to the gloomy night.*

*When in the East the evening stars burn clear*
*We know our time of toil is drawing near;*
*For as the evening deepens in the West*
*It brings an ending to our day-long rest.*

*One after one we slip into the gloom,*
*And through the dusk like great cockchafers boom;*
*High in the stars you hear our mournful cry*
*As we sail onward through the sapphire sky.*

*The twilight shadows welcome in our day;*
*The silver dawn will hurry it away.*
*The golden stars act as a changeless guide -*
*The gloomy skies our wanderings will hide.*

*The Rhenish cities hear our throbbing hum,*
*And o'er the Belgian coast we go and come.*
*From Zeebrugge to Metz our name is cursed,*
*At every township where our bombs have burst.*

*The cunning searchlights haunt the midnight skies,*
*Where chains of emerald balls of fire rise,*
*To mingle with the spark of bursting shells -*
*High in the darkness where the bomber dwells !*

*Across whole countries we move to and fro*
*As on our restless pilgrimage we go:*
*With tanks filled up with petrol and with oil,*
*With loaded bomb-racks—all the night we toil.*

*We know the meaning of the lights which shine*
*Upon the world beneath—each is a sign,*
*Which tells us of some dim and frightened town,*
*Which dreads to hear our bombs fall*
    *whistling down.*

*Or of some railway junction full of dread,*
*Whose workers hear us thunder overhead,*
*And darken every lamp—that we may pass*
*And leave no twisted rails and shattered glass.*

*We know the meaning of the sudden glare*
*Of dazzling light which blossoms in the air.*
*For us the green and scarlet rockets blaze*
*And whisper urgent secrets through the haze.*

*The dials with their phosphorescent face*
*Record our passage through the star-lit space;*
*Our height, our speed, the lapse of time is told*
*By steady fingers, calculating, cold.*

*Above a strange and darkened world we ride*
*And over dim mysterious forests glide;*
*When we are silent, we can move unknown;*
*Our only warning is our engines' drone.*

*Dusk is our dawn, and midnight is our noon;*
*And for the sun we have the silver moon;*
*We love the darkness, and we hate the light;*
*For we are wedded to the gloomy night.*

An outstanding example of a night bomber pilot was Cecil Hill Darley. Born on April 23, 1889 in Shropshire, England, he was living in Montreal when the war broke out. As an engineer he could have had his pick of army jobs. Instead, he applied for the Royal Naval Air Service. It was early in the war and Canadians had to go through a rite of passage—securing a pilot's certificate from a private school. Darley attended the Curtiss School in Toronto, passed his tests on September 1, 1915. He was commissioned as a Probationary Flight Sub-Lieutenant in the RNAS, which also refunded him the costs he had incurred at Curtiss. He subsequently trained at Hendon and Eastchurch, then was assigned to No.4 (Naval) Wing in May 1916. One of his earliest sorties ended ignominiously, when he ran into telegraph wires in a B.E.2c. Other trips were more successful (details from Hitchins):

August 2, 1916
    1 Farman, 5 Caudrons, bomb raid on St. Denis Western Aerodrome. Caudron 9123; 4 hours 21 minutes; left at 11.42 a.m. Reached objective at 1.40; four 65-lb bombs from 9,000 feet; exploded near 4 aircraft on ground.

September 15
    5 Caudrons made bomb run on Tirpitz battery [a German coastal artillery position]. Caudron 9123 left at 4.49; 2 112 lb. bombs from 4,500 feet. Machine picked up by searchlights but only 1 anti-aircraft gun fired.

Canadians also flew the 85-mph Armstrong Whitworth F.K.8 "Big Ack" bomber/observation plane. This early model (B856) had oddly arranged radiators forming an "A" atop the engine, and had been rebuilt from salvage. On March 27, 1918, 2Lt Alan A. McLeod (No.2 Squadron) and observer Lt Arthur Hammond, while bombing and strafing in F.K.8 B5773, shot down a Fokker Dreidecker. Now seven more Dreideckers attacked. Both men had to climb from their flaming cockpits, yet McLeod effected a crash-landing. McLeod, bearing at least five bullet wounds, pulled Hammond toward friendly lines, as German fire fell around him. Australians came to the rescue. McLeod received a well-deserved VC, Hammond a Bar to his MC. McLeod was invalided back to Canada. (CF PMR82-603)

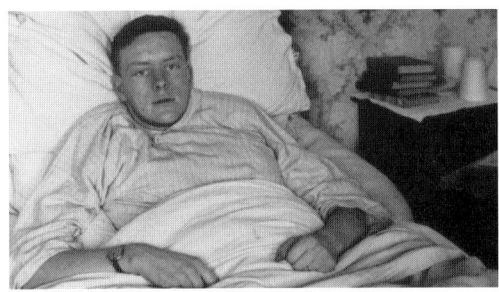

Lt McLeod recuperates. Having survived the caldron of fire amongst the Hun, he fell to influenza in Winnipeg in 1918. Hammond, although he lost a leg, survived. (NAC PA6736)

September 23
*6 Caudrons; raid on Handzaeme Aerodrome; extremely unfavourable weather; heavy ground fog. On Caudron 9123. Left about 5.15; reached objective about 5.45. Dropped 4 65's, 6,000 feet. Could not see aerodrome owing to thick ground mist; saw village though.*

November 15
*Combined raid on Ostende by Nos.4 and 5 Wings; almost ideal weather conditions; 22 aircraft; great damage; large fires; great explosion. Short bomber 9321 (Gunlayer A.M. Kirby). Was hit by anti-aircraft fire and forced landing on beach near Dunkirk. (4 Short bombers, 7 Caudron, 9 Sopwith, 2 Nieuport).*

November 17
*6 a.m. Perfect weather conditions; most successful raid on Ostende by 20 aircraft of Nos.4 and 5 Wings; 171 bombs; damage even greater than on November 15; direct hits on Atelier de la Marine [naval repair facilities]; 4 Short bombers, 6 Caudron, 9 Sopwiths, 1 Nieuport. Short 9313 (Gunlayer Kirby).*

February 3, 1917
*Early morning raid on Torpedo Boat Destroyers in Bruges harbour by 12 aircraft of Nos.4 and 5 Wings; extreme cold; only 6 reached objective; others had engine trouble due to cold and had to land. Darley reached objective. Short 9337 (Gunlayer Bager).*

February 10
*Early hours; 2 Shorts raided shipping at Bruges. Visibility hazy; 12 65-pound and 4 112-pound bombs dropped by 2. Bursts in corner and on Torpedo Boat Destroyers. Short 9338 (Gunlayer Kirby).*

April 30
*Four aircraft left to attack shipping at Mole, Zeebrugge; second pair encountered thick mist and returned. First pair dropped 4 100-pound and 12 65-pound bombs; exploded on objective but results not seen. On return, F/L Darley encountered thick mist and lost his bearings. After flight of 7 hours and 17 minutes he returned safely with Observer.*

The wing's old Short and Caudron bombers now gave way to Handley-Page O/100s. On August 13, 1917 the *London Gazette* announced that F/L Darley had been awarded the Distinguished Service Cross: "For conspicuous skill and gallantry on the night of the 2nd July 1917. One of his engines having seized whilst he was over Bruges, he dropped his bombs on the objective and managed to fly his machine home on one engine and effected a safe landing on the aerodrome." Darley continued in night bombing. On August 18, 1917, raiding Belgian docks, he came low enough to strafe railway sidings. Raiding Bruges on September 5, 1917, he scored direct hits on ammunition stores and started many fires. On a further raid (September 25, 1917), flying Handley-Page No.3129 (one of 10 engaged), he bombed and strafed rail lines. He attempted to fly two sorties that night; bad weather forced him to abandon the second. On February 18, 1918 Darley completed two raids on St. Denis Westrem aerodrome, the longest double trip successfully performed to date by any of the Handley-Pages at Dunkirk. He was awarded a Bar to the DSC on April 17, 1918, by when he was on No.214 Squadron. On July 2, he was awarded a DFC; the citation paid tribute to his night work. Unhappily, on September 28, 1919 Darley's Handley-Page O/400 hit a tree while taking off near Lake Bracciano (Italy) for Cairo; fire consumed the bomber and ended the life of a much-decorated Canadian hero.

Another notable Canadian on bombers was Arthur Barlow Whiteside of Inverness, Quebec, who interrupted his studies at McGill University to enlist in the Canadian Expeditionary Force. He served with the Canadian Army Medical Corps, attached to the Princess Patricia's Canadian Light Infantry, and was wounded in June 1916. Whiteside switched to the RFC, trained as a pilot, and was posted to No.102 Squadron (Bristol F.2B Fighter), serving from October 20, 1917 to June 7, 1918. One scarcely can grasp the man's courage and endurance by reading the citations to his gallantry awards:

Military Cross - announced in *London Gazette* of March 26, 1918.
*For conspicuous gallantry and devotion to duty. He carried out several night bombing raids with great success, attacking enemy aerodromes, trains and billets, often from a low altitude. On one occasion, he attacked a train with his machine gun from a height of 100 feet. He showed splendid skill and initiative.*

Bar to Military Cross - announced in *London Gazette* of September 16, 1918.
*For conspicuous gallantry and devotion to duty. This officer has taken part in over fifty night bombing raids, many of which, carried on at heights considerably under 1000 feet, and in adverse weather conditions, were only successful through the skill and energy displayed by him in discovering and attacking*

*his objective. On one night ... having successfully bombed a large ammunition dump, which was set on fire and blown up, he ... drop bombs on a town which held large numbers of the enemy, also firing from a low altitude with his machine gun on the roads leading to it. Returning to his squadron he obtained more bombs and ammunition, and with the same observer proceeded to drop bombs on a train behind the enemy's lines. On several occasions his machine was badly knocked about by enemy fire from the ground.*

Summaries of Whiteside's work consist chiefly of dates, bomb loads and target names. On November 7, 1917 he released a single 112-pound bomb on hangars from 750 feet, then went down to strafe. On several occasions he flew two sorties in one night, three on November 30, 1917. On January 26, 1918, by the light of parachute flares he and another pilot observed three aircraft on an enemy airfield. They descended to 300 feet, scattering 25-pound bombs over the hangars and parked aircraft. His own machine took a bullet through the main fuel tank; he struggled back to Allied lines on a reserve tank and force landed before reaching base. At the conclusion of his tour Whiteside won three months leave before becoming a bombing instructor. He was killed in England on April 22, 1919, when his Handley-Page bomber failed to clear a building on takeoff.

An interesting account of bomber operations (preserved in the Creagen Papers) is that of Capt James Gray, a former University of Toronto student. After training in the RFC (Canada), he instructed until the autumn of 1917. After advanced training he was posted to No.27 Squadron (D.H.4), reporting to Cerby aerodrome in late December. Excerpts from his narrative pick up the story:

*The winter of 1917-18 was a cold and snowy one in northern France, there being snow on the ground when I arrived there and for a month afterwards... I went up on a few familiarization flights after arrival at 27 Squadron. On one ... I took Lt Ronald, one of the more experienced observers, to have a look at the "line" which was a series of dirty looking trenches and shell holes in a white snowy landscape. When returning to our aerodrome we ran into a snow flurry ... I had lost track of the location of the aerodrome ... The navigator also was lost, so we flew west until I was able to see the ocean, so as not to get on the German side of the "line". I landed in a field where I saw a farmer ploughing and my observer, who spoke good French, was able to get our position... We got back to our aerodrome as darkness came on.*

*The weather remained unsuitable for bombing for quite a while, although we were able to make a bombing raid on January 4th, on the railway sidings at Denain, this being my first "show"... We had a formation of about eight or so planes and, as it was only the third time I had flown in formation, I was kept rather busy keeping my place. We ran into plenty of anti-aircraft fire or "Archie"... and were attacked by a formation of enemy aircraft... I did not see any too much of the fight as I left the shooting to my observer, Lt [J.A.] McGinnis, a fellow Canadian, and concentrated on holding formation, but when we returned to the aerodrome we had one plane missing. Our bombing altitude on this flight was 13,500 feet, while the duration was 2 hours 35 minutes. Our planes always took off singly for these "shows" and climbed to height and closed formation over the aerodrome, it taking about an hour to complete this before crossing the line, which was usually at ... 12 to 15,000 feet. The importance of keeping a good close formation was not realized at this time and pilots were being sent to the "active" squadrons with very inadequate training...*

*Early in March the squadron moved to an aerodrome near Villers Bretneux on the Somme, where we continued our bombing raids chiefly on the different railway sidings behind the German line, such as Roulers, Inglemunster, Markkquain, Friedmont, Menin, Dezne, Basinay. I also did a couple of photography reconnaissance shows to Bachant and Audinard. On these ... we went over alone at the highest altitude we could on D.H.4s, which was from 17 to 20,000 feet, as we depended on our altitude for freedom from attack by Huns. We were seldom "archied" at this altitude but were occasionally jumped by Hun scouts. The only time I myself was jumped was by a flight of Bristol Fighters, which I presume had been sent over on an offensive patrol. Fortunately, they recognized my machine before opening fire, but they gave me and my observer a fright as we were well behind the line...*

*I was given command of a flight and promoted to Captain ... On my first show as formation leader to Basinay we ran into Hun scouts over the target. I believe they were "Richthofen's Circus" as each plane was painted a different fancy colour with red noses. Our formation was broken up in the ensuing "dog fight" and I ended up with a Hun on my tail, who managed to put a few bursts in our machine until my observer, Lt McGinnis, was able to drive him down with the rear Lewis gun. On the way back my engine failed before we got over the line. However, I had enough height to glide across the line, although we were heavily "archied" on the way. I managed to make a forced landing ... on our side of the line without further damage...*

*I found on landing that there was a bullet hole in the oil tank, among a few others... the engine had seized when the oil ran out. The machine was in "dead ground" and was not under observation from the German side of the line. I found an Army Service Post not far away and got a message through to my squadron to send a truck to dismantle the plane and take it away before it was observed by the Germans. However, they only sent a tender to pick up Lt McGinnis and myself and we heard later that ... the Germans shelled the machine and destroyed it... At this time the Germans were building up their forces ... our bombing raids were usually met by German planes whose favourite trick was to ... attack as we were dropping our bombs... We then had to fight our way back to the line, where they usually turned back... On one of these attacks I had just crossed back over our line and was looking around for the other planes of our formation when I spotted one well below me with a Hun ... firing at him. I pointed them out to my observer, Lt McGinnis, and put the plane into a sideslip to enable him to bring his Lewis gun to bear on the Hun. After he had fired a drum of ammunition I started to kick off rudder to get into position to dive on the Hun and bring my Vickers forward gun to bear on him, when I found my rudder was jammed... I shut off the engine and shouted to McGinnis, as our cockpits were about four feet apart on the D.H.4s ... to check the controls in his cockpit, where there was a duplicate set. I saw him stoop down and then throw an empty Lewis drum overboard. I then found my rudder was free. After this he threw all empty drums overboard; he had dropped the drum in the cockpit after it was empty, where it had got under the rudder bar and jammed it. By this time we were unable to see anything of the other two planes so returned to our aerodrome. The other plane ... which I had seen being attacked, did not return and we heard later that it had been shot down on our side of the line...*

*The Germans mounted a strong attack on this part of the line (the Somme) about the middle of March and we were forced to evacuate our aerodrome on Villers Bretneux. The front line eventually passed right through this aerodrome after the German push had been brought to a halt. We moved to Candas aerodrome for a few days and it was from there that 27 Squadron was assigned to low altitude ground attack... Each plane went across the line alone with bombs and plenty of machine gun ammunition with orders to attack any target we saw, such as troop concentrations, transport, aerodromes, artillery, observation balloons, etc. Such attacks were made from ... 1000 feet or less. It was very exciting work, but the danger from ground fire was great and we usually returned with some holes in the machine. On one of these attacks I thought I had been shot myself, as there was a loud crack and a sharp pain in my right leg. I immediately started back to our aerodrome feeling my leg for blood without result. On landing I found that a bullet had come up from below through the floor of the cockpit, which was of thin hardwood, and a piece of the wood had hit me on the leg.*

*After a few days at Candas we moved north to a new aerodrome at a small village named Russeauville, not far from ... Agincourt... We were billeted in the village here until our tents could be erected, as the British front seemed to be in a bit of a mix-up after the retreat on the Somme.*

*Several days after our arrival at Russeauville, the village was bombed one moonlit night by German night bombers and several houses destroyed with casualties to the mechanics of our unit. The Germans now started another offensive on the northern part of the British front opposite Armentières and we were put on to bombing railway yards and ammunition dumps on this part of the front. I remember us setting off a tremendous fire and explosions in one dump ... We also bombed the railway yards at Bruges where the "archie"...was particularly heavy. Our squadron did not lose a single machine from "archie" on high altitude bombing while I was with them, although machines often returned with shrapnel damage. I had my auxiliary gas tank punctured by this means on one occasion.*

*My observer, Lt McGinnis, was sent back to England early in April to take his pilot's course... After trying a few others I selected Lt Gooding... He had been a machine gunner in the infantry and was a cool hand with the machine gun and not as excitable as Lt McGinnis who, I remember, put a burst into my tailplane once when we were dogfighting a Hun. Lt. Gooding was an artist in civilian life and used to take his sketch book along with him...*

Gray took leave in England in early May. While there he was invited to lecture to aviation cadets at Cranwell. In the end he ferried a D.H.9 to France and resumed operations:

*Early in June, 27 Squadron was moved to an aerodrome near Beauvais opposite the Compiègne part of the French front, where the Germans were mounting an attack against the French. We made a good many bombing raids there with the railway yards behind the German lines as targets and also some low level ground strafing attacks. We were night bombed by the Germans several times when we were on this aerodrome and suffered some casualties among our personnel.*

*On June 21st we returned to our former aerodrome at Russeauville and continued our bombing on this part of the line until July 15th. On this date we were moved down to the French front again in the Champagne district, near Chailley, where the Germans were attacking and had crossed the Marne River. We were immediately put onto low level attacks on the bridges that the Germans had put across the Marne River... had the satisfaction of seeing one of my bombs hit the end of one of the bridges and seeing it collapse into the river with artillery transport on it. We ran into "flaming onions" here which was an incendiary projectile that the Germans used and appeared as a series of balls of fire in bursts of about a dozen or so coming up at you. Rather "windy" but not too accurate. My own method of putting off the aim of the enemy on these low level attacks was never to fly straight, but to be turning, diving or climbing all the time I was over the enemy lines...*

*We had pretty heavy casualties in 27 Squadron during the spring and summer of 1918. It was said that the average life of a pilot or observer over the line at this time was only six weeks, and I can quite believe it. The pilots coming out to the squadron were very inadequately trained with only about 20 hours solo. We had no parachutes or flame-proof fuel tanks and ... planes with their crews going down in flames was a frightening sight. We heard that the Germans were using parachutes and armour plate on their machines by the summer of 1918... It must have been a boost to the morale of their airmen to have a means of escape from a "flamer".*

*On August 3rd we moved from Chailley on the French front to an aerodrome called Beauval on the British front not far from Amiens... This attack started from Villers Bretneux, our old aerodrome which we were driven off in March. The attack started on August 8th and we did a low show, attacking aerodromes and communications behind the German lines. This attack was a success and was the beginning of the end for the Germans... Shortly after ... I was laid up with hepatitis and a bad case of jaundice, so was invalided back to England after nine months with 27 Squadron. I was senior pilot with the squadron by some five months by the time I left and can only attribute my survival to some 150 hours flying experience before going to France, plus good shooting by my observers, plus a whole lot of luck, with emphasis on luck.*

Gray's combat report for March 8, 1918, found in the National Archives of Canada, illustrates the importance of the gunner. Gray and his gunner (2Lt McGinnis) were in D.H.4 B2088 at 14,500 feet when German fighters appeared:

*The formation of four D.H.4s dropped bombs on objective (Busigny). When turning we were attacked by 12 ... Albatros Scouts. One of our machines (Lieutenant Perkins) turned to engage 3 enemy aircraft with his forward gun, but was seen to go down in flames. Before he went down, Lieutenant Foley, his observer, appeared to hit one of the enemy aircraft, which went down in a spin, apparently out of control. Lieutenant McGinnis put a long burst into another enemy aircraft which stalled, turned over on its back and went down out of control. Altogether, Lieutenant McGinnis fired 480 rounds. We did not see what became of the third machine as we were engaged with enemy aircraft. We were forced to land northeast of Peronne with engine trouble.*

Two encounters by Canadians of No.57 Squadron further illustrate bomber-fighter conflict. On August 14, 1918 Lt Eric M. Coles was piloting D.H.4 B8377 at 15,000 feet; Sgt J. Grant was his rear gunner. Coles reported: "Whilst taking photographs of enemy aerodrome south of Roisel, the formation was attacked by 8 Fokker biplanes. Two climbed above D.H.4's tail and Observer fired about 5 bursts of 12 rounds each into the nearest machine. The enemy aircraft fell over on its side and immediately went into a spin. Observer ... saw it crash into the ground. The pilot jumped out in a parachute. The other enemy aircraft ... flew off."

Lt Herbert Leonard Rough was born in Britain, but was living in Victoria when war broke out. He went overseas with the 48th Battalion, Canadian Expeditionary Force, but transferred to the RFC in April 1917. Trained as a pilot, he served in No.49 Squadron (D.H.4) from November 1917 to June 1918. Shortly afterwards he was awarded a DFC with the following citation:

*A keen and gallant officer possessing great skill and judgment. He has carried out twenty-six successful bombing raids and five photographic long distance flights. On a recent occasion, when low bombing, having obtained a direct hit on a limber, he dived to seven hundred feet and, by skillful manoeuvering, three enemy aeroplanes on the ground were set on fire, one by a direct hit with a bomb, and two by observer's fire. During the whole time his machine was subjected to heavy machine-gun fire. With the assistance of his observer, he brought his machine safely back, though he was shot through the leg.*

Rough and his gunner, Lt V. Breschfeld, had their share of aerial combats. On January 13, 1918 he was on reconnaissance at 16,500 feet, when enemy aircraft appeared:

*While doing reconnaissance I saw 5 enemy aircraft below, 3 of which were biplane two-seaters and 2 triplanes. The 2 triplanes climbed above us, whilst the biplanes attacked from below. We turned and fired ... 20 rounds from the Vickers gun at 200 yards into one of the biplanes, all of which then turned away climbing. One of the triplanes then dived at us and we fired a burst of 20 rounds from the Lewis gun; the triplane then turned over when we fired another ... 20 rounds with the Lewis and it went down in a spin completely out of control. The remainder of the enemy aircraft then broke off ... but continued to follow us from a distance as far as the lines.*

Rough filed another report on May 17, following a reconnaissance at 14,000 feet. Although neither he nor his gunner made any claims, the mission was remarkable. At no time was the photography abandoned, although D.H.4 C6093 was attacked three times. Each time the enemy was driven off. More decisive was an encounter on May 19 during a bombing raid at 13,500 feet in C9063:

*After having dropped our bombs, the formation turned south, and I noticed 10 enemy aircraft... immediately dived upon us out of the sun. I ... pulled up ...in a stall and put a burst of 100 rounds Vickers into the enemy aircraft near to us at a range of 40 yards. This machine turned over and nearly crashed into us... was unable to observe it further as two machines were attacking the rear of our machine. My observer fired 100 rounds Lewis at the one nearest at a range of 30 yards, when it broke off the combat. The second enemy aircraft sat on our tail and my observer fired a further 100 rounds, when it broke off...*

Christian Burgener (right) with two of his 214 Squadron crew at their stations in an O/400. Then, another view of Burgener. These photos show many of the H.P.'s details. Burgener's last night on ops (November 10/11, 1918) also was his last, and the greatest such raid of the war— 31 bombers took part, Burgener's being C9646. (Burgener Family Col.)

## Christian Burgener: Dairy of a Bomber Pilot

Another Canadian on Handley Page bombers, Lt Christian Burgener, kept an excellent diary; in 1978 he loaned it to the author, who was researching for his first book, *Aviation in Canada*. A photographer with a studio in Picton, Ontario before the war, Burgener had gone overseas as an infantryman in the Eastern Ontario Regiment, but transferred later to the RFC. He began flying on October 12, 1917, going aloft in a Maurice Farman for 15 minutes. By November 26 he had logged 7:50 hours. In December he was flying at Narborough in Norfolk, where he and his friend, W.J. Dalziel of Saskatchewan, did advanced training on the B.E.2e. On December 11, having flown 11:55 hours, Burgener soloed. His 5-minute flight was not auspicious. He stalled on take-off and nearly came to grief, then crashed into a tree on landing: "Smashed my nose and spent a few days in hospital. Had a beautiful black eye for about two weeks ... Dalziel also crashed on Dec.17th."

The winter weather in Norfolk was poor, assuring the student aviators their share of boredom. Mail was important during such spells. On January 18 Burgener wrote: "At last, mail from Canada. Got five letters. Raining practically all day, so had a good chance to read them." On January 21 he took some instruction on the D.H.6, soloing the same day. Four days later he remarked: "Fly my old D.H.6. Like them very much. Also had a ride in an R.E.8. Think I will like them as well." Training grew in intensity.

February 18, 1918:
> Got a new Bus today, little A.W. [Armstrong Whitworth]. Russell takes me up and loops eight times. Do not like the sensation after the third or fourth attempt, but on the whole think they are a fine Bus.

February 21
> Very windy but put in one hour and 15 min. in the air ... very bumpy. Have developed a wonderful confidence in the air now, and simply love flying. What a contrast to about six weeks ago. Passed all tests now and expect to graduate soon. Poor Shaw and Low are buried today. [They had crashed two days earlier.]

February 23
*The third fatal accident occurred today. One of our best pilots, Lt Law, a Canadian who had done a lot of flying, crashed a D.H.4.*

March 21
*Fine trip today on photographic work. Remained in the air for two hours. Stayed over the water over ½ hour taking pictures of English Zeppelin.*

March 22
*Put in some more time in R.E.8, "graduate", and get permission to take up passenger ... Anderson of 121 Squadron as first passenger. Am now supposed to be full-fledged pilot ... do not feel like it, although have done 48 hours solo flying.*

March 28
*Received my pilot's certificate today. Needless to say, very proud of it.*

May 30
*... a real good game of baseball between our American boys ... it gives us Canadians all a feeling of home again to hear their typical expressions.*

On June 15, 1918 Burgener was posted to the No.1 School of Navigation and Bomb Dropping at Stonehenge. He arrived two days later, but to bad news: "A black day for the squadron. My friend Lt Aldertin, with whom I was in Lynn last evening and had dinner with, cracked up today in a '4' and paid the price." Burgener soon was busy checking out on new types and learning night flying. His notebooks cover the F.E.2b:

Running on the ground
*Run the engine at about 600 revs until thermometer needle commences to move. Then run it up to 1000 revs and test both mags separately. Open throttle fully and note revs ... Do not run engine full out for more than ¼ minute on ground ... engine should be looked over for water leaks while it is running on the ground. Never let a mechanic climb about on the machine while the engine is running, with his cap on, as he may knock it against a wire, in which case it goes through the prop and may break same, causing a lot of damage...*

Fire
*If the machine catches fire in the air, turn off petrol at once. Close throttle. Switch off. Land.*

Draining water:
*In frosty weather care must be taken to drain off all water. To do this, the tail of the machine must be raised just above flying position and the prop rotated...*

Cross country
*White rockets are sent up when a machine is due back and it is in sight 18-20 miles away ... Permission to land must be asked for by firing the colour of the day. This must be replied to by a like colour...*

Engine trouble
*... If you have to have a forced landing and you have fired Michelin Flares and found no suitable landing space, fly over and fire Very's lights. They give considerable light and a landing place may then be discovered.*

At first Lt Burgener was not excited about bombers. On June 21 he entered: "Saw 'Windy' and he kind of got me to console myself with the idea of becoming a night pilot." Burgener soon adjusted. After a few nights on the Maurice Farman and F.E.2b he advanced to Handley Pages. Meanwhile, there was always time for a bit of fun. On July 18 Burgener visited London where he took in two plays—*Man from Toronto* and *The Naughty Wife*.

June 27
*Go on H.P.s for the first time and this makes my eighth Bus... Am not very stuck on H.P.s, but will try and stick them.*

June 28
*Since I did all my H.P. flying required to pass my day tests, I won't have anything to do for about a week.*

June 29
*... Take a walk to the Stones which make Stonehenge famous. The weather is very hot now and it is none too pleasant hanging around aerodrome and attending lectures.*

July 5
*... From now on am a night bird. Had to put in my first three hours during the night. All went off successfully. At the same time do not find any great fascination in the work and so far would prefer day flying.*

July 11
*Finished all tests on F.E.2b, so didn't have to fly last night.*

July 13
*Moved to H.P. today, and even after staying up all night didn't get a chance to make a good start. First a heavy ground mist came up and then we did get a start. Burst two tyres.*

July 19
*After arriving back from London last night, found that the severe rain here had blown my tent over. Luckily, however, my batman moved everything into a new tent and no damage was done. Had to stay up all night and finish my time on H.P. Got along fine. Did my bombing, cross country, and landings OK. [Six flights in aircraft 9671 for 3.8 hours.]*

On July 24 orders came to report to 214 Squadron at St.Inglevert, near Dunkirk. Burgener arrived the same evening: "Find out here that this is a famous squadron and consider myself very lucky." August 8: My first start today. Go up as rear gunner on raid to Bruges ... do not reach objective on account of dud weather.

August 10
*Another raid as rear gunner with Russell, and this time a real one. Raided Bruges with 1660 lb bombs. Found it quite exciting and saw there a sight which I will for a long time remember. Everything turned out OK. Arrived back at Dunkirk. Land on beach and return from here in the morning.*

August 12
*Am quite sure of getting a Bus of my own almost any day. Would, however, rather put in more time as passenger, but am afraid will not get the chance.*

August 13
*... am on a raiding list for tonight ... Reach my objective OK ... attacks by e/a ... nothing serious ... took the whole procedure very coolly and, after meeting no further troubles, managed to drop bombs and return to Dunkirk ... followed by a Hun but he did not attack ... had the misfortune to run into my hangar and damage Bus, so it will be out of action for a few days.*

August 21
*A very long and tiresome raid. In the air five hours. Got it very hot over Bruges and Zeebrugge. Came back with one engine dud. The night was as light as day and we could see everything below. Thought once or twice that we would not survive the barrage and made for Holland, but we were safe once out to sea again. [On this raid Burgener flew aircraft 3489 carrying 16 112-lb bombs that were dropped on a canal.]*

August 24
*Raid Zeebrugge ... got attacked by three e/a off Ostend ... port engine went dud on return. Again afraid I might have to come down in Holland but managed to get back.*

August 25
*A great surprise today. Old Dalziel came over in his Bus before I was out of bed this morning...*

September 13
*Raining again and the inactivity nearly driving us crazy...*

September 23
*Go to Calais to do some shopping. At night getting settled down nicely writing letters, when orders come to turn out for raid... suddenly cleared and off we went. On my return had the misfortune to come down without warning ... landed OK in water off beach at Clipon. Have to wait an hour before tide goes down and go ashore. Took us two hours of walking around in night to find a telephone. Found one at Casino at Clipon 10 km from Dunkirk. Found out at daybreak that machine is OK. Engines cut out from lack of petrol ... Relief came at 9 o'clock. Got her filled up again, left for home at 10:15. Arrived 30 minutes later. Spent the rest of the day in bed having a well-earned rest. The only thing I had to eat or drink since the night was a cup of*

coffee at a French farmhouse. Fortunately for me, my observers could both speak French nicely. [Soon afterwards, Burgener's observer, Cpl Thomas, who had spent time in the water after the ditching, took pneumonia and died.]

October 4
And another raid tonight ... weather very bad and had a hard time finding our way. However, succeeded in bombing obj. On return, surprised both engines cut out over aerodrome and unable to land on account of another Bus ahead of us getting the right-away signal. On trying to land on upper end of drome, the machine became unmanageable and crashed into woods ... The Bus is a total wreck and we narrowly escaped injury ... again must consider myself exceedingly lucky. Stay in bed all day trying to recuperate from shock.

October 6
Big doings today. Have our RAF band from England, on tour in France. Here for a concert in the afternoon. Got a bunch of the doctors and nurses from the Canadian hospital up.

October 11
War news is continuing good. Everybody begins to think that we won't have to do so very many more raids.

October 14
Another raid tonight. All goes well again ... weather is very bad ... saw tonight the greatest sight I have ever experienced in war flying. Almost the whole of the Belgium front seems to be on fire.

October 18
... signs of us having to move up soon as the Huns are evacuating the Belgian coast and we will soon be able to go and see our own bomb holes. Our C.O. [H.G. Brackley, DSO, DSC] has the honour today of taking the King and Queen of Belgium over to Ostend in a Handley.

October 20
Another day which I will not forget so soon. A bunch of us decided to take a run up to the lines ... Never had even the remotest idea of the devastation and state of the country and the largeness of the whole battle front. I have flown over the district before, but at night and consequently not seen it the same as I have today.

October 30
A perfect day for our moving [from Quilen to Chemy]. Start off at 10 o'clock and after 1:20 hours of very interesting trip arrived at the new place... What a sight ... when crossing No Man's Land. Have fairly good quarters in a chateau where the Huns had been 10 days ago.

November 10
Hate to go tonight, as every hour we expect an armistice and would hate to meet with a mishap... Have engine trouble at first, therefore a long time behind the rest... Raid Louvain siding, which we find all ablaze from previous bombing... The greatest news when I open my eyes in the morning. The armistice has been signed and our work is over, so expect to have done my last raid ... A fine finish and completely satisfied to pack up. Everybody almost wild with joy and celebrate all day.

November 12
All have thick heads today ... it hardly seems possible that the show is over.

The H.P. bomber's 1650-lb bomb—an omen of things to come in WWII, when the 4000-lb "cookie" and even bigger bombs became commonplace in the RAF. H.P.s also carried bombs of 20-, 40-, 112-, 250- and 550-lb. (NAC PA6294)

The H.P. O/400 had twin 360-hp engines, compared to the O/100's 250s, the top speed comparison being 98 mph versus 76 mph. D9689 was from a production batch of 46 O/400s. (David Thompson Col.)

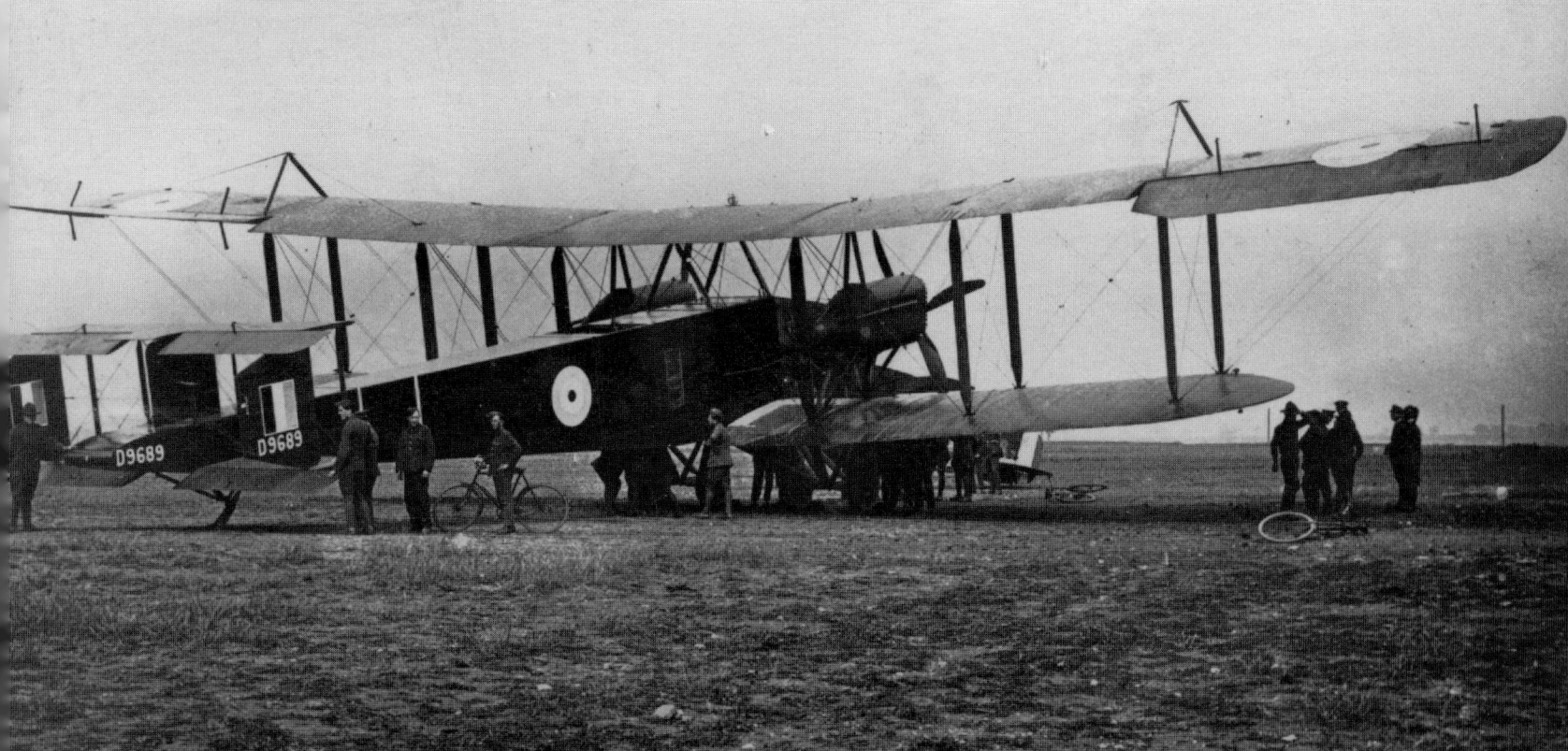

Redford H. "Red" Mulock. The first Canadian to qualify as a pilot in the British services (March 9, 1915), he rose to head the RAF's strategic bomber force. (CF PMR71-404)

## Bombing Leader

From all the stories of the air war, one Canadian stands out—Redford Henry Mulock of Winnipeg. Ron Dodds, in *The Brave Young Wings*, described him simply as "The Man Who Did Everything". The National Archives of Canada has copies of correspondence demonstrating how senior British officers almost quarrelled as to whom would have Mulock on their staff. His rise from private to colonel in four years was even more meteoric than that of Billy Bishop, and in every respect he made a far more significant contribution to the air war effort than any other Canadian. Mulock had begun with the First Contingent, Canadian Expeditionary Force, but transferred to the RNAS (whereas most air-minded CEF members escaped to the RFC). By May 1915 he was bravely (though ineffectually) chasing Zeppelins. On September 28, 1915 he bombed a Zeppelin shed near Brussels, a remarkable feat of navigation, relying on compass and watch. From late 1915 he was routinely spotting for naval guns and artillery along the Belgian coast, breaking off occasionally to pursue German observation planes, or to harass enemy submarines. From March 1916 he was developing the tactic of night spotting for big guns. He also tested parachute and landing flares.

Mulock was awarded a DSO in June 1916. The following spring he took command of No.3 (N) Squadron. Although an active fighter pilot, his scoring was modest. The true mettle of his courage—and modesty—was shown during a fire on July 10, 1917 at an ammunition rail head. For a while his role was unknown, but it slowly emerged. He was commended for "courage and devotion to duty displayed... in rescuing the wounded and salving ammunition under shell fire". Another officer, investigating the affair, reported that Mulock, with a doctor and three enlisted men, had gone to a burning ammunition train: "Surgeon Panter and one Driver took the first wounded man they found away in an ambulance, and after this I was unable to get any definite statements, as Squadron Commander Mulock is very reticent on the subject. I gather from the remaining drivers, however, that he went alone and searched the Dump for wounded and would not allow the drivers to accompany him."

Mulock took command of the Dunkirk Aircraft Depot in November 1917, but soon was brought into more important planning and operational roles, culminating in promotion to colonel and assignment to command the Handley-Page V/1500s that would have bombed Berlin, had the weather broken before the Armistice. At war's end Col Mulock was the second highest ranking Canadian air officer (Brigadier Arthur C. Critchley, who had worked mainly in training, outranked him). Mulock had been made a Commander in the Order of the British Empire, a Chevalier in the Legion of Honour, had two DSOs and was thrice Mentioned in Despatches. He could have had his pick of RAF appointments. Instead, he returned to Canada as a businessman, member of the Air Board, and eventual president of Canadian Airways. He died in Montreal on January 23, 1961, just before his achievements were about to be rediscovered by a new generation of historians.

## The Fighter Pilot Mystique

Although WWI fighter pilots were supporting players in the air war, their actions have received disproportionate attention in published history. In part this was due to wartime propaganda (more particularly in France and Germany). Fighter pilots, who made better copy than artillery observers, were turned into aerial knights, practicing the ancient concept of the duel. Yet, fighter pilots knew that the object of combat was to gain an advantage—in height, numbers, experience, or aircraft performance—and then exploit it for a kill. If the odds were adverse, combat should be avoided. There was little of the chivalry of a duel in the fighter pilot's world.

George Shiras, writing in the American edition of *Cross and Cockade* (Vol.5, No.4, Winter 1964) addressed this same question. On examining contemporary records (e.g. diaries, combat reports) he concluded that the "knights of the air" were a figment of others' imaginations—politicians seeking to mask the war's ghastliness, and reporters who refused to admit the brutalizing effects of battle. One such scribe, after a frank interview with the ace Mick Mannock, tore up his notes and exclaimed that Mannock was a "monster". Pulp writers, more inclined to copy from each other, than to investigate the true nature of the air war, followed suit, mindlessly repeating myths that had been born of denial and propaganda. Shiras concluded that chivalrous conduct could be found—but not in combat. Once an enemy aircraft had been shot down, it was easy to treat its crew in a civilized manner—a fine funeral if they were dead, a mess party if they were well enough to attend, a visit to the hospital if they were wounded. But in the heat

No.3 (Naval) Wing formed early in RNAS days and included many Canadians, a dozen of whom eventually were aces. The first Canadian to score a kill during the war was F/S/L Arthur Ince of N°. 3. On December 14, 1915 he downed a German seaplane off Belgium, while flying as F/S/L C.W. Graham's gunner in a two-seat Nieuport. This No.3 Sopwith 1½ Strutter, a type that was both fighter and bomber, is shown at Ochey in 1917. F/S/L Percy G. McNeil of Toronto is leaning on the wing. He died in combat on June 3, 1917, flying a Sopwith Triplane with No.10 (N) Squadron. (Stewart Taylor Col. via J.S.T. Fall)

F/S/L Edward R. Grange of Toronto is noted as Canada's first ace, being credited with his 5th kill on January 4, 1917, while flying a Pup with No.8 Squadron, RNAS. Then, William W. Rogers of Alberton, Prince Edward Island, the first to shoot down a Gotha bomber. This action was on December 12, 1917, while he was flying a Nieuport 24 of No.1 Squadron: "Just after climbing through the clouds, I saw ... Gothas coming West about 7500 feet... observed one E/A turn back East, so attacked it, firing ¾ drum at 30 to 20 yards range. E/A burst into flames, fell to pieces and crashed North of Frelinghien." Rogers received an MC a few days later. (CF 64-2932, CAHS Col.)

of battle, while doing one's job (which included staying alive and killing the other fellow), it was cruelly practical. One might feel sorry for an opponent who was going down in flames, but compassion did not prevent one from shooting him down in the first place.

The most fundamental tactic of aerial warfare was achieving surprise—by sneaking up on one's enemy, using cloud cover or the sun, and delivering a killing burst of fire from behind. How could such manoeuvres be chivalrous? Fighter pilots did not deliberately square off in one-on-one combats—they tried to achieve superiority (hence, the evolution of the wingman). When the German ace Werner Voss was shot down in September 1917, he was trying to get out of a fight at six-to-one odds, but none of his opponents broke away to allow a more equal fight. Stragglers from enemy formations were attacked without hesitation—no sporting chance for them. Enemy aircraft with jammed guns were as much fair game as any others, and why not? Who could say if the enemy would not clear his guns. Aircraft that had crash-landed were shot up, and observers parachuting from burning balloons were attacked. Shiras quotes a French officer: "Why destroy a balloon that can be replaced and let an experienced observer get down with his observations?" There was nothing personal about such ruthlessness; it was war!

Canadian preoccupation with WWI fighter pilots was remarkable. Chief culprit in this must be George Drew, whose 1930 book *Canada's Fighting Airmen*, became the most widely read aviation work in the country. Writing with no analysis of tactics or events, much less of his heroes' significance, Drew sparked an interest in his subject that had its full impact a decade later at RCAF recruiting offices. Unhappily, *Canada's Fighting Airmen* was a bad book. Its tunnel vision with respect to fighter pilots amazed even reviewers for *The Aeroplane*, the most respected of aviation periodicals. The book incorporated myths as readily as facts. It stated that Raymond Collishaw had taken part in a pre-war Antarctic expedition. This myth took such firm hold, that an artist did a portrait of Collishaw, adding the ribbon of a Polar Medal to which the great pilot was not entitled. Drew named G.E.H. McElroy (an Irishman) as a Canadian ace. The cult of the "aerial knight" as expounded by Drew and others distorted truth in the strangest possible ways. Nowhere was this more evident than in the controversies that have swirled around W.A. "Billy" Bishop. An icon in Canada, Bishop's reputation and fame eclipsed even Collishaw's and Barker's. For a time it appeared that the only criticism one could ever make about Bishop was that he distracted attention from so many other worthy contemporaries.

That began to change when Canada's National Film Board released a Paul Cowan film, *The Kid Who Couldn't Miss* (1974) which suggested that Bishop had faked his VC action—itself remarkable above all else for having been the only Victoria Cross awarded solely upon evidence provided by the recipient himself! Cowan's film was

LCol W.A. "Billy" Bishop, was one of several leading WWI Canadian fighter pilots. Close inspection of the historical record has made his story a controversial one. (CANAV Col.)

a second-rate mish-mash of facts, legends and allegations. Its footage included actors playing Bishop and others, clips of interviews with aging First World War aviators (none of whom had personally flown with Bishop), snippets from 1916 newsreels and a 1927 Hollywood epic, *Wings*. The producer also had leaned heavily on Arthur Bishop's biography of his father, *The Courage of the Early Morning*. The contention that Billy Bishop was a great shot, but a bad pilot, was hardly borne out by the fact that his 1917 Nieuport fighter still was flying with the RAF (in Egypt) in 1921.

*The Kid Who Couldn't Miss* stirred a hornet's nest of opposition. In a committee of the Canadian Senate, critics condemned the film, although many stumbled over the facts themselves. Some wanted the film suppressed, but the final recommendation was that the NFB label it as a "docudrama". Clifford Chadderton, a spokesman for veterans groups and opponent of all "historical revisionism" (whether or not based on fact) wrote *Hanging a Legend: The NFB's Shameful Attempt to Discredit Billy Bishop*, which roundly condemned such iconoclasm as unpatriotic. It defended Bishop's reputation with character references (but no contemporary evidence). Chadderton's principal argument was that, if George V thought a VC was warranted for Bishop, then there must have been substantiating evidence for the honour.

The first volume of the RCAF's official history, *Canadian Airmen in the First World War* (1980), did not question Bishop's account of his attack on an airfield, that led to a VC recommendation, but did include a disturbing footnote (disturbing for the historically orthodox): "The location of the airfield Bishop attacked has never been definitely established... By Bishop's own admission he didn't know where he was". The footnote constituted a time bomb, which would explode half a decade later. Brereton Greenhous, a member of the DND staff which compiled *Canadian Airmen and the First World War*, had been intrigued by Bishop's VC. The British previously had granted aerial Victoria Crosses with political shading (e.g. Warneford, Leefe-Robinson); the Bishop case had such shading—a high award at a time of great losses, coupled with paying respect to a growing colonial component in the air forces. Greenhous, who described Bishop as "a flamboyant extrovert", twice questioned the Bishop legend, first with an entry in a revised edition of the *New Canadian Encyclopedia*, then in a short article, "The Sad Case of Billy Bishop", which was printed in the June 1989 issue of *Canadian Historical Review*. Both proposed that, since Bishop's account of his VC action was dubious, his entire record might be. Chadderton now questioned the suitability of Greenhous for employment at the Directorate of History. To his credit W.A.B. Douglas (who succeeded S.F. Wise at the Directorate of History) would have nothing to do with this attempt at thought control, nor was the Minister of National Defence prepared to censure Greenhous.

Next on the scene was a retired RCAF wing commander, Phillip Markham. He came to the controversy with unique assets. His dedication to Canadian aviation history was unquestioned (he is widely regarded as a founder of the National Aviation Museum). He understood the importance of research, reliance on primary documents (rather than second hand accounts or dated recollections) and he knew how to conduct that research. His work at investigating the circumstances by which Robert Hampton Gray had won a posthumous VC (1945) had been a model of investigative history. Above all, he had no axe to grind. If anything, he undertook to investigate Bishop with a view to *contradicting* the revisionists. Markham wrote a fact sheet about Bishop which was published by the Canadian War Museum in 1989. He did not attack the legend, but endorsed it cautiously.

Markham had pulled his punches. Even as the fact sheet appeared, he had come to the conclusion that Bishop's VC action was unwarranted. After research in Canada, England, France and Germany, poring over maps and contemporary documents—in short, having searched everywhere to try substantiating the VC—he concluded that there was no evidence even hinting that Bishop had done what he had claimed. Uncertain that his findings would be published in Canada (and not wanting to diminish his circle of friends), Markham published them in an American journal, *Over the Front* (Fall 1995). The article, coming as it did from a man of his reputation, was the most credible indictment of Bishop. Yet, its American publication went unnoticed in Canada. It failed to attract the thunderous denunciations, which would have assured wider knowledge of its contents.

But could it be that Bishop told the truth—that the supporting documents have simply been lost or destroyed? Like a trial lawyer testing the credibility of a witness, Markham had earlier investigated another episode in Bishop's career—his account of an inconclusive dogfight with Manfred von Richthofen. He traced this account from a combat report, to a letter to his fiancee, to his autobiography, demonstrating that, at each step, Bishop had expanded the event—one that did not happen; records showed von Richthofen was not even flying that day. This study, titled "A Flight of Fancy", appeared in the Summer 1996 issue of the CAHS *Journal*.

Although published in a Canadian journal, "A Flight of Fancy" did not trigger the orchestrated outcry that greeted *The Kid Who Couldn't Miss*. Markham continued to dig into Bishop's claims which could be substantiated by independent evidence, as compared to claims by aces such as Collishaw, Barker or McLaren. In private conversations he revealed that, while in the case of most such pilots, the ratio of claims *confirmed* to claims *submitted* was usually very high (in Barker's case, perhaps 100 percent), Bishop's ratio was below 50 percent. In 1995 Phillip Markham died of a heart attack while skiing, before he could publish his analysis. His research notes were acquired by an experienced historian, who is expected to write with an eye on history, rather than hagiography.

The Royal Aircraft Factory S.E.5 (top speed 126 mph) was the Camel's chief rival for the honour of best British fighter of WWI. First flown in December 1916, it had a side-mounted Vickers and an overhead Lewis. No.56 Squadron, replete with Canadians, introduced the type in France in April 1917. The S.E.5a was the mount of several Canadian aces, Billy Bishop included. Some 5200 were built (478 are estimated by Henshaw to have been lost on the Western Front). (CF RE19742-6)

Nieuport fighters were well-known to Canadians on British squadrons. This Nieuport 17, a pristine replica constructed in 1962 by Carl R. Swanson of Sycamore, Illinois, was photographed at Rockcliffe in June 1968. Today it may be seen in the National Aviation Museum in the 60 Squadron colours of Billy Bishop. (Larry Milberry)

Canada's great WWI ace, W.G. Barker, VC, DSO, MC poses with a friend beside a captured D VII at Hounslow, England in April 1919. Barker was long overshadowed by Bishop, until the 1997 publication of Wayne Ralph's fine Barker biography. (NAC PA6311)

Barker taxis his 28 Squadron Camel while fighting in Italy in 1917. (NAC C59854)

The ungainly Royal Aircraft Factory F.E.2b could hit only 72 mph and took nearly 45 minutes to reach 9000 feet. One Lewis gun usually was carried, the observer/gunner standing in the nose (without a parachute). Spent shells were spit into a leather bag, lest they damage the prop. Nearly 2200 of these kites were built. Generally reliable and durable, they served into 1917 before being pulled from the Western Front. The great German ace Max Immelmann met his doom against an F.E.2b. (CANAV Col.)

The F.E.2b also was a bomber, often operating from 1000 feet or lower. With a speed of only 70-75 mph, this was hazardous work. Trenches, road transport, artillery, anti-tank gun emplacements, bridges, rail yards, supply depots, etc. all were fair game. This 58 Squadron F.E.2b is ready for a bombing sortie from Clairmarais, France in February 1918. Lt H.T. Leslie of Toronto is the pilot, his observer is Brock. W.A. Leslie, "HT's" brother, also was a pilot on 58 Squadron. On May 16, 1918 "WA" went down behind German lines, becoming a POW till December 30. (Penhold Col.)

## A Selection of Notable Fighter Pilots

This book is intended to bring fresh stories to the public discourse, not replicate the work of others. Consequently, for the same reason that "Red" Mulock's story has been summarized rather than repeated, we must turn from these famous names to lesser known ones. In passing over the honourable careers of a few to concentrate upon their less renowned (but no less deserving) comrades and contemporaries, it was necessary to choose among many candidates. It would be wrong to describe these as "lost heroes". Their records have been publicly available for decades and those who might claim to have "discovered" them merely would be taking credit for looking beyond the obvious. The so-called lost heroes have more often been ignored in favour of other men, whose stories have already been told around the historical campfire.

The tunnel vision of fighter mythology, concentrating as it does on certain high scorers, has not only obscured many deserving men, but also numerous interesting aircraft. Consider the case of the F.E.2b and a notable early Canadian fighter pilot, Lt Reginald M. Makepeace of Montreal, who served in No.20 Squadron from June 29 to November 17, 1917. The F.E.2b was not what we consider a fighter—certainly not like a Camel, SE.5 or even a Bristol F.2B Fighter two-seaters. It was a two-seat pusher with the pilot in the rear, the gunner in the front "pulpit". The latter not only fired forward; he also was expected to stand up in the airstream, firing a gun mounted on the upper wing at whatever might be overhead and to the rear—all without a parachute!

Makepeace claimed his first victim in June 1917 (the squadron shot down 20 enemy aircraft that month). On July 27 No.20 was the bait for German fighters; other squadrons were to bounce once the enemy was committed. Things went awry; the escorts were late and the F.E.2bs had a 45-minute running combat against superior odds before help arrived. Yet the old pushers put up a good fight, claiming 9 enemy aircraft (6 with certainty). Makepeace and his gunner, Pte S. Pilbrow, claimed three. The pair received an MC and MM respectively. Pilbrow was killed with another pilot on August 15. Makepeace managed one more kill on the F.E.2b (August 17) before No.20 Squadron re-equipped with F.2Bs. He then achieved the unit's first victory on "Brisfits" (September 3, Lt M.W. Waddington, observer, one Albatros scout crashed).

Arnold Jacques Chadwick was born in Toronto on August 23, 1895. He studied music there, then took courses in music and modern languages in Germany. Newspapers later reported

Capt A.E. McKeever (right) and Capt Donald R. MacLaren of Ottawa while with the Canadian Air Force (England). Flying Camels with 46 Squadron, MacLaren was credited with 54 aerial victories. After the war he helped lay the groundwork for commercial aviation in Canada. (NAC PA6023)

As a two-seater in a mêlée the Bristol F.2A and F.2B Fighter proved the equal of most single seaters. Note the proximity of the cockpits, a vital feature in crew communication during combat. This robust type reached the front in April 1917; 3100 were built. F.2 pilot Capt Andrew E. McKeever, DSO, MC and Bar of Listowel, Ontario was dubbed "King of the Two Seaters"—he and his observers downed more than 30 of the enemy between June and December 1917. McKeever survived only to die in an automobile accident in Canada in 1919. (NAC PA6023)

Once in the UK fresh Canadian pilots usually completed their training on vintage types like the Maurice Farman S.11 Shorthorn. An example may be seen in Canada's National Aviation Museum. (David Thompson Col.)

that he had escaped Germany five weeks after war was declared. Upon returning to Canada, he applied to the RNAS, which demanded that applicants first clear a hurdle by obtaining an Aero Club of America certificate, the cost of which would be refunded on acceptance into the RNAS. Chadwick began flying training at the Curtiss Flying School in Toronto on September 28, 1915 and had logged only 76 minutes on flying boats when the school closed for the winter. The RNAS waived its ACA requirement and on December 30 Chadwick was appointed a Probationary Flight Sub-Lieutenant. Two weeks later he reached England.

Excerpts from his log, preserved in the Creagen Papers, show an interesting career. His first RNAS flight was a 3:50 minute hop from Chingford in a Maurice Farman Longhorn on January 24, that ended with engine trouble. He flew again the same day, this time for 27 minutes, which he described as "a little bumpy". On January 27 he was checked by Flight Sub-Lt Alcock, then soloed. Much of his subsequent flying was figure 8s plus "circuits and bumps". On January 31 he logged 16 minutes in a JN-3: "First trip in a tractor. A little unconfident at first". By February 2 he was much more at home in the JN-3, practicing spirals and steep banks. Although he occasionally piloted the JN-3 or an Avro, most flying was on Longhorns and Shorthorns until the end of June, when he began flying Avros almost daily, switching to B.E.2s on July 18. Occasionally he force landed with engine trouble and on July 20 he wrote of a crash landing after his engine quit.

Chadwick moved to the Gunnery School at Eastchurch, dropping bombs and shooting at balloons and surface targets. On August 17, 1916 he reported to Dover where armament training continued. Finally, on August 28 he flew his first operational patrols—45 minutes in Nieuport 3165 and 35 minutes in Nieuport 3966. He still had a few things to learn; on August 31 he flew Nieuport 9210 for 25 minutes and recorded, "First trip on machine with Clerget engine. Landed side wind and bent axle. Poor showing!" By September 9 he had flown 67:15 hours. He was now ready for more advanced work. On September 14 he was posted to No.5 Wing, RNAS, Dunkirk, from where he recorded his first trip over the lines on September 23—two hours at heights to 10,000 feet in Sopwith 1½ Strutter No.9658, during which he became lost, was archied at Ostend, and finally reached base.

His first great adventure came on October 2, when a formation of four Sopwiths raided the Zeppelin sheds at Evère, near Brussels. Chadwick bombed his target but was archied and had to force-land near Tirlemond. He escaped before the Germans arrived and contacted Belgian civilians. He sheltered first at a priest's house, where he switched to civilian garb. He was taken to meet an agent, code-named Charles Lapin, staying with him three days. He then hid in a chateau near Meldert for three weeks, while Lapin made arrangements. German troops and agents visited the chateau eight times, yet Chadwick remained undetected. At the end of this period he left, disguised as a woman, for Liege, then Bassing,

"Archie" was a colloquial term for anti-aircraft fire. Here a British Army mobile unit stands by on the Western Front in May 1918. Archie was important in keeping at bay German trench fighters like the Junkers J.I. (NAC PA2594)

An unknown aircraft brought down during the Canadian advance on the Arras front in August 1918. (NAC PA3182)

hiding in a farmer's house for a day. With 17 others he attempted to sneak into Holland, succeeding on the third try. He then travelled to The Hague, where passage to England was arranged. How he evaded internment is not clear, but his disguise may have done the trick.

Chadwick reached London on November 10, then returned to his unit. He had leave in Canada, but by January 22, 1917 was back on operations, this time on bombers with No.5 (Naval) Wing at Dunkirk. He flew Sopwith 9394 during a raid on Bruges, but was turned back by intense cold (it was on this mission that F/L Darley succeeded). Bruges was attacked on February 7, Ghistelles on the 9th, enemy ships on the 13th, Bruges on the 14th and 16th. On the latter occasion the formation (8 Sopwiths) reported a large explosion, while facing intense, accurate anti-aircraft fire.

Chadwick's log for March 15, 1917 notes a 30-minute flight in Pup N5184: "tried to loop six times but failed". A few days later he was posted to No.4 (N) Squadron, first at Petit Synthe, then Bray Dunes. He reported his first "spinning nose dives" on April 8; an attempt to intercept enemy aircraft on April 21 failed ("could not pick out enemy machines"). His luck changed—on April 24 he drove an enemy aircraft away from a French photographic machine under his protection. On April 26, after a "fighting patrol" in Pup 9899 (115 minutes), he wrote, "Shot down Hun near Bruges at 6,000 feet—reverse loop". Now, instead of bombing, he was escorting bombers as well as flying offensive patrols. On April 30 he piloted one of four aircraft (3 Pups, 1 Belgian Nieuport). They ganged up on a German two-seater, which force-landed half a mile behind German lines. Further log entries (all with Pup N6176) describe the intensity of operations:

May 19, 1917, 8.15 a.m., 1:35 hours.
Reconnaissance patrol; Flight Commander Shook landed in sea. Picked up by French destroyer Oriflamme.

May 22, 3.30 p.m., 1:00 hour.
Attack on kite balloon at Ghistelles. It was pulled down.

May 25, 6.15 a.m., 1:20 hours,
12 aircraft. With Flight Commander Newberry, Flight Sub-Lieutenant Enstone. Shot down hostile aircraft near Zydecoate; [6.45 p.m., Pup 6176, 1:10 hours] attack on hostile aircraft bombing Dover. One shot down off Ostend.

May 26, 8.05 a.m., 2:00 hours,
15 aircraft, up to 15,000 feet. With Flight Sub-Lieutenants Enstone and Hodges. Met hostile aircraft (Albatros) when returning. Driven down in Foret d'Houthulst.

June 3, 4.10 p.m., 1:56 hours,
15 aircraft, up to 18,000 feet. With Flight Lieutenant House, Flight Sub-Lieutenants Hemming and Ellis. Was attacked by Albatros Scout following formation. Driven down to 8,000 feet without engine. Got engine, stalled under hostile aircraft which crashed from 4,000 feet. Right top elevator controls shot away...

In service since May 1916, the Sopwith Pup (112 mph, all-up weight 1225 lb) was praised by its pilots. With a Vickers gun and light bombs, it was a versatile fighter. This example (N6181—No.3 Naval Squadron) was flown by F/S/L L.S. Breadner, later RCAF Chief of the Air Staff. (Stewart Taylor Col. via J.S.T. Fall)

June 5, 4.25 a.m., 2:45 hours.
Fleet patrol. Shelling Ostend. Eight machines. Flight Sub-Lieutenant Smith shot down. Hostile kite balloon. [It is not clear if Smith shot down a kite balloon or if Smith was shot down and a kite balloon was sighted]; [7.15 p.m., 1:30 hours, 17 aircraft, to 18,500 feet] Patrol to intercept hostile aircraft reported off Nieuport; Flight Commander Newberry, Sub-Lieutenants Busby, Hodges landed in England. Flight Commander Shook shot down two hostile aircraft, Flight Sub-Lieutenant Enstone shot down one...

Now the squadron switched to Camels, but Chadwick reported gun jams during inconclusive combats on June 12 and 14. He flew five inconclusive Camel patrols between June 20-24 and chased an LVG without result on the 21st. His next few sorties make chilling reading; that for the 27th demonstrates the ruthless nature of the war:

June 25, 1917, 10.00 a.m., Camel N6345,
19 aircraft, up to 14,000 feet. Met Albatros over Roulers. I shot one two-seater down in flames. "Travelling Circus" supposed to be here now.

June 27, 1917, 6.25 a.m., Camel N6345,
35 minutes, 5,000 feet. Attack on kite balloon at Dixmude. Shot observer descending by parachute. Machine gun fired from trenches at 900 feet; 9.40 a.m., Camel N6370, 1:00 hour Attack on kite balloon anchored to trawler off West Deep. Could not see trawler there so attacked kite balloon at Dixmude. Did not see observer descend, but was very heavily shelled both ways. Kite balloon hauled down very suddenly.

On July 3 Chadwick fired on an enemy machine; it went down in a steep dive, but probably survived. On July 6, however, there was no question as to the outcome. With Flight Sub-Lieutenant S.E. Ellis, he attacked a two-seater

A 208 Squadron Camel after force-landing in France in May 1918. The fighter pilot faced many hazards. This chap may have been hit by archie, been shot up by an enemy aircraft, had mechanical trouble, or run out of fuel. On April 9, 1918 the Germans overran 208 Squadron's base at La Gorgue. The squadron, which included many Canucks, escaped by road in rotten weather, having been obliged to burn all 17 of its Camels. (NAC PA3894)

Albatros, spotting for enemy artillery. It went down smoking, then shed its right wing and crashed. He was flying Camel N6369 on July 10, when 5 RNAS aircraft intercepted 3 enemy two-seaters. As they engaged, 10 Albatros attacked. One Camel, chasing a German, was shot down by 4 Albatroses: "I shot down one in flames in canal at Nieuport (confirmed) and other out of control; Flight Sub-Lieutenant R.M. Kierstead shot another down ... seen by the army." RNAS communiques described the first victory as "shot down completely out of control" and indicated that Kierstead's victory was, in fact, the same machine attacked by Chadwick in his second combat. On July 11-12, 1917 he was engaged in "Y" patrols —a new term which meant looking for German bombers that had been raiding London. His log entry for July 14 was unusually chatty in describing such an operation (3.45 a.m., Camel N6369, airborne 2:40 hours): "With Flight Sub-Lieutenants Enstone and Hodges, chasing early morning Hun bombers over Dunkirk. Flight Sub-Lieutenant Enstone shot down one bomber over Ghistelles ... I attacked a Halberstadt two-seater over St. Pierre Cappelle which disappeared suddenly... I think I got him, but it was not confirmed."

On July 20 Chadwick test flew Pup N6476 and exuberantly remarked, "It looped perfectly". Later that day, however, he was in Camel N6369 on an offensive patrol and had only bad things to report: "Poor formation. Flight Sub-Lieutenant Akers and I were attacked over Westende by four Albatros scouts. Both guns jammed, *as usual*, so I was unable to assist Akers. When last seen was going down in a vertical dive with two Huns on his tail." Such trouble with guns was making Chadwick wary. On July 21 he tested a new Camel (B3853). "Guns jammed, so refused machine", he wrote. Next day he was part of a four-plane offensive patrol, but two machines left the formation (at least one with engine trouble) and he turned back. Later that morning he took off to pursue another intruder, but landed after 10 minutes (engine trouble). He switched to another Camel, took off with Flight Commander Shook, and finally got a shot (no results) at an Albatros at 18,000 feet.

On July 25 shore batteries were engaging British warships near Nieuport, an Albatros directing their fire. No.4 (N) Squadron despatched a patrol to deal with the spotter; Chadwick, flying N6369, was airborne two hours. His log reported the destruction of the Albatros (RNAS communiques confirmed this). Chadwick's last combat—an attack on a two-seater over Ostend on July 27, 1917—was inconclusive. He now had over 210 flying hours. Next day he went missing; his body was washed ashore on August 8 (he had drowned). His commanding officer, B.L. Huskisson, subsequently wrote: "... an excellent pilot and officer who, if he had lived, would have made a great name for himself. He was with me from the time this squadron was formed and his loss is greatly felt by all his brother officers and myself." What "score" might be assigned to Chadwick can be disputed. The citation to his DSC, published in August, gave no figure: "For exceptional gallantry and remarkable skill and courage whilst serving with the RNAS at Dunkirk during May and June, 1917, in repeatedly attacking and destroying hostile aircraft."

Reports by his superiors are contradictory. Shortly after Chadwick's death, Huskisson wrote, "During the last four months he has led his flight with conspicuous success and certainly caused six machines and several Kite Balloons to descend." The phrase "to descend" is vague. A report of June 11, 1917, prepared by the wing commander at Dunkirk, credited him with "the destruction of four enemy machines". This, of course, was composed before several of his victories. The same officer declared in a July report, "He has destroyed eight enemy machines and has on three occasions forced down balloons." Chadwick's score is immaterial. From evading capture to his final combats, he had shown courage of the highest order; neither medals nor artificial scores could embellish this.

Alfred Clayburn Atkey was born in Toronto on June 18, 1894. He was farming in Saskatchewan from 1913-15; in the winter of 1915-16 he was a reporter for the Toronto *Telegram*. The nature of the land war was apparent by then, and he chose to approach RFC recruiters. He was accepted, commissioned as a 2nd Lieutenant on Probation in October 1916 and sailed aboard the SS *Olympic*. Once in Britain he followed a program that had been developed by trial and error and which would become familiar to thousands of other young men. This began with attendance at the School of Military Aeronautics (Oxford), followed by posting to No.27 Reserve Squadron (December 23, 1916), No.28 Reserve Squadron (Gosport, February 7, 1917), No.9 Reserve Squadron (Norwich, February 22, 1917), and No.9 Training Squadron (date uncertain). Along the way Atkey moved from Longhorns and Shorthorns (the most common primary trainers) to B.E.2 and F.E.2 two-seaters. Finally, on September 8, 1917, he reported to No.18 Squadron (D.H.4s). Almost immediately, 2Lt Atkey began to figure as one of the unit's more active bomber pilots. He subsequently was posted to No.22 Squadron (F.2B, May 9, 1918), ultimately returning to England to instruct.

Atkey's first aerial victory was on February 4, 1918. RFC Communique No.125 reported that he and his observer, Lt C. Ffolliott, were returning from a photo and bombing mission, when attacked by about 10 e/a. His companion fired a burst at the leader, who went down, shedding part of a tailplane. Atkey fired on another opponent, who also lost control. The remainder broke off. His "Brisfit" had been damaged—one machine gun, plus the observer's ammunition drum shot through and the control wires to one elevator severed. It was the beginning of a scoring career which, on closer examination, is almost a model to demonstrate the confusion and ambiguity surrounding First World War fighter pilot statistics. He was *associated* with 35 claims, but several were credited to his gunners (he had at least five) and not all were mentioned in communiques.

If one subtracts *gunner's claims*, the number drops to 22-25; if one subtracts claims *submitted but not mentioned in communiques*, it drops to 15 or 17 kills. Yet such arithmetical juggling assumes that gunner and pilot were separate entities (when they were a team). Failure to have many victories listed in RFC communiques reflects an apparent policy, from mid-May onwards, to exclude certain ambiguous claims (notably "out of controls") from the communiques. Indeed, many of the victories *not listed* in communiques (themselves published by RFC and then RAF HQ) are nevertheless listed in a document compiled by 10 Army Wing HQ (an inferior formation, but one apparently concerned with detail). The latter document clearly enhances the credibility of Atkey's claims.

Thanks to W/C Fred Hitchins, as well as the copying of Air Ministry records on behalf of the Directorate of History, the texts of many combat reports—Atkey's included—are available in Canada. The following is typical of his 1918 engagements:

April 21, D.H.9. 2Lt A.C. Atkey, Lt P. Anderson. Photography at 10,000 feet—one driven down out of control:

*While taking photographs over Aubers we were approached by five Pfalz scouts, three of which we attacked. After firing front gun on nearest scout, rear gun was brought to bear on same machine. This machine was seen to spin down obviously out of control, and clouds prevented us from following him to the ground. The others immediately dispersed into the clouds.*

May 7, Bristol Fighters B1164, B1253. 2Lts A.C. Atkey, J.E. Gurdon, C.G. Gass, A.J.H. Thornton. 6.45 p.m., offensive patrol at 15,000 feet, 10 miles northeast of Arras:

*While doing an offensive patrol in pairs, 2nd Lieutenant Gurdon and myself dived on a formation of seven enemy aircraft. In the first dive both machines shot down one enemy aircraft in flames and on coming out of the dive, 2nd Lieutenant Thornton (observer) fired at one which was on his tail which also burst into flames. One enemy aircraft almost collided with the tail of 2nd Lieutenant Atkey's machine and was shot down in flames at a range of a few feet. During this fighting the enemy aircraft were reinforced by two other formations which brought their number up to about 20. We were fighting with them for about half an hour during which time many of them spun away, possibly out of control or just breaking off the engagement. Only four were actually seen to crash and are claimed by 2nd Lieutenant Atkey (two), 2nd Lieutenant Gurdon (one), 2nd Lieutenant Gass (one). We had then run out of ammunition for the back guns, so broke off the fighting. As we left we counted the enemy aircraft, which did not follow us, and there were only seven left.*

May 19, Bristol Fighter C4747. Capt A.C. Atkey, 2Lt C.G. Gass. 6.45 p.m., offensive patrol:

*While patrolling in the vicinity of Lille, in pairs, we met about 12 enemy aircraft [LVG two-seaters] slightly below us. I dived firing a burst of 100 rounds intermittent at a range terminating at about 20 yards and one enemy aircraft was observed to dive vertically for a long distance and must have been out of control. 2nd Lieutenant Dunster also dived and saw this machine turn on to its back when close to the ground. My observer afterwards fired at a machine following on our tail; he fired about 200 rounds at long range and this machine fell out of control with thick black clouds of smoke issuing from the fuselage. My observer having expended all his ammunition we were obliged to ... discontinue the combat.*

May 20, Bristol Fighter C4747. Capt A.C. Atkey, 2Lt C.G. Gass. 10.45 p.m., offensive patrol at 15,000 feet:

*While doing an Offensive Patrol in pairs, in the vicinity of Lille and Armentières, we sighted 10-12 enemy aircraft. They were 1,000 feet below when we attacked. I fired a burst of about 100 rounds at the nearest machine which dived vertically ... apparently out of control... I dived on a second enemy aircraft which also dived vertically ... a third machine immediately climbed under our tail ... Having outmanoeuvred him, my observer fired a long burst at a range of about 10 yards. This machine rolled over on its back ... then disappeared in a vertical dive completely out of control. Lieutenant S.F. Thompson also observed these machines go down. Seven machines were counted in the air after the combat when, owing to a possible lack of petrol and ammunition, we were obliged to discontinue the combat.*

Scrutiny in Atkey's reports reveal some discrepancies in 10 Army Wing's list of victories credited to him. It would seem that both kills of March 25 were scored by his gunner, that of April 21 was a Pfalz sent down out of control, rather than in flames (which would explain its not being mentioned in communiques). Of the 3 victories of May 19, one owed as much to Gass as to Atkey, while one was clearly a gunner's kill. On the other hand, one of the 2 aircraft described as "out of control" on May 22, was seen to crash into a tree. Of the 3 claimed on May 27, the "crashed" was clearly Atkey's while the 2 "out of control" were by Gass. In the combat of May 30, Atkey's share was a Pfalz shot down in flames (not merely out of control) in a head-on pass; the "out of control" claim went to Gass, who saw his victim "rolling wing over wing". Similarly, the honours for May 31 and June 2 were evenly divided between pilot and gunner. Not counting victories by gunners, then, Atkey's "score" could thus be calculated as 7 kills (the number he shot down burning or saw crash) or 15 (victories mentioned in communiques) or 22 (victories tallied by 10 Army Wing). Atkey was awarded the Military Cross on June 22, 1918. The citation had been drafted weeks earlier, whilst he was still with No.18 Squadron: "For conspicuous gallantry and devotion to duty. When engaged on reconnaissance and bombing work he attacked four scouts, one of which he shot down in flames. Shortly afterwards he attacked four two-seater planes, one of which he brought down out of control. On two previous occasions his formation was attacked by superior numbers of the enemy, three of whom in all were shot down out of control. He has shown exceptional ability and initiative on all occasions."

The *London Gazette* of September 16, 1918 announced that Atkey had been awarded a Bar to his Military Cross, the citation noting: "...he proved himself a brilliant fighting pilot, and displayed dash and gallantry of a high order." Atkey was demobilized in May 1919. He returned briefly to the *Telegram*, then tried barnstorming, but was bankrupt by the fall of 1920. When the Canadian Air Force offered refresher flying courses, he attended Camp Borden in 1920. Curiously, he did not impress all the staff; F/L N.R. Anderson wrote: "This officer has flown heavy machines overseas for a considerable length of time and seems to have found a fixed way of flying which contains many erroneous ideas. These we have tried to correct and also have given him the main points of the Gosport System."

W/C Scott Williams considered Atkey "a fairly good pilot on Two Seater Machines, very keen and conscientious." Early in 1922 he applied to join the CAF full-time, then lost interest. He enquired about possibly buying a surplus Air Board seaplane for a commercial enterprise, but none were available. Between 1922 and 1929 he was a Customs Inspector in Saskatchewan and a railway switchman in California. He finally settled into farming and teaching music at Lloydminster, Saskatchewan. When the Second World War began, Atkey applied to the RCAF. The Recruiting Officer handling his forms wrote, "A keen type—civilian record not impressive. War record good. Sincere and honest." He was commissioned in August 1941 as a Link Trainer Instructor, in which capacity he served until May 1944, when BCATP reductions commenced. He went back to farming and teaching, ultimately retired to Toronto, and died there on January 29, 1971.

Captain Ernest James Salter was born in 1897 at Greenbank, near Ottawa. He was living in Mimico, Ontario when he joined the RFC. He trained in Canada and sailed overseas in October 1917. After further instruction he embarked for France, joining No.54 Squadron (Camels) on March 19, 1918. His tour was interrupted by a stay in hospital in April and May, but on rejoining No.54 he proved skilled and aggressive. His luck faltered on September 2, when he was wounded, but until then he had registered the following claims:

July 4 (Camel D1948)
*Hanoveraner driven down out of control;*

July 5 (D1948)
*Albatros Scout driven down out of control;*

July 21 (D1946)
*2 Halberstadter 2-seaters destroyed;*
(D9497) - *one Fokker scout destroyed;*

August 22 (D1946)
*2-seater Albatros destroyed (shared with 3 other pilots);*

August 25 (D1946)
*Fokker biplane driven down out of control.*

Perhaps Salter's most interesting combat was the last, when he was leading 6 Camels on low level bombing. They were at 2000 feet when they spotted a formation to the northeast. Salter led his group to 10,000, gaining a height advantage which enabled him to stalk the enemy. He caught up to and surprised 7 Fokkers. He dived on one white-tailed e/a, firing 100 rounds at 50 yards. The Fokker stalled, fell over on a wing, and was last seen at 500 feet, out of control. A general dogfight followed, with No.60 Squadron joining the fray; one other enemy aircraft was sent down out of control.

Not all pilots could be aces, yet fame is accorded only to a few. Salter was successful enough to be accorded the French Chevalier of the Legion of Honour, and a Croix de Guerre. Nevertheless, he appears on none of the traditional "ace" lists—something which may simply be due to bad timing! In his case the RAF ceased to publicize "driven down out of control" claims after May 1918—yet such annotations in earlier RFC and RAF communiques often were the basis of fighter pilot scores compiled by popular writers. Thus, instead of 6 victories (3 destroyed, 3 out of control, and a share in a fourth), communiques acknowledged only his successes of July 21. If he had been in combat even six months earlier, Salter might have been accorded ace status by the unofficial arbiters of that rank. Ernest Salter returned to Canada, flew briefly as a barnstormer, then moved to other things. In WWII he joined the RCAF as a bombing instructor; he died in Oakville in March 1959.

Some men may have been condemned to relative obscurity because they were flying the wrong type of airplane. Aviation "buffs" are familiar with the Sopwith Camel and SE.5; but who remembers the Martinsyde Elephants and the men who flew them? The Elephant was not famously successful; too stable and heavy for dogfighting, it spent more time in reconnaissance and bombing. Its armament was also unusual—a Lewis gun atop the wing firing over the propeller, supplemented with a swivelling Lewis on a port side bracket to provide some protection to the rear. The Elephant's very obscurity demands that its pilots be recognized.

# A selection of German fighters encountered through the war by Canadians

The Fokker Dr I *Dreidecker* entered service in the summer of 1917. This nimble fighter became famous on *Geschwader 1*, commanded by Manfred von Richthofen— the "Red Baron"; and on *Jasta 10*, where Werner Voss flew it with distinction. Only 320 Dr Is were built; several replicas are flying today. (Stewart Taylor Col. via J.S.T. Fall)

The Fokker D VII was Germany's best scout, even though it did not enter service until 1918. With a 160- to 185-hp, it hit 125 mph, and was highly manoeuvrable at altitude. This D VII met misfortune inside Canadian lines near Amiens. Note how its fabric has been peeled off—everyone wanted a souvenir patch to bring home. The scene has one saving feature— it shows the D VII's welded steel fuselage frame, an innovation for 1918. The wing was of traditional wood and fabric. (NAC PA3606, '2953)

The Albatros D-V was a prominent German fighter, but was plagued by structural failures. Like the Dr I and D VII, armament was a pair of Spandau machine guns. (Tom Dietrich Col.)

Although not a hot performer the diminutive Fokker *Eindecker* had one important feature when it entered service in 1915—its machine gun fired *through* the propeller with the aid of an interrupter gear. Until the Allies developed such a system, the Germans had a vital edge. This captured E-III was being evaluated by the British. (CF RE13510)

The DFW C V was the most successful German two-seater. F/S/Ls Anderson, Bayne, Beamish and Harrower, all Canadians on No.3 (N) Wing, took part in the action on September 10, 1917, led by Flight Commander Ronald F. Redpath of Montreal, which saw this example forced down with its observer wounded. (Stewart Taylor Col.)

Percy Clark Sherren, born at Crapaud, Prince Edward Island in July 1893, secured a commission in the 26th Battalion, Canadian Expeditionary Force. He transferred to the RFC in May 1916; by August he was piloting Martinsydes with No.27 Squadron. On November 25 he was awarded a Military Cross—fast work! The citation was tantalizingly brief: "He led a successful bomb raid, collecting and landing his formation with great skill. Later, he dropped bombs on an ammunition train from 500 feet, causing much damage." Two of Sherren's combat reports are held in the National Archives of Canada. That of September 23, 1916 (Martinsyde A1567) makes exciting reading, even if the outcome was inconclusive. A German fighter attacked Sherren's flight commander; Sherren drove it away. A multi-coloured biplane approached; he attacked it, firing 30 rounds before his guns jammed. His return to base was the most dangerous part of the sortie:

The Martinsyde G.100 Elephant of 1915 was intended as a long-range escort, but proved itself mainly in bombing. (Jack McNulty Col.)

"On return journey from Cambrai pilot was attacked by several machines, but managed by diving at them to drive them off, being assisted by two other machines of the formation. Pilot was able to get his side gun going eventually and fired ¾ drum at close range at one enemy machine which was attacking one of the formation. The tracers were seen going through the fuselage, but enemy was lost sight of. Pilot saw one machine going down under control with smoke coming from fuselage." Sherren was in A1567 again on September 26, 1916 when he engaged a Roland at 12,000 feet. "Pilot noticed hostile machine on tail of another Martinsyde; pilot turned and fired ½ drum from top gun at enemy machine which was then only five yards above him and going in the same direction. Tracers were seen entering bottom of fuselage by pilot's seat. Enemy dived steeply in front of pilot, who followed and finished drum. Enemy was last seen sideslipping and nose diving."

Sherren, rising quickly to captain, was awarded a Bar to his Military Cross on June 4, 1917. There was no published citation, so it is unclear what deeds won him this honour. He completed his tour with No.27 Squadron in March 1917 and was at the Central Flying School, probably learning to instruct, when the announcement was made. He returned to Canada that summer, serving with No.92 Canadian Training Squadron, RFC Canada, from August 1917 to April 1918. At war's end he was commanding No.98 Squadron. He remained with the RAF until June 1936. He died in a flying accident in the King's Cup Race on September 10, 1937.

One of the most formidable fighters of late 1918 was the Sopwith Dolphin. It carried two Vickers machine guns firing through the propeller arc, plus two Lewis guns firing obliquely upwards. The latter posed a serious threat if an aircraft turned over on landing; almost all operational Dolphins reduced the Lewis guns to one, and some pilots removed them altogether. The most successful exponent of the type was Capt Frederick I. Lord, an American in the RFC, who had trained in Canada. He destroyed at least 12 e/a while on Dolphins. Maj Albert D. Carter of Point de Buts, New Brunswick also scored well on Dolphins.

No.23 Squadron had a cluster of successful Canadian Dolphin pilots including Lt Harry N. Compton, DFC (Winnipeg and Vancouver Island), Capt Arthur B. Fairclough, MC (Toronto) and Lt Harold Albert White (British born, home in Brantford, Ontario) plus an American trained in Canada, Capt James W. Pearson, DFC (Nutley, New Jersey). From many combat reports one may be taken as typical. It was filed by Compton on October 28, 1918 after a fight with a dozen D VIIs camouflaged grey with yellow noses.

Pearson (F3961) was leading a patrol and attacked an enemy formation, accompanied by Compton (C4130) and four others; further aircraft gave cover. He sent one Fokker out of control (not seen to crash), another crashed into a tree. Apart from witnessing his commander's success, Compton's role was also significant: "While on offensive patrol I dived on a formation ... in company with Captain Pearson... I singled out a Fokker, which I attacked. I got in about 150 rounds at very close range (finishing at about 20 yards). I saw my fire going through his planes and some appeared to hit the engine, as it emitted puffs of black smoke... I saw that his propeller had stopped... I did not see it crash as I had to turn to engage another, but without decisive result."

A rising star was Lt Frederick Joseph Stevenson, whose war record was honourable, yet blown out of proportion by admirers. Frank Ellis, in *Canada's Flying Heritage*, described his hero as having 18 victories. The truth is more prosaic.

Fred Stevenson after the war as a bush pilot. He died in the crash of a Fokker Universal, while landing at The Pas, Manitoba on January 5, 1928. (CAHS Col.)

The speedy, manoeuvrable Dolphin played an important role in fierce battles that raged through 1918. On May 20, for example, Lt C.A. Crysler of Toronto (23 Squadron) became embroiled with seven Dr Is at Le Hamel, shooting one down, then ramming another and falling to his death with it. October 30, 1918 is noted as the war's heaviest day of air fighting, the official RAF history listing 67 e/a downed for 41 lost. Returning from a raid on Mons that day, 98 Squadron D.H.9s, escorted by 19 Squadron Dolphins were ambushed by Fritz. At least 9 RAF machines were lost, 5 being Dolphins. Dolphin pilot Lt C.M. Moore of Montreal downed a D VII in this mêlée. (Tom Dietrich Col.)

Born in Parry Sound, Ontario in 1896, Stevenson was in Winnipeg when he enlisted in the 196th Battalion, Canadian Expeditionary Force, and went overseas. He transferred to the RFC in September 1917, trained as a pilot, and was posted to Dolphins on No.79 Squadron in June 1918. He was wounded on August 8, 1918 (the first day of the Allied offensive at Amiens), but remained with his unit until April 1919.

Records show that Stevenson filed exactly three combat reports, all for decisive engagements involving the same machine (C8189). The first was on November 3, 1918 when he dived on a Halberstadt 2-seater at 1500 feet, poured in 150 rounds, and saw it crash into a tree. The following day he attacked an LVG 2-seater at about 1000 feet; it burst into flames and crashed. On a 4-plane offensive patrol on November 9, he attacked another Halberstadt at low level. He fired two long, close-in bursts; the German went down out of control, crashing into a hedge. In June 1919 Stevenson was awarded a DFC, but the citation gave no details other than "in recognition of distinguished services rendered during the war".

There were subsequent anomalies to Stevenson's story. Ellis wrote that he had joined an RAF mission to southern Russia—yet RAF documents show him as a Fairey seaplane pilot at Murmansk in northern Russia in July 1919. It was there he apparently received a White Russian decoration, the Order of St. Stanislas, 2nd Class, which anti-Bolshevik generals were awarding lavishly to members of Allied forces intervening in the Russian Civil War. Few (Stevenson's included) were officially acknowledged by British authorities. How did Stevenson's war record come to be so exaggerated? It seems that Ellis, a business partner and friend, may have been taken in by some "line-shooting". Stevenson was known to embellish. When applying to join the Ontario Provincial Air Service in 1924, he supplied an inflated figure for his seaplane flying hours—no doubt to better his chances of getting a job. As a pioneer bush pilot he was without peer; he just followed what others did with their military records—he took a good story and made it better.

## Lt Louis Bennett

So vast a war inevitably was filled with tragic, heroic and hilarious stories—sometimes narratives that included all three elements. One such, encompassing three nations, involved Louis Bennett. Born in Brooklyn, New York in 1894, he enlisted in the Canadian Expeditionary Force in September 1916, joining the 213th Battalion at Camp Borden. The unit moved to St. Catharines, Ontario in December 1916, but Bennett chose not to go. A Board of Inquiry declared him a deserter. By the usual method—Part II Orders (meaning the second part of published daily orders)—he was struck off the CEF rolls as of January 4, 1917. His case was not unusual.

Early in 1929 the Department of Militia was approached by Bennett's mother, requesting a Memorial Cross (a medal issued to the mothers and widows of soldiers who died on active service). Departmental records indicated that Louis Bennett had been a deserter and that a Memorial Cross could not be issued. Mrs. Bennett's letter started a chain reaction. British authorities were asked if they knew anything of the man. They did. Bennett clearly had reconsidered his role in the 213th Battalion. He joined the RFC, but without using normal channels. He trained as a pilot, being commissioned in January 1918. Overseas he was posted to No.40 Squadron in France, serving there until his death on August 24. (Bennett was hit by ground fire while attacking kite balloons with his SE.5a.) He was posthumously Mentioned in Despatches. Air Ministry concluded its report: "I am to add that according to the records of this Department, Lieutenant Bennett during his service in France was officially credited with eight enemy balloons destroyed, four of which were destroyed in one day (19th August 1918); two enemy machines crashed, and one shot down out of control." His story was complicated, however, by an obituary that had appeared in *Flight* (September 4, 1919) which seemed to be a product of reports, misunderstandings and possible deliberate misinformation that Bennett himself may have generated before his death:

*Lieutenant Louis Bennett, 40th Squadron, RAF, BEF, reported "missing" on August 24, 1918, is now officially reported as killed in action, having been shot down in flames after destroying two enemy observation balloons. He was an American and the only son of Hon. and Mrs. Louis Bennett of West Virginia. When the United States declared war he left Yale, and raised and trained at his own expense the West Virginia Flying Corps which he offered to his government to serve as a unit in France, like the Lafayette Escadrille. This being refused, he joined the RFC, and came over with the Canadians, in order to get into action. His record between August 15 and 24 was three enemy planes and nine balloons destroyed—four in one day—for which he was congratulated and recommended for the DFC.*

The Creagen Papers include copies of several combat reports filed by Bennett; they do not add up to any of the scores quoted, but may not be complete. Another item in *Flight*, this one of January 9, 1919, added greater poignancy to his story: "Mrs. Louis Bennett, whose son Louis lost his life flying with the RAF, has offered £100 to the Aero Club of America as a prize for a competition to develop parachutes to be used in escaping from aeroplanes which are out of control." Whatever contradictions and anomalies may have surrounded Bennett, there was no doubt that the CEF absconder and the RAF hero were the same man, and that his combat career was as brilliant as it had been brief. In August 1929 the Canadian Army formally cancelled the Part II Orders declaring him a deserter, substituting an entry to the effect that he had been struck off CEF strength for purposes of enlisting in the RAF (although in 1916 it had been the RFC).

## Canadian Air Organizations

Upon the outbreak of war Canada's eccentric Minister of Militia, Sam Hughes, was approached by Ernest L. Janney, who suggested that an air arm accompany the Canadian Expeditionary Force, being assembled at Valcartier, Quebec. The Minister abruptly scribbled a note commissioning Janney in something called the "Canadian Aviation Corps" and authorizing its sole officer to spend $5000 on an airplane. It seemed an odd move, but not unusual for Hughes, who had scrapped prewar mobilization plans on the spur of the moment in favour of building a "Canadian Expeditionary Force" from scratch. The force was being thrown together in such chaotic fashion that the impulsive authorization of an aviation corps was but one small incident. Janney's commission consisted only of Hughes' note; it was never formally proclaimed in the *Canada Gazette*, so is of doubtful legality.

Artist Robert W. Bradford created this work showing Canada's first military aircraft—the $5000 Burgess-Dunne tailless, built in Marblehead, Massachusetts in 1914. (CAHS)

Janney hastened to the United States where he bought a Burgess-Dunne biplane. He had no piloting skills (only four Canadians had Aero Club of America flying certificates at that time). The company pilot ferried the aircraft to Quebec City (Janney picked up a few tips en route); it sailed aboard a steamer in October 1914 with the first contingent, CEF. Stowed on deck, it may have suffered damage. Once in England, apparently it never flew. Janney, having recruited another officer and a sergeant, abandoned his post, although he did find time to write a report proposing a substantial CEF air arm. Within a year after Janney's appointment, the Canadian Aviation Corps had disappeared. Back in Toronto, he promoted a flying school, which never graduated a pupil, and walked about flaunting his rank and claiming to be a veteran flier. A senior Militia officer instructed a subordinate to "clip the wings of Captain Janney".

Flying the Burgess-Dunne was a case of sitting right out in the wind! (via Carl Mills)

The Canadian Aviation Corps had been a comical incident; more serious attempts had to wait until 1918, when two Canadian air forces were authorized. German submarines had operated off the New England coast in 1916, but in 1918 they threatened to become a serious problem. The Royal Canadian Navy was revitalized and a Royal Canadian Naval Air Service was established on paper, the latter encouraged by American authorities, who set up naval air stations at Halifax and Sydney, pending activation of the RCNAS. This force had taken only the most tentative steps towards formation—preliminary plans for a uniform, modest aircrew recruitment, and training of some technicians in the United States—when the war ended. It was disbanded before it got off the ground. Curiously, one of its officers was E.L. Janney, who received a generous official letter of thanks for his services, as he was let go.

Proposals for an overseas Canadian Air Force had been floated several times during the war, but never passed the memorandum stage. However, the public became increasingly aware of Canadian participation in the air war, and senior Canadian officers (Sir Arthur Currie and Sir Richard Turner included) began to push for a CAF to support the CEF. Formation of the Australian Flying Corps provided an example. Prime Minister Sir Robert Borden came on side in May 1917. Yet there was a war going on, and it seemed a bad idea to start pulling Canadians out of British units and spreading confusion in the established air forces, when events everywhere were in a state of crisis. Borden was not in a position to push for a CAF until he had grasped another nettle—conscription—and fought an election on the issue. In the total pattern of military and political events, creation of an overseas CAF was small beans.

Even small beans germinate, however, and a distinct CAF was approved on May 29, 1918. A technical school for Canadian mechanics was formed at Halton, England on August 22; but it was not until November 20 that Nos.81 and 123 (RAF) squadrons were designated CAF units. Based at Upper Heyford, they became Nos.1 and 2 Squadrons, CAF, equipped respectively with Dolphins and D.H.9s. Air crew selection was conducted by LCol W.A. Bishop, who stacked the units with some of the most decorated pilots and observers of the war. Nevertheless, the Armistice condemned the CAF to irrelevancy. It reverted to being a paper force, its business more that of demobilization and repatriation of personnel and shipping aircraft from Britain to Canada. When the CAF was formally disbanded on February 5, 1920, scarcely anyone noticed.

No. 1 Squadron, Canadian Air Force in England, 1918-19: standing are Lt W.L. Rutledge, AFC, MM, Lt P.F. Townley, Lt G.R. Howsam, MC, Lt. E.A. Kenny, Lt F.V. Heakes, Lt C.M. McEwen, MC, DFC, Lt H.A. Marshall, Lt J. Whitford, Lt R.W. Ryan; seated are Capt D.R. MacLaren, DSO, MC, DFC, Capt G.O. Johnson, MC, Maj A.E. McKeever, DSO, MC, Lt J.F. Verner and Capt C.F. Falkenberg, DFC. The gongs show that many of the group had been in hot action. In peacetime, some would play leading roles establishing the RCAF. Others, like McEwen, became top RCAF men in 1939-45. (CF RE17474)

Dolphins of No.1 Squadron, CAF. Since a Dolphin pilot sat in the slipstream above the wing, nobody wanted to flip this fighter. Some squadrons installed roll bars to counter this danger. No.1 Squadron also flew the Pup, Snipe, Bristol Fighter and Fokker D VII. (CF M816Q)

D.H.9a bombers of No.2 Squadron, CAF at Shoreham-by-Sea, England in 1919. (National Museum of Science and Technology 001561)

The flightline and flying field at Camp Borden in 1918. One of the key RFC/RAF (Canada) bases, Camp Borden remains prominent to the present. Several of its original "temporary" hangars survive. (CAHS Col.)

## Training in Canada

In 1917-18 the RFC and its successor (from April 1, 1918 forward), the RAF, directed an ambitious flying training operation in Canada. The scheme (which included a stint in Texas) had no precedent, but inspired later schemes—the British Commonwealth Air Training Plan (1939-1945) and a general program to train NATO aircrew in Canada from 1952 onwards. The importance of air power had been growing, and led to successive expansions (and casualties) affecting all belligerent air forces. Late in 1916 an RFC plan suggested 35 new training squadrons. Most would be outside Britain itself, due to lack of space for airfields and the need to tap overseas aircraft production capacity. Those requirements were the genesis of the RFC/RAF Canada training program.

Although Canadians had been joining the British air services since December 1914 (some by enlistment in Canada, most by overseas transfers from the CEF), the Borden government did not hinder British recruiting efforts in Canada, but also did not promote air training at home, either directly or through subsidies to the few existing Canadian flying schools. Unlike Australia and South Africa, Canada was indifferent to having its own air force, so had no incentive to organize the air training that would have accompanied such a force. Faced with this official Canadian apathy, yet anxious to secure Canadian resources for the RFC, British authorities adopted a policy best described as "If you want it done—do it yourself".

The program that emerged was developed with minimal Canadian government participation, but with the assistance of the Imperial Munitions Board. Initially established in 1915 to co-ordinate shell production and other military contracts in Canada, the IMB was composed of Canadians—yet was essentially a British organization, paid for chiefly by British taxpayers (yet supplemented by loans from the Canadian government). Historian F.J. Hatch described it as a unique example of the flexibility in economic and military affairs within the British Empire. The IMB secured land for air training in southern Ontario, arranged for construction of barracks and hangars, and established Canadian Aeroplanes Ltd. of Toronto to manufacture Curtiss JN-4s for the program. The RFC/RAF provided the direction, including

When freeze-up in Toronto threatened, Curtiss set up in Long Branch. Many young men journeyed there to take up flying on the Curtiss Jenny biplane. Licence in hand they then sailed for England to join the RFC or RNAS. Here is Curtiss' first Toronto class. Standing are D. Hay, E. MacLachlan, Homer Smith, J. Day (mechanic), Clarence MacLaurin, Cornelius I. Van Nostrand, D.G. Joy. In front are C. Grant Gooderham, A. Strachan Ince, V. Carlstrom (instructor), C. Geale and Warner H. Peberdy. Peberdy and Ince were the first Toronto graduates. MacLaurin went on anti-submarine flying boats, was in the short-lived RCNAS, then commanded RCAF Station Jericho Beach. Gooderham and Ince also were RNAS. On December 14, 1915 Ince was a gunner with F/S/L C.W. Graham when they shot down a German seaplane off Belgium. No sooner had the e/a crashed than Graham and Ince were treading water—their engine quit, but they were rescued. Van Nostrand was the first from Curtiss in Toronto to get into the RFC (July 1915). He would have some hairy experiences on 12 Squadron. On March 14, 1916 he and Air Mechanic T. Parkes were on a reconnaissance around Douai. German fighters intervened; Parkes fought well, scaring them off. Next their B.E.2c was hit by flak—the engine faltered, the pilot looked for a place to land. An Eindecker now attacked, killing Parkes, both of whose Lewis guns were out of action. Somehow, Van Nostrand made it home. On July 1, 1916 he became a POW. (CAHS Col.)

Canadians hoping for a wartime career at first had to pay for their flying lessons, usually in the US. In 1915, however, Glenn Curtiss of Hammondsport, New York established a flying boat school on Toronto Bay. Shown is a replica Model "E" (length 25', span 40') built by the Curtiss Museum and flown from its birthplace in 1998. (George Stewart)

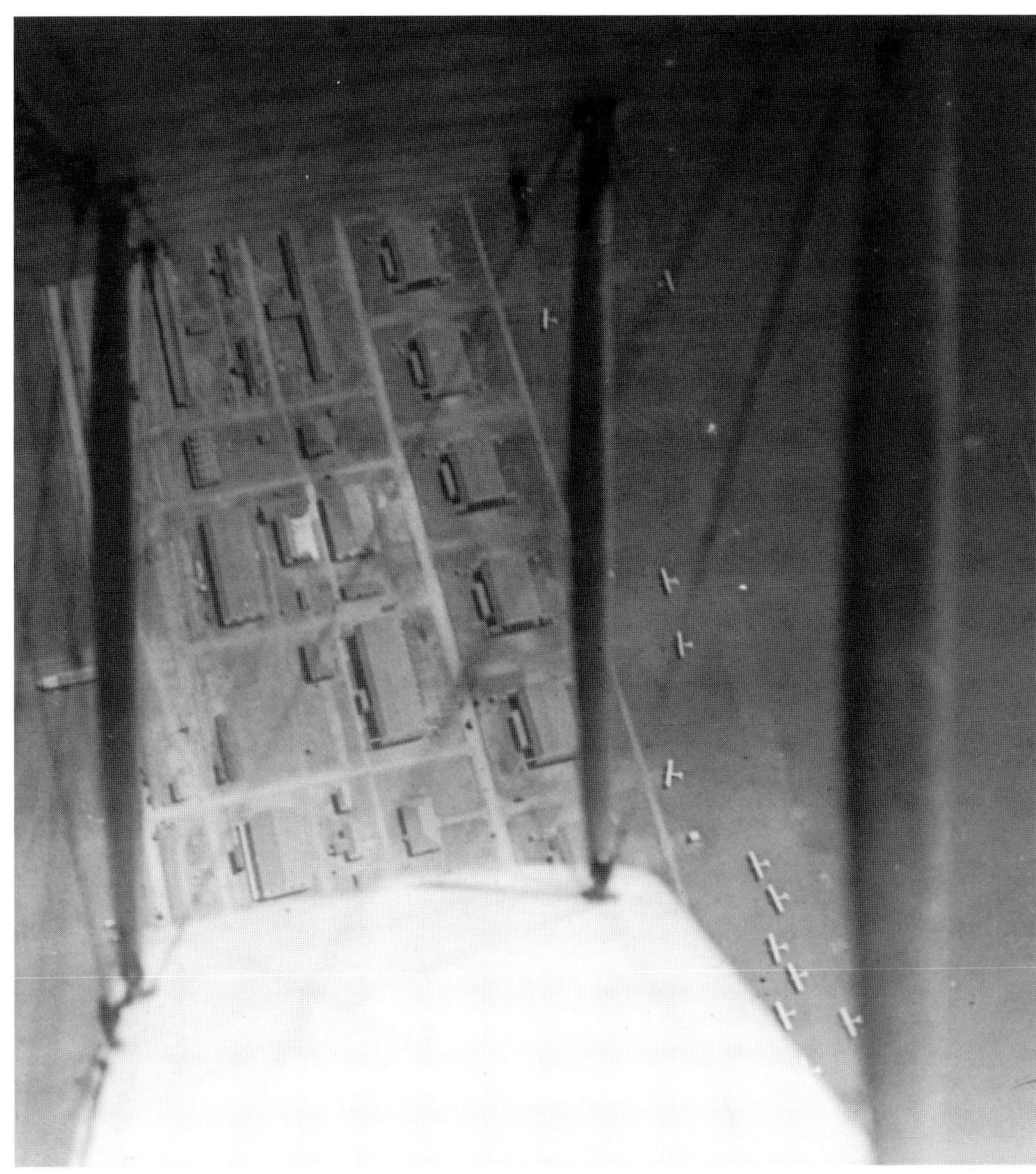

A different sort of aerial view, looking down the wing of a JN-4 trainer over Camp Rathburn, an RFC (Canada) drome near the Lake Ontario shore between Belleville and Kingston. (CF RE14089)

syllabi, uniformed managers, and instructors in dozens of specialist trades—armament, gunnery, aircraft maintenance, navigation, flying etc.

LCol (later Brigadier) C.G. Hoare, an Imperial officer, arrived with his advance staff in January 1917. Only one officer was a Canadian. With buildings under construction and the first JN-4s accepted only on February 22, Hoare ordered that flying commence at Long Branch, west of Toronto, on February 28, 1917. The largest school, Camp Borden, launched training on March 30. The Canadian enterprise included a vigorous recruiting campaign, including newspaper advertisements. A comparable scheme in Egypt was instructional only, without active enlistment efforts.

The program grew monthly, as schools opened and pupils arrived. At war's end the organization occupied quarters at Hamilton (Armament School), Toronto (School of Military Aeronautics, Recruiting Depots), Long Branch (cadet ground training), Beamsville in the Niagara Peninsula (School of Aerial Gunnery, renamed School of Aerial Fighting), Armour Heights (pilot training, School of Special Flying to train instructors), Leaside (pilot training, Artillery Co-operation School), Camp Rathburn and Camp Mohawk (Deseronto, east of Belleville, pilot training), and Camp Borden (pilot training). Armour Heights, Leaside and Long Branch were suburbs of Toronto. Facilities used included schools, a prison, and much of the University of Toronto. Camp Borden alone had room for 122 officers, 496 cadets and 1014 other ranks.

The plan is best viewed in the context of global RAF operations. The Canadian organization was equivalent to what the British would have called a Training Brigade. It provided training up to the advanced level, where pilots were almost ready for combat. The finishing touches would be applied at advanced schools in Britain or France. Training grew more sophisticated with Canadian experience, "feedback" from the Western Front, and revised British procedures. The most important changes came with adaptation of the Gosport System, developed in Britain by Maj R.R. Smith-Barry and in general use by 1918. The name derived from the school where he devised and propagated his theories. Originally, flight training had told pupils little about why an airplane behaved as it did, and instruction concentrated more on what to avoid. The Gosport System taught the dynamics of flight, then progressed to how to use the airplane. For example, earlier pupils had been warned to avoid spins; those of 1918 were taught how to get into a spin, then recover from it.

At first the RFC envisaged recruiting ideal candidates, described as "the clean bred chap with lots of the devil in him, a fellow who had ridden horses hard across country, or nearly broken his neck motoring, or on the ice playing hockey". What they got was more mundane— a keen, healthy specimen of middle-class Canadian youth. Rough studies reveal that those enrolled were more likely to be from a city or large town, than from the countryside. This was most apparent from the occupations of recruits. Although 34% of working Canadians lived by farming, only 4.9% of wartime Canadian fliers identified themselves with agriculture. Those from industry represented 18.4% of workers, but only 5.4% of aviation recruits. The "professions" (including accountants, engineers and teachers) constituted 4.4% of the labour force, but 22.7% of the fliers. Those in commerce were 10% of the labour force and 22.2% of Canadian aviators. However, the largest group of air-minded applicants were students (28.2%), representing an unknown proportion of the populace. The average age on enlistment (or transfer from another service) was 23. Of thousands of volunteers, fewer than a dozen mentioned having "previous aeronautical experience".

The Curtiss JN-4C "Canuck", powered by a 90-hp Curtiss OX-5, was the chief RFC/RAF (Canada) trainer. (Molson Col.)

A snowbound Canuck, propeller broken, location unknown. Canadian Aeroplanes Ltd. of Toronto manufactured more than 2000 JN-4Cs. (via John Wegg)

Some accident scenes tickle the imagination. The reader might enjoy musing about what this JN-4 pilot was explaining after his prang. (National Museum of Science and Technology 4809)

## Action, Adventure, Romance

AERIAL warfare stirs the imagination of the keenest intellects. It satisfies the wildest longing for adventure. It offers the young man free play for all his talents.

In the air he may win fame single-handed, for the air is an element that youth dominates.

The very finest of our young men have taken up the study of military aeronautics in the R.F.C. The work requires a clear brain, a sturdy physique and a fair education.

Men between the ages of 18 and 30 entering as cadets receive a thorough training to fit them for commissions. During the training the cadet receives $1.10 per day. Class 1 men under the M.S. Act are eligible.

Those wishing to enroll as cadets should apply in person or in writing to one of the following addresses:

### Imperial Royal Flying Corps

Recruiting Office: 93 King Street East, TORONTO

HAMILTON—C. G. Booker, Mayor;
LONDON—Capt. W. S. Smith, 402 1-2 Richmond St.;
BRANTFORD—W. C. Livingston, Police Magistrate;
KINGSTON—Lieut.-Col. R. E. Kent;
PETERBORO'—G. H. Logie, Manager Bank of Nova Scotia;
WINDSOR—T. C. Ray, Secretary Border Chamber of Commerce;
ST. THOMAS—W. Trott, Mayor;
STRATFORD—J. D. Monteith, Mayor;
OWEN SOUND—R. D. Little, Mayor;
SAULT STE. MARIE—C. F. Farwell, Registrar;
SARNIA—L. M. MacAdams, Editor "Observer";
BELLEVILLE—A. R. Walker, Public Library;
WOODSTOCK—W. J. Taylor;
BARRIE—Judge G. M. Vance;
COBOURG—John T. Fields, Registrar;
ORILLIA—Mayor Curran;
CHATHAM—J. A. Kerr, Mayor;
BROCKVILLE—W. J. Cairns;
OSHAWA—C. M. Munday;
CORNWALL—E. J. Roth;
SMITH'S FALLS—J. C. Knox;
PORT HOPE—E. A. Mulholland, Mayor;
WELLAND—M. Vaughan, Mayor;
PRESCOTT—F. S. Evanson, Mayor;
TIMMINS—G. A. Macdonald.

Recruiting posters appeared everywhere around the country, whether on billboards or newspapers. The sketch was by Fred Varley, later one of Canada's famous Group of Seven landscape artists. (National Museums of Canada 76-2094)

Daily accidents kept carpenters, fabric workers and other tradesmen (sometimes coffin makers) busy. These Canucks came to grief near Fort Worth over the winter of 1917-18. (David Thompson Col.)

JN-4Cs being built in Toronto, then a look at the fabric shop, where many seamstresses found good jobs. (David Thompson Col.)

Canadian Aeroplanes Ltd. also built the giant F.5 flying boat for the US Navy. One is shown under construction, then the first is seen in flight at Philadelphia. Such contracts gave a young Canada valuable experience in aircraft manufacturing. (David Thompson Col.)

The basic flight trainer was the Curtiss JN-4 (Can), an American design modified by Canadian Aeroplanes Ltd. to meet military training needs, e.g., by removal of a wheel control and substitution of a joystick, adding camera guns, reconnaissance cameras and machine guns. Those flown in Canada sported a variety of distinctive markings, including maple leaves, terriers, black cats, shamrocks, and the Jolly Roger. Some were named for cities; at least six had carried lettering commemorating battles of the War of 1812. Had the war continued, JN-4 production would have been superseded by the Avro 504K, but only two were built by CAL before the Armistice.

William Hector Ptolemy was a typical trainee of the late war. His first flight on December 3, 1917 was a 10 minute "joy ride" with an instructor at No.88 Canadian Training Squadron, Armour Heights. Two days later he flew for 25 minutes and took the controls for the first time. Bad weather occasionally interrupted his progress and on December 16 he broke a propeller, while landing in snow. He smashed another on December 22, and generally had difficulty with turns. On January 3, 1918 he flew for 40 minutes, executed seven landings, and made an emergency landing, when his engine failed. He reported his first landing on skis on January 29. Finally, on February 5, having flown 7:25 hours dual, Ptolemy made his first solo circuits (most pupils soloed after five hours).

Thereafter, Ptolemy regularly flew alone. His terse log entries hint at excitement. On February 11 he was airborne 70 minutes and described the trip as "Up to Newmarket—went for a joyride". In mid-February he switched to No.90 Canadian Training Squadron and in late March to No.91 CTS (both at Leaside), where more advanced manoeuvres were taught, notably formation flying and the first aerial photography exercises. On April 10, 1918 he first reported dropping bombs. He subsequently attended the School of Aerial Gunnery at Beamsville for advanced gunnery and photography (11:10 hours between April 27 and May 3). He was then posted overseas and, after further training in Britain and France, reported to No.201 Squadron (Camels) on October 4, 1918. Following the war Ptolemy flew as a bush pilot.

Aerial instruction was supplemented by intensive ground training in class, at gun butts, and even with training aids that included battlefield models and primitive flight simulators. By any standards the program was sophisticated, dealing even with such topics as aviation medicine and psychological screening of candidates. With no experience in severe cold weather flying, RFC authorities feared that training might be shut down entirely for the winter of 1917-18. Consequently, much of the program relocated to Fort Worth, Texas, where it also trained many Americans and led to mutual exchanges of training information. Squadrons left in Canada adapted their JN-4s to cumbersome skis, devised formulae for lubricants and kept the system operating at least as well as in Texas, where mud proved as frustrating as snow.

There were many aspects to RFC/RAF (Canada) training other than flying. Here cadets learn basic airframe. (CF RE19065-13)

Aerial combat was not easy to simulate. In this view a cadet rides in a crude motion simulator at the School of Aerial Gunnery in Beamsville. The gondola is being winched along a cable, while the student fires at a stationary target. (CANAV Col.)

Cadets take a class in aerial observation. (CF RE19065-14)

Overall, the scheme enrolled 9200 cadets. Of these 3135 completed pilot training and more than 2500 went overseas. The balance were retained as instructors or were awaiting postings overseas when the Armistice was signed. In addition 137 observers graduated, of whom 85 got overseas. The plan also turned out at least 7400 mechanics. Some American personnel (navy and army) were trained in Canada, as were a few White Russians. These results were achieved at some cost. At least 129 cadets and 20 instructors were killed in flying accidents. A nasty instance was a head-on collision at Beamsville on May 2, 1918. One instructor was shaken up, the other had a broken hip; the two pupils in the front cockpits died. Yet the safety record improved with time; in April 1917 there was one fatality for every 200 hours flown; for December 1917—one for every 1500 hours; for October 1918—one for every 5800 hours. The most publicized accident involved no injuries. A JN-4, force-landing on Oshawa's main street on April 22, 1918, became entangled with telephone wires and wound up pinned to a large store front. There it remained for several hours. Photographs of the bizarre crash turned up in every history of the plan.

Reminders of the RFC/RAF Canada are scattered throughout North America, including plaques in Fort Worth and Toronto, as well as cemetery headstones. An instance of the latter is at Deseronto. A square grey monument, eight feet high, has a representation of RFC and RAF wings, together with an inscription: "To the memory of the officers, non-commissioned officers, cadets and airmen of the Royal Flying Corps and Royal Air Force who died while on duty in Canada 1917-1919." Grouped around the memorial are six graves; a seventh headstone associated with the 1918 RAF is in a local family plot. There are some anomalies. It is unclear how Cadet C. Bender (died June 10th, 1918) came to be in a family grave at Deseronto. Newspapers confirm that Cadet John Robson (a Scot) died there in a flying accident on July 3, 1918. The drowning of Lt C.J. Humphreys following a flying accident (July 15, 1918) is also confirmed by a headstone (Humphreys, aged 21, was from Victoria). Three other deaths (those of Pte Frederick W. Grand on October 3, 1918, Air Mechanic George D. Marshall on October 19, 1918 and Sgt John R. Holland on December 16, 1918) could be attributable to the 1918 influenza epidemic.

Behind the headstone of Lt C.G. Coleridge (age 29, killed in a crash on July 23, 1918) lies an interesting story. *Aeroplane* of October 23, 1918 carried an extensive obituary. A native of Norfolk, England, he had been made a Member, Order of the British Empire for "saving life at the risk of his own while in great personal suffering". This accident occurred on February 9, 1918 and involved JN-4 C463. *Aeroplane* noted:

*When stationed at Fort Worth, Texas, last February, Mr. Coleridge had gone up with a cadet, to whom he was giving instruction. The cadet pulled the control pillar over too hard,*

This monument to the RFC/RAF (Canada) is in the village cemetery just east of Deseronto. In the foreground is the grave marker of Lt Humphrey. That of Lt C. Coleridge, MBE, killed July 25, 1918, is beyond. (Larry Milberry)

*and being too near the ground (about 500 feet) for an attempt to right the machine to succeed, it crashed and immediately caught fire. Mr. Coleridge was thrown out with four ribs broken and a severe cut on the head, but he crawled into the burning aeroplane and rescued the cadet, who was strapped in and pinned under some burning wreckage with a broken leg. Mr Coleridge, who was severely burnt about the face and hands, then dashed loose earth over the cadet "whose legs were soaked with gasoline and burning badly", and put out the flames.*

RFC Canada graduates began sailing for Britain in June 1917. Probably the most famous was Lt A.A. McLeod, who trained at Long Branch and Camp Borden, received his wings in July 1917, and reported to No.2 Squadron (Armstrong-Whitworth F.K.8 army co-operation aircraft) on November 29, 1917. His brilliant career culminated in an action on March 27, 1918 for which he was awarded the Victoria Cross. Other distinguished alumni included Capts D.R. MacLaren and W.G. Claxton (54 and 31 estimated aerial victories).

While the organization was dedicated to training, it made news in ways that heralded future developments. The first airmail in Canada was carried by Capt Brian Peck from Montreal to Toronto on June 24, 1918, and four additional airmail flights (Toronto-Ottawa return) were conducted by RAF instructors between August 15 and September 4, 1918. The Ottawa terminus was the Rockcliffe Rifle Range (an area now occupied by the National Aviation Museum). An unexpected development was the recruitment of women.

This was a direct consequence of introducing army conscription in late 1917. Shortages in civilian manpower created opportunities for females. Thousands volunteered; more than 1200 were accepted, chiefly as mechanics and drivers. Their incorporation was orderly and aroused little comment. One wonders what difficulties the authorities expected; Alan Sullivan subsequently wrote in *Aviation in Canada, 1917-1918*, "... in spite of many predictions, they have caused no trouble whatsoever, and submitted themselves apparently without effort to the necessary discipline of the Force."

Mechanic R.B. Brock, DFM of Brockville, Ontario is buried in Hamilton, Ontario. Brock, who died on July 12, 1919, was the only Canadian to receive the DFM in WWI. His gallantry is thought to have been connected with work in airships or balloons. (Larry Milberry)

Although the RFC/RAF Canada plan had begun with negligible Canadian direction, it came to include Canadians at all levels. The Canadian Militia assigned paymasters, doctors, and other non-flying personnel to schools and headquarters. Increasingly, Canadian pilots and observers joined the instructional staff. Some were recent graduates, others were veterans of the Western Front. By November 1918 Canadians commanded the School of Aerial Fighting, 2 of the 3 training wings, and 12 of the 16 squadrons; roughly 60% of all instructors were Canadians.

Lists of personnel decorated or commended for services associated with the scheme include many administrative and support personnel seconded from the Militia to the flying stations. Several instructors were Canadians with service on the Western Front. In the former category (administrators seconded by the Militia) we may identify such individuals as Lt George John Blackmore (Commended for Valuable Services in Connection with the War, *London Gazette* of January 22, 1919), a Toronto hotel inspector, who served in headquarters as a quartermaster. Maj Breney Rolph O'Reilly, Canadian Army Medical Corps, was made an Officer, Order of the British Empire (*London Gazette*, March 29, 1919) for services rendered with the program in Canada and Texas. As of February 1918 he was Senior Medical Officer at Headquarters in Toronto. His contributions were considered sufficiently valuable that he was sent to France in April 1919 on a special mission relating to aviation medicine. Unhappily, no citation has been found describing his achievements.

At a lower level, 2Lt Edward Alan Sullivan was "Brought to the Notice of the Secretary of State for Air in Respect of Valuable Service" (August 29, 1919). A graduate of the University of Toronto (Applied Science, 1889), he had been an engineering and technical officer with the RFC/RAF Canada scheme. His honour was described only as being for "valuable services rendered". As author of *Aviation in Canada 1917-1918* he was the first chronicler of the program, his book being a valuable source for subsequent researchers.

At least 10 Canadian and 2 British RFC/RAF (Canada) instructors were awarded the Air Force Cross in connection with the training program; several more were commended. The best known was Acting Major Albert E. Godfrey, MC, a fighter pilot who, at wars' end, commanded the School of Aerial Fighting at Beamsville. Godfrey's most treasured memory of his experiences was taking Brigadier Hoare for a flight into the turbulent Niagara Gorge, a trip that momentarily terrified Hoare, who nevertheless forgave Godfrey. Another decorated instructor was Lt John Owen Leach (Toronto) who had won the Military Cross as an infantry officer, transferred to the RFC, became a fighter pilot in France, then lost a leg through wounds. This did not ground him. He persuaded his superiors to keep him on the flying list as an instructor. Posted to Canada, he served at Armour Heights, then Deseronto. Hoare sometimes used him as a personal pilot; at least once he was called as an expert witness at an inquest investigating the death of a student. A newspaper described Leach as "the best flyer in Canada".

Historian S.F. Wise has described the RFC/RAF Canada scheme as "the single most powerful influence in bringing the air age to Canada". JN-4s left over after the war were less important than the pool of men determined to fly and service them. The public, at least in the Niagara-Hamilton-Camp Borden-Toronto-Deseronto arc, became accustomed to aircraft, no longer viewing them as novelties or menaces. The RFC/RAF Canada organization proved the feasibility of year-round flying in Canada and even resulted in special winter clothes being developed. Postwar barnstormers, usually RFC/RAF (Canada) graduates, quickly gave way to aerial forestry surveyors and bush pilots operating on northern frontiers. They, in turn, trained others to fly. The plan proved a foundation on which to build the saga of Canadian bush flying, and the RCAF of wartime and peacetime fame to come. Perhaps no man better exemplified its continuity with succeeding events than Capt Murton A. Seymour, a Toronto staff officer in 1917-18, later a founder of the Canadian Flying Clubs Association, and an organizer of RCAF training during the Second World War. With many others, he bridged two wars and two generations in Canada's flying heritage.

Just how many Canadians served in the First World War flying services? The records are incomplete and contradictory. The appropriate text panel of the Memorial Chamber in the Peace Tower, Ottawa, quotes 22,182. The Canadian Armed Forces Directorate of History and Heritage came up with only 11,160 *associated* with Canada, of whom 1736 were known to be non-Canadians (chiefly Americans). To this one might add 7453 mechanics recruited in Canada for the flying program—which leaves 18,653, including the non-Canadians. Given the uncertainties as to enlisted numbers, the accepted figure for Canadian aerial losses—about 1500— is probably a bit high. As shown by the casualties, the first air war had been waged bravely and fiercely. When tallying the costs, the RAF estimated that, as of November 1, 1918, it had sustained the following losses:

| Killed | Wounded | Missing | Prisoners/ Interned |
|---|---|---|---|
| 5972 | 5813 | 1873 | 1093 |

In 1893 Rudyard Kipling wrote, "If blood be the price of admiralty, Lord God we ha' paid in full" (later adapted, by himself for another context, to "If blood be the price of the Arctic, Lord God we have paid in full"). By 1918 a new form of "admiralty" had appeared—and it too demanded its monstrous fee. Air casualties in 1914 had totalled 37 (all categories of which 14 were killed); for 1916—1418; for 1918—5368. Not surprisingly, the worst losses were in France, followed closely by the British home front, which exacted a terrible toll in flying accidents. The following summary by theatres (which seems to omit training accidents in the United States during the RFC sojourn in Texas) illustrates the pattern:

|  | Killed | Wounded | Missing | Prisoners/Interned |
|---|---|---|---|---|
| France | 3287 | 3855 | 1764 | 985 |
| Great Britain | 2147 | 2261 | 35 | 25 |
| Middle East | 347 | 341 | 32 | 69 |
| Italy | 47 | 76 | 42 | 14 |
| Canada | 133 | 278 | - | - |
| East Africa | 11 | 2 | - | - |

After the war surplus JN-4Cs flooded onto the civilian market. Many were bought by barnstormers in Canada and the US. This one, still in RFC colours, was joy riding at the Calgary fair of July 5, 1919. Pilot Fred McCall had engine trouble and came down on one of the carnival rides. (Glenbow-Alberta Institute NA1451-27)

On May 20, 1983 a group of WWI airmen visited 410 Squadron at CFB Cold Lake. Steve St. Martin of Milwaukee took this photo showing Lt William W. McGill (23 Sqn), unknown, Lt Alfred Koch (70 Sqn), unknown, Lt George A. Riley (57 Sqn), Capt Russell N. Smith (54 Sqn), Lt W. Dunbar (254 & 236 Sqns), Lt C. Victor Scaife and Lt Ewart T. Griffith. By 1999 only one Canadian WWI aviator remained alive.

The D.H.4 bomber first flew in England in August 1916. More than 6000 were built. Able to operate above 15,000 feet, it also could hold its own with German fighters. Canada's Imperial Gift D.H.4s were converted to single seaters with air-to-ground wireless sets, mainly doing forestry. F2714, still in RAF colours, is seen at High River in May 1922; it was on RCAF strength to November 1928. Note the wind-driven generator (between the cockpits); it provided electricity for the wireless. D.H.4 specs included: engine 375-hp R-R Eagle VII, wingspan 42' 4⅝", length 30' 8", all-up weight 3472 lb, top speed 143 mph, endurance 3¾ hr. (K.M. Molson Col.)

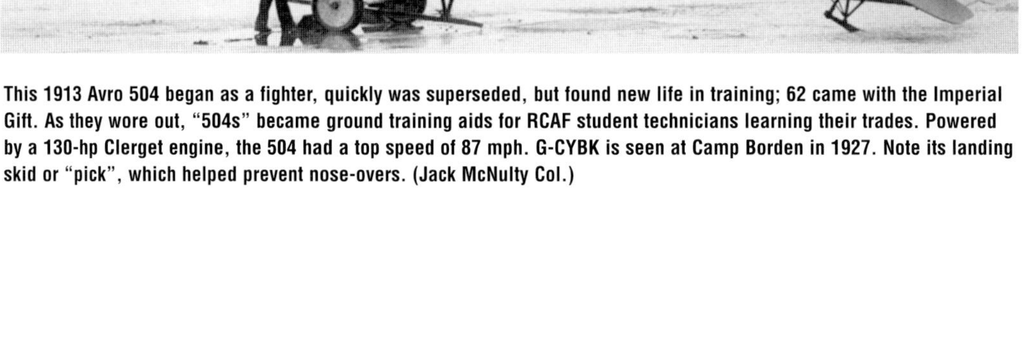

This 1913 Avro 504 began as a fighter, quickly was superseded, but found new life in training; 62 came with the Imperial Gift. As they wore out, "504s" became ground training aids for RCAF student technicians learning their trades. Powered by a 130-hp Clerget engine, the 504 had a top speed of 87 mph. G-CYBK is seen at Camp Borden in 1927. Note its landing skid or "pick", which helped prevent nose-overs. (Jack McNulty Col.)

# ❖ BETWEEN TWO WARS ❖

The Canadian Air Force / Royal Canadian Air Force should never have happened. Before the war Ottawa had shunned aviation. It had tolerated formation of the RCNAS only because of a grave submarine threat, then hastily disbanded it. Creation of a distinct overseas air force had come late, then it disappeared. As to government support for commercial aviation, scarcely any existed in early postwar Canada. Meanwhile, Europe was quickly criss-crossed with subsidized airlines, and the US instituted air mail. Canadian air services struggled without such help. Although airmail contracts flourished from 1927-32, they were slashed in the austerity of the Depression. A government airline with national aspirations was established only in 1937. Given this atmosphere, the birth of the RCAF, then its survival, were little short of miraculous.

Formation of a CAF resulted from a peculiar combination of circumstances and personalities. Firstly, there was the so-called Imperial Gift. At war's end Great Britain had thousands of surplus aircraft. Its self-governing Dominions—Australian, Canada, New Zealand, South Africa—each received about 100. This was partly Britain's gratitude for wartime assistance, partly a device to encourage aviation in the Empire, partly a promotion for later sales. This last motive was blatant in similar (though smaller) gifts of British aircraft to nations outside the Empire, e.g. some South American countries and new European states created by the Treaty of Versailles. For its part in the Imperial Gift, Canada selected: Avro 504—62 aircraft, D.H.4 and D.H.9—23, SE.5a—12, F.3—8, Curtiss H.16—2,

**Avro 504 'BL with two well-bundled pilots about to go flying. 'BL remained on strength into 1930. Skis were among the early Canadian mods for keeping aircraft on the go in winter. (CF RE75821)**

**Flexistowe F.3s at Victoria Beach, Manitoba in 1922. Although valuable in war, F.3s were too cumbersome for daily work in the bush, so had a brief CAF career. Their usual end was to be stripped of components, then burned. (CANAV Col.)**

Bristol F.2b—2, Sopwith Snipe—2, Fairey C.3—1. There were also airships, kite balloons, engines, spares, tools, etc.. Some lighter-than-air equipment was scrapped or re-cycled as patching materials for hangar roofs. Canada also inherited a few JN-4s from the hundreds in the RAF (Canada), and 12 ex-US Navy HS-2Ls abandoned in Nova Scotia. Most of Canada's airplanes eventually found employ-ment, thanks principally to J.A. Wilson, Deputy Minister of Marine.

**Various interest groups claim this or that individual as their pioneer Canadian aviation hero. However, considering his impact on the Air Board and enthusiasm in bringing aviation to the Canadian scene, none compares to J.A. Wilson. He came to Canada in 1905, following a career in India, supervised formation of the RCNAS, then was Control-ler of Civil Aviation from 1920. From earliest times he pushed for development of a trans-Canada airway. (NAC PA64561)**

After the war some officers had grandiose schemes for the Canadian Militia and the RCN. None came to fruition—as now, governments were indifferent (even hostile) towards large, permanent military establishments. The army and navy were retained almost by force of habit, then starved of funds, manpower and purpose. However, a cadre of civil servants, led by Wilson, was convinced that aircraft could be tremendously useful in civil roles. They sold this idea to Cabinet. What transpired was an air force with almost no military duties. Simply by virtue of existing, however, it might become a true fighting service.

### The Organization: 1919-23

In 1919 almost every nation was enacting legislation to regulate aviation. Canada's *Air Board Act* received Royal Assent on June 6, 1919. Under it, a board of civilians (some with distinguished military careers) was to study aeronautics, regulate safety, license pilots, air engineers, aircraft and air harbours. It also would administer the Controller of Civil Aviation (regulating private operators), the civil Flying Operations Branch (operating aircraft for federal departments), and the Canadian Air Force. Although most aircrew went by their wartime titles, only the CAF was "military", and only its members wore the distinctive dark blue CAF uniform.

The CAF was large on paper (5245 officers and men). In practice fewer than 300 were in uniform at any time. HQ was in Ottawa; there was only one base—Camp Borden. The principal task was to give refresher training (flying and technical) to officers and men who became CAF members for a month or two, then returned to civil life (although "civil" could mean working for the Flying Operations Branch). Lines were blurred, since personnel were traded among branches. The weakness of this scheme was that the CAF was simply retraining veterans (many of whom soon disappeared into civil life). No new blood was introduced. Refresher courses were suspended in mid-1922, pending a re-organization of government services.

### The Organization: 1923-36

So far the situation governing Canada's armed forces had been confusing. The RCN was under the Department of Marine and Fisheries; the army

Aviation and forestry both knew that war surplus aircraft were stop-gap. New designs were needed, one being the Canadian Vickers Vedette. Before it flew, a scale model was wind-tunnel tested at the University of Toronto, a first in Canadian aeronautics. On December 2, 1924 S/L Basil Hobbs was the first in the RCAF to fly a Vedette, his passenger that day being J.A. Wilson. Here prototype Vedette G-CYFS climbs from Shirley's Bay, Ottawa on October 20, 1925. Of 60 Vedettes 44 went to the RCAF. A batch sold to Chile was the first export sale of a Canadian production aircraft. No original example remains, but a replica exists in the Western Canada Aviation Museum in Winnipeg. (NAC PA62459)

Barnstorming seemed to be the first popular application of war surplus JN-4s. In this 1923 view G-CADF of Laurentian Air Services (St. Jovite, Quebec) is about to give a ride to a chain of thrill-seekers. Earlier it had flown with Price Brothers of Chicoutimi. (Wheeler Col.)

was under the Department of Militia and Defence; the CAF was under the Air Board. From January 1, 1923 a newly-created Department of National Defence controlled all three (an early example of armed forces unification). But the RCAF (formed April 1, 1924) was subordinate to the Army, the Director of the RCAF reporting to the Chief of Staff. The RCAF was responsible for aid to the civil power as well as licensing and regulating civil aviation. Instead of all aviation (including the CAF) being administered by a civil body (the Air Board), the "military" RCAF called the shots for commercial aviation.

Matters changed in 1927, when the RCAF split into four divisions. A service aviation branch reported to the Chief of Staff, so was under direct military control. A newly-created Civil Government Air Operations Branch (CGAO—manned largely by the RCAF) flew on behalf of the government (photography, forestry patrols, etc); an Aeronautical Engineering Branch dealt with technical developments; and the Controller of Civil Aviation (largely civilian) handled pilot licenses, aircraft registrations, etc. These last three divisions reported to the Deputy Minister of National Defence. With no enemy in sight military aviation atrophied, while CGAO flourished. Aid to the civil power was to the RCAF then, what "peacekeeping" became much later—a visible, politically popular reason for having an air force.

### The Organization: 1936-38

With a few changes in nomenclature, the RCAF remained much the same until 1936. With war looming in Europe, it was clear that the air force should re-equip and train for more belligerent duties. At the same time non-military aviation grew more complex. In a general "clearing of decks" the government formed the Department of Transport, and transferred to it the RCAF's former responsibilities for civil licensing and regulation. The force was now almost wholly military (although it remained the principal supplier of air services to federal departments such as Indian Affairs or Interior). However, it was not until 1938 that the RCAF achieved true independence from the Army; that year the Senior Air Officer (soon renamed Chief of Air Staff) reported directly to the Minister of National Defence, equal in status to the Army and Navy service chiefs.

### The Air Board Organizes

Establishment of the Air Board from scratch was a remarkable achievement. Appointments had to be made; the first regulations drafted for licensing and regulating aircraft, personnel and bases; Air Board sites selected. How this was done can be illustrated with a case study. Maj Ambrose B. Shearer, a Canadian still in England and expecting to be sent to North Russia, heard of the Air Board's creation. On July 25, 1919 he wrote to enquire about a position. Col O.M. Biggar (first chairman of the Air Board) replied, enclosing application forms. Shearer briefly accepted an RAF commission, then resigned it and returned to Canada. On February 7, 1920 he wrote the Air Board again, seeking a position as an Air Station Superintendent. Living in Vancouver, he hoped to obtain a position in Manitoba. On May 18 he was appointed a Pilot-Navigator (Aeroplanes) at a yearly salary of $2460.

Robert Leckie, Director of Flying Operations, had never met Shearer. But, knowing of his war service and ambition, he wrote that, although an assignment to the air station at Morley, Alberta was being offered, there was hope of a better position. "Your name also appears on the eligible list of Sub-Station Superintendents," read Leckie's letter, "and you should consider your present employment as merely a stop gap until such times as the service is extended to allow of posting you to a Sub-Station." Shearer's luck was better than he could have hoped. On June 11 he was instructed to report to Leckie in Ottawa. Confused as to whether he was now going to Morley or elsewhere, he asked for clarification. Leckie cabled that a sub-station appointment in the east had become available. Again, Shearer asked for details. Leckie replied on June 21: "You will be located temporarily seaplane station Halifax. Will endeavour to place you in west when opportunity arises. Imperative if you accept you should report to Ottawa at once."

Shearer cabled Leckie on June 22: "Sorry cannot accept Halifax. Too far to move. Will accept Morley pending opening of new sub station in west." Leckie must have been anxious to secure Shearer's services, for he replied with a cautionary telegram: "You are making mistake in turning down sub station superintendent position. Halifax only temporary for a few months. No immediate prospects for new sub stations in west. Your expenses refunded from Vancouver to Halifax." Shearer immediately accepted unconditionally.

Reclassified as a Sub-Station Superintendent (salary $2940), he reached Halifax on July 6. He apparently arrived with instructions to close the former US naval base, in RCN hands for some months. This entailed assembling and testing all HS-2Ls there, and others being shipped by rail from North Sydney. He was also to arrange

**Major Ambrose B. Shearer rose to be A/V/M Shearer and AOC of No.2 Training Command 1940-43. Here he is in 1918 wearing his RNAS uniform. (Halliday Col.)**

shipment to other Air Board stations of useful and serviceable equipment at Dartmouth Air Station. His stores officer, F/O Peter J. Moloney, did not arrive until October. When he did, Shearer instructed him to take on charge all serviceable items, inside or out, that might be useful at other stations. No inventory had been taken since late March 1920, and some confusion developed, aggravated by Moloney, who proved troublesome.

From September, Shearer was fitting cameras to and flight testing HS-2Ls (he logged 58 hours in less than 3 months) and shipping tool kits to Ottawa and Winnipeg. On September 9 he asked if he might participate in the forthcoming Trans-Canada Flight. This was refused (personnel had been selected), but when LCol Leckie crashed the Fairey seaplane in New Brunswick early in that flight, Shearer and an airman named Dowell went to assist him. Meanwhile, Ottawa decided to keep Dartmouth Air Station open. On June 25, 1921 Leckie requested a promotion for Shearer to Air Station Superintendent (salary—$3360), and listed the staff to be supervised—1 additional Pilot Navigator, 2 Air Foreman Mechanics, 2 Air Riggers,

**Dartmouth Air Station as it was when Shearer took command. (CF RE19537-A)**

3 Air Engine Fitters, 1 Motor Transport Driver, 1 Fabric Worker, 1 Air Boat Builder (seaplane repairman), 2 Clerk Stenographers, 1 Camp Cook, 1 Storekeeper and 2 Watchmen.

When he had arrived, Shearer had found his station a virtual hay field, which he considered a fire hazard. He allowed a local farmer to graze cattle on the premises; in return the station purchased milk at a slight bargain. The base had some leftover HS-2L packing crates. Shearer gave them to whomsoever would remove them, keeping only two. There were also many old barrels, dating to the construction of Dartmouth in 1918. These were collected, sold, and the money used to buy a second-hand gramophone for the base mess. Such minor details came to light because of a dispute between him and Moloney. The latter had been representing himself as an RAF flight lieutenant, but enquiries in Britain showed that he had held no RAF rank higher than acting lieutenant. Faced with this, he became indignant, arguing with Shearer on several points of administration. Shearer finally succeeded in having him let go. Moloney's reaction was a long letter to W/C Ernest C. Stedman (head of RCAF engineering), accusing Shearer of incompetence, waste, and using government resources for personal ends, notably in servicing and fuelling two private motor cars. A court of inquiry was convened in February 1924. It quickly was clear that Moloney had mounted a vicious personal attack and that many of his charges were off the mark. What most occupied the inquiry was the vehicle state at Dartmouth—the line between private and government issue was blurred. Shearer explained:

*Mr. Moloney also neglected to mention the number of times my cars were used in hauling the Reo Ambulance, Ford and even the Kelly 3½ ton truck back to the Station when they broke down... I also always used my car during Artillery Co-operation practice, to carry the wireless gear to and from the F.C. Post at Sandwich Battery and the Air Station... transport supplied to the Station was so bad that it was mostly under repair, and could not be relied on for important trips. At different times when the Cleveland Tractor was out of action, one of my cars was used to haul HS-2L machines up and down the slipway.*

The board concluded that, although Shearer had been administratively careless, he was no wrong-doer. Many of the problems had been inherent in the Air Board process of rushing air stations into service in 1920. Although Shearer was not formally disciplined, he was re-assigned to Camp Borden. When the RCAF was organized as a fully professional force in April 1924, he was not given a permanent commission. Instead, he was granted a commission in the Non-Permanent Force (a means of placing him on probation) on the understanding that he would receive a Permanent Force commission, if he proved capable. What followed was absorbing, not only in terms of his personal career, but as an illustration of how "personnel management" was handled in an age of diversity and practicality, rather than of excessive specialization. At Camp Borden Shearer was placed in charge of the Motor Transport Section (the field where his Dartmouth administration had been most muddled). His handling of the section was described by W/C L.S. Breadner in a letter to the Deputy Minister of National Defence of August 25, 1924 (With Breadner's recommendation, Shearer was let out of the dog house—he was confirmed as an officer in the Permanent Force, effective April 1, 1925.):

*Shortly after assuming command of this station, I placed Squadron Leader Shearer in charge of Mechanical Transport, which for some considerable time had been operated under the sole supervision of an NCO... Immediately improved conditions were noted... with the result that this Section, which frequently is the cause of much trouble and inconvenience, has operated in an entirely satisfactory manner, in spite of the fact that, in nearly all instances, the vehicles are old.*

*On April 28th, 1924, Squadron Leader Shearer was placed in charge of the Aeroplane and Engine Repair Sections and was charged with turning out sufficient machines and engines for the summer Cadet Training. Work progressed in a highly satisfactory manner. By May 26th five machines were serviceable and placed in No.1 Hangar for Flying Training. At about 0100 hours on the 29th May, No.1 Hangar and contents was entirely destroyed by fire, which left this Station without a single serviceable machine and with some 24 Cadets here for instruction and a further seven due in a month's time... All Mechanics on the Station were immediately turned over to Squadron Leader Shearer...with the result that at the present time there are about eight serviceable machines and the number of hours flown by the 1st term Officers approximated four-fifths of that originally intended and 2nd Term Officers slightly more than two-thirds. Squadron Leader Shearer has demonstrated his marked ability in handling men and his technical knowledge has been of the utmost value to this Station. His discipline is good and it is felt that the services of this Officer are such that the RCAF can ill afford to lose them.*

## The Trans-Canada Flight

In 1920 the Air Board conducted a trans-Canada survey flight to study the feasibility of coast-to-coast air operations. Although many countries were involved in such trials, for Canada the venture was especially daunting, the country being so vast, and weather and terrain so challenging. The flight commenced at Halifax on October 7 when LCol Leckie and Maj Hobbs took off in a Fairey seaplane. Mechanical trouble forced them down near Saint John, but they carried on in an HS-2L as far as Rimouski, landing on the St. Lawrence in darkness. From there they switched to an F.3 for the legs to Ottawa, North Bay, Sault Ste. Marie, Kenora and Selkirk, on the Red River downstream from Winnipeg. From there two D.H.9s took over, the land of lakes now being behind. On October 11 one D.H.9 reached Calgary. Two days later the crew of Tyee and Thomson bucked stiff winds to reach Revelstoke, where they were weathered in till the 15th. That day they pushed on to Merritt. On the 17th they made Vancouver. Eleven days had passed, 45 hours had been spent aloft and 3265 miles covered. Although the trans-continental flight had been a success and much had been learned, many years would pass before Canada would be traversed routinely by air.

**Trans-Canada Flight aircraft:** Felixstowe F.3 G-CYBT/N4016, (Rimouski-Selkirk); and D.H.9A G-CYAJ (Winnipeg-Saskatoon). The F.3 was at St. Vital on the Red River; the D.H.9A, at Camp Borden. (CF AH-88, RCAF RC553/Jack McNulty Col.)

## Aircraft of the Trans-Canada Flight

| Type | Registration | Crew | Route |
| --- | --- | --- | --- |
| Fairey IIIC | G-CYCF | Maj B. Hobbs, LCol R. Leckie | Halifax-Saint John |
| HS-2L | G-CYAG | same | to Riviere du Loup |
| F.3 | G-CYBT | Hobbs, Leckie, Heath, Capt G.O. Johnson | to Selkirk |
| D.H.9 | G-CYAJ | Capt C.W. Cudemore, Sgt Young | Winnipeg-Saskatoon |
| D.H.9 | G-CYAN | Capt Home-Hay, LCol Tylee | Winnipeg-Regina |
| D.H.9 | G-CYAN | Capt C.W. Cudemore, LCol Tylee | Regina-Calgary |
| D.H.9A | G-CYBF | Tylee, Capt G.A. Thompson | Calgary-Vancouver |

Camp Borden's hangar line facing the landing ground in a photo dated 1927. These buildings, described as "temporary", were erected hastily in 1917 to serve the RFC (Canada). By the year 2000 several remained in use. (CF)

## Camp Borden: 1920-24

The CAF took over Camp Borden in July 1920. Early in this era it was as much a storage depot as a flying unit. From the summer of 1921 three Felixstowe F.3 and two Curtiss H.16 flying boats were on charge and rotting away. There also was an assortment of German war trophy aircraft. Much of this would be junked, yet some would be surprisingly durable. A gantry crane, used to assemble Imperial Gift aircraft in 1920, helped assemble Canada's first Fairey Battles in 1938. Compared to the heady days of 1918, Camp Borden seemed a ghost town. By May 1921 a provisional establishment called for only 13 officers and 70 other ranks. Station HQ alone encompassed 14 (all ranks); flying instruction required 17 (5 officers, 6 aero engine mechanics, 6 airframe riggers); ground instruction needed 3 officers and 6 other ranks; the balance were assigned to transport, equipment, maintenance, messing and medical duties. The establishment was subject to constant negotiation with the Air Board and, later, with the Chief of Staff; but in the RCAF's infancy, its principal flying base remained small, occupying quarters that would have served a much larger force. The station filled as refresher flying courses arrived and trained, then shrank to its semi-permanent establishment, awaiting the next trainees. The main aircraft was the Avro 504K.

No.1 Wing and the School of Special Flying (SSF) were organized on September 7, 1920. No.1 Squadron (the flying component of No.1 Wing) was formed on September 29, F/L Keith Tailyour commanding. In January 1921 he was promoted, then commanded the School of Special Flying; S/L G.A. Thompson took charge of No.1 Squadron. Over the next few years the base and its sub-units went through various name changes, but the essence of Camp Borden remained. Personnel also constantly changed, reflecting the uncertain status of the CAF. Few officers held a post more than six months, before transferring to the Flying Operations Branch, thus reverting to civil life. Occasionally, a member returned to duties with the RAF.

The Officer Commanding, CAF visited Camp Borden on July 20, 1921. His report appears to have stimulated a modest expansion of the staff, particularly among mechanics and ground instructors. For September 22, 1921 the base had an establishment of 97—still pitifully small. A proposed expansion to 131 was cancelled in January 1922. About 15 civilians were regularly employed as plumbers, carpenters, postal clerks, etc. The highest paid civilian—a Supervising Engineer—earned $2200 annually; the lowest—a school teacher—received $900. The night watchman, at $1000 a year, was better off.

A January 6, 1922 report describes the duties of all supervisory personnel and identifies many of those serving. Some later attained high rank; most had only brief military careers. In charge of flying instruction was F/L A.L. Cuffe. Flying instructors were F/L Roy S. Grandy and G.G. Wakeham (G.E. Brookes may also have instructed). S/L R.A. Logan headed ground instruction. This must have been a jack-of-all trades position, given the description of his assignments:

"1. To be in charge of all Ground Instruction. 2. To be in charge of all practical work on the station in connection with wireless, photography, navigation and armament. 3. From time to time to make recommendations to the Officer Commanding Station for improvements to the syllabus of training, changes in or improvements on the methods employed and work done in Armament, Wireless and Photography in practical operation as well as for training purposes." The duties of Flight Sergeant Fitter Instructor—teaching all trainees, regardless of rank "the theory and practical overhaul and assembly of aero engines", were performed by Sgt C.S. Caldwell. F/O L.F. Stevenson was in charge of motor transport.

Some accidents at Camp Borden suggest a marginal airfield. On March 15, 1921 an SE.5a came to grief when F/L B.H. Windsor taxied into a sand hole, putting the aircraft on its nose. JN-4 G-CYDD was wrecked on June 18, 1921, a report noting: "The above machine piloted by Pilot Officer A. Tapping with pupil pilot E. Delorme when landing after doing a wireless flight at 0725 hours hit a small stump on the aerodrome and the axle broke close to the starboard wheel. After sliding on the ground a short distance the machine turned over on its back, breaking the propeller and rudder. No injury was done to the occupants."

An example of those going through the refresher training system was Lt Ortho A.C. Gibbons, whose log (preserved by the Atlantic Aviation Museum in Halifax) gives an idea of his flying, although it does not clearly distinguish between dual and solo flying, nor name his instructors. Nevertheless, it is a rare look at the course as reported by one who took it. Gibbons had flown 310:45 hours during the war, e.g. on D.H.6s (75 hours), Avros (60) and D.H.9s (73). His first CAF entry was on December 20, 1920—25 minutes of instruction on Avro 'BG. Next day he flew twice in Avro 'AU—25 and 15 minute hops described as "landings"—before doing 40 minutes on "loops, spins, half rolls".

After the Christmas - New Years break Gibbons resumed flying on January 5, 1921 (25 minutes on 'AU); he made two flights on the 6th in 'BR and 'AU. This was followed by 30 minutes dual on the 7th in 2-seat SE.5a 'CQ. On January 8 he was again up in 'CQ—"stunting". He piloted a JN-4 on January 10 on a "night flight" (10 minutes). The same day he flew SE.5a 'AY cross-country for 65 minutes (to and from Guelph). Gibbons flew twice on January 11—50 minutes in

Capt Roy Grandy was one of the seasoned flying instructors at Camp Borden in the early 1920s. Here he wears the CAF's dark blue uniform and distinctive wings. (CF RE13434)

'AY "stunting" and 75 minutes in SE.5a 'BQ, flying to Barrie and return at 6000 feet. His final flight was on January 12—30 minutes stunting at 3000 feet in SE.5a 'BQ (another pilot wrote off 'AY at Guelph on this day). He was signed off by W/C J. Scott Williams, MC, AFC on January 13. Gibbons (who had no subsequent CAF or RCAF experiences) may have been unusual among those given refresher flying for the amount of time he spent in the natty SE.5a.

The example of F/L C.M. McEwen may have been more characteristic. He reported to base on December 8, 1921, began training next day (his instructors were F/Ls Wakeham and Cuffe) and completed his course on January 5. The subjects in which he was tested, plus their assessments, suggest a comprehensive course, although the forms from which this was compiled do not indicate what standards were applied, e.g. Morale—Very good, has initiative, Ability as a Pilot—Good, Landings—Good. S/L Logan supervised a ground school; it included wireless (2-4 hours), artillery observation (4 hours), photography (4½ hours, but McEwen seems not to have taken this at the time), gunnery (4½ hours), engines (3-8 hours), aeroplanes (i.e. airframe technical studies, 2 hours), and navigation (4½ hours allocated, McEwen took 8). Although the rapid turnover of personnel caused problems, a report of February 18, 1922 listed 40 airmen and NCOs on strength for a year or more. Every effort seems to have been made to retain efficient airmen, while discharging those who had grown stale, careless or lost interest.

The continuing scarcity of personnel was evident in early 1923. The Commanding Officer (W/C W.G. Barker) was without an adjutant, the Flying Training Depot was short two instructors, and numerous other vacancies existed in technical and clerical offices. On April 23 Barker suggested that a sergeant be sent to Toronto or Montreal to recruit people. He was emphatic about the type of person to send: "It is especially recommended that an NCO be sent, owing to the fact that he can talk to these men, as well as trade test them, and make clear to them the conditions that will be encountered at Camp Borden. This should ensure getting men who would be satisfied with the living conditions on this Station."

On April 27, 1923 a letter went across the country authorizing recruiting expenditures from $50 to $250 per station. The sort of personnel sought reflected the curious times. In another memo Barker noted the need for batmen (personal servants): "It is pointed out that when the Cadet Training takes place, there will be somewhere in the neighbourhood of 45 officers to be looked after, and the request for one batman for eight officers appears to be very reasonable. In this connection, it is pointed out that the RAF [has] one batman per Squadron Leader and one to every three officers is allowed." Barker was looking ahead to the greater "professionalism" he expected in a reorganized Canadian Air Force. The same letter addressed the status of manning the Officers' Mess, an institution he deemed more than just a centre for dining: "The conditions, owing to lack of establishment, in the Officers Mess are disgraceful ... this situation should be remedied. It is... doubly necessary to have a first-class mess, in view of the fact that these Cadets should be taught a good deal in the Officers' Mess, and impressions they take away with them will either be for the good, or to the detriment of the RCAF." HQ agreed and immediately sanctioned an enlarged establishment, including 5 batmen and 4 mess orderlies. Camp Borden was the largest element in the RCAF, just as it had always been the biggest Air Board base. For July 1923 overall RCAF establishment was:

| Unit | Officers | Other Ranks |
|---|---|---|
| Headquarters | 10 | 11 |
| Vancouver | 6 | 21 |
| Winnipeg (including High River) | 16 | 43 |
| Ottawa Air Station | 4 | 20 |
| Camp Borden | 18 | 133 |
| Dartmouth | 9 | 25 |
| Photographic Section | 2 | 6 |

On the eve of the permanent force being established (April 1924) Camp Borden was authorized for 19 officers and 140 other ranks. Barker had left (to become temporary commander of the RCAF). His successor was W/C L.S. Breadner, another man with a good war record. Camp Borden, despite its small staff, was busy updating inventories and cleaning out dead storage. In 1920 (the effective "working year" being only six months) 86 officers and 111 airmen completed courses; 934 hours were flown, or roughly 11 per officer taking refresher training. In 1921, 375 officers and 835 airmen took courses. This was Camp Borden at its peak. Not all flying was instructional. Commencing January 1, Lt Allan G. McLerie operated a 17-week Camp Borden—Toronto trial mail service, using an Avro 504. He also carried 86 passengers. Aircraft at Camp Borden performed limited military duties—spotting for militia training groups, and doing aerial photography.

**This was the scene at Camp Borden on November 30, 1921 as the CAF ensign was inaugurated. Unless this photo has been doctored, the 504s were doing some impressive flying. (CF RE15524)**

Yet the station's principal task remained training. The trouble was that this was a dead end—no new men were being recruited. The program was suspended in 1922, then replaced in 1923 by the Provisional Pilot Officer scheme, wherein university students would train for a service flying career.

Camp Borden witnessed its share of triumphs, perhaps none so poignant as a ceremony of November 30, 1921. The CAF had received permission to use the ensign of its RAF parent. There were discussions about adapting the ensign to a Canadian design, but, on both sides of the Atlantic, sentiment favoured a common pattern. Authorities wanted to inaugurate the CAF ensign with a ceremony "as pompous as possible", but all cabinet ministers were engaged in a general election campaign. Senior army officers also absented themselves, and the Governor General merely sent a congratulatory telegram. The event ended as an Air Board—CAF "family affair" with 45 officers and 169 airmen on parade, a chaplain's blessing, and Avro 504s circling the flag staff. The lack of officialdom seemed not to faze anyone, and the *Borden Flyer* of November 30 described the event as one to be long remembered. In the midst of flux and transition, the CAF looked to the past (the RAF ensign) in preparation for the future.

Avro 504Ks on inspection at Camp Borden. (RCAF RC1501/Jack McNulty Col.)

The CAF's last 10 JN-4 Canucks were on strength 1920-23 at Camp Borden. This one is seen with undercarriage trouble. (RCAF RC539/Jack McNulty Col.)

Long before Ski Doos, Camp Borden converted the gondola car of an Imperial Gift airship into an snow-borne ambulance. In case of an accident it could plough across the snow to the rescue. The stretcher was loaded through a detachable nose cone. Powered by an aircraft engine, this contraption served for some years. (RCAF RC1419/Jack McNulty Col.)

A 1927 Camp Borden class watches as a mechanic "props" an Avro 504K. Wheel chocks are not in place, although the 504 had no brakes! (RCAF RC1613/Jack McNulty Col.)

Canuck 'DC (previously C240) was modified for armament training at Camp Borden, even though the CAF had no service role in this era. (CF RE15446)

Camp Borden weather observations were made from atop this tower. The Stevenson Screen, a box containing met instruments, is near the tower base. (CF RE15838)

## Aircraft and Incidents from the Early Days

As the CAF grew, a great deal of paper work was generated. Much eventually found its way into the National Archives of Canada, where it remains a resource for researchers. Many Camp Borden documents survive. Typical is file No.HQ1021-3-44, dealing, in part, with the CAF's small 1920's fleet. For example, there is a May 6, 1921 memo dealing with JN-4A No.C502, S.E.5a G-CYBQ, and Avro 504Ks G-CYAI, 'BD, 'CH, 'CJ, 'CS and 'CY. These are noted as "to be reduced to salvage", their engines to be returned to Air Board stores. A look at other records shows that most of these machines had been involved in accidents. Much of HQ1021-3-44 deals with accidents, the data showing that the CAF took flight safety seriously. Wrecks were analyzed, witnesses interviewed, and technical people brought before enquiries. Conclusions and recommendations came forth.

On February 15, 1921 P/O J.H. St. Martin of the School of Special Flying took off from Camp Borden in Avro 504K 'CS. He was taking fitter S.E. Calver to nearby Shelbourne, to help repair

**W/C/ J. Scott Williams, who headed several accident investigation boards at Camp Borden. (CF PL117510)**

'CT, stranded there. P/O Martin reported: "I landed at the fair grounds... which was the best place I could see." He let off Calver, then departed, this time with P/O Clearwater. On lift-off, however, his engine failed. Later, F/L A.L. Cuffe reported: "There were obstacles (high poles) on both sides, and the pilot stalled the machine on the roof of the grand-stand which, under the circumstances, was the only thing he could do..." A three-man Court of Inquiry was convened under W/C J. Scott Williams (OC of the SSF). Documents entered showed that 'CS had a weight (less gas and oil) of 1230 lb, and was permitted to carry 600 lb. As to St. Martin (later a renowned Quebec bush pilot), he was described as experienced, having 150 hours on the Avro 504, D.H.9: 120 hours, JN-4: 60, and D.H.6: 10. He was not blamed, the cause being listed: "Engine failure due to loss of air pressure when air pump came loose." It was recommended that the air pump be bolted, not screwed, in place.

A more serious accident is filed in HQ1021-3-44. Avro 504 'CL of the SSF went down soon after take-off on January 3, 1921. W/C Williams again headed the enquiry. The final report noted engine failure: "The pilot [P/O H.D. Wiltshire] made a half-turn before putting the machine into a glide and, consequently, fell into a spin, striking

**S/L Tailyour's funeral in Barrie. (CF RE15472)**

the ground left wing first... I cannot account for the behavior of this machine ... there was plenty of aerodrome in every direction, no wind, and sufficient height to get out of a spin." Wiltshire could not help: "I have no recollection of what occurred." He, too, was experienced, e.g. 1700 hours on the JN-4. On March 3 authority was given to rebuild 'CL, but this seems not to have happened.

There was an even worse accident on April 21, 1921, when S/L Keith Tailyour of Edmonton crashed Avro 504 'CY. Several witnesses appeared at the enquiry. F/L J.H. "Tuddy" Tudhope, for example, swung Tailyour's prop, then helped extricate his body from the wreck (Tudhope later was the DOT's senior pilot). F/L P.C. Garratt (later president of de Havilland Aircraft of Canada) saw the accident. A flying instructor, he reported: "At 1650 hours on April 11th, 1921 I was coming out of "B" Flight hangar, when I saw a machine at a height of about 500 feet diving... slightly over the vertical. The engine was full on. As far as I could see, no effort was made either to bring the machine out of the dive, or to close off the engine. The machine hit the ground in that position."

F/L S.H. McCrudden saw 'CY take off, fly up and down the hangar line, climb to 500 feet, perform loops and half-rolls, then descend to about 50 feet: "He then zoomed to about 300 feet and attempted a side roll." The most remarkable account was that of FSgt E.E. Moore: "The machine was stunting around the office and sheds and as I did not feel safe I remained outside the office. Before the accident the machine made an Immelmann turn and came out on its back. It came down some distance in this position until the engine suddenly opened full out and the machine nose dived straight into the ground. It appeared as if Squadron Leader Tailyour was trying to get out of the position on his back, but had not enough height."

Tailyour died instantly. His funeral in nearby Barrie, held on the 13th, was complete with band and firing party. An enquiry concluded: "The pilot of the machine was to blame for disobeying orders in stunting at too low an altitude."(Tailyour's was only the second fatal accident at Camp Borden since flying had recommenced there on June 2, 1920. The other involved Capt J.A. LeRoyer, killed the week before Tailyour.) Another Avro 504K dealt with in HQ1021-3-44 is 'EE, written off after 288:40 hours. It had been converted to a seaplane in 1924, so instructors could train on floats on nearby Kempenfeldt Bay (Lake Simcoe). It later was reconverted to a landplane by the Ottawa Car Co. A memo of February 21, 1930 from F/O E.D. Dawson, Camp Borden's technical officer, states of Avro 504 'CM: "This aircraft is in such a dilapidated condition as to warrant a complete write-off, involving no salvage of any spares."

HQ1021-3-44 reviews the history of other types. D.H.4 G-CYBW, for example, crashed at High River on May 31, 1921. A/P/N Albert W. Carter and WOp Wilfred E. Beattie were injured (Carter had more than 500 flying hours). They had taken off at 0925 to patrol the Bow River Forest Reserve. One witness reported: "He appeared to me to be too slow to complete a bank turn." Carter later stated: "After Air Forman Mechanic Govett had run up the engine ... I took my place in the pilot's seat, and examined the controls of both machine and engine. I also tested the speaking tube, talking to Wireless Operator W.E. Beattie, who occupied the observer's seat. The next thing I remember happening was awakening to find myself in the High River Hospital ... I can throw no light whatever on the cause of the accident." In the end Maj G.M. Croil, superintendent at High River, blamed Carter for attempting to turn too low.

HS-2L G-CYAG in the backwoods. Such machines took a beating. Built to fight, they were never intended for long use. While overseas units had huge servicing echelons to keep such complex machines going, in the bush they were maintained by an ingenious little crew, with a few spares and tools in a satchel. It often was season's end before an HS-2L get home for overhaul. In October 1920 pilots Hobbs and Leckie flew 'AG on the Trans-Canada Flight. Although no HS-2L survived, Canada's National Aviation Museum has a gorgeous replica. (CF RE13546)

In January 1924 W/C Stedman reported on aircraft at RCAF Station Winnipeg. All its HS-2Ls were in poor condition, mainly with rotting hulls. Only 'AF could be serviceable for the 1924 season: "DT is beyond repair and 'ED is damaged to such an extent that the repairs cannot be undertaken at this station." Meanwhile, Viking 'EU required hull and wing repairs. Stedman suggested to the station superintendent that he keep his hangar floors wet to help slow the drying process and reduce cracking in the all-wood aircraft through the winter. He explained: "Examination of the Viking wings at this station shows, as was anticipated, that machines built in Montreal shrink up slightly when stored in this climate." Thus, all wings had to be opened to tighten their fittings and bracing wires. 'EU was the last RCAF Viking and the correspondence found in HQ1021-3-44 indicates that the air force wished to get rid of it. In April 1931 Stedman decided to scrap it, since it required so much repair. S/L N.R. Anderson, OC at RCAF Station Winnipeg, concurred in a report of March 27:

*They are difficult to land, and pilots are always breaking tail planes and elevators. This often involves expensive transportation flights to supply replacements... They will not hold their rigging ... The two Vikings in service during 1930 received more rigging attention than any other type ... on the station, but their performance remained so poor that they were very little used ... A number of flights had to be carried out at a dangerously low altitude. The wind-driven fuel pumps are always giving trouble and no serviceable pumps remain on hand. The Rolls-Royce Eagle X engines used in this type of aircraft are now seven years old and considered most unreliable ... years of wear and tear have made the Viking ... unsafe for further flying.*

Many other insights into early RCAF days may be found in archival files. For example, HQ1021-3-44 reports about the crash on October 8, 1930 of Viking 'EV at Cormorant Lake. F/O A.D. Ross and Cpl Cooper (fitter) escaped. 'EV was a dead loss and Ross (later Air Commodore Ross) was blamed. In the accident Cooper lost his watch, so filed a claim for $14 for a replacement. R.P. Brown, RCAF financial superintendent, concluded: "If the DCO [Director of Civil Government Air Operations] will state that Corporal Cooper was required to carry a watch, special application might be made to Privy Council to grant him compensation."

## Civil Government Air Operations

With no wars to fight, it would have been difficult for the CAF/RCAF to justify a purely military *raison d'être* to Parliament. Thus, for much of this period, aid to the civil power was the most realistic air force role. Yet, the RCAF lacked even the limited political patronage enjoyed by the peacetime Non-Permanent Militia (populated by weekday politicians and weekend colonels). The near-demise of the RCN was an object lesson in what might befall the RCAF, lest it fail to prove its usefulness. Nevertheless, some RCAF officers opposed CGAO involvement. The most senior and vocal critic was G/C J.S. Scott, MC, AFC

G/C J.S. Scott, MC, AFC strove to give the RCAF a military semblance, but most interwar flying remained civil in nature. (CF HC10381-B)

during his tenure as Director of the RCAF (May 1924 to February 1928). He opposed linking service and civil air expenses in one financial vote. Late in 1926 he suggested to Chief of Staff Major-General J.H. MacBrien that other government departments form their own flying service, so the air force budget (which he did not imagine being cut) might be devoted to the creation of a small, efficient military air arm. This was vigorously opposed by J.A. Wilson (by now Secretary of the Air Force—virtually its chief administrative officer), who supported the existing policy of putting all flying appropriations under one parliamentary vote.

In 1927 Scott opposed the purchase of special airplanes for agricultural experiments, urging that such flying be done by commercial

The RCAF's eight Vikings, new in 1923, were heavily used each summer. Inevitably, they wore out, the last being disposed of in 1931. Here 'ET is readied at Victoria Beach in 1924. It was lost in July 1927, when it broke up in flight. The Viking had a wing span of 50', length of 34', weighed 5600 lb all-up, and cruised at 80 mph. Readers wishing to know more about Canada's early bushplanes should consult Ken Molson's *Canadian Aircraft since 1909*, and the *Journal of the Canadian Aviation Historical Society*. (Wheeler Col.)

companies. MacBrien replied: "Owing to the experimental nature of this work it is improbable that any civilian company would undertake it unless it was offered on very favourable financial terms... all experimental work should be carried out by the RCAF." The clash between Scott and his staff on one hand, and MacBrien, backed by Wilson, on the other, was emphasized early in 1927 during a policy study centred on aerial surveying. Starting from the premise that the proper function of the peacetime air force was preparing for war, Scott declared that the existing policy was inadequate because it retarded the development of commercial civil aeronautics, civil government air operations were restricted by the money voted for service flying, men were being commissioned and enlisted in the RCAF to carry out civil functions, and it was impossible to maintain a high RCAF training standards and discipline when civil operations required intimate contact between officers, men and civilian employees. Scott recommended that the RCAF discontinue flying for other government departments and concentrate on establishing a military organization. Regarding aerial photography, he recommended that this be done by commercial firms or provincial air services, except where these found the task beyond their means.

Scott's arguments found little sympathy with MacBrien, who pointed out that certain federal responsibilities in the Prairie provinces soon would be ending with the handing over of natural resources to the provinces. As the first task of the RCAF in wartime would, in his opinion, be in conjunction with the Surveys Sections of the Royal Canadian Engineers, extensive aerial photo operations constituted the most valuable RCAF training. MacBrien also pointed out that, at the last Imperial Conference, Canada had been congratulated for using its military air arm for such useful and rewarding purposes. The RCAF therefore continued to operate as the principal air arm of the federal civil power until the mid-1930s, when international developments brought drastic changes.

Although Scott's arguments may ring oddly in our ears (e.g. his objections to chumminess between officers, airmen and civilians), one might observe that some of his subordinates held similar views and had practical arguments to counter MacBrien's ideas of how civil operations could prepare the RCAF for war. On November 24, 1926 F/O C.C. Walker, Photographic Officer at No.2 (Operations) Squadron at High River, wrote a long memo relating to photography and war requirements, pointing out that control of aerial photography was in the hands of civil servants, whom he considered indifferent to military needs. Although the methods employed were suitable for the Topographical Surveys Branch, it was uncertain that they would be applicable in war.

Walker was critical of a system that relied on camera operators. Wartime attrition would quickly lead to personnel shortages, if such specialization continued. He felt that pilots should be responsible for the photography and mentioned that in 1914-18 observers had merely changed film magazines for their pilots, their principal task being to fend off enemy fighters. He also argued that Canada's system of a centralized photo development centre also was inapplicable in war. Local development of pictures at corps or divisional levels was most common. While admitting that field photo labs might be impractical in Canada, Walker argued that the matter of such units being established for wartime should be studied now, rather than waiting for a war to break out.

## Other Historic Firsts

It would require another book to describe the duties of the interwar RCAF and recount the adventures of its small, enthusiastic pool of men. The broad outlines of that story are found in such books as *There Shall be Wings*, *Creation of a National Air Force* and *Sixty Years*. Here, attention is paid to a few activities which did much to justify the RCAF in the eyes of the public, thus saving it from budgetary annihilation. Occasionally, these activities caught the headlines. On September 11, 1926 Dalzell McKee, an American adventurer accompanied by RCAF S/L Albert Earl Godfrey, left Montreal in a Douglas MO-2B seaplane. They reached RCAF Station Jericho Beach on the 19th after numerous stops for fuel, weather and rest. This epic was the first trans-Canada crossing by a single aircraft. McKee then created the Trans-Canada Trophy (usually called the McKee Trophy). He planned further exotic flights, but died in a crash before his plans bore fruit. The McKee Trophy remains Canada's premier aviation prize.

In September 1927 the RCAF inaugurated experimental air mail between Ottawa and Montreal, thence down the St. Lawrence to Rimouski. There, mail bags were delivered to and taken from trans-Atlantic liners. Service began on September 9, using a Canadian Vickers Vanessa (the only example). Pilot J.H. "Tuddy" Tudhope and engineer Gerry LaGrave were the crew. The mail was received from the *Empress of France* at Rimouski, but the Vanessa cracked up on takeoff. Its cargo reached Montreal *after* its liner had docked, but the service soon was working well, using an HS-2L and Fairchild FC-2. (Of the wreck Tudhope noted: "To Rimouski wharf to take on first overseas mail, rough sea at wharf, tied up with difficulty... 502 lbs. mail. No flight. Mail saved and aircraft towed to wharf... Engine only salvaged.")

**Dalzell McKee's Douglas MO-2B, the first Canadian-registered aircraft with Pratt & Whitney power, was in the RCAF flying the mail till 1929. RCAF Fairchild FC-2 'YT is in the distance. (CF HC1548)**

**This famous photo is one of the few that shows the ill-fated Vanessa to advantage. (NAC PA61996)**

## Aerial Photography

It is now a cliché to claim that the airplane helped map Canada. Looking at the 1918-19 and 1919-20 annual reports of the Topographical Surveys Branch, Department of the Interior, one is struck by how detailed surveys, including mapping, evaluation of resources, magnetic readings and elevations, remained largely in the Prairie provinces. The boundary between Alberta and British Columbia was still being traced. The outlines of northern lakes appeared on maps, but the islands dotting those lakes were largely absent. The 1921-22 report shows systematic northern exploration only beginning, and that this was directed at the Mackenzie River basin, owing to reported oil strikes. Even so, every piece of equipment and ounce of supplies moved by steamer, canoe, toboggan, pack animal or backpack. This limited when surveys could begin, how much could be covered, and when the season ended.

The camera in mapping was not so new. Commencing in 1886, Canadian surveyors had practiced what they called "photo topography" in mountainous areas. The airplane, however, gave new possibilities, particularly in rolling, forested and isolated regions. The Air Board recognized this, and in 1919 an Aerial Survey Committee formed with representatives of various surveying departments. The Topographical Surveys Branch made its dark rooms and map-making equipment available to the Air Board, until the new organization could establish its own facilities. Meanwhile, the committee studied the applicability of aerial surveying, and the Air Board was asked to co-operate by performing a practical experiment.

On May 20, 1919 the Minister of the Interior wrote to the Minister of Militia and Defence, requesting a test of aerial surveying and suggesting that $2000 might be provided from Interior's budget. This was considered at an Air Board meeting. There Mr. E.G. Deville, the Surveyor-General, considered that the test should be made near Ottawa, where there would be great interest and where results could be checked with existing maps. Although some feared failure under the public eye, Deville was optimistic. An experiment was planned, provided that Interior supplied the funds, and that an HS-2L be secured from Halifax or Sydney. The project did not materialize that year, presumably because no suitable aircraft was available. However, it was reconsidered in May 1920. The scheme (whereby 700 pictures would be taken) was recommended by W/C Leckie and approved with a budget of $2500.

The flying field selected was the danger area behind the rifle butts at the Rockcliffe rifle range. Deville supervised, but the day-to-day photography was directed by F.H. Lambert, a veteran Dominion Lands Survey officer. F/O J.B. Mulvey, who had done RAF reconnaissance work, was in charge of the flying. Preparations were in order by the end of August. The experiments went on until November, using an Avro 504K and Bristol F.2b. The camera was a wartime L.B. (Mark I),

A typical scene (location unknown) during RCAF photo detachment work. Life usually was a bit rugged. Pilots Bill Weaver and Gordon F. Mason-Apps are on the right. (CF RE13131)

operated by hand or by wind-driven propeller. The Avro proved unsuitable, owing to its low ceiling and vibration. The Bristol was suitable, although its controls were overly sensitive (the D.H.4 and D.H.9 were suggested for future work). The camera had been bolted to the aircraft but, when the mosaics were produced, there were evident distortions. This was due partly to the aircraft crabbing in the wind—pointed away from its line of flight and thus flying partially sideways. Moreover, the camera had not been absolutely level at all times (as when climbing or banking). Distortions lessened as altitude increased. One thing was clear—pilot and photographer would need great skill and training, and close co-operation between them was vital.

In scientific terms these were the Air Board's most important 1920 operations, having even greater long-term significance than the Trans-Canada Flight. Photo operations were not restricted to the Ottawa area—some were conducted along the main river routes around Lac St. Jean. Experiments continued from Ottawa in 1921, including testing a camera developed by Professor R.H. Cooke of Princeton University. Photography was done over London, the Gatineau Hills, the Welland Canal and the St. Lawrence River with its associated canals, and around Vancouver, High River, Winnipeg and Halifax, the principal Air Board bases for forestry and marine patrols. On behalf of the Militia, the CAF also flew 6 hours photographing around Petawawa. To date there was no co-ordinating authority beyond the Air Board itself. It was left to each department to request its own work. Asked to carry out a variety of tasks, the Air Board sometimes combined one with another. Sometimes departmental requests could not be met due to weather, aircraft shortages, etc. This would be alleviated when the Topographical Surveys Branch assumed control of aerial photography in 1925.

Typical of early photographic projects was one requested in July 1921 by the International Boundary Commission. Ownership of an island in the St. Croix River near St. Stephen, New Brunswick was disputed by Ottawa and Washington. Which branch of the river constituted the main channel was the point to resolve. To prepare a model of the St. Croix, the Commission asked that the river be photographed from St. Stephen to Milltown. The operation was flown in October using an HS-2L piloted by S/L A.B. Shearer of Dartmouth Air Station. He was to photograph the river from 5000 feet, but his Liberty engine overheated and he could not exceed 3000. Adding to this, he could only communicate with his camera operator by passing notes. Yet the operation succeeded, the island being awarded to Canada.

The Topographical Surveys Branch was quick to appreciate the value of aircraft, although Air Board resources were meager, compared to needs. The 1922-23 annual report noted work in progress to define the Manitoba-Ontario boundary from a point 69 miles north of the Winnipeg River to one 111 miles further on. This was done with three surveyors and 21 other men (many were packers carrying supplies) proceeding on foot through the bush, pausing to build survey monuments (markers), take magnetic readings, and sketch shorelines. The document notes: "A study was made of a series of vertical photographs taken by officials of the Air Board of Canada over a portion of the line, and, after being reduced to the scale of the plan, these were used to supplement or amend the topography obtained by the surveyor. The photographs showed the courses of streams not traversed and the outlines of a few lakes not discovered by the surveyor. The large scale photograph showed greater detail of topography, but this was rendered much less conspicuous upon reduction to the scale of the final plan."

In September 1923 boundary surveyor R.D. Davidson accompanied an RCAF flying boat doing oblique photography in northern Manitoba. This was to investigate how, and to what extent, aircraft could help in mapping the northern Prairie provinces. Standard methods could map only the strip of land which a surveyor walked. There were no mountains to provide high points for greater geographical studies. Davidson travelled 375 miles along one line and 140 miles down another, taking 745 oblique photos from 2700 feet. About 500 showed identifiable features that would enable ground parties to locate themselves—or which might enable aircraft to photograph along charted flight lines. This boded well for the future.

Some photography was far in advance of its final application. In May 1921 the Canadian Secretary of the International Joint Commission, Mr. M.J. Burpee, wrote to the Air Board requesting a familiarization and photography flight over the St. Lawrence from Kingston to Montreal. This was motivated by plans to build a deep waterway connecting the Great Lakes with the sea, a project not undertaken for a generation. The Air Board complied by assigning an HS-2L, then a D.H.4 from Ottawa; 35:55 flying hours were logged in July and August. Another operation was an attempt to prepare a photographic mosaic of London. This was initiated by Maj D.H. Nelles, Supervisor of Topography, Geodetic Surveys Branch, Department of the Interior. He had been in contact with the Eastman

**H.L. Holland, pioneer aerial photo pilot. (CF RE17955-1)**

Kodak Company, which had done such work over Rochester, New York. The London operation was to fix the position of topographic details preparatory to a large scale city map being produced. Nelles suggested that the work be done before the trees were in leaf and that, in accordance with Kodak's experience, be flown at 4000 feet. In April 1921 the CAF sent D.H.4 G-CYDE from Camp Borden to London with the crew of Capt Hubert L. Holland (a wartime reconnaissance pilot and veteran of RE.8 operations with 34 Squadron in Italy), and Lt E.R. Owen. They began photographing from 3000 feet, but the overlap from photo to photo was insufficient. They completed the operation at 5000 feet. Further photography was planned for the spring of 1922, but was curtailed when Holland died in a crash. In November 1922 London again was covered, this time by Bristol F.2b G-CYDP piloted by F/L R.S. Grandy with photographer Cpl A.J. Le Sueur.

One important question still had to be resolved—relations between the air force and the civil flying firms wishing to do photography under contract. The matter had been raised in 1920 with respect to air force operations on behalf of provincial governments. On a few occasions the Air Board referred government departments to civil firms, but the Geodetic Survey was dissatisfied with Canadian Aerial Services Limited. Having solicited and obtained a contract to photograph over Montreal, it failed to do any work. Thereafter, the Department of the Interior dealt exclusively with the air force, although private organizations such as timber firms and universities were referred to commercial companies.

Although selected areas were photographed for various departments in 1922, operations were largely experimental, not so much to prove to the Air Board that they were feasible, as to demonstrate to clients and parliament that aviation was practical. The area mapped in 1921-22 totalled only 1775 sq. mi. By now, however, the technique had been proven, and more literature was becoming available about attaining maximum coverage, and transposing photographs, particularly obliques, into accurate maps. A more ambitious survey was to have been flown in 1923, but was hampered by delays in the delivery of aircraft and cameras. Nevertheless, 1965 sq. mi. were covered.

By 1924 the RCAF was ready for a sustained aerial photography program. The Department of the Interior drew up a proposal for oblique photography in western Nova Scotia, the Cross Lake, Oiseau and Fort Alexander areas of Ontario, and north of The Pas in Manitoba. Vertical photography around Edmonton, Wainwright Park and Vermilion, Alberta also was requested. This program was modified and enlarged following a conference of April 18, attended by officials of the DND and Department of the Interior. The intention was to review forestry and survey operations done so far, relating them to each other when possible. The meeting produced an inter-departmental committee composed of W/C J.S. Scott (Acting Director of Flying Operations), D.R. Cameron (Acting Director of Forestry) and A.M. Narraway (Controller of Surveys), charged with drawing up a 5-year plan. It contemplated developing forestry operations in Manitoba in 1924 with survey operations pushing into Saskatchewan.

The most ambitious operation of 1924 was done from the Victoria Beach Air Station between July 18 and August 14. It covered the Reindeer Lake and Churchill River districts. S/L Basil D. Hobbs was in charge, accompanied by Cpl Alex J. Milne (mechanic), F/O James R. Cairns (photographer) and R.D. Davidson (Dominion Lands Survey, acting as navigator). Their route through Manitoba and Saskatchewan was indirect. Leaving Victoria Beach on July 18, they went to The Pas, then to Pukkatawagan (July 20), Reindeer Lake (July 23), Pelican Narrows (August 1), Stanley Mission (August 9), Ile-à-la-Crosse and Prince Albert (August 11), back to The Pas (August 12) and Victoria Beach (August 14). Hobbs' report details the execution of a single expedition, illustrating the problems and conditions encountered. The original document is available in archival and library sources, notably the National Archives of Canada; major excerpts are found in Vol.1 of *Air Transport in Canada*.

A total eclipse of the sun was to occur the morning of January 23, 1925. In what perhaps was an aviation first, two Avro 504s took off from Camp Borden to photograph the phenomenon. One was piloted by F/L George E. Brookes with F/O Arthur L. Morfee as cameraman; the other carried F/L Roy S. Grandy and Frederick C. Griffin of the Toronto *Star*. Grandy's machine force-landed in Newmarket with engine trouble. Brookes and Morfee encountered cloud—although they climbed to 9400 feet, they missed most of the eclipse. Morfee succeeded in shooting a number of exposures through gaps in the cloud just as the sun began to emerge from behind the moon. At times it was so dark that Brookes was unable to read his instruments.

In 1925 topographical survey techniques were revised. Some 48,000 sq. mi. were covered by vertical and oblique photography (40,000 the previous year). For 1925 aircraft were assigned to photography without the expectation that they also would do forestry patrols. Not being tied to existing bases for such duties, crews could fly to areas not hitherto covered, as S/L Hobbs had done the year before. A new method of oblique photography was introduced. Previously, 3 pictures had been taken (1 ahead, 1 on either side). The method introduced in 1925 called for 5 photographs (1 ahead, 2 on either side); these would be taken at 3-mile intervals from 5000 feet. A set of 5 photographs covered an arc of 180° and considerable overlap was achieved. While parallel flight lines had been 8 miles apart, now they were flown at five.

Although some photography was done by Ottawa Air Station (mainly scenic views of historic sites), most work was done from Dartmouth, High River and Winnipeg. From Winnipeg, Vikings 'ET and 'EZ were used. F/O L.R. Charron (pilot), F/O J.R. Cairns (photographer), Cpl A.J. Milne (mechanic) and R.D. Davidson (DLS, navigator) crewed 'ET. F/L F.C. Higgins (pilot), AC1 R. Marshall (mechanic), AC2 D.O. Craig (photographer) and John Carroll (surveyor/navigator) crewed 'EZ. Separate orders were issued, which expressed hopes rather than final accomplishments. To be covered were: about 30,000 sq. mi. in eastern Manitoba and western Ontario directly east of Lake Winnipeg on both

Vedette G-CYGA on 1928 summer photo operations near Cypress Lake, northern Quebec. Cpl H.S. Diller, MM occupies the front cockpit, his pilot being F/L J. Sadler. Bert Green, flying with F/O A.L. Johnson in Vedette 'ZL, took the photo. 'GA served to September 1934, then joined the Saskatchewan government as CF-SAE. On May 27, 1936 it got into turbulence near Stony Lake, obliging the pilot to take to his parachute. The Vedette had a 42' wingspan, was 32' 10" long and weighed 3200 lb. all-up. 'GA had a 220-hp Wright J-5 engine. Top speed was 92 mph. (H.S. and Gordon Diller Col.)

A Roberval-based CAF Curtiss HS-2L in 1921. One can imagine the clatter from the 400-hp Liberty engine and its huge wooden propeller. Note the radiator and the oil cooling tank (one per side). There are two cockpits behind the camera man and two pilot cockpits (windscreens folded forward). Hundreds of HS-2Ls were completed, mostly in Buffalo, NY. The RCAF used 30. (CF RE19533)

HS-2L engine change scenes from Roberval. The men had to be careful handling a Liberty engine—it weighed 900 pounds. Then, the sub-station's mobile workshop, where repairs and field mods could be made. (CF RE19523, '24, '13)

Varunas and Fairchilds await breakup at Sub-station Cormorant Lake, Manitoba in the spring of 1929. (Arthur Fleming Col.)

The Fairchild FC-2 series proved its worth in forestry, photo, air ambulance and anti-smuggling work from 1927-38. 'XT was at Island Lake, Easter Sunday 1929. Apart from photography, in 1927 RCAF pilots were credited with several illicit Manitoba stills "going into hiding"—their operators apparently feared that the photo planes were searching for them. (Molson Col., Arthur Fleming Col.)

Bellanca Pacemaker 'VG and Vedette 'WS at a gas cache on Aylmer Lake, NWT during the 1931 survey season. (CF A4379-1)

The 1928 photo detachment at St. Donat, Quebec. Fairchild 'YU and two Vedettes are tied up near the hotel, where personnel stayed. The tents served office, workshop and photographic needs. (H.S. and Gordon Diller Col.)

In the RCAF's early years countless aerial images were produced by low-level oblique photography. This 1920s view shows downtown Winnipeg. (RCAF/Halliday Col.)

Some of the key men on aerial photo operation from Victoria Beach in 1924: Eric Frye, mechanic Bob Marshall, pilot G.F. Mason-Apps, pilot Leo Charron, mechanic Alex J. Milne and navigator Bob Davidson. (Wheeler Col.)

Viking 'ET at Prince Albert on August 12, 1924. Here pilot Basil D. Hobbs (inset) faced tricky conditions—a strong current, shallows and drifting logs. This was during the famous Reindeer Lake-Churchill River survey, a story told in the CAHS *Journal* (1983-84), and the book *Air Transport in Canada*. Hobbs, from Sault Ste. Marie, and several other Canadians had begun flying at Orville Wright's school in Dayton, Ohio in 1915. From there he obtained an Aero Club of America certificate, was accepted into the RNAS and flew flying boats from the channel port of Felixstowe. On June 14, 1917 he and his crew spotted Zeppelin L.43 near Terschelling and shot it down in flames. (CF PMR70-214, Wheeler Col.)

sides of the interprovincial boundary (oblique photography); 30,000 sq. mi. in central Manitoba from the northern end of Lake Winnipeg to Pukkatawagan, Nelson House and Split Lake (oblique); routes between Cormorant Lake, Pelican Narrows, Waskesiu Lake and back to Cormorant Lake (oblique); and vertical photography of 85 sq. mi. around the southern end of Lake Manitoba between Delta and Lake Francis.

The Vikings were together at the beginning of the season, while covering the Lac Seul-Little Grand Rapids area, flying from Minaki, Sioux Lookout and Pine Ridge (Lac Seul). This began on June 10, but was interrupted frequently by camera breakages and bad weather. On June 15 F/O Charron operated between Minaki and Little Grand Rapids (despite clouds) to get pictures urgently needed by surveyors in the area. Because of overcast the flight was made at 3000 feet, more than 1000 feet lower than desirable. Nevertheless, the photos were usable. On July 12 F/O Charron departed to photograph the water routes east and northeast of Lake Winnipeg, while F/L Higgins continued around Lac Seul and Little Grand Rapids. Higgins remained in this area for most of the season, flying from Little Grand Rapids, Matheson Island, and Pine Ridge, where he frequently was held up by camera failures and cloud—between June 14 and August 31 he reported only 27 of 79 days suitable for photography, but many non-photographic days coincided with returns to Victoria Beach for maintenance. In one such period (August 14-19) Higgins transported G/C J.S. Scott, then touring sub-bases at Norway House and Cormorant Lake. On August 31 Higgins moved to Victoria Beach for operations near Delta. This required the Viking to use amphibious gear, lest it force-land in a farm field. A vertical camera was also installed. A small inland lake at Delta was used as a base, rather than the shore of Lake Manitoba. To obtain best performance, only three-quarters fuel was carried and the mechanic was left behind. The operation was done in three flights (September 2-6), after which 'EZ went to Victoria Beach, was reconfigured as a flying boat, then returned to Little Grand Rapids. Between September 8 and October 16 there were only two suitable photography days, and the approach of freeze-up forced Higgins back to Victoria Beach.

F/O Charron operated alone from mid-July. On July 12 he had left Higgins to survey water routes in the Little Grand Rapids, Island Lake, God's Lake, Oxford House, Norway House area. That day, en route to Island Lake, he encountered storms north of Deer Lake, which cut short photography. On July 15 he flew south from Island Lake to fill gaps from his earlier sortie. These films were urgently needed by survey parties. Charron waited for good weather on July 21, then flew to Norway House, photographing one strip en route, then he shipped his films to Ottawa. Clear skies demanded that more work be done that day, but the wind had dropped to a dead calm. In order to get the heavily loaded Viking airborne, Charron put his three passengers in the nose, then taxied in circles to make waves. After a prolonged run he got the hull up on its step and lifted away to photograph another line between Norway House and Island Lake, then moved to God's Lake. Over the next two days he shuttled back to Oxford House and Norway House, broadening earlier coverage.

A fuel pump failure on July 24 forced Charron to Victoria Beach. By August 6 he was back at Little Grand Rapids; on the 10th he flew to Island Lake. Photographs taken earlier revealed the existence of a hitherto unknown river and on August 15 Charron turned his camera on it for extended charting. Next he was to photograph the region along the Hudson Bay Railway and lower Nelson River (the railway had not yet been extended to Churchill and was then to terminate at Port Nelson). Charron arrived at The Pas for this operation on August 27 and proceeded from there to Wabowden (September 3), Split Lake (September 7) and Port Nelson (September 12). At Port Nelson he faced a tidal mud flat shoreline—the Viking had to moor on the mud at the mouth of the river. Fuel had to be man-handled almost two miles, and for the last few hundred yards had to be carried down a steep slope in small buckets. Thus was 'ET refuelled on the 13th, while the tide ebbed and the crew pushed the machine out to prevent being stranded. That day a flight to York Factory was made. On the 15th the expedition returned to Split Lake, carrying on to Wabowden on the 21st. On the last few flights the engine delivered less than full power; on the 22nd 'ET refused to take off with a full load. The mechanic, surplus kit and exposed films were left at Wabowden, as Charron flew a photo sortie to Cormorant Lake, where Cpl Milne and the baggage subsequently arrived on a gas-powered rail car.

On September 26 Charron took off for Pelican Narrows with forester Mr. C. McFodgen. Milne was photographer (F/O Cairns was left behind). The flight extended to Waskesiu Lake and Prince Albert that day. The return to Pelican Narrows next day was uneventful. Some additional work along the Hudson Bay Railway was done on September 29, then Charron flew to Norway House. On the 30th he carried on to Wabowden, but poor weather prevented further photography. The approach of freeze-up forced the detachment to Victoria Beach on October 8. Viking operations for 1925 were significant for the large areas covered, but also for lessons learned. They had re-emphasized that a skilled NCO could double as mechanic and photographer. Viking crews also learned that their presence at temporary bases was not always welcome. Isolated trading post employees were hard-pressed to conduct their normal business, while cooking for the fliers (this also sapped a post's limited supplies). In future, personnel on detachment would carry their own tents and food, and do their own cooking.

## Standardizing Photo Operations

Thus far aerial surveying was simple but comprehensive. Once an area was selected for photography, RCAF HQ issued an Operational Order to the flying units responsible. A map, however rough, was provided on which a series of numbered, parallel lines was drawn. Preliminary ground surveys established control points at either end of the lines and, if they were long, at points between. Fuel caches were established in the preceding year or in the spring. Thus prepared, aircraft were despatched to their temporary bases. Once films were shipped from a detachment to Ottawa, and processed, a telegraphic report was sent to the field, advising whether results were satisfactory. If any film was badly-exposed, or unsuitable for other reasons (static on the exposures, poor definition, etc.), lines could be re-photographed. Lines were flown according to prevailing weather, rather than in strict sequence. Each detachment was assigned more work than it could ever complete on the understanding that, if one operation be delayed, another might commence. Nevertheless, ideal weather was uncommon—sometimes aircraft were grounded for weeks. When this occurred, they might fly reconnaissance, transport ground survey parties and supplies, or check fuel caches. Most operations called for oblique photography, a simple method which covered a wide area with a few pictures. Ideally, this was done from 5000 feet AGL. Heavily laden Vikings and HS-2Ls had difficulty reaching that height, so 4000 was more common. Vertical photography was best done at about 10,000 using the D.H.4 and D.H.9.

## Further Operations

Photographic flying increased rapidly. In 1926 59,000 sq. mi. were covered—50,300 oblique, 8700 vertical. Work over Vancouver Island was hampered by smoke, but at High River a D.H.4 flew 132:40 hours and covered 2550 sq. mi. by vertical photography in southern Alberta and northern Saskatchewan. The most significant development of 1926 was the increased use of photo detachments, by now self-contained with provisions, tents and a cook. Two Vikings from Winnipeg operated in eastern—northeastern Manitoba, and northwestern Ontario. Ottawa Air Station established a similar flight with 2 Vedettes and a Varuna. Operating from Larder Lake, Ontario and around Senneterre, Quebec, they photographed 6500 sq. mi. near Rouyn. Aircraft from Ottawa covered the Ottawa Valley and eastern Ontario.

For 1927 High River covered about the same area as in 1926, using two D.H.4s. Winnipeg had 2 flights, each with 2 Vikings, 1 flight in Saskatchewan; the other in Manitoba and northwestern Ontario. The Manitoba/Ontario detachment finished 1927 with Vedettes. Bad weather hampered both detachments. Although 36,456 sq. mi. were photographed, only 45% of planned work was completed. Unhappily, 'ET crashed near Hilbre, Manitoba on July 11, 1927 killing P/O W.C. Weaver, AC1 J.T. Eardley and Mr. F.H. Wrong (DLS). An observer on the ground watched the aircraft break up, as it emerged from cloud. Investigations, including strength tests directed by W/C Stedman, indicated

Varuna 'GV comes to the dock. An outgrowth of the Vedette, it operated 1927-30, then was replaced by the Vancouver. (NAC PA20099)

that the hull had failed near the pilot's cockpit. In his memoir, Stedman noted: "At Lac du Bonnet we rigged up a Viking and loaded it with sandbags to simulate the weight of the crew and equipment. Art James assisted... The load was gradually increased until failure occurred at a figure somewhat lower than was desirable, and reproduced the type of failure that had occurred in the air. The hulls of all Vikings were afterwards fitted with an extra mahogany plank on the outside of each side of the hull at the level of the top longerons... no further difficulty was experienced from this cause."

Ottawa had two detachments in 1927. Two Vedettes photographed in Muskoka from mid-June until mid-September, then proceeded to Sudbury, where they were held by weather for 5 weeks, before starting an operation that was only half-done by freeze-up. The other flight (Vedettes and a Varuna) operated in Quebec from La Tuque, Baie St. Paul, Lac St. Joseph and Roberval. Dartmouth Air Station, using a Varuna, completed vertical coverage around Shelburne, Nova Scotia. Although areas covered in 1926-27 were substantial, the Topographical Surveys Branch of the Department of the Interior was not wholly satisfied. On June 2, 1927, for example, the Surveyor-General, Mr. F.H. Peters, complained to G/C Scott:

*Reports ... concerning the two machines which have been detailed for this work [aerial surveys near Prince Albert and Wood Buffalo Park] indicate that ... there seems to be doubt as to their being in proper condition to carry out the work ... Reports being received from the other stations indicate that we are facing a serious situation in connection with all the aerial photographic operations... At High River a number of rolls of film have been exposed, but the prints received indicate that the personnel are either inexperienced and new to the work... The quality of the pictures is good, but the flight lines are very much out of position... At Winnipeg we know that there has been a delay in getting the equipment ready and that considerable time has been lost... At Ottawa a serious condition has apparently developed. Having ended the past season with three planes in operating condition, we cannot understand why at least three machines were not in operation early in April. Yet we have seen photographic days go by with no work at all being done and other days during which only one or two machines have been in operating condition... there exists little hope of any improvement... In Nova Scotia the conditions seem to be the same... I trust that some satisfactory explanation may be made for the serious delays... and that you may be able to give some definite assurance that some immediate remedy is in sight...*

Scott replied, cataloguing the problems. The chief cause of delay was the unreliability of the Lynx engines in Vedettes and Varunas. He also pointed to increasing demands for vertical photography. This taxed machines and personnel more than oblique work. While financing had

**A May 1931 north-looking view of Sub-station Lac du Bonnet, Manitoba, a key RCAF forestry/photography base. (NAC PA62734)**

Rockcliffe on a summer's day in 1929. This was home to RCAF's photo detachments. In WWII major infrastructure was added, especially above the bluff, and runways crisscrossed the flying field. The area beyond is now modern-day Ottawa. (CF)

G-CYXS, the first of six Canadian Vickers Vancouvers, at Rockcliffe for acceptance trials on October 28, 1929. (NAC PA62530)

Another Rockcliffe scene, this time with Vedette 'YC (on beaching gear), Puffer 'ZI, Moth 'YP and FC-2 'YT. (NAC PA62413)

W/C W.G. Barker, VC, MC and 2 Bars, DSO and Bar died in the crash of this Fairchild KR-21 at Rockcliffe on March 12, 1930. (CANAV Col.)

RCAF personnel at Rockcliffe over the summer of 1932: Rear—Sgts Poulin, Macauly, Dearaway, Silsby, Bradford, Roberge, Ewart, Fleming, Gear, Howard, Cousins, Millar, and Rathwell; front—FSgts Anderson, Diller and McClatchie, WO2s Colp and Black, W/C A.E. Godfrey, WO2s Raymond and Palmer, FSgts Greene, Bennett, Winship and Dasey. (H.S. and Gordon Diller Col.)

A scene at Norway House with RCAF and civil Fairchilds awaiting changeover to floats. (NAC PA20371, RCAF/Halliday Col.)

been disrupted in 1926 by a political crisis and general election, Scott admitted to inexperience concerning some personnel. Aircraft were not the most suitable for vertical photography. The Vedette and Varuna had been designed for forestry. Some hopes were expressed that a new Canadian Vickers design might resolve many problems. Instead, the company was about to unveil the Velos—an aeronautical monstrosity, perhaps the ugliest aircraft designed in Canada. When the RCAF examined the prototype at Montreal in November 1927 it found the Velos weighing 5875 pounds—1691 over design weight. This was so unreasonable that examining officers F/Ls Alan Ferrier and Roy Grandy insisted that the scales be checked.

The Velos was redesigned with more powerful engines. By September 1928 it was ready for testing. Grandy found the controls dangerously sluggish and the whole concept impractical. Visibility forward was excellent, but poor otherwise. The large cabin was cluttered with bracing tubes and there was fear that groundcrew conducting normal duties could be hit by a propeller. The sole Velos came to an ignominious end the night of November 30, 1928, sinking at anchor under the weight of heavy snow. It was later reported that this program had cost Ottawa $43,100, although most of this was later recovered as credits on Vedette orders. Precisely why the Velos turned out so badly was never clarified, but F/L Ferrier had a theory: "Unfortunately the specification was prepared by a government committee on which certain topographic mandarins had a weighty influence. The result was an overweight mongrel toward which [T.R.] Reid [Chief Designer for Canadian Vickers] had difficulty in concealing his disgust." Happily for the RCAF, late in 1927 it began receiving Fairchild FC-2s; by mid-1928 it had 13. The FC-2 had been designed by the Sherman Fairchild Corporation, which had begun as an aerial survey firm. Dissatisfied with existing camera planes, Fairchild branched into aeronautics. The result was a family of superb high wing monoplanes adaptable to wheels, skis or floats. With these the RCAF could extend deployment of its photographic detachments.

In 1928 no fewer than 8 units were deployed, e.g. No.1 Photographic Detachment—FC-2Ws seaplanes 'XN and 'XQ in BC. Pilots F/Os A.L. Morfee and E.J. Burke, with 4 airmen and 1 navigator from Topographical Surveys. Oblique photography near Victoria and Esquimalt for Department of Public Works, vertical photography at Squamish and Mud Bay. Flew 4 hours of customs patrols for Department of National Revenue; No.7—FC-2Ws 'YU and 'XP under F/O W.M. Emery with F/O Graham. Vertical photography from Ottawa to Montreal, north around headwaters of Gatineau, Rouge and Mattawin rivers. Emery reconnoitered near Lac Claire, 35 mi. north of St. Michel des Saints, carrying John Carroll of the Geodetic Survey. Oblique photography along St. Maurice River. Unexpected freeze caught detachment at Ste. Agathe des Monts (one Fairchild towed back to Ottawa behind a truck).

An unnumbered detachment active in 1928 operated from Deer Lake on Mile 437 of the Hudson Bay Railway. Formed for the Department of Railways and Canals, it transported personnel to Churchill and flew ice patrols along the west coast of Hudson Bay from Churchill to Port Nelson. Commanded by F/O A.L. James, the unit (2 pilots, 4 fitters, 2 carpenters and a cook, all living in tents) began work on June 28, 1928. FC-2Ws 'XL and 'XM flew 524 hours before the station was temporarily closed and its gear stored for the winter. All this was part of a larger scheme begun the previous year. Its reports on ice conditions near Churchill were compared with those from three similar Hudson Strait detachments of 1927-28. Indeed, these latter were not temporary, but components of the ambitious Hudson Strait Expedition, the story of which has been told in several Canadian books. This was the largest interwar deployment involving RCAF aircraft, personnel and expenditure. It also was one of the most significant in its findings. The task was to determine ice conditions to assist in research about navigation into Hudson Bay. The matter had been raised as early as 1920, when the Department of Railways and Canals requested a survey. The Air Board favoured the project, but the civil department, having raised the idea, let it fade.

During the next seven years little interest was shown in the eastern Arctic. The only RCAF involvement was in the summer of 1922, when Ottawa despatched an expedition to the Arctic archipelago to explore, and establish police posts to assert sovereignty. On the urging of the Department of the Interior, the Air Board assigned S/L R.A. Logan to investigate flying conditions in the areas visited. He had been a Dominion Lands Surveyor before the war, had Arctic experience and was an expert in meteorology, navigation and wireless (he taught them all

Obliques showing two key tations: Ladder Lake, Saskatchewan (note the many features—wharf, shops, cabins, tents, fuel cache); and below, Cormorant Lake (the presence of so many aircraft suggests the start or end of a season). (RCAF/Halliday Col.)

at Camp Borden). The expedition sailed from Quebec on July 18, returning to the same port on October 2, having been to the north end of Baffin Island, Bylot Island, Ellesmere Island and North Devon Island. Because of the short season and the priority of establishing police posts, Logan did relatively little work, but did determine that certain local resources such as coal might be available to any air stations. He noted that ski-equipped aircraft could be used for more than half the year, and recommended that an Eskimo be present on long distance Arctic operations, noting that such people "can find food and direction where a white man would be lost, starved or frozen to death". He also discussed the difficulties with supplies, aero engines and navigation. Logan was concerned that, before any major flying operations in the area, aircrew spend several months familiarizing themselves with conditions and making meteorological observations. Logan's report to the Air Board mentioned the possibility of Arctic development relative to any defence scheme involving Russia:

*Whether war with such a country as Russia would ever come or not, should not affect the determination to develop flying in the Canadian Arctic and sub-Arctic regions, because Canada, if it considers itself worthy to be called a Nation, should have enough pride and spirit to take at least ordinary precautions and be prepared to defend itself in any emergency.*

*Canada proved in the Great War that her men in the Royal Air Force were equal to any in the world, and it now remains for her to show the rest of the world that she can defend herself, and the whole British Empire if necessary, from all comers from the cold countries in the north of Asia—or Europe—by having men trained and proper material and information available through actual practice within her own boundaries.*

Isolated Wakeham Bay, main base for the Hudson Strait Expedition, on a bright winter's day. (H.S. and Gordon Diller Col.)

## The Hudson Strait Expedition

Construction of the Hudson Bay Railway focused government attention on Hudson Strait. In December 1926 the Deputy Minister of National Defence requested that the Senior Air Officer organize an aerial expedition to the area. Order in Council PC.85 established an Advisory Board to organize the expedition, members including N.B. McLean (Department of Railways and Canals) and G/C J.S. Scott (later succeeded by W/C J.L. Gordon). Before establishing the Advisory Board, the RCAF considered using kite balloons for ice reconnaissance. The idea was rejected, not as impractical, but on the grounds that balloons would cost too much (about $45,000 each). In February 1927, when the Deputy Minister of Marine and Fisheries insisted that personnel of his department handle wireless during the expedition, Scott protested. The Royal Canadian Corps of Signals had hitherto provided the RCAF with such services. Scott felt that the civil department would be less experienced and co-operative than the RCCS. Supported by the Chief of Staff, he threatened to withdraw RCAF participation. The matter was settled by making the Department of Marine and Fisheries responsible for ground stations, while the RCCS would install (but not operate) aircraft radios. Several British and American aircraft were considered, evaluation being supervised by W/C Stedman, the chief DND aeronautical engineer, and F/L T.A. Lawrence, a staff officer in HQ. Lawrence eventually was promoted to squadron leader and placed in charge of flying operations under the general supervision of Mr. McLean. The Fokker Universal, a sturdy, single-engined monoplane with a 200-hp. Wright J.5 air-cooled engine, was selected. Six were purchased—G-CAHE to 'HJ, along with D.H.60 Moth 'HK (to do reconnaissance in selecting bases).

Personnel trained intensively at Camp Borden in flying, meteorology, navigation, engines, first aid, seamanship, snowshoeing, skiing, shooting, dog-handling (with RCMP instructors), welding, carpentry, rigging, photography and instrument servicing. Attention was paid to emergency supplies, even to the smoking and reading needs of the men, who would be living in isolation for six months. Loading aboard the CGS *Stanley* and SS *Larch* commenced in Halifax on July 4, 1927 and the little convoy sailed on the 17th. Ten days later, at the eastern entrance to Hudson Strait, the Moth was lowered from the *Stanley* and S/L Lawrence, accompanied by F/L A.A. Leitch, reconnoitered to the first base site. A construction party was aboard *Larch* to erect the first camp. On August 26 the Moth was caught in a storm while moored. It demonstrated its seaworthiness by riding out the tempest for 12 hours before capsizing, but it already had served its purpose, noting ice conditions ahead, and helping choose base locations.

On November 11 the ships weighed anchor. Camp construction had not been finished, so the RCAF was occupied for some time putting the final touches to temporary homes and workshops. Three bases were created—"A" (Port Burwell), "B" (southern corner of Nottingham Island) and "C" (Wakeham Bay, between "A" and "B"). Thus, air patrols would report conditions all along Hudson Strait. Two Fokkers were at each site. A system of normal and special patrols was drawn up and approved before flying began. During certain periods routine sorties would be flown daily (weather permitting) by aircraft of each base. Aircraft from different bases could rendezvous at selected points to ensure that information for a particular period was collected through-

Universal 'HE is readied at Wakeham Bay on June 30, 1928. The Universal, one of Anthony Fokker's American designs, was built at Atlantic Aircraft Corp., Teterboro Airport, New Jersey. The Universal accommodated a pilot in an open cockpit, and 4-6 passengers (max. 1200 lb payload). Wing span was 47' 9", length 33' 3", gross weight 4000 lb, speed 105 mph, engine: 200-hp Wright J-4, fly-away price on floats $16,650. (NAC PA202541)

out the Strait. Special patrols were laid on if considered essential. Hand-held oblique cameras were used, with 60% overlap between frames. Pilots filed detailed reports on ice conditions following a patrol. Air-to-ground radio communications were maintained using CT.21A transmitters with trailing antennae. Voice and key methods were used and a remote control installation enabled pilots to use the radio themselves. Crews on patrol communicated with base every five minutes.

Float operations were limited during late summer and autumn. Weather was involved, but so was the fact that the flying staff was helping assemble the bases. The key task was ice reconnaissance, and ice did not appear until November 16. Regular patrols began on September 29 from Wakeham Bay; 53 days elapsed before freeze-up. Float patrols were flown on 10 of those days. For the rest fog, wind, snow and shore ice prevented flying. For three weeks conditions were unsuitable for either floats or skis, but on December 12 ski patrols resumed. This prevailed until June 18, 1928. Now break-up curtailed operations, but on June 29 flying from Wakeham Bay resumed on floats, continuing until the expedition completed its work in July.

Regular patrols from Nottingham Island began on October 11, 1927, continuing until freeze-up on November 16. It was colder there than at the other bases and the interval between float and ski operations was short, with regular flying resuming on November 23. At the end of May 1928 the break-up allowed a return to floats. At Port Burwell regular float patrols began on October 23 and continued for 31 days. Ski operations began on December 13 and the aircraft reverted to floats on May 23, 1928. Ice conditions during initial freeze-up were of vital interest. At Nottingham Island the problems were not acute, owing to rapid freeze-up. Elsewhere it was a different story, as S/L Lawrence explained in his final report:

*Conditions made the launching and removal of aircraft to and from the water impossible, and previous experience had taught us never to leave machines at their moorings for more than a few hours and then ... under continuous watch... 18 days passed from the time of Wakeham Bay freezing over until it was possible to get a machine on to the bay ice and commence flying, using skis... Some idea of the conditions to be contended with in eventually getting a runway over the rough shore ice to the bay ice can be visualized... Tons of ice and snow were chopped down, filled in, levelled off and packed into a runway, using the tractor as a roller. At each change of tide, this runway, which extended across a beach about 200 yards long, would heave up and crack until ... after much labour it became a solid bridge of ice, rising and falling with the tide, but immune to serious damage.*

Throughout the expedition meteorological observations were recorded. At Port Burwell sub-zero temperatures were regularly reported from December 8, 1927 until April 9, 1928 and low temperatures frequently were accompanied by high winds. On February 9, 1928 minimum and maximum temperatures were -14.1° and -2.8°F, while winds varied from 12 to 27 mph. The lowest temperature there was -23.6°F (February 17, 1928). Temperatures were more severe at Nottingham Island, which recorded lows of -30°F or colder on 11 days in January and February 1928. Conditions at Wakeham Bay resembled those at Port Burwell. Flying continued at every occasion until January 25, 1928. By then it was apparent that there would be no significant changes in the ice for many weeks. On instructions from Ottawa fortnightly patrols commenced. After February 21 even these were limited to specific routes, to enable a Fokker in trouble to glide to the coast.

On three occasions aircraft became lost. On December 15, 1927 F/L Leitch was returning with 'HJ to Nottingham Island, when he encountered a snowstorm. Not sighting land, he alighted on a thin ice floe. The crew drained its engine oil, then made themselves as comfortable as possible. That night the temperature dropped to -16°F, causing minor frostbite. Next day Leitch calculated his navigation error. Using emergency equipment, the fliers warmed their oil, poured it back into the engine and started up. They reached base with barely a quart of gasoline remaining. Next, on January 8, 1928 S/L Lawrence, flying between Wakeham Bay and Nottingham Island, ran into snow about 20 miles east of Cape Digges. He turned back and landed at Suglet Inlet, tried to reach Nottingham Island next day, met further storms, so put down at Deception Bay. For nine more days storms battered the area, while the crew camped in the Fokker. On January 16 F/O Carr-Harris, flying 'HE from Wakeham Bay, located the missing plane, and landed. The rest of the day was spent digging out 'HE and getting it airworthy. Both Fokkers overnighted at Deception Bay, returning to Wakeham Bay on the 17th.

On February 17, 1928 F/O Lewis took off from Port Burwell in 'HG, accompanied by FSgt N.C. Terry and an Inuit named Bobbie. Having flown as far as southeastern Baffin Island, they were returning to base when the engine began sputtering. Visibility deteriorated and Lewis feared that, if he descended too soon looking for landmarks, he might be unable to regain altitude. At last he thought he recognized some feature. He laid on a course which, he thought, would lead to Port Burwell. At regular intervals the crew radioed base, where messages were recorded, some of which, in the last two hours, seem a bit peculiar, e.g. at 1319 hours: "Engine missing badly, can just see land, losing revs, think will make it. Engine missing badly, will just make it, if at all." (1325) "Should be home in about 45 minutes if this blasted engine keeps going all out. Buttons [Button Islands] look very cheerful after the past exciting half hour." (1334) "Just had another drink... flying over Akpatok Island, wind must have drifted us but will head across." (1340) "Have another drink." (1407) "Just now on our way to the old country." (1409) "I suppose will land on the Thames." (1416) "Hello, engine vibrating and losing revs, above the clouds, no sign of land." (1425) "Should see the coast of Ireland soon." (1443) "Believe I see land, petrol nearly gone, don't know where we are". (1514) "V.C.H. [an emergency code], landing on ice in Ungava Bay."

**S/L Lawrence, F/L Leitch, F/O Carr-Harris and Mr. Valiquette gather survival kit before leaving in 'HI to search for Lewis and Terry, February 19, 1928. (NAC PA202545)**

Lewis alighted on hummocky ice, but 'HG sustained only minor damage. The men began walking eastward with their emergency kit. After a day they realized that they were on an ice floe in the Atlantic off Labrador. The jests about arriving in Ireland had, ironically, been too close to the mark. They doubled back. For the next seven days they travelled west, frequently crossing icy water in their inflatable raft. Their emergency rations ran out, so they resorted to raw meat from walruses shot by Bobbie. Finally, they struck Labrador. Over the next four days they saw neither birds, animals nor humans. Search aircraft from Port Burwell were grounded by weather for much of the period. On the fifth day after reaching the coast, the party met an Inuit hunter and his wife. Through Bobbie they explained their needs. Food was provided, as well as transportation by dog team to Port Burwell. They reached base on March 1, bringing to a close a northern Canadian saga.

Meantime, Ottawa had been pondering the future of the expedition. On March 2 the Hudson Strait Committee recommended that the operation be extended for one or two years. The Department of Marine and Fisheries, satisfied with results, informed the DND on March 10 that there was no need to prolong matters. Ice observations ceased after August 3, 1928. Personnel prepared to evacuate the bases. The aircraft would fly back to Ottawa, using fuel caches at Cape Smith, Povungnituk, Port Harrison, Great Whale River, Fort George, Eastmain River, Rupert House, Moose Factory, Remi Lake and Trout Mills. The Fokkers rendezvoused at Erik Cove and attempted to get away on August 25. Three became airborne, one failed to takeoff due to engine trouble, and the fifth, piloted by F/L Coghill, crashed with a broken pontoon. F/O Carr-Harris immediately landed and rescued the three from Coghill's machine. S/L Lawrence radioed SS *Larch* and CGS *Montcalm* for assistance. The other four aircraft were examined; the undercarriage mounts and pontoon fittings on two were deemed unfit. The fleet was disassembled and stowed aboard ship.

As to statistics, Base "A" machines flew 83:46 hours, made 47 patrols, and took 227 photographs; for Base "B" 134:10 hours, 82 patrols, 756 photographs; for Base "C" 151:48 hours, 98 patrols, 1302 pictures. The RCAF learned much about ice conditions, the potential for a shipping season in Hudson Strait, ice-breaker requirements, winter flying, Arctic clothing, and establishing semi-permanent bases in an extreme climate. Equipment and supplies issued had proven appropriate, as shown by the ordeals of Lewis and Lawrence. The final report noted, however, that messing had been deficient in fresh meat (one had to acquire a taste for seal, white whale and walrus). A 21-day expedition by Constable Murray from "C" Base in search of game had failed dismally; he and his Eskimo companion travelled inland for 200 miles, never fired their guns, and returned with only five fish.

## Survey and Winter Flying

In 1922 the Geodetic Survey was attempting to chart more accurately the boundary between Alberta and British Columbia north of the Yellowhead Pass. At the suggestion of F.H. Lambert, surveyor in charge, A.O. Wheeler requested an aircraft from High River. On July 11 Capt J.H. Tudhope flew with a mechanic in D.H.4 'DM to Henry House, nine miles from Jasper. Next day he took Wheeler for a short flight to acquaint him with daily conditions. On the 13th they flew over the boundary from the Yellowhead Pass, following the Miette River to Colonel Mountain, thence through Moose Pass, Jackpine Pass, and along Jackpine Creek. North of this, clouds shrouded the peaks. In attempting to edge away from them Tudhope followed the wrong divide and emerged in the Fraser Valley. With fuel low they landed at McBride. Tudhope and Wheeler secured food and gas, then took off for Henry House. Wheeler was thrilled by the experience and impressed by how he had been able to trace routes which his surveyors would have to follow. In one instance he garnered trail information that would save considerable time.

Wheeler left next day, but Lambert wished to follow the interprovincial boundary from Jasper to 120°W. On the 14th Tudhope flew him to Moose Pass, where cloud forced their return. Next day the weather remained poor and the aircraft narrowly escaped being wrecked by wind. Finally, on July 16 Tudhope flew Lambert as far as Jarvis Pass, dropping a bag of mail to a survey camp on the return. They had been handicapped with camera problems and Lambert wanted to make another reconnaissance. However, a pall of smoke over the area made the sortie unwise. Tudhope returned to High River on July 19. Relatively little was done for the Geodetic Survey over the next six years. Forestry and photography had cornered the greater number of scarce aircraft. One operation requested in 1923, transporting supplies and personnel for the survey between Yellowhead and Jarvis passes, had been rejected, terrain being too rugged, potential landing fields too few, and equipment inadequate. Early in 1929, however, the Department of the Interior expressed renewed interest in aerial reconnaissance on behalf of ground parties working for Geodetic Survey, particularly in winter, when aircraft and personnel were more available. This coincided with the RCAF which was investigating

**Sgt Arthur Fleming wearing the sort of winter flying gear that the RCAF developed in the interwar years. (Arthur Fleming Col.)**

J.H. Tudhope was one of those RCAF types who was always involved, no matter what the aircraft or task. Having made his mark in the air force, he became a key man in the newly-formed Department of Transport. His story is told in *Air Transport in Canada*. (CF PL117529)

further operations by aircraft far from civilization.

In considering Geodetic Survey support flights in winter, the RCAF assigned FC-2 'XK with S/L R.S. Grandy to fly reconnaissance between Barrière and Senneterre, carrying J.L. Rannie and J.E. Ross of Geodetic Surveys. The first phase was completed February 23-26, 1929, being flown at 300-500 feet; in some cases hills forced Grandy as high as 2000. Rannie reported: "The test showed that aeroplane reconnaissance in this type of country was successful and very quick. The actual reconnaissance was nearly com-

Officers at a Camp Borden wings parade in August 1929. First is the station commander, W/C G.M. Croil, AFC (later A/V/M Croil, Chief of the Air Staff); then P/P/O G.F. Kimball, winner of the Sword of Honour, S/L George E. Brookes (later A/V/M Brookes, OBE), S/L C.M. McEwen (later A/V/M McEwen CB, MC, DFC, AOC No.6 Group) and F/O J.T. O'Brien. (CF PL117109)

Student riggers learn their craft in 1930, practicing on a superannuated Avro 504. (NAC PA62793)

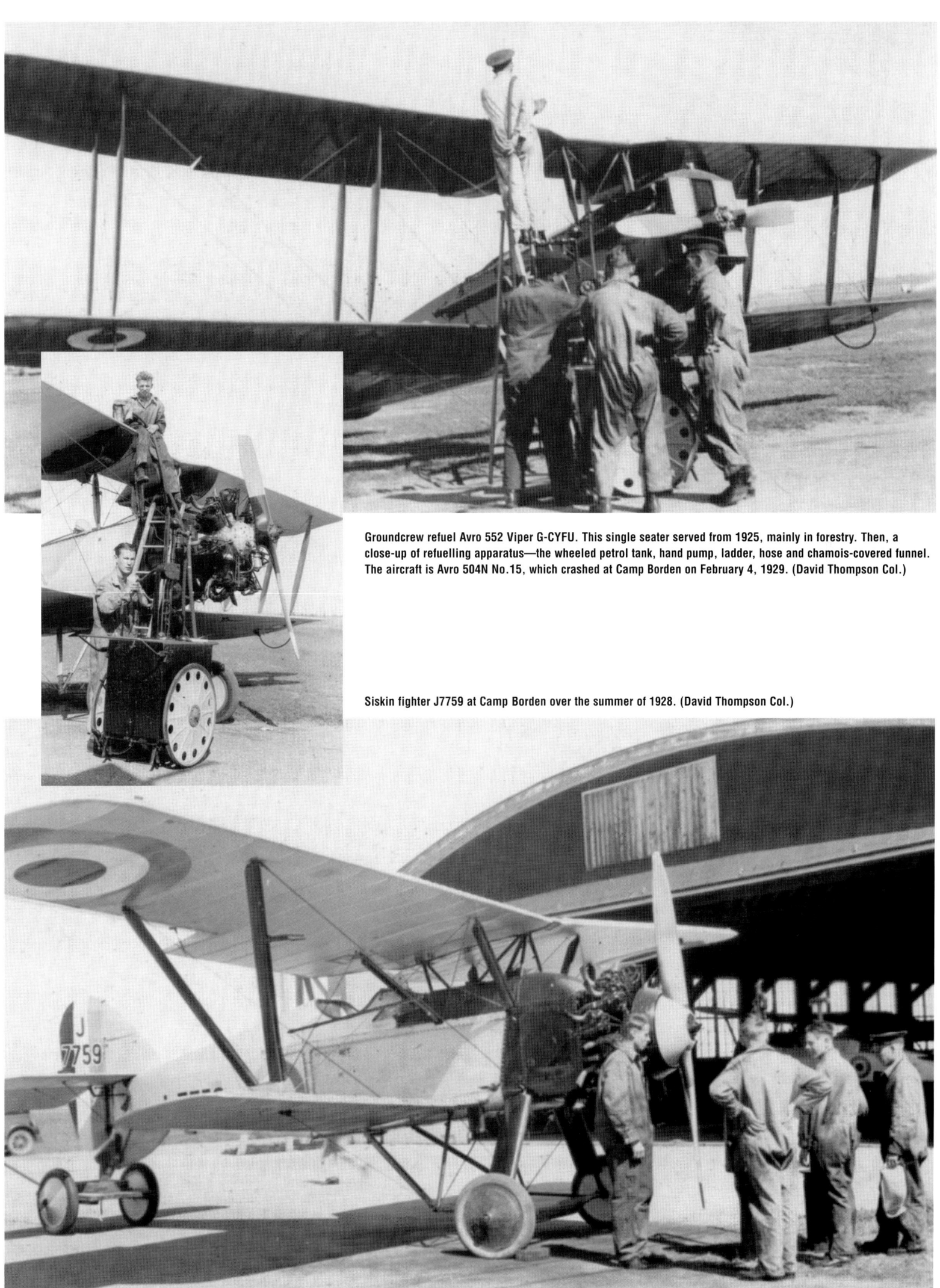

Groundcrew refuel Avro 552 Viper G-CYFU. This single seater served from 1925, mainly in forestry. Then, a close-up of refuelling apparatus—the wheeled petrol tank, hand pump, ladder, hose and chamois-covered funnel. The aircraft is Avro 504N No.15, which crashed at Camp Borden on February 4, 1929. (David Thompson Col.)

Siskin fighter J7759 at Camp Borden over the summer of 1928. (David Thompson Col.)

**The tent-and-blow pot kit referred to by Roy Grandy. The plane is an FC-2. (NAC PA62629)**

pleted in about nine hours flying time, including about 4 hours going to and coming from the area ... it is estimated that it would have taken at least 2 months to have accomplished the same result by canoe." A second operation occurred March 8—13 with Grandy and Rannie in FC-2W 'XU. Ross was replaced by F. Steers, and Grandy was accompanied by crewman Sgt S.A. Greene. The area lay between Clova and Windigo in northern Quebec. On one flight (March 11), oil pressure dropped and Grandy landed on Menjobaques Lake. Discovering congealed oil in the lubricating system, he drained his oil, re-heated it over a fire pot, then poured it back into the engine (a delay of 4 hours).

Rannie again was satisfied with the reconnaissance. His only complaint was that a better idea of the lines of sight for the surveyors might have been obtained by flying level with or slightly above the surrounding hills. This, he admitted, was not always consistent with safety in a single-engined plane, and he suggested that a tri-motor might be better. The other objective of Grandy's flights had been to consider the feasibility of detached operations in winter. On his February sorties he had used blow torches and stove-pipes to heat his engine, before starting in sub-zero temperatures. This had failed—blow torches kept going out and the stove pipe effect proved disappointing. In March, however, Grandy successfully used a small stove (fire pot) and a tent pitched over the engine. He also concluded that existing skis were too small for good handling in slush. The FC-2 had performed poorly; in some cases a mile-long takeoff run was needed. The Wright engine had also faltered with carburetor icing. The FC-2W, on the other hand, could unstick in 100 yards, performing admirably with its P&W Wasp. Overall, however, Grandy felt he had not learned enough; the operation had exploited good weather to help the Geodetic Survey at the expense of investigating the full complications of winter flying. Grandy concluded that the policy of combining experimental with other work was unsound.

It was thus that the Transportation Flight in Manitoba was so much engaged in flying triangulation survey crews. In eastern Canada, No.6 (P) Detachment operated so well on non-photographic days in reconnaissance and transport on behalf of triangulation parties east and southeast of Doucet, that one surveyor near Senneterre wrote, "The Air Force is marvelous, don't know where we would be without them". In British Columbia No.1 (P) Detachment flew 13 hours in reconnaissance for the Geodetic Survey in the Parsnip River area. As the 1929 summer season drew to a close, the Department of the Interior evaluated the results. Noel J. Ogilvie, Director of the Geodetic Surveys, wrote a lucid report to his superior, J.D. Craig, saying in part:

*Since last February some 270 hours of flying has been done by the Royal Canadian Air Force to explore some of the possibilities of the use of aeroplanes on operations of the Geodetic Survey Branch of the Department of the Interior. It is now possible to judge the results which have been obtained. Planes can be used with great success for three types of triangulation operations:*

1. *Preliminary reconnaissance to determine the most suitable routes for triangulation.*
2. *Detailed reconnaissance for actually selecting triangulation stations.*
3. *Transport of triangulation parties—tower building, angular measurement and light keeping—and other parties...*

*For detailed reconnaissances the plane has been found a great success with certain present-type planes ... Reconnaissance was done in four areas of northern Quebec which fully demonstrated the success and economy of the method. This work can be carried out effectively in winter, a considerable advantage, as the planes have a better performance with skis than with pontoons. Perhaps the most spectacular success of the season was in the experiment of transporting all triangulation parties by plane in the district south and east of Senneterre, Quebec, by the two photographic planes operating from that base. Aerial photography took precedence in the use of the planes, but there was ... little conflict between the photography and transportation due to the small number of good photographic days. This country was a very difficult one to work over by land methods, as transportation would have been entirely by canoe and man-packing involving long delays and much loss of time. Aeroplane transportation changed this condition entirely ...*

*The season's experience shows ... that low power planes must not be used for detailed reconnaissance and transportation of triangulation parties... a cabin monoplane with as good or better performance than the Wasp Fairchild is essential even for well watered areas. A plane with better visibility, such as the Bellanca, would be a greater improvement, while for drier areas a powerful tri-motored plane of at least the performance of the Wasp should considerably extend the zone of the information obtained. For transportation of triangulation parties power is also essential to permit landings and take-offs from small lakes.*

*Specially designed canoes for being carried under the fuselage between the pontoon struts are a frequent necessity ... these should be provided for next season... Radio communication between plane base and triangulation parties is also a desirable and economical provision for next season.*

The upshot was that Ottawa was assigned an extensive winter flying program, including experimental oblique photography between Chelsea and Maniwaki, reconnaissance in northern Ontario and northern Quebec, and winter equipment trials. The reconnaissance (areas as far north as Chibougamau, Dore Lake and Oskelaneo) was conducted between January 29 and March 1, 1930 by F/L A. Carter in Pacemaker 'WT. Mr. F.P. Steers (Geodetic Survey) accompanied him as observer, AC1 Oldridge was the mechanic. Apart from reconnoitering, the operation enabled the RCAF to further test winter flying clothing, as well as the engine tent and heater used previously by Grandy. While engine heating equipment proved successful, cabin heat was inadequate (Grandy had complained of this with both Fairchilds). Carter found that in extreme cold his feet were comfortable only when he was wearing two pairs of heavy woolen socks, one pair of duffel socks, light moccasins and flying moccasins.

The 1930 season saw increased contact between the RCAF and the Department of the Interior, with transport and reconnaissance done

Fort Fitzgerald, a place on the Slave River about half way between its source (Lake Athabaska) and destination (Great Slave Lake). This frontier post often was visited by bushplanes. Note the paddle-wheeler—in these days the river was the chief means of transportation and communications in the region. (RCAF/Halliday Col.)

on behalf of Geodetic Surveys. Regular aerial photography involved 11 special detachments. The year was notable for the first truly extensive work in the Northwest Territories, e.g. two Fairchild 71s involved in oblique/vertical photography around Fort Resolution, Great Slave Lake, Lesser Slave Lake, Hay River, Lockhart River, and the Slave and Taltson rivers. The most dramatic operation was one by No.2 (GP) Detachment under F/L F.J. Mawdesley (Fairchild 71 'VX and Vedette 'WS). The aim was to inspect fuel caches in the NWT, photograph selected areas and, if possible, make a reconnaissance to Herschel Island. Accompanying Mawdesley were FSgt W.J. "Harry" Winny (pilot), Cpl S.C. Dearaway (photographer), two mechanics and Colin S. McDonald (Department of the Interior). On July 2 the Ottawa *Citizen* reported: "What aeronautical people regard as promising to be the greatest aviation feat of the year, in Canada at least, is to begin in a short time when two aircraft ... take off from Ottawa on an extended expedition into the Northwest Territories." The detachment left on July 4, arriving at Ladder Lake, Saskatchewan on the 9th and Fitzgerald, Alberta on the 11th.

Flying down the Mackenzie River, the aircraft paused frequently for photography and to inspect caches. On July 12 they checked those at Resolution, Hay River and Providence. Next day they were at Fort Simpson, and on the 19th reached Aklavik at the mouth of the Mackenzie, becoming the first RCAF aircraft to visit the shores of the Arctic Ocean. A flight to Herschel Island was attempted on July 21, but fog intervened. On the 22nd the expedition returned up the Mackenzie as far as Norman, where it turned eastward to Hunter Bay on the east shoreline of Great Bear Lake. It arrived there on August 5, then proceeded via Dease Bay to Coronation Gulf, then traced the Coppermine River to Point Lake. On August 18 it reached Reliance (Great Slave Lake), then back-tracked to Lake de Gras (Coppermine headwaters, August 23), returning to Reliance on the 29th. Flying east on September 2, it reached Chesterfield Inlet on the 7th, swung northeast to Cape Fullerton, Wager Inlet and Repulse Bay, then returned to Chesterfield Inlet on the 12th, from where it flew down Hudson Bay with stops at Mistake Bay and Eskimo Point. On September 21 the expedition reached Churchill, and on October 1 returned to Ottawa, having covered 12,000 miles in 193 flying hours. Some 3100 photographs were taken, mostly obliques (the weather deteriorated after the Thelon, so photography ended). The expedition covered unmapped territory and regions where magnetic compasses were unreliable.

Naturally, not everything had gone perfectly. Differences arose between Mawdesley and McDonald. In July the latter had threatened to leave the expedition. On July 22, however, Mawdesley telegrammed Ottawa that McDonald's grievances, due to personal misunderstandings, had been discussed, and their friendship for the remainder of the operation was assured. This illustrates the problem of a small group of men working in strenuous semi-isolation. It is not surprising that occasional cases of "cabin fever" arose. W/C F.H. Hitchins, describing the operation, cited another incident (although he did not mention his source):

*A member of the party recalls that even F/L Mawdesley, who seems to have had the navigational instinct of a homing pigeon, on one occasion became lost because of ... compass variation... he followed a river for some time until the dwindling fuel supply made it imperative to lose no more time in getting on course. Landing on the river, "Mawdy" went ashore, climbed to the top of a hill, surveyed the landscape and, with the help of his watch, estimated the correct direction to fly. Then he returned to the Fairchild and took off again but, to his crewman's surprise, headed in the opposite direction along the river. His instinct—or luck—held, and he finally reached his destination "with scarcely enough fuel left to fill an eye-dropper".*

**Frederick Joseph Mawdesley was one of the great prewar RCAF figures. His impromptu ideas and pranks became legendary. Once he was forced down in his Vancouver on the St. Lawrence River. Concerned that the mail might be late, he transferred it to a passing rum runner. The mail was dropped at Rimouski, but Mawdesley's CO was not impressed. Another time, "Mawdy" had some reporters aboard his Stranraer. Along route he left the controls and went back into the cabin. This worried the passengers who, in those years, didn't know about auto pilots. Here is Mawdy in 1929 wearing summer dress, while CO at Cormorant Lake. (Arthur Fleming Col.)**

## Ad Hoc Taskings

Apart from regular photographic and general purpose operations, *ad hoc* surveys were conducted. Typical was one requested of Ottawa Air Station at short notice by the Department of the Interior, for a reconnaissance of Lake Mistassini. Bellanca 'VJ, Fairchild 71B 'VY were assigned with pilots F/L J.E. Jellison and F/O J.G. Bryans, mechanics Sgt W. Keighley and LAC G.W. McCrea, and Messrs P.E. Palmer, John Carroll (Surveys Branch men) and N. Cauchon (Chicoutimi Harbour Commission). At 1425 on August 30, 1930 the expedition left Ottawa for Parent, Quebec via Blue Sea Lake, arriving at 1710. Next day it flew via Gouin Dam to Dore Lake, arriving at 1105. Accommodation was at the Chibougamau Prospectors' Camp at $3.50 per day per man. Heavy rains had raised water levels, flooding sandy beaches that would otherwise have been used for pulling up aircraft. The weather stayed poor, and, although 47:55 hours were flown, no photography was done. Visual recces were made, but the most significant accomplishment seems to have been measuring lake elevations near Chibougamau and Mistassini. The detachment returned to Parent on September 12, thence to La Tuque on the 16th.

Another *ad hoc* operation illustrates RCAF flexibility. No.9 (P) Detachment (Fairchild 71s 'XO and 'PO) was at Orient Bay in northern Ontario. On May 31, 1930 an Ontario Provincial Air Service cook there fell ill and had to be rushed with a doctor to Port Arthur. But the OPAS Moths could carry only one passenger. F/L C.R. Slemon took 'XO, picked up patient and doctor, and flew them to destination. June 2-9 the same detachment was on fire patrol near Port Arthur on behalf of Ontario. While at Orient Bay 'XO and 'XP also worked for Geodetic Surveys, dropping two surveyors with 900 pounds of equipment along the Albany River. No.9 later was despatched to Alberta.

Although aerial photography had become routine, it was not without difficulties. On October 13, 1934 F/O R.C. Hawtrey, commanding Nos.6 and 7 (GP) Detachments, submitted a report detailing loads. This ranged from 1719 pounds to 2264 pounds. Moving from base to base regularly took more—3 people plus baggage, while operational flights carried only 2 crew and less baggage. The breakdown of only one type of operation—multiple oblique photography—illustrates the load:

| Personnel | Weights (lb) | Total (lb) |
|---|---|---|
| Pilot | 190 | |
| Photographer | 190 | 380 |
| **Emergency Equipment** | | |
| Aircraft | 20 | |
| Ration kits (2) | 58 | |
| Personal kits (2) | 16 | |
| Axe | 3.5 | |
| Shotgun | 9 | |
| Ammunition | 7.5 | |
| Sleeping bags with ground sheets (2) | 30 | |
| Tent (7 x 7 feet) | 9 | 153 |
| **Fuel and Oil** | | |
| Gasoline (90 gallons) | 639 | |
| Oil (7 gallons) | 66.5 | 705.5 |
| **Camera Equipment** | | |
| Outside cameras with loaded magazines (2) | 106 | |
| Centre camera | 48 | |
| Magazines loaded (3) | 51 | |
| Camera Operator's Seat | 14 | |
| Magazine shelves (2) | 12 | |
| View finder with box | 11 | |
| Intervalometer with leads | 18 | |
| Thermometers (inside and outside) | 2 | |
| Altimeter | .75 | |
| Triple mounts | 32 | |
| Camera tool kit | 43 | 339 |
| **Miscellaneous** | | |
| Anchor and rope | 50 | |
| Stone bag | 4 | |
| Paddles (2) | 5 | |
| Tool kits | 40 | |
| Spares (engine parts) | 16 | |
| Ropes | 22 | |
| Parachutes with harness (3) | 60 | |
| Perrins Life Belts (3) | 9 | |
| Gas pump with hose | 53 | 259 |
| | Sub-Total | 1835.25 |

An RCAF Fairchild 71. This example, typically in overall yellow, served 1930-41. There were 61 RCAF Fairchilds, the last not being disposed of till 1946. In post-RCAF years, many flew for civil operators. (David Thompson Col.)

## The Late Thirties

Through 1938 No.8 (GP) Squadron under S/L W.W. Brown was the only RCAF unit doing photography on behalf of the Interdepartmental Committee on Air Surveys and Base Maps. Nos. 1, 2, 3, 6 and 7 Detachments operated from Halifax to Vancouver and from the American border to 66°N lat. Operations commenced March 24, 1938 and all detachments had returned to Ottawa by October 2. They had exposed 589 rolls, photographing 39,900 linear miles (76,015 sq. mi.) in 1729:50 flying hours. Vertical photography included 326 linear miles for the Department of Public Works, 4502 for the Geographical Section, General Staff (Militia), and 29,862 for the Interdepartmental Committee on Air Surveys. Oblique photography included 5210 linear miles for the Interdepartmental Committee. The squadron worked under the authority of 23 separate Operation Orders (not all of which were fully executed). Each detachment submitted a report. No.1 operated a Bellanca Pacemaker, No.7 had 2; the others flew the new Northrop Delta with 2 Norsemen for transportation and support. Excerpts from their reports throw light on difficulties encountered:

*Northrop Delta aircraft were used entirely throughout the season, both on wheels and floats, and were subjected to a variety of conditions which undoubtedly tested them completely. Generally speaking, the airframes are satisfactory, except for certain items... The workmanship of the manufacturer is not all that is to be desired and in many instances may be attributed to lack of adequate inspection during manufacture... The type of aluminum finish on the airframe is far from satisfactory. Because of its porous nature, it readily retains oil that cannot be cleanly removed by mere wiping, hence it is necessary to wash the airframes frequently... The interior finish of the cabins is entirely unsatisfactory.*

*As a landplane, the performance ... is satisfactory, except for manoeuvring by hand on the ground and in the hangars... Generally speaking, the Northrop Delta is not a good seaplane for the type of operations carried out. No really rough surface conditions were experienced during the summer, yet '672' developed three slight wrinkles between Nos.1 and 3 frames on the port side and other Delta airframes were wrinkled similarly. They are relatively difficult to handle on water, due, no doubt, to the inefficiency of the water rudders... Because of the distance required for take-off and landing, their scope of operations is restricted... The supply of hot air to the carburetor is deplorably inadequate and all pilots at one time or another had disconcerting and distressing moments of concern at faulty running of their engines, caused by ice formation in the throat of the carburetor while flying through rain. It is an inexcusable condition that should not be tolerated. (From F/L H.H.C. Rutledge, commanding No.2 Detachment, photographing around Regina and Swift Current, Camsell Portage, Reliance, Winter Lake and Hill Island Lake, NWT.)*

*The seven-man camp cooking kit was not adequate, but the supply of culinary equipment was augmented by utensils drawn from Stores at Fitzgerald. This did not prevent the civilian cook ... from coming forward daily with grievances which found their origin in the necessity for making one pot do the work of three. Constant contact with service equipment, however, finally won him over to a more philosophical outlook. Nevertheless, table-ware contained in the seven man kits is made of a very inferior grade of material. Knives, forks and spoons had a habit of breaking when personnel applied the amount of pressure necessary to sever the type of sinew normally found in such foods as boiled dinner... Blistered lips were the inevitable result of any attempt to take a hot drink from the aluminum cups provided. Personnel were compelled either to provide themselves with earthenware cups, or drink tepid tea...*

*When detachments are operating on the Prairies, where the weather is usually extremely good and photography can be carried out nearly every day, it is necessary for the Detachment Commander to carry out photography, run his detachment, make out the multitude of reports and returns required of him, and also compile his film returns each night. Frequent compilation of travel claims under these conditions is out of the question. It does not take long to exhaust the small advance of $500.00 provided for travelling expenses. Even when claims are submitted regularly, the delay in passing and paying them is so excessive that the Detachment Commander's bank balance is exhausted long before he receives notification that the money has been replaced in the bank. During this period it has been the custom in the past for the Detachment Commander to provide his detachment personnel with food and lodging.*

*When a detachment is operating in the Northwest Territories, food shipments are obtained periodically from Edmonton. Supplementary orders of food and incidental supplies must be obtained from time to time from posts at Fort Smith or Yellowknife. In all cases, it is necessary to run accounts. Statements of these accounts are mailed to the Detachment Commander, who mails a cheque in payment. Receipts must then be received before claims can be submitted. All this takes time. Claims mailed to Ottawa take a week to reach the accounting unit. After that, anywhere from two weeks to a month may elapse before payment of these claims. As a result, the Detachment Commander is again thrown upon his own resources... such a state of affairs is not in the best interests, either of the Service or the Officer in Charge of a detachment. A junior Officer should not be expected to finance a detachment operated by and for the Dominion Government, simply because the financial organization is unwieldy and the amount of money placed at his disposal so small... It is strongly recommended that the advance ... allotted to detachments operating on the Prairies and in the Northwest, be increased to $1000...*

*A civilian cook was hired while the detachment was based at Yellowknife. Although his skill with the skillet left much to be desired, all ranks were relieved from kitchen duties which gave them full time to devote to recalcitrant engines and ailing cameras... The detachment was compelled to leave Ottawa for operations before personnel had had the opportunity of obtaining more than a sketchy knowledge of the Delta aircraft. When the detachment left RCAF Station Ottawa, neither pilot had completed ten hours' flying on the type and the Fitters possessed only the knowledge of their engines imparted by an extremely short and, as a result, not-too-informative course at No.1 Aircraft Depot. The trouble experienced with the engine in Delta '670' was enough to*

**The Northrop Delta was the RCAF's first all-metal design. The RCAF operated 20, mainly in aerial photography. Delta 668 is shown in the St. Lawrence River at the factory. After its photo career it was a ground instructional airframe till 1945. (CANAV Col.)**

*call forth a great deal of effort from Fitters well versed in the maintenance of these engines. It made things doubly difficult for men who had so little previous knowledge of that type of engine. Being keen in their work, proud of their aircraft and very desirous that the detachment's record in the field should be a good one, these men "worked like Trojans" to maintain both aircraft in a serviceable condition. That they were able to do so, in spite of the difficulties attendant upon servicing aircraft in the bush, is greatly to their credit. Voluntarily working, as they did, 20 to 22 hours a day for several days, without once complaining, their record is worthy of praise. The Wireless Operator rendered invaluable assistance during these periods of stress. He helped the Fitters in their work, loaded and unloaded camera magazines, helped repair cameras, and nearly achieved a monopoly on the gasoline pump...*

*The crew of the Norseman transportation aircraft [aircraft 678, Sgt W.G. Pate, fitter LAC G.F.G Gayton] are also to be commended upon the manner in which they performed their work. The task of keeping four separate units*

*supplied with food, mail, etc., is a job in itself, particularly when each unit is separated by from 100 to 200 miles and the whole are operating in an area 200 miles from the source of the required supplies. In addition to these supply flights, each party of surveyors must be moved to a new control point every three or four days. The co-ordination of all these requirements calls for intelligent planning on the part of the pilot, as he is on his own a large part of the time. His plans must be sufficiently flexible to overcome delays occasioned by bad weather, changes in plans of any one of the four units, etc. He is frequently forced to fly in bad weather over poorly mapped country remote from inhabited areas. The Fitter, for his part, is just as important as the pilot... The crew of the Norseman tackled their duties with enthusiasm. They worked very hard, used their heads, flew continuously, slept in queer places, ate doubtful food and were never known to lose their highly developed sense of humour...* (F/L S.S. Blanchard commanding No.3 Detachment, photographing drought areas around Medicine Hat, Lethbridge, Calgary and Edmonton, May 15 to 30, 1938 and August 16 to October 2, 1938, around Lake Athabaska, May 30 to June 13, 1938, and around Yellowknife, June 11 to August 16, 1938.]

Not all RCAF photography was civil. With the approach of war, the force began looking for new airfield sites. Thus, in September 1937 No.5 (FB) Squadron, Dartmouth, received a K-3/5 camera to photograph potential sites around Sydney, Yarmouth and Dartmouth. Two years later the situation was more urgent. On May 31, 1939, AFHQ wrote to Eastern Air Command Headquarters:

*In considering defence requirements of the East Coast, it is essential that we have accurate information respecting the Labrador coastline.*

The RCAF kept tabs on the latest products. Sometimes industry offered off-the-shelf imports, such as the Bellanca, sometimes, new types like the Vedette. These advertisements from 1928-29 editions of *Canadian Aviation* and *Canadian Air Review* give a sampling of the hot products of the day. (David Thompson Col.)

*Existing Admiralty Charts are admitted to be very sketchy, and in many instances the information ... is unreliable... little or no indication is given in existing maps and charts of what lies a few miles from the coastline. It is proposed to make a detailed reconnaissance this summer in an attempt to discover localities which would be of use to an enemy desirous of establishing operational or temporary bases from which submarines or aircraft could operate against trade routes converging on the Gulf of St. Lawrence and Atlantic Seaboard...*

Operation Labrador, commanded by W/C G.E. Brookes, was done with Deltas 676 (FSgt R.I. Thomas, pilot, Cpl C.H.C. Hoseason, engine mechanic, LAC J.R. Fraser, rigger), 677 (Sgt R.A.W. Gilmour, pilot, Cpl N.E. Harvey, engine mechanic, Cpl J.F. Schultz, rigger) and Norseman 678 (Sgt S.D. Turner, LAC D. Cameron). Cpl J.W. Newbrigging was base wireless operator. The Operation Order included a secret appendix, for Brookes' eyes only, stressing the military nature of the task. The party arrived at St. Mary Harbour on July 20, where it contacted HMS *Franklin*, a Royal Navy survey ship on local duty since 1933. Photography commenced on the 23rd. The Norseman had engine trouble, so a Delta was diverted for reconnaissance and transport. "Labrador" involved 45:45 hours on photography, 42:00 on reconnaissance, 98:10 on transportation. Brookes took a Delta as far north as Port Burwell. All work was concluded by August 9. The expedition was ordered back to Ottawa on the 13th—mobilization for war was imminent. The operation had been aided by civilians including Moravian missionaries. There was no suggestion that anyone's loyalties were suspect, but Brookes' report contained a paragraph indicating a hitherto unappreciated German interest in the region: "Moravian Missionary at Nain, Mr. Hettasche, informed OC Labrador Flight that he sends half yearly reports to Hambourg Germany and has been sending these records for many years. He also stated that he has sent similar reports to Toronto and to St. John, Newfoundland... It should be noted that Mr. Hettasche has the following instruments only: barometer, thermometer maximum and minimum, rain gauge."

Such data may have been used by the *Kriegsmarine* in 1943, when a U-Boat installed a robot weather reporting station on the Labrador coast. Nevertheless, Brookes concluded that there was only slight danger of infiltration: "It is submitted that a power unfriendly to the Newfoundland Government might establish a base or bases on the Newfoundland Labrador coast ... establishment of such bases would be very liable to detection unless skillfully cloaked under the guise of lumbering, whaling operations, or trading posts." If anything, nature's defences would prove more formidable—Brookes noted: "Mosquitoes and black flies apparently were at their worst during the period under report and the use of head nets was found to be essential during calm weather and during the evenings. Mosquito oil is absolutely necessary ... when head nets cannot be used."

Viking 'EU over its natural habitat—the waters and islands of Rainy Lake in Northwestern Ontario. From his cockpit on many such flights Francis Vernon Heakes dreamed his dreams and composed his poetry. (RCAF/Halliday Col.)

F/L (later A/V/M) George E. Brookes headed Operation Labrador. In this 1920s snap he wears typical flying gear. Such casual-looking get-up was scorned by some officers, whose sense of decorum exceeded their appreciation of daily life in the bush. Note the rubber life belt which the wearer had to inflate by puffing into a valve. (CF RE18223)

## Forestry and Fire Patrols

*Over lake and pine-clad forest,*
*Over river flashing by,*
*Sails the white-winged forest watcher*
*Softly in a cloud-swept sky.*

*Softly drones the distant engine,*
*Over virgin timberland,*
*Speaking peace unto the forest*
*Where the mighty giants stand.*

*Onward then o'er tracts scarce charted,*
*Over cataracts asweep,*
*Over mountain, plain and valley*
*Over glades that lie in sleep.*

*Far into the western twilight,*
*Flashing wings against the sun,*
*Hums the softening song of engine,*
*Throbbing until day is done.*

"The Forest Watcher" by F/L F.V. Heakes, RCAF
*Canadian Aviation*, July 1928, p.42

The Air Board, CAF and RCAF long were active in forestry protection. In part this was due to interest in promoting aviation, a concern shared with private companies. However, there also was a constitutional angle. The provinces united by the British North America Act in 1867 had retained control over natural resources, but the prairie provinces, carved from lands purchased from the Hudson's Bay Company, had no such powers. Their natural resources came under Ottawa, hence the 10-year presence of federal aircraft over those forests. In 1930, however, a constitutional amendment transferred them to the provinces, which, thereafter, continued aerial patrols using either commercial or provincial air services. Federal responsibilities were limited to patrols over national parks.

Experimental forestry patrols were carried out by the Air Board late in 1920 using High River and Roberval. The costs of the latter were shared with Quebec, which contributed $20,000 and assisted in putting up permanent buildings. A detachment also was established at Haileybury, Ontario to survey areas affected by insect infestations. This was completed over 14 days. Roberval and Haileybury used HS-2Ls and were concerned only with surveying. Not until 1921 were air- (fire patrol), High River (fire patrols and some reconnaissance in Jasper Park), Victoria Beach, (fire patrols between Lake Winnipeg and the Ontario border and over Manitoba's largest lakes), Sioux Lookout (mainly surveys) and Roberval. Following Quebec's lead of 1920, BC and Ontario contributed towards such operations.

In order to extend its surveys, the Air Board established the Northern Ontario Mobile Unit at Sioux Lookout under F/L H.S. Quigley. The unit

HS-2L 'AH at Sioux Lookout in 1921. While some of the locals look on, the crew tinkers with the plane. Then, an HS-2L engine change at Whitney House in 1922. (Hutt/Molson Col.)

planes used as part of a general scheme of fire detection. The experimental work of 1920 was greatly appreciated by the Forestry Branch, particularly following the Haileybury job in an area where little was known of the extent and nature of timber resources. As early as December 1920 plans were made in the Department of the Interior for surveys in northern Ontario astride the Canadian Transcontinental Railway. Thus, in 1921 operations were conducted from Vancouver (fire patrols, photography, survey), Kamloops was a rolling air station. Two HS-2Ls and an F.3 flew from any large lake along the railway, while stores, crew quarters and a photographic darkroom followed in rail cars. Patrols were conducted from early June into October using landing sites at Sioux Lookout, Banning, Minaki and Allanwater. The cumbersome F.3 proved inefficient and was replaced by a third HS-2L. Three incidents marked this operation. On June 6, 1921 F/L W.G. Boyd was on a long patrol when he ran out of gas. Forced down on Davies Lake, he was soon missed, but an air search that evening found nothing. Late on the 6th a telegram arrived at Sioux Lookout reporting his whereabouts. Next day Quigley landed as close as possible to the downed machine and portaged gasoline to the site. Boyd was refuelled, but because his HS-2L was in a confined part of the lake, it had to be lightened. Kit and rations were portaged back to Quigley's machine, then Boyd got off.

In mid-June F/L Albert Carter took over the unit, Quigley having taken a job with Price Brothers of Chicoutimi. Plans were to establish the Minaki detachment early in July, but this was postponed when Ontario sought help with a forest fire near Sioux Lookout. July 1-3 five flights transported rangers and equipment. The Air Board was thus associated with the conflagration from detection to extinction. Later that month similar assistance was given near Eileen Lake. The third incident involved an HS-2L force-landed in Lake Minnitaki on August 27. It was beached and had a new engine installed. It was not back at base until September 4. Meanwhile, there were difficulties in Alberta, where several de Havilland aircraft were wrecked landing and taking off; Capt W.E. Shields was killed on August 1, 1921 (G-CYBV), apparently after striking a hillock on takeoff. F/L Carter, who had served there before and after his Ontario assignment, wrote of one of the hazards to D.H.4 pilots in this region—"bumps", or what today might be called an updraft, downdraft or microburst: "I have found that the most vicious bumps have been experienced at a considerable distance east of the mountains, particularly when following the CPR line from Carstairs to High River. However, on one occasion whilst turning to the right over the Red Deer River at the Gap, I got a bump which threw my machine into a left vertical side slip and I lost 800 feet ... before regaining control." High River completed 201 patrols in 1921; 57 fires were detected, while several photo and publicity flights were made. Operations were scrutinized over the winter. The Forestry Branch was anxious to extend coverage in BC, Manitoba and Alberta. By now no one doubted the value of airplanes in forestry. Indeed, a Manitoba forester was outspoken about the economy of air patrols, compared to the old system of ranging by canoes. The latter, he declared, did not justify the expense of paddles which the rangers broke.

The Forestry Branch remained concerned with technical questions involved in timber surveys and patrols. Alberta District Forestry inspector C.H. Morse had some rangers take elementary wireless training to facilitate experiments in air-to-ground communications. Director of Forestry R.H. Campbell was anxious to determine the optimum altitudes for photography of wooded areas and to compare the relative merits of vertical and oblique photography regarding the nature and extent of timber. All this meant even more serious work in 1922, when the Air Board flew several photo surveys over Alberta. Operations were similar to 1921, although the Mobile Unit was replaced by fixed bases at Parry Sound and

Whitney. Ontario paid the men's wages, the cost of consumables, interest at 5% on the equipment used, and depreciation on ground and air equipment. F/L C.M. McEwen, a veteran of Haileybury and the Mobile Unit, was mainly in charge of these patrols. When he was absent, e.g. ferrying aircraft, F/L G.E. Brookes took over. P/O T.A. Lawrence was at Parry Sound much of the time; other pilots involved were P.F. Townley and R.M. Smith. Operations began on May 23, continuing to October 4, during which 242 patrols (575:45 hours) were made. Usually 3 pilots were on duty, plus 2 supply clerks, 7 mechanics and 1 labourer. An Avro 504K and several HS-2Ls proved handy on smaller lakes, although the latter, showing their age, suffered engine failures, leaky hulls and rot. F/L Brookes' log for 1922 HS-2L flights gives some interesting literary snapshots of service life at the time, including an informal—even undisciplined—approach to air force life:

*(June 24, G-CYED; Brookes, Watson, O'Gorman, 1:15 hours flown from Cedar Lake) Commenced fire patrol but had to land at Pembroke on account of fearfully hot day and engine overheating. Weather did not improve and we had to stay the night. (June 25, 'ED, same crew, 1:25 hours) Returning to Whitney from Pembroke, landing at Golden Lake for fuel. Also landed at Victoria Lake on account of low clouds and stayed there for an hour or so. (June 27, 'AG, same crew, 3:40 hours, from Parry Sound) Fire patrol. Landed at Dorset and left McEwen and then landed at Wawa for luncheon and fuel. Held up by bad weather until 1650 hours and then toddled along to Skeleton Lake where we again forced to land on account of heavy rain and low clouds. Finally reached Parry Sound by 2015. (June 30, 'DT, Brookes and Watson, 3:30 hours flown. Parry Sound) Fire patrol. Had to do patrol without observer because Courtrage left us after we ran EJ on dead head, and I did not feel like hunting all over town for him. Landed at Wawa for fuel. (July 29, 'ED, Brookes, Watson, Courtrage; 2:35 hours, from Whitney) Fire patrol. Covered western and northern area of patrol and had forced landing at Cedar Lake, owing to valve breaking off and falling into cylinder. Had spares and McConnell brought up by Grandy within four hours of breakdown. Stayed overnight with the Bargers. McConnell, O'Gorman and I fired up engine and had her running by 1900 hours. (September 4, 'DT, Brookes and Delahey, 1:50 hours, from Whitney) Cross-country to Parry Sound to relieve Lawrence. Darkness settling very quickly, catching us at Lake Rosseau, but we went in and landed by moonlight.*

These operations were appreciated by the local rangers and provincial authorities, who expressed their gratitude and interest. The district forester in the Algonquin area reported:

*In no case did the weather hinder flying operations during the season to any extent, except during short periods when heavy, low clouds and low visibility rendered flying ineffectual... During such periods the accompanying precipitation or high humidity considerably lessens the fire hazard, and in no case during the season did any fire fail to be reported within 24 hours of its inception...*

*The particularly outstanding feature of this department that cannot be too highly commended are its accuracy, its reliability and its efficiency. Anyone who can read a good map intelligently should have no trouble in determining the location of a fire when viewed from the air to within a few rods... The certainty with which the fire may be classified with regard to size and possibilities enables the observer to determine the number of men and the amount of equipment required to combat it... When considering its efficiency a point which should not be overlooked is the moral effect on the people who are responsible for possibly 50% of the fires. The sight of the plane each*

The 1926 summer photo detachment at Larder Lake, Ontario. Varuna 'GV and Vedette 'FS are awaiting their next patrols. (NAC PA20103)

*day is a constant reminder of the fact that the Government is spending vast sums of money in fire prevention and tends to make them more careful... It is noteworthy that during the latter part of September, when so many small fires were burning in the surrounding districts as a forerunner to the disastrous fire of October 4, this district was practically free from any fire.*

In Manitoba S/L B.D. Hobbs directed 37 men in forestry. There the F.3s proved awkward for mooring on small lakes and rivers; but the establishment of temporary bases at The Pas and Norway House succeeded in extending coverage. The idea of the detachment was to be adopted in subsequent operations. In 1923 Ontario and Quebec assumed the responsibility for forestry patrols in their respective jurisdictions. Ottawa now could concentrate on its western forests. BC, though having power over its forests, continued to rely on RCAF patrols until 1927, when contracts were let to commercial firms. In 1923 the CAF gave special attention to Manitoba, with operations from Victoria Beach, The Pas (Cormorant Lake) and Norway House. Compared to earlier years, where the commanding officer had also been in charge of a base, Headquarters was in Winnipeg, for closer co-operation with the District Forest Officer. Aircraft were: 1 H.16 and 4 HS-2Ls available immediately, 4 new Vikings to be delivered. The flying season would prove frustrating. An HS-2L was blown from its moorings and destroyed, leaving only one for each sub-station. The H.16 did not come until July and the Vikings were delayed—by September 16 only 3 had arrived. Worse, a number of experienced mechanics joined commercial firms, meaning that the level of maintenance fell. The Forestry Branch had priority in air operations, so work for other departments was curtailed.

Although forestry was reduced in 1924, High River flew 382 hours on fire patrols and 19:20 on forest surveys; Winnipeg logged 939:45 hours on fire patrols, 81 on related duties, including timber cruising and transportation of forestry personnel. In Alberta the western forestry reserves and Waterton Lake National Park were regularly patrolled, with occasional flights over Rocky Mountain Park. One-way radio enabled aircraft to contact High River, which was linked by telephone to the Forest Service. Sub-bases at Eckville and Pincher Creek widened coverage. In Manitoba 4 Vikings patrolled. The 1925 introduction of Avro 552s (named "Vipers" for their engines) eased the shortage of machines. On wheels in Alberta and floats in Manitoba, Vipers were easy to handle and service. As single-seaters they could patrol, leaving Vikings free for suppression, photography and transportation.

Although coverage gradually increased, difficulties remained. Radio communications in Manitoba were incomplete, leaving many fires unreported until patrols returned to base. Performance of the Vipers on floats was disap-

A 1923 overview of RCAF Station High River looking northwest. Big canvas Bessonneau hangars, part of the Imperial Gift, are foremost—a 4th was ruined in a hail storm. (Museum of the Highwood 977-080-011)

D.H.4s at High River. G-CYBV met a tragic end, crashing on the aerodrome on August 1, 1921, killing Capt W.E. Shields. (David Thompson Col.)

Many RCAF pilots trained in parachuting. Here C.M. Anderson is ready at High River for a practice jump. On May 7, 1926 F/L A. Carter set a British Empire parachute record jumping at High River from 20,000 feet from a D.H.4. (Arthur Fleming Col.)

A High River Avro 504N. (Arthur Fleming Col.)

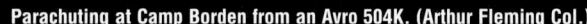
Parachuting at Camp Borden from an Avro 504K. (Arthur Fleming Col.)

In 1924 several Canadians took parachuting at Chanute Field, a US Army base in Illinois. Jumpers went aloft clinging to a wing. When ready, a man would deploy a small drogue 'chute (which would pull out his "main"), let go and be swept off the wing. Here F/O Arthur Fleming and F/L Carter are in position for takeoff atop a D.H.4. (Arthur Fleming Col.)

Keystone Puffer 'ZH at High River in 1927. Two Puffers had been purchased from Keystone of Bristol, Pennsylvania ($8295 each, less engine). They were used in experimental crop dusting. (H.S. and Gordon Diller Col.)

On September 4, 1926 pilot Jack Caldwell and engineer Irenee Vachon brought Viking 'EB to High River for storage. It remained until late in 1928, then was shipped to Vancouver for overhaul. In 1932 it again changed hands, the new owner having in mind an expedition north. On a September 16, 1932 test flight 'EB was lost. (H.S. and Gordon Diller Col.)

One of the Viper-powered Avro seaplanes that saw forestry service in Manitoba and BC from 1925. (CANAV Col.)

pointing, although the RCAF had adopted a more efficient American naval float over the cumbersome flat-bottomed British type. Forestry was not happy with the Vikings' small loads and suggested larger machines for suppression. Finally, although such bases as Cormorant Lake had been ideal, Victoria Beach was too exposed to the elements. CGAO Director W/C J.L. Gordon agreed—steps were taken to transfer its functions to Lac du Bonnet. However, he objected to a new design as too costly—perhaps $115,000 for development and $72,000 per machine.

As early as June 1925 a survey had been ordered as to using aircraft in Saskatchewan. The provincial District Forest Inspector favoured such patrols, but did not wish to prejudice Manitoba operations. He pointed out that the late spring breakup on northern Saskatchewan lakes would delay flying, but recommended that, if a new base was built, it should be at Ladder Lake. G/C Scott (Director, RCAF) was reluctant to push into Saskatchewan for fear of over-taxing existing service, although he estimated that patrols then being flown cost only half a cent per acre. The matter was settled at an interdepartmental meeting of February 3, 1926. It was pointed out that patrols in Manitoba could be reduced, as much of the timber was of doubtful value. To maintain the relatively low cost per acre, it seemed reasonable to extend into Saskatchewan. This was begun from Cormorant Lake, and in 1927 Ladder Lake was a full-fledged air station.

The same year saw the introduction of the Vedette and Varuna, the latter to replace the Viking. In Manitoba, Vedettes supplanted Vipers (which continued in Alberta), but W/C Gordon then advised the Director of Forestry that the Vipers be sent to Camp Borden for training. With a proposed replacement (Vickers Vigil) delayed in development, he suggested Keystone Puffers for patrol. The Director of Forestry replied that a system of stationary towers being built in the Rockies probably would replace most air patrols and touted the cheaper D.H. Moth. Gordon persisted with the Puffer, until he witnessed Ontario's successful use of Moths. These replaced Vipers in the RCAF in 1928.

The Varuna was the type of specialized aircraft that Gordon had opposed two years earlier. Its service career justified his doubts. It never attained the status of the Vedette or the Moth. On June 19, 1927 F/L R.S. Grandy, with W/C Stedman, a mechanic and a full load, thoroughly tested Varuna 'ZS. Grandy concluded that it was generally satisfactory, although some aerodynamic matters needed attention. The gravity fuel tanks projected above the top plane, spoiling lift on the centre section, while engine mounts presented a large flat surface to the wind. Grandy liked the water handling and concluded that performance would improve with modifications, e.g. metal propellers.

Two Canadian Vickers low-cost forestry designs were the aluminum-hulled Vista, and Vigil biplane. S/L Roy Grandy flew the 60-hp Vista in Montreal in October 1927, and the Vigil in March 1928. Neither went anywhere, especially since the D.H.60 Moth proved ideal for RCAF needs. (CANAV Col.)

G/C Scott evaluated a Varuna a week later. He agreed that it took off well, but climbed slowly (undesirable in forestry) and its top speed was only 72 mph (Grandy and Stedman had achieved 82 mph). Cleaning up the design would help, but it was clear that the Varuna was 700 pounds overweight. More evaluation followed in July, as the Varuna was being rushed into western service. There it was found that the hull needed strengthening with additional stringers. Since Vedettes needed the same mods, it seemed that the RCAF had under-estimated the operational stresses placed on their machines. Other mods were needed, e.g. shields to deflect water that splashed into cockpits. At Lac du Bonnet F/O Charles E. Kelly had several propeller pitch adjustments made to improve takeoff with a 900-pound load. He still found the run lengthy, even in ideal conditions. On July 11, 1928 he submitted a disturbing report on Varuna 'ZQ:

*I was given instructions to proceed to Little Grand Rapids ... and transport back to base two rangers with their kit and some fire fighting equipment. Three attempts were made to take off ... I found it impossible to get the aircraft on its step. The Hamilton air screws were then removed and replaced by two standard air screws. At 1420 hours of the same day and with the same load I made a take off and the aircraft appeared to develop the necessary speed fairly quickly... climbed very slowly, taking ... 35 minutes to 2000 feet. At no time on my way to Little Grand Rapids [160 miles] was I able to reach over 2100 feet.*

*Owing to this poor performance I did not feel justified in carrying more than two rangers and their kit, with which, pilot and mechanic and small equipment amounted to 900 lbs... On taking off at Little Grand Rapids I found climbing conditions better, due perhaps to a rainstorm which had just passed over, and the climb to 1000 feet was made in 12 minutes ... 2400 feet being the maximum height... flying had to be carried out at maximum RPM and a fair amount of engine trouble has been caused ... through this fact. Shortly before reaching Lac du Bonnet ... the oil pipe was broken and a considerable amount of oil was lost.*

Ottawa felt that experiments with propeller pitch had aggravated the Varuna's performance on this occasion. Nevertheless, as soon as the RCAF reduced forestry operations, the sluggish, overly-specialized Varuna faded, all but one of the eight built having been struck off strength by December 1930.

## How the System Worked

Forestry patrol, as it had evolved by 1927, was simple yet comprehensive. In Alberta, Vipers and Moths merely reported fires by radio or message bags. Rangers with no wireless communicated with pilots by laying out signal strips. In Manitoba and Saskatchewan, Vipers and Vedettes (usually two per base) patrolled and reported fires. Local foresters, weighing the

**The versatile Moth was a blessing to Canadian aviation. It served the RCAF (91 aircraft), commercial operators, flying clubs and private flyers. Here is one at the RCAF's St. Donat detachment in 1932; another, on skis, was at Cormorant Lake over the winter of 1929-30. Fairchild 'XT runs up beyond. (H.S. and Gordon Diller, Arthur Fleming Cols.)**

fire hazard, could recommend flights over particular areas, although the base commander (responsible for the aircraft) decided if a sortie was feasible. With small fires, a pilot might land and take suppression action. Most outbreaks, however, were handled by rangers. A Viking (later a Varuna) was held ready at each base to speed pumps and personnel to a blaze.

Hitherto, forestry air ops had been restricted to summer. Early in 1928, however, an attempt was made in Manitoba to locate fires using ski-equipped Moths. The district forester was enthusiastic about results, suggesting that Fokkers on skis be used to expand the program. Early in 1929 orders were for the detachments at Lac du Bonnet, Cormorant Lake, Norway House and Ladder Lake to start patrols as early as March, using spotter Moths and fire suppresser Fairchilds. The RCAF again won high marks: "Flight Lieutenant Mawdesley has been having a hectic time flying men to fight their fires. The RCAF have done wonderful work and have greatly enhanced their reputation... They were flying as late as May 11th, landing on ice well out in the lake, the ice adjoining the shore not being safe for the planes." Due largely to such spring patrols and the use of far-flung detachments, the two final years in which the RCAF performed this work were the most active. In 1929 some 5819 forestry hours were flown, in 1930—5316. This was also the peak year for aircraft deployed in summer: High River— 5 Moths (2 reserve); Grande Prairie—2 Moths; Lac du Bonnet—2 Moths, 1 Varuna; Norway House—1 Moth, 1 Vedette, 1 Varuna; Cormorant Lake—2 Moths, 1 Varuna; Ladder Lake—1 Moth, 1 Vedette, 1 Varuna; Berens River—1 Vedette; Thicket Portage— 1 Vedette; Lac la Ronge—2 Vedettes. The RCAF patrolled 60,385,820 acres and detected 368 fires. Some problems remained; in Manitoba and Saskatchewan radio communications between aircraft and base were so disappointing that wireless sets were carried on only a quarter of patrols. Even then 75% of messages were not received. Pigeons were more useful in reporting incidents like forced landings.

Aircraft were damaged or lost, when striking submerged logs and rocks. Smoke from fires could blot out the horizon, impeding navigation. A 1929 incident, while unusual for its drama, illustrates the dangers. Sgt J.M. Ready, a pilot at Lac du Bonnet, took off on the morning of August 26, 1929. He was to land at Gordon Lake, but in the haze became lost. He returned to Lac du Bonnet and began a descent through smoke. With no horizon for reference he let the nose drop. Ready heaved back on the controls, but the floats, and his own airspeed, had stabilized the aircraft and it refused to respond. At 500 feet, with his Moth at 125 mph, he baled out, leaving the aircraft to plunge into Lac du Bonnet. He landed in the lake, but his life preserver failed to inflate. He swam ashore, walked five miles to Davis Lodge, procured a motor boat, and returned to base.

Details of aerial camera installations: port side and belly cameras on Bellanca No.603 (May 1935); and the tri-camera set-up in Fairchild Super 71 No.665 (December 1936). (NAC PA63076, '181)

The busy printing room at the RCAF Photo Establishment, Rockcliffe. (NAC PA63220)

By the fall of 1929 operations near High River could be reduced, as the tower system in the Crowsnest and Bow River areas was almost complete. By then, too, the transfer of natural resources from Ottawa to the Prairie provinces was imminent. In the spring of 1930 the RCAF suggested that, regardless of the date of transfer, it should patrol through that season and that the CGAO Branch should continue on behalf of the provinces on contract. The cost was estimated at $577,000 for Manitoba, $207,000 for Saskatchewan. The provinces considered this excessive, so declined. In 1931 they contracted with commercial firms. The RCAF was not completely relieved of fire patrol, however, for there remained several western national parks needing protection. The Department of the Interior requested patrols. The DND replied that the RCAF had been sharply reduced, that no suitable base existed near the Rocky Mountain parks (High River having closed in 1930), but that assistance could be given in the cases of Riding Mountain and Prince Albert National Parks, provided that the Parks Branch pay. From 1931-36 the RCAF provided an airplane and crew for each park during the summer. In 1936 it advised the Department of the Interior that service training commitments made it impossible to continue, thus ending the RCAF's role as fire ranger.

## Vancouver: Snapshot of a Station

The affairs of each RCAF station appear in its daily diary, and related documents. Such papers are available in the National Archives of Canada, and at the DND Directorate of History. Vancouver Air Station (later known as RCAF Station Jericho Beach) is a case study for such early establishments. It opened in June 1920 under Maj Claire MacLaurin, DSC, Lt William Templeton and Earl L. MacLeod being his pilots. Five HS-2Ls and F.3s were designated (HS-2Ls G-CYBA, 'BB, 'EA, 'EB and 'DX soon appear in the diary pages along with F.3 'DI). Superintendent MacLaurin made his first test flight in September. The first service flight carried Dominion entomologist Mr. Hurle to inspect and photograph a mosquito swamp (film was processed and prints made at the station). The next flight was to Victoria on October 4 to show the HS-2L to the BC premier. Then came a demonstration at a lumberman's convention on the 9th. Air Commodore Tylee, and S/Ls Hobbs and Leckie were flown on the 20th and 21st in 'BA. Many technical snags were experienced in this period. November 5 to 15 'BA was as far inland as Kamloops, doing forestry. On December 13, 1920 the station diary notes that it was at Esquimalt "for coastal defence reconnaissance."

For 1921, 507:32 hours were flown from Vancouver. In January wireless tests and instructional flights were made. The first customs patrol operated on March 1, the first fisheries patrol on the 9th. On the 11th HMCSs *Aurora*, *Patriot* and *Patrician* were escorted into Vancouver harbour. On the 17th 'BB made its first flight. On April 27 machine gun practice was carried out from 'BA

Jericho Beach in 1921, not long after operations began. Two Bessonneau hangars are in use, four large flying boats are in view. (CF RE19539)

(the particulars of this as yet are unknown). In May a pigeon loft was built. Mr. Dradon of Victoria provided 50 pigeons. Alec Dickie, a former Army pigeon handler, was in charge. Vancouver soon began raising messenger pigeons for other stations. In his report of June 4, 1922 to the Air Board, MacLaurin talked about pigeons: "Our pigeons have been doing very well. We have given them a great many practice flights during our flying, and every one so far has been successful in returning to the loft in good time. In experimenting the other day with different wrappings for throwing them from the machine in flight, one pigeon failed to clear itself from the wrapping and dropped 1000 feet into the water. However, in some miraculous way it managed to shake free and returned to the station two days later." In reporting on February 21, 1922 MacLaurin described typical seasonal maintenance:

*At present I am concentrating my men as much as possible on the preparation of hulls and wings for the spring operations. I am putting two hulls at a time in the carpenters' workshop, turning them upside down and drying them out as thoroughly as possible. I have two stoves in this building which I keep going all the time, and keep the temperature up as high as conditions will permit.*

*I am stripping all the wings and shall dry these out thoroughly also before re-covering. I found that the wings (in fact, all the machines) have suffered greatly, owing to being exposed to the weather so long. All the woodwork is saturated with water, and signs of mildew are everywhere apparent. When the wings have been thoroughly dried and cleaned up, I shall make perfectly sure that the strength of materials has not been affected, before re-covering. From a careful examination of our equipment it is quite obvious that flying operations cannot be carried out indefinitely without proper hangar accommodations.*

Pigeons were important in RCAF communications. Here is the pigeon loft at Rockcliffe about 1931; then, the standard way of releasing a bird while in flight. (NAC PA62835, '830)

On May 16, 1922 Capt Roald Amunsden visited to consult with MacLaurin about using aircraft on a summer trip to the North Pole. On May 30 news came that E.G. Fullerton (a Canadian war veteran who had been in the headlines during Imperial Oil's 1920 Mackenzie River aerial expedition) would be one of Amunsden's pilots. In his weekly report of June 30, 1922 to the Air Board, J.L. Gordon of Vancouver Air Station discussed ways in which BC could make better use of CAF resources, citing 1921 operations in Ontario and Quebec. He suggested a BC forestry man be attached to the station and that sketching be more widely used: "This sketching [in Ontario and Quebec] includes topographic details of the country together with burnt and green areas, types of timber and a rough estimation of quantity. The sketches were also supplemented by both vertical and oblique photographs." On July 5, 1922 MacLaurin and Mr. Munn of the Provincial Forestry Department flew in 'EA with a pump and 2000 feet of hose to a fire in Hardwicke Island. In all such contracted work, the air station billed its client. For July 31 the station diarist recorded:

*The provincial government put in a hurry-up call for the big machine to take a firefighting crew, together with a pump and various other fire fighting equipment and provisions to a fire reported at Buttle Lake in the heart of Vancouver Island. Two men were sent out from the city office and the machine left immediately for Campbell River. Here four more men were taken on board and a large quantity of provisions. The total load amounted to 4895 pounds. Some difficulty was experienced in getting the machine off the water, as there was practically no wind at the time. However, this was eventually accomplished and the machine was successfully flown through the smoke over the extensive forest fire west of Campbell River and into Elk Valley. The party and equipment were unloaded and put ashore at the fire. The machine returned to Comox for more men and equipment. Another party with outfit were moved in the following day and the foreman in charge reported that he expected to have the fire out by Wednesday night, August 2nd.*

*This operation is of interest as the fire could not have been attacked by any means other than by air without a great deal of time and expense.*

Much effort was required to keep the fleet serviceable. A July memo from MacLaurin discussed engine oil, noting that Castrol was difficult to remove, once it splashed on any surface. On the other hand, he felt that Mobil oil was an excellent lubricant and easy to remove: "The keeping of machines clean and presentable

P/O Earl MacLeod, S/L A.E. Godfrey and P/O Tom Cowley at Jericho Beach in 1923. Godfrey recently had taken command, following the death of Maj MacLaurin. (CF PMR79-288)

is important and, when only a few men are available, the use of an oil which may be easily removed from different exposed surfaces is important." A July 23, 1922 note describes a typical job: "G-CYDX has been laid up for a few days for a general overhaul. The metal fittings on the lower planes have been corroded considerably due to the action of the salt water. These are being wire-brushed and repainted. The wings are being thoroughly washed off and will be given a new coat of varnish. The hull is being thoroughly cleansed and given a coat of paint. The engine will be tuned up generally. The work of overhauling and assembling G-CYEB is now complete. This machine is in very fine shape and will be tested and ready for service today."

Also in his July 1922 report to the Air Board, MacLaurin complained about the decrepit shape of F.3 'DI: "If there is any possibility of the machine being replaced with a more modern and efficient type for future operations I would certainly recommend that this hull be destroyed." (A note from March 23, 1923 suggests that this likely was done: "Machines and parts condemned by Board of Survey have been destroyed by burning and metal parts salvaged.") Tragedy visited the station on September 11, 1922, when 'EA crashed in English Bay. Superintendent MacLaurin drowned, mechanic Hartridge and J.R. Duncan (a civilian engineer) were injured, Duncan fatally. MacLaurin's funeral on the 14th included a ceremonial flypast of 'EB by Lt MacLeod. S/L A.E. Godfrey was named to command the station. On September 17 an F.3 carried a crew of 3 and 10 lumbermen from Harrington Lake to Vancouver. The station put on an air photo display in September at the Exhibition of International Photographers in Vancouver. On October 8, 1922 F/L MacLeod reported that he had only 'DX serviceable. 'DI was awaiting servicing for propeller vibration. The same day F/Ls Cowley and MacLeod carried 6 officials in 2 aircraft to study the white pine blister rust problem around Squamish. There was engine trouble, so half the inspectors returned to Vancouver by steamship, the others in the good HS-2L. At year's end the diarist boasted: "Several seizures were made by the Customs, which could not possibly have been made without the co-operation of the pilots and aircraft from the station. Many fires were located through the use of aircraft, and fire equipment, forestry officials and fire fighters were transferred to the scenes of the fires. During the year ... 244 hours 10 minutes flying were completed, most ... for the Customs and Forestry Departments... At the end of the year the mechanical staff numbered 11." On March 25, 1923 there were three flights "made in an effort to locate various whiskey runners for the Department of Customs." This covered all islands and inlets between Vancouver Island and the mainland from Victoria as far north as Qualicum Beach and the west coast of Vancouver Island from Victoria north, including Barclay Sound and the Alberni Canal. Customs officers were pleased with the work, "particularly the seizing of the launch Truscilla."

On June 30, 1923 Vancouver Air Station made its longest flight: Vancouver-Thurston Bay-Alert Bay-Thurston Bay-Vancouver—595 miles in 9:55 hours. The aircraft was HS-2L G-CYDX, carrying customs inspector Barton. On July 25 S/L Godfrey and F/L Cowley, on behalf of the Minister of Public Works, flew newspapers and a message to President Harding in Campbell River. The station's first flight to Prince Rupert (a fisheries patrol), via Alert Bay and Bella Bella, came on July 22 (7:45 hours, 478 miles). A visit by W/C Stedman in January 1924 taught station superintendent S/L Godfrey something new. To this time Viking 'EW, on strength at Vancouver, was thought to have a ceiling of 4000 feet. Stedman recommended some carburetor changes—on a test 'EW reached 8000 feet! In February 1924 Col E.L. Broome arrived at Vancouver to consult about a proposed British round-the-world flight. On the 23rd he was flown to Bellingham to arrange for caches in Alaska. A report about the pigeons from March 9 singled out one bird: "Bird AB #17 flew 930 miles in 1923... This bird flew from Bella Bella to Vancouver, 330 miles. Its longest previous flight was 130 miles."

In May 1924 'EW made two machine gun flights. The gunner was a Fisheries Department employee hoping to reduce the Hair seal population in the Fraser River estuary. The diary notes: "Owing to the number of fishing vessels in the vicinity and the cloudy condition ... the operation was not successful." On June 21 S/L Godfrey in 'EW escorted HMSs *Hood*, *Rodney* and *Adelaide* to Victoria. F/O C.J. Duncan took aerial photos of the fleet, and several RN officers had flights that squadron of His Imperial Japanese Majesty's naval fleet, accompanying them into the inner harbour." Later, S/L Tudhope and F/O Mercer visited the warship *Asuma* and met Admiral Saburo Hyatoutake. Japanese naval officers later visited Jericho Beach. In mid-March the station began its first flying boats training course with seven pilots. In April F/O Carter, MM made a demonstration parachute jump, leaping from an HS-2L at a mere 600 feet ("a good landing was made on the golf course"). The motorboat *Express*, a 40-footer with a 100-hp motor, was purchased to support seaplane training (re-designated "M.96"). On May 23 eight officers completed flying boat training and were posted to Winnipeg for summer ops. In mid-June two aircraft were despatched from Jericho Beach to Prince Rupert (sub-base operations were not carried out this year by the RCAF." The next entry was not made until September, a note of several men, station commander S/L Tudhope included, making a few parachute jumps. In November there is mention of photography over Victoria. Yearly flying was a paltry 203:15 hours.

In January 1927 three Avros arrived. Pilot training carried on, including the occasional civilian or other non-RCAF, such as F/L J.A. Sadler, RAF. On January 8 the diary notes: "Sgt Flewelling and LAC Tall injured by low flying of Moth G-CAHS piloted by E.C.W. Dobbin. Admitted to hospital." In February Avro 'GJ arrived to augment training (later that year it was shipped to Winnipeg). The first instructional flight was on February 14. Fairchild 'XN arrived

One of Gordon S. Williams' grand Jericho Beach scenes. While a Vancouver sits at the buoy, Vedette 803 has just returned from a flight (delivered in July 1929, it was wrecked on November 4, 1935).

week in 'EW. In August flights were made in co-operation with the RCN, unsuccessfully seeking a ship smuggling goods from the Orient. On September 10-11 there were flights to Esquimalt supporting coast artillery and naval firing. Aircraft wireless proved useful. In October HS-2L 'GA was received from Laurentian Air Services. In December 1924 S/L Tudhope took over the station. Flying hours for 1924 totalled 381:15.

A January 1925 entry in the station's handwritten diary notes: "A new HS-2L aircraft 'DU was received from Ottawa and work has been started to assemble same." By year's end 'GA, 'GM, 'GN, 'GQ, 'GR and 'GS had been added. In February F/L MacLeod and F/O Hull flew HS-2Ls to Port Atkinson, "to escort a Case Cove). In July pilots F/O Morfee and Lt McCullagh returned to Vancouver for the trials of several people caught fishing illegally. Fines were assessed from $5 to $25. The same month mechanical trouble in 'DU forced F/O Morfee down in the Queen Charlottes. 'DU was dismantled and sent by steamer to Vancouver. For 1925 Vancouver logged 474:46 hours.

For January-February 1926 it was back to training, some officers even taking an RCMP equitation course. In May the diary reports a sudden cutback: "Station personnel reduced to 2 officers and 9 other ranks... A number of NCOs and aircraftsmen not wishing to be transferred from Vancouver purchased their discharges." Others took posting to Camp Borden and Ottawa. The final note about this is a terse, "Fisheries in May for photography, being test flown on June 10. P/O Van Vliet took command of the station from S/L Tudhope on June 1, 1927. There were artillery co-operation flights at Victoria on July 2-3. The diary makes no mention of fisheries flying, but there was customs and forestry work. 1927 totalled 355:48 hours; 1928—714:10 hours. In January 1929 Vedettes 108, 109 and 110 arrived for assembly, and so began a busy season. Students included Barclay, Brookes, Costello, Dunlap, Ewart, Godwin, Gordon, Heslop, Hickson, Miscampbell, Ready and Spradbrow, all of whom would be known in Canada's aviation heritage. In April 6 Moths were taken on strength for training. The CP "Empresses" *Asia*, *France* and *Russia* were escorted for customs purposes this year.

Vancouver II No.904 over the still-incomplete Lion's Gate Bridge. The old flying boat has been modified for service duties—the nose and rear cockpits have Scarff rings to take light machine guns. Underwing bomb carriers also were fitted. (Bill Dunphy Col.)

hours. On February 9, 1931 F/O Gobeil had an accident in 123. In April and May seal patrols were flown. In July, Vancouver 'VS (later re-registered "904") arrived from Winnipeg with F/L Morfee and Sgt McNee. The Vancouver, developed as a solution to the Varuna crisis, first flew in April 1929. Trials showed that, in spite of its size, it was more manoeuvrable than a Vedette. The Vancouver would have a lengthy West Coast career. In July 1931 the Trans-Canada Air Pageant reached Vancouver. Festivities included the opening of the new airport at Sea Island; 2138:35 hours were flown for the year.

In the cutbacks of 1932 Vancouver again was reduced. Flying fell to 372:05 hours. At year's end seven Moths were shipped to Winnipeg. Hours in 1933 rose to 766:40. On September 19, 1934 two more Vancouvers arrived. A blizzard of January 20, 1935 led to the hangar roof collapsing, damaging Fairchilds 617, 'XN and 'XQ, and Vedette 123. Associated with a severe winter, there were flights to survey relief camps cut off by snow in lower BC, in some cases supplies being air-dropped. On November 4, 1935 Sgt N.E. Small and AC1 A.P. Whalen crashed in Vedette 803, but survived. Further details of Vancouver's early pre-WWII years are related in Chris Weicht's excellent book, *Jericho Beach and the West Coast Flying Boat Stations*.

For February 1930 mention is made of Vedette amphibians 115 and 123 arriving. A few days later M. E. Cooper of Canadian Vickers visited, probably in connection with the Vedettes. On June 17 Vedette 115 crashed with Sgt Fleming. For 1930 flying totalled 1738:40

That Jericho Beach thrived in WWII is evident in this June 1944 view. The old Bessonneau hangars were gone, replaced by 6 modern structures and the PBY was the dominant aircraft. Note how the city had encroached in the intervening 23 years. (RCAF via Chris Weicht)

## Training and the "Let's Pretend" Airforce

The CAF had formed in 1920 to give refresher training to veterans; but it had to renew its personnel through recruitment and instruction. *Ab initio* training for new men began in 1923, first with the Avro 504. The D.H. Moth was the next generation trainer, but the RCAF was constantly on the watch for other such types, the Avro Avian being one. In October and November 1929 several pilots, including F/O W.I. Riddell, F/L D.A. Harding and W/C G.M. Croil, assessed Avian 921 at Camp Borden. While it had well-balanced controls, pilots generally were unimpressed. The cockpits were too small (parachutes tended to snag on entering or exiting). Harding was blunt: "I do not recommend this light type of aircraft... construction is not heavy enough to warrant a prolonged and heavy instructional campaign, nor can it carry out strenuous aerobatics... An aeroplane is needed that will carry the average pupil through his elementary instruction into higher training and stand up under the heavy hand of an inexperienced pilot..." Nevertheless, the RCAF took 28 Avians from Ottawa Car. Avians also were used extensively by flying clubs (which operated with substantial RCAF assistance, including maintenance), but by 1931 the type required wing and cockpit modifications. In January 1932 Croil wrote about the Avian: "performance is very poor and the controls have a different feel from other aircraft, which makes them unsuitable ... maintenance requirements ... far in excess of what is necessary for other types of modern aircraft..." On October 22, 1932 the RCAF declared the Avian obsolete, designating its remaining examples suitable for "ground instructional purposes only and not to be flown".

## Entering the RCAF

Being a small force, the RCAF could set high qualifications for those aspiring to commissioned rank. Most pilots came up through the Provisional Pilot Officer (PPO) scheme - single men, aged 18 to 21, British subjects and enrolled either in Royal Military College or in science programs at Canadian universities. They received three summers of instruction before being commissioned. For 1929 a new pilot officer would receive $2100 plus a uniform allowance of $100, rations allowance of $200 and quarters allowance of $250 (per annum). Commercial pilots also were considered eligible, even without a degree, provided they had practical experience, a high school diploma, were British subjects and medically fit. Married applicants or those past age 25 were considered only with wartime experience. Scarcely anyone entered the RCAF by this route. Non-flying officers (e.g., stores administrators, technical officers) faced fewer hurdles; they had to be British subjects, age 20 to 30, have a high school education, and experience in their field. Upon commissioning, a non-flying pilot officer earned $1300 with rations and quarters allowances as per flying personnel.

The RCAF looked to another pool of flying talent—its own non-commissioned ranks. In 1926 it selected six promising airmen and gave them elementary flying training in the winter, when the PPOs were back in university. Nevertheless, NCO pilots were not given the quasi-combat courses, e.g. Army co-operation, but flew photo and fisheries patrols, thus freeing some commissioned pilots for more career-enhancing flying. When war came, these NCOs also received commissions and rose quickly. One such was Arthur Fleming. Born in December 1906, he enlisted in the RCAF as a labourer in March 1927. Posted to Winnipeg, he took courses in engine maintenance and photography and became a Varuna and Fairchild crewman. F/L F.J. Mawdesley reported in July 1929 that Fleming had flown 335 hours in two years—more than most pilots—and recommended him for pilot training. He was posted to Camp Borden on a course running from October 7, 1929 to March 28, 1930. With nearly 61 hours of instruction on the Avro 504N, Moth and Consolidated Courier, Fleming earned a ringing endorsement from W/C G.M. Croil: "Corporal Fleming has been a keen pupil and he has shown aptitude for flying. He is smart in appearance and very amenable to discipline. During his training he has had one crash whilst executing a forced landing on soft ground. His flying has been very satisfactory and steady improvement has been made."

Promoted to sergeant, Fleming went to Vancouver for a course on Vedettes and Moths, commencing April 12, 1930. He excelled in everything from navigation to signals to handling pigeons. On June 17, with eight hours solo on Vedettes, he took No.115 cross-country from Jericho Beach to Comox, where he alighted and refuelled. Returning, his engine quit between Texada Island and Blind Bay. He tried to glide for the shelter of islands in Blind Bay, but could not make it. At 500 feet he turned into wind to land, although eight-foot waves were running. He subsequently reported: "When about to make the landing I struck a large swell ... The machine turned over ... myself and mechanic [Cpl L.S. Thompson] were thrown into the water and submerged. We managed to get back on machine... After about ten minutes we were picked up by a Forestry boat, and we proceeded to tow the aircraft upside down to a small island in Blind Bay... I communicated with my Commanding Officer through the wireless equipment on Forestry boat."

No.115 was a write-off, although its hulk was towed to Jericho Beach for salvage. Injured and temporarily grounded, Fleming resumed his course in January 1931. His instructor, F/O G. Jenkins, reported: "Acting Sergeant Fleming is a good average seaplane and flying boat pilot. He appeared to be a little afraid of the water after his crash, but is regaining his confidence and should be OK with a little more practice." Fleming now was ready for CGAO work. He spent 1931 in Ottawa on transport (roughly 219 hours), and 1932 through 1935 at Dartmouth on transport and anti-smuggling patrols (915 hours). He returned to Ottawa at the end of each summer to help recondition aircraft and take courses in navigation, rigging and engines. W/C A.E. Godfrey routinely praised him as a pilot and fitter. Commanding a patrol detachment in 1933, 1934 and 1935, he showed administrative as well as flying skills. Promoted in June 1936, he spent that summer in Ottawa. When Governor General Lord Tweedsmuir undertook a tour of northern Canada in the summer of 1937, F/L Dave Harding and FSgt Arthur Fleming piloted the two Vice-Regal aircraft. Fleming received a letter of thanks from Lord Tweedsmuir and later that year received a Coronation Medal.

Early in 1938 Fleming studied photographic techniques, navigation and automatic pilots, before spending that summer and the next flying

The three types on which Fleming trained: Moth, Avro 504 and Courier. (David Thompson Col.)

Arthur Fleming (left) in a Varuna. His companion is the famous Mawdesley, then probably a flight lieutenant, but later a group captain. Then, the demise of Vedette 115. Although deemed a Category "B" mishap (repairable) it was cannibalized. (Arthur Fleming Col.)

survey Deltas. Senior officers recommended him for a commission, but only when war broke out did this occur. He spent most of the war with Eastern Air Command, rose to wing commander, and received the Air Force Cross. In January 1940 Fleming filed a document which listed his flying times. They show the life of a prewar RCAF NCO pilot engaged only in training and civil duties: Avro 504—20 hours, Moth—100, Vedette—50, Bellanca—215, Fairchild—1200, Delta—250, Bolingbroke—15.

Year to year training figures are revealing. Numbers for the 1935-36 fiscal year give some idea of the scope of such training. The most significant number was for trainees given initial instruction—53 (48 officers, 5 NCOs), of whom 41 graduated. Nine men took courses to train instructors, 11 received the RCAF Army co-operation course and 21 underwent instrument flying training; 13 NCOs took aerial wireless training. Nevertheless, the largest numbers were "Apprentices", whose trades were as mundane at they were vital—carpenters (32), fitters (29), armourers (13), ground based wireless operators (28), motor transport mechanics (8), fabric workers (7), motor boat crewmen (7), instrument makers (5), coppersmiths (2) and machinists (2). The same year 10 airmen were attached to various engine and aircraft manufacturers to learn details of their trades. Ten officers went to Britain for specialist courses, including one to the Imperial Defence College (London) and two to RAF Staff College (Andover). The balance attended such units as the RAF School of Army Co-operation (Old Sarum) and the Armament School (Eastchurch). It was still an Imperial heyday—the RCAF had no formal contacts with the US Army Air Corps. Amid all the statistics were a few others—these for courses given by the RCAF to Militia personnel—21 PF and 24 NPF officers were trained in RCAF methods and communications. Of these 24 took an Intelligence Liaison

Officers course. No matter what doctrines were evolving in Britain or elsewhere, the RCAF was still training essentially for an Army co-operation war similar to that of 1914-18, but it had no other model to follow.

Although the CGAO had been the RCAF's bread and butter, the force aspired to a semblance of being military. But there was never anything resembling a "war game", such as the RAF staged, with real bombers and fighters in mock campaigns. Anti-smuggling patrols might be deemed practice for future anti-submarine patrols, but little else seemed military. The chief difficulty was that there was no independent role for the RCAF in any of the rudimentary defence plans circulating in NDHQ, and the air force's natural partners in any war—the Militia and the RCN—were even more strapped. The navy came near to being disbanded altogether. There were, however, a few opportunities for the RCAF to practice military flying. There were always requests for aircraft to assist with artillery shoots and militia familiarization flights. Not all could be accommodated—aircraft might have forestry or fishery work, and there were often scarcities of officers trained in artillery co-operation. Some Militia taskings consisted merely of taking up officers to let them see a mock battlefield. On July 26, 1923 at St. Jean, Quebec, F/L George Owen Johnson took up 11 officers, each for 20 minutes. Similarly, on August 14, 1923 at High River, S/L G.M. Croil and F/L A.A. Leitch co-operated with "C" Battery, Royal Canadian Horse Artillery (4.5-inch howitzers, Sarcee Camp). The pilots subsequently were thanked, but a letter from Maj J. Crossley Stewart (RCA) to Croil suggests that people were losing their touch for the deadly trades practiced five years earlier: "It was certainly a most interesting and instructive shoot and I think, taking everything in consideration, such as the error of the gun and the rustiness of both the pilot and myself in this type of work, also the limited amount of ammunition available, that it was a most successful shoot and that the number of effectives obtained were above that as laid down in the probability table of the gun."

The RCN also benefited from air support. On October 12, 1923 a shoot was arranged for the Halifax Fortress guns, aiming at a moored target representing an 300-foot ship lying 7500 yards distant. An earlier trial had been unsuccessful "as an inexperienced pilot was flying the plane". This day HS-2L 'EL (Maj A.B. Shearer) with Lt O'Brien (Royal Canadian Corps of Signals) and Capt Bishop (Royal Canadian Artillery), directed the Sandwich Battery, the guns of which each had 16 rounds of 6-inch shell. Their guidebook was a pamphlet laying down coastal artillery procedures for 1918. The first round fell about 600 yards short and 3 degrees left. The 11th, 12th, 13th and 14th were hits, knocking off the superstructure and splitting the target—a tow boat had to return for a new target.

Another example of coastal artillery co-operation was on September 10, 1924. S/L A.E. Godfrey flew Viking 'ES to Victoria for a shoot with the 12-pounders at Fort Rodd Hill, firing at a target towed by HMCS *Armentieres* at 7000 yards. The shoot was directed through wireless. Godfrey returned the same day, but force-landed 15 miles from Vancouver, low on fuel. On alighting he found the gasoline feed pipe had cracked close to the carburetor, 25 gallons being lost. He effected repairs and continued to Vancouver. On September 11 he returned to Esquimalt and, from 5000 feet, directed by wireless a shoot for the 9.2-inch guns at Signal Hill. The Battery Commander considered the exercise a great success. A Militia Order of June 23, 1926 describes how other exercises were arranged and their objectives:

1. *Information*
   *The Royal Canadian Dragoons and the Royal Canadian Regiment are in camp at Niagara for annual training. The Royal Canadian Air Force have been requested by the Officer Commanding Military District No.2 to carry out Army Co-operation Practices with these units.*

2. *Intention*
   *To Co-operate.*

3. *Method of Execution*
   *This co-operation is to consist of close reconnaissance practices such as co-operation in attack, co-operation with an advance guard, concealment from the air, etc. and is to be carried out under instructions issued by District Officer Commanding, MD No.2. The Officer Commanding, No.1 FTS, Camp Borden, is to detail the following personnel and aircraft for the operation:*

*Personnel*
   *Flying Officer Coghill as pilot observer, one other pilot, one mechanic*

*Aircraft*
   *One Lynx Avro aeroplane equipped for radio telephony*

Accordingly, F/O F.S. Coghill and P/O D.H. MacCaul flew to Camp Niagara in Avro 504 'FY. Work on June 24 was handicapped by the lack of a radio telephone, and little could be done except take up officers to show them how troops appeared when deployed. On the 25th there was R/T (working poorly), so they resorted to signals by strips laid out by the troops and message drops from 'FY. The aircraft was damaged on landing late that day; the 26th through 28th involved repairs. 'FY was tested on June 28 and exercises resumed on the 29th and 30th (recces, message drops) until noon on the 30th, when the aircraft returned to Camp Borden.

Exercises had to be reconciled with peacetime conditions. S/L J.H. Tudhope, reporting on Esquimalt garrison artillery exercises in October-November 1925, described delays of up to two hours occasioned by shipping proceeding across the line of fire. Technology also complicated exercises; it was generally recognized that "wireless telephony" (i.e. radio voice communications) was less clear and shorter ranging than "wireless telegraphy" (i.e. Morse code). Laying out cloth strips in patterns was still an accepted means for troops to communicate with aircraft. This meant time wasted in circling back to the guns, and the strips were not always clearly visible. The 1925 exercises at Esquimalt also stretched resources at Jericho Beach, which was so hard-pressed with fisheries that artillery "schemes" had to be postponed from August to October. In 1927 Vancouver had difficulty again, this time because its wireless facilities had closed as an economy measure, much of its portable equipment being diverted to the Hudson Strait Expedition. The Militia was frustrated as requests for aircraft were turned down on various grounds. What made it worse was that officers genuinely respected the RCAF and what it could do. Major-General J.H. MacBrien (Chief of Staff and himself a pilot) was an enthusiastic booster. In June 1928 F/L A.H. Hull, Commanding Officer, Vancouver Air Station, recognized the problem of saying "no" to friendly superiors, when he urged that requests by the General Officer Commanding Military District No.11 should be met: "General McNaughton is, apparently, a great believer in the usefulness of aircraft in connection with Military and Naval manoeuvres and has mentioned ... several tentative dates that he hopes to ... have aircraft available for operating with both the Army and the Navy ... if aircraft could be spared for this kind of work the personnel carrying out the flying would benefit from the experience."

Until 1925 a few Sopwith Camels and SE.5as lingered at Camp Borden, flown periodically by veterans, who might fantasize about former adventures. Nevertheless, they played no part in *ab initio* training. For war training the CAF/RCAF had to make do with exercises in conjunction with Army and Navy units. The 1927 reorganization with its CGAO emphasis made it difficult even to maintain a representative military stance. Writing on September 10, 1927, G/C J.S. Scott declared:

*Due to reorganization, the RCAF is left very deficient of officer personnel. Every effort is being made to remedy this deficiency by appointing non-permanent officers to fill permanent force officer vacancies. Suitable types are most difficult to obtain, and when obtained, require considerable training, and even then do not come up to the standard required in the RCAF. If the RCAF continues to be depleted of its qualified officers, the whole Air Force policy and training programme will be impossible of attainment, and these Headquarters cannot be held responsible for the efficiency of training and the safety of personnel under instruction... Camp Borden is already reduced far below the minimum requirements of qualified officers.*

The RCAF was virtually unarmed until 1928, when it ordered 9 Siskin fighters and 6 Atlas army co-operation aircraft. The 1927 Atlas had resulted

The Armstrong Whitworth Atlas and Siskin were the RCAF's first true service aircraft (6 Atlases, 12 Siskins acquired 1926-29). In 1934 10 further Atlases were acquired (No.409 seen here). By then there were 28 service aircraft: 15 Atlases, 8 Siskins, 5 Vancouvers. 1935 brought more orders—6 Westland Wapiti bombers and 4 Blackburn Shark torpedo bombers. The deal for the Wapitis was a give-away by the RAF—$13,914. (CF HC7283)

from an RAF project to replace the Bristol Fighter. With its 450-hp Jaguar engine, top speed was about 140 mph. The nimbler Siskin, also with a Jaguar, could hit about 150. Both were obsolete by the time they reached the RCAF. Even so, grouped into specialist flights at Camp Borden, they were welcomed. No more RCAF warplanes would be ordered until 1934. A report submitted by F/L E.G. Fullerton on August 26, 1929 described exercises held earlier that month at Petawawa with Atlas 18 and Fairchild 30 (F/O Davey). The latter took officers aloft to familiarize them with aviation, demonstrate the importance of troop concealment, and observe artillery fire as directed by Fullerton. His report was one of the most detailed relating to army co-operation exercises, as a few excerpts, including one about message drop/pick-up, show:

*All Army Co-operation work called for by the Directing Staff was carried out with complete success and the Directing Staff comment very favourably upon the part played by the Air Force... The work included Close Reconnaissance, Artillery Reconnaissance, Message Picking Up, Message Dropping, and Passenger Carrying. The total flying time was: Atlas—10 hours 55 mins, Fairchild—13 hours 30 mins... The Army Co-operation aeroplane flew over the area where Battalion Headquarters was supposed to be located at ... 3000 feet, and then fired a green Verey Light (being a pre-arranged signal indicating that the aeroplane had a message to drop and requiring that the Battalion Headquarters in question should indicate the location of its Headquarters by displaying the requisite signal). As soon as the ground signal was displayed, the aeroplane dived down and threw out a message bag containing a fictitious message. A few minutes later the ground signal "M" was displayed indicating that a message was ready to be picked up. The hook ... was then lowered and the aeroplane swooped down and picked up the message.*

*The message was prepared by being suspended between two service rifles, the fixed bayonets of which were stuck in the ground about ten feet apart, in accordance with F.T.M. [Field Training Manual] Part IV for message picking up.*

Fullerton emphasized the need for high wing monoplanes for this work, given that the lower plane of a biplane interfered with observation; but almost any aircraft would do for exercises. Between June 28 and July 1, 1930, F/O R.C. Gordon (Vedette 109) and FSgt E.P.H. Wells (Vedette 110), based at Esquimalt, joined in manoeuvres around the Strait of Juan de Fuca and Shawnigan Lake (Nanaimo). Patrols were flown in conjunction with HMCS *Dauntless*. On June 29, 80 militia officers made reconnaissance flights over exercises in progress. On June 30 a Vedette co-operated with "attacking" forces and one with "defending forces". Gordon reported: "Wireless was installed in both aircraft and was used very successfully throughout the exercises... signalling by Aldis Lamp was also carried out from both aircraft." One sortie (0735-1035 hours, June 30, FSgt Wells, Vedette 109) gives the flavour: "Took off with Lt. Vincent as observer officer and Sgt J. McCulloch W/T operator. Co-operated with the British Columbia forces, observing movement of enemy force and position of artillery, contact being established with Headquarters by W/T and Aldis Lamp. Owing to temporary breakdown of W/T, message bags were twice dropped on Headquarters."

The log of W.W. Brown gives insights into Army Co-operation on a February 6-May 25, 1933 training session. He headed one page of his log "First RCAF Course—School of Army Co-operation". During this time he flew 34:15 hours as pilot, 15:10 as passenger on various exercises, mostly on Avro Tutors. The course began with "Pinpointing" (i.e. precision navigation and location of specific sites), followed by photography, puff shooting and numerous wireless exercises. Towards the end there were various recce flights, and three message pickups.

The formation of Auxiliary squadrons seemed to stimulate Army concern for the RCAF. On April 19, 1934 Brigadier W.G. Beeman, District Officer Commanding, Military District No.10 (Fort Osborne, Manitoba), wrote to AFHQ,: "I might say that a great deal of interest has been aroused in the local Non-Permanent Active Militia by the creation of No.12 (Army Co-operation) Squadron of the Non-Permanent Active Air Force, and a practical demonstration of some of the work performed by this type of squadron would prove of considerable instructional value to the Non-Permanent Forces and to the Permanent Forces in this District." RCAF squadron reports just before the war offer a picture of men striving to organize meaningful training with inadequate equipment (this was complicated in the spring of 1939 by a Royal Visit, which provided welcome pomp, but distorted flying schedules).

Air Commodore W.W. Brown, who did extensive photo work as a young RCAF pilot. (CF PL135067)

## Adapting the Atlas

The RCAF's new ground attack plane could not match training syllabi requirements and its effective ceiling was only 11,000 feet. Supply dropping apparatus did not fit existing bomb racks, and age and construction demanded excessive maintenance. The Atlas had such short range that even an Ottawa-Camp Borden flight had to be interrupted for refuelling. Aircraft were as apt to travel long distances by rail. In August 1929 Atlas No.16 was tested on British floats. Performance was satisfactory, but the RCAF noted that the floats needed padded nose bumpers for mooring, and water rudders to improve manoeuvrability. The manufacturer declared that its "up to date" Short Brothers floats had been designed especially for the Atlas.

Until 1935 the Atlas had no winter flying hood and its engine was difficult to run in the cold. The condition of the rear gunner was especially trying, yet the presence of a Scarff ring (i.e. machine gun mount) complicated the fitting of a hood. The situation was underlined on February 11, 1936 when W/C G.O. Johnson, Commanding Officer at Trenton, submitted a report: "Recently a flight of one hour and forty-five minutes was made in an Atlas aircraft with an atmospheric temperature of minus 20 degrees Centigrade at 3000 feet. On landing, the pilot and air gunner reported that ... they were ... unable to carry out army co-operation duties—writing, using the wireless and so on... it is strongly recommended that the matter of hoods and heat for Atlas cockpits be given early consideration." Some gear to improve the Atlas was procured. On February 25, 1936 NDHQ issued two Test and Development Orders, "Streamline" and "Comfort", by which Trenton would test a transparent hood, heaters and skis. Ski trials on Atlas 406 by F/L W.D. Van Vliet on February 6 and March 2 showed good results. All such efforts reconfirmed the understanding that a combat design rarely meets expectations.

## The Atlas in Service

In the summer of 1938 No. 2 (Army Co-operation) Squadron had Atlas detachments on Militia exercises at St. Jean, Camp Shilo, Petawawa and St. Catharines. A report of the summer's work described the activity. "Enemy" troops, weapons pits, barbed wire, etc. observed by the pilot were reported to the ground by radio. No.1 Sub-Detachment re-broadcast the pilot's messages on a public address system, so that officers could check troop positions and ascertain why camouflage may not be effective. Anti-aircraft aiming and ranging was conducted at he end of a recce sortie. Aircraft flew at 3000, 2000, 1500 and 500 feet in succession, acting as targets for troops using camera guns. Aircraft might dive to give troops practice in close deflection shots. As an aircraft changed height, the troops were informed by R/T, to assist their training in estimating altitude. Camera gun films were developed and distributed so units could appraise results.

In low flying attacks troops followed a specific route, taking them through various countryside—on a road through heavy woods, then on one lined with trees, across open country, etc. Atlases attacked, dropping flour bag "bombs", if troops did not take advantage of cover or keep dispersed. At St. Jean this type of attack was made against transport on the move, using camera guns. This training proved so useful that the Militia asked that more be scheduled. On August 20, 1938, with the Munich crisis looming, the Atlas detachments prepared for a move to Halifax. The aircraft assembled at Camp Borden for repairs, then flew to Ottawa on September 4. Seven reached Halifax on September 29, spares and mechanics following by train. In Halifax aircrew took lectures on coastal artillery shoots, several such drills being undertaken with the guns at Sandwich Point.

Atlas 408 demonstrates message pick-up at Rockcliffe in 1938. Note that in the snatching view, the undercarriage is damaged. This had happened earlier in the flight (the pilot was either unaware of the damage, or sporting enough to finish his demonstration in spite of it). Then, the aftermath of the landing. (NAC PA63299, '298)

The squadron soon returned to Ottawa, where it resumed training. Three new officers were initiated to night flying with the help of the Link Trainer, then No.2 practiced night flying. Crews detached to the School of Army Co-operation (Trenton) followed a rigid syllabus that owed more to WWI tactics than modern warfare. Exercises included forced landings, message pick-ups, camera gunnery, and photography; ground school training centred on signals, photography, airframe and engines. When Atlas 403 had engine failure during night flying on December 15, 1938, the crew parachuted. Although it seldom had more than four aircraft serviceable, No.2 Squadron's progress report for January 1939 proudly stated: "Service Training ... consisted of intensive practice in two and three plane formations. The pilots of the squadron ... have now attained proficiency in all types of formation work for Atlas aircraft." The relationship between Militia training and the work of an Army Co-operation Squadron was described in another progress report. Having moved to Petawawa on May 30, 1939, No.2 worked first with the Permanent Force artillery until June 24, when Non-Permanent Active Militia units (the reserves) began to arrive:

*Six aeroplane observation shoots were conducted during the Permanent Force Artillery Brigade Training. The first shoot was with the 60-pounder guns and difficulty was experienced in observing the rounds because of the extreme range of 9000 yards, the low ceiling (1700 feet) and the fact that shrapnel shell was being used. Subsequent shoots with the 4.5 howitzers and H.E. shell were successfully observed... One ... had to be conducted by ground observation from an elevated observation post, owing to low ceiling and heavy rain. Also the deep pools of water on the aerodrome after rain made it dangerous to attempt take-offs.*

The situation at Camp Petawawa was mirrored at other bases where No.2 operated that July. The field at Camp Shilo was described as "just large enough for the operation of Atlas air-

A Siskin Scout running up at High River in September 1926. Holding down the machine are LAC Osterbauer (far side), Cpl Davey, AC1 Elliot (in cockpit), Cpl Richards and Cpl Diller. This was one of a pair of Siskin IIIs—J7758 and J7759, the RCAF's first modern fighters—evaluated at High River. J7758 crashed at High River on June 28, 1927, killing P/O C.M. Anderson. J7759 lasted to 1935 as an instructional airframe at Camp Borden. (H.S. and Gordon Diller Col.)

craft". Camp Dundurn was worse: "too small and rough in its present condition for satisfactory operations..." The surface was so rough that undercarriages and airframes were stressed and radio tubes broken in hard bounces. In January 1939 No.3 (B) Squadron at Calgary was faring no better with the Westland Wapiti. Radio aerials could not bring in signals at greater than 40 miles; blueprints for improved aerials were "in the mail". A report by S/L A. Lewis, however, underlined how inadequately prepared the RCAF was for training, much less for defending the nation's coasts: "It is unfortunate that Night Flying Training cannot be proceeded with. The aerodrome at this Station, owing to its height above sea level (3460 feet) is not large enough to provide a margin of safety for Wapiti aircraft when landing and taking off from the flare path at night. It would be large enough if telephone wires and high power lines were taken down. It is considered that Night Flying Training is the most important part of a Bomber Pilot's training and should not be carried out spasmodically with long intervals in between, but should be proceeded with as a matter of routine at least once or twice a week throughout the year." Problems were alleviated when No.3 moved to Ottawa. Meanwhile, squadrons anxiously awaited new aircraft to make up for the lean years.

## A Fighter at Last: The Siskin

Canadian reliance on British military aircraft was understandable, if unfortunate. It meant that, in acquiring fighting machines, the RCAF ignored a range of good, relatively cheap US designs. Nevertheless, it was expected that in any future war the RCAF would operate closely with the RAF—familiarity with British aircraft, procedures and personnel was important. Even so, the choice of one type or another was not inevitable. S/L G.E. Brookes would have preferred another fighter to the Siskin. While on RAF exchange in 1927, he flew several types, including the Gloster Gamecock. Writing in his log on February 10, 1927, he was ecstatic: "The Gamecock is quite the nicest aircraft I have flown. Controls light and quick to respond, wonderful climb, slow glide under full control and a reasonable landing speed. Side slips ... beautifully and tail-down landings were OK each time on this flight. Am full out for Gamecocks." (The last of the RAF's all-wood fighters, Gamecock was a favourite of all who flew it. But the age of the wooden airplane was dying. The Siskin, with its metal-frame fuselage, was a better choice.)

The Director of the RCAF, G/C. J.S. Scott, was disappointed to learn that Siskins could not be fitted with floats. "The only suitable way for aircraft to travel in this country during the summer season is on floats", he wrote, although the type would surely have been useless as a fighter if so equipped. F/O W.A. Jones tested a Siskin ski undercarriage in 1933, while F/L B.G. Carr-Harris and S/L R.S. Grandy made similar trials in the winter of 1934-35. Carr-Harris carried out the most extensive ski trials in February 1936 with Siskin 60. Although some "drooping" persisted, the fighter could become airborne in 45 yards, and remained aerobatic.

Siskins 22 and 59 commemorated a historic event. On June 12, 1929 Mr. Benjamin F. Wilson of Toronto, wrote to Senator George Graham, suggesting that the 10th anniversary of Alcock and Brown's flight across the Atlantic be noted. "Ottawa has named a flying field for Lindbergh", he declared, "but no notice has been taken of those boys who had not the improved machine that Lindbergh had". Graham forwarded Wilson's letter to Prime Minister King. By June 27 it had reached W/C. L.S. Breadner, who suggested that two aircraft be named for the fliers. No.22 was painted on both sides of the cockpit with an inscription, "Captain Arthur Whitton Brown, First Trans-Atlantic Flight, June 15th, 1919"; No.59 similarly honoured Captain Sir John Alcock, DSC. On July 19, 1929, Brigadier A.H. Hill (District Officer Commanding, Military District No.2) visited Camp Borden. After a short

This grand scene from High River on July 13, 1927 shows the heaviest of RCAF trucks. The crate contained a Siskin, shipped from the UK aboard SS *Comino* via Camp Borden. From the left are LAC Stu Dearaway, Cpl A. Richards, FSgt R. McGrandle (below), then (atop the crate) Sgt C.M. Anderson, AC J. Telfer, AC Duncan, LAC McDonald, LAC A.V. Green, LAC S.N. Green, Cpl P.N. "Phil" Green, Cpl Diller and Sgt R. "Bob" Eadie. (H.S. and Gordon Diller Col.)

J7758 after its fiery demise. One of High River's Bessonneau hangars is beyond. (H.S. and Gordon Diller Col.)

A Siskin trio ("John Alcock" nearest) on August 24, 1929, location not recorded. With a 425-hp Armstrong Siddeley engine this model could top 150 mph. (CF HC3070)

Siskin 20 runs up at Rockcliffe. (CF HC2093)

This view of a Siskin cockpit illustrates well the spartan and confined nature of aircraft design during the 1930s (NAC 145985)

One of the Siskins dedicated in honour of the Alcock and Brown flight. (NAC PA62471)

Skis were developed for the RCAF's Siskins. These aircraft came to a sad end. On May 11, 1940, G/C S.G. Tackaberry noted in a memo that all Siskins were at technical training schools. He asked if one might be saved for the "National Research Museum". This was not done, apparently due to lack of storage space. A Jaguar IV engine was retained for some time, but disappeared during WWII. Canada's Siskins are remembered only in archival records, photos, paintings and models.
(RCAF RC1954/Jack McNulty Col.)

speech he broke a bottle of wine over each machine's propeller. That done, they put on a flight demonstration.

The Siskins were best known for their air displays, either singly or as a team, although they lacked carburetors for extended inverted flight. In October 1930 the RCAF attended an air meet at London, Ontario in force—9 machines and 19 airmen from Camp Borden, with a Siskin, a Fairchild, a Tomtit and six Moths. The star was the lone Siskin, piloted by F/L F.V. Beamish (RAF exchange pilot), who "sent the spectators into ecstasies with a display of aerobatics." A group appearance at the Montreal Air Pageant on September 6 and 7, 1930, combining airmanship and melodrama, thrilled those present (today, parts of their show would be criticized as politically incorrect):

*Perhaps outstanding in the many events and contests that took place during the two days was the performance of the Royal Canadian Air Force Siskin fighting planes from Camp Borden... piloted by Flight Lieutenant F.V. Beamish, leader, Flying Officer Gobeil and Flying Officer McNab, these speedy machines gave a brilliant display ... flying in a V, one on top of the other, and one after the other, so that sometimes it seemed they were almost touching. Inside loops, all precisely together, drew loud appreciation from the spectators ... An attraction on the Sunday afternoon was the mock storming of an Arab stronghold by the Siskins, aided by two of the Montreal Club Moths and a troop of Boy Scouts disguised as Arabs. Pilot Bernard Martin, in a club machine, made a supposed forced landing with engine trouble, whereupon the "Arabs" dashed out shooting wildly. Ere they could reach the stranded pilot, Miss Daphne Patterson, who had been flying in another club machine, picked him up, leaving the plane in the hands of the Arabs. Flying off to a supposed British air base, nearby, Miss Patterson soon returned with the three Siskins, which wreaked havoc with the village by dropping bombs which tore the stockade apart and finally set it on fire. Then the stolen machine was recovered...*

The log of Fowler M. Gobeil is part of a document collection in the Canadian War Museum. He joined the RCAF in 1927, receiving his wings in 1929. Assignment to the Siskin Flight in April 1930 was a dream come true. He first took instruction from F/L Beamish on dual control Siskin 63 (April 22, 1930). After 2:15 hours, he soloed on April 25 and did his first formation flying next day. Practice followed on every suitable day, with two or three sessions daily, sometimes going back to No.63 to try new manoeuvres (e.g. inverted flying) with Beamish. On August 18 the Siskins (Beamish, McNab and Gobeil) flew to Ottawa via Deseronto. On the next three days they performed before Central Canada Exhibition crowds. A newspaper account noted: "The final display by the three pilots left the large gathering in Lansdowne Park breathless, and it seemed that all three officers had gone out of their way to make their last show in Ottawa as spectacular ... as possible... Daring in the extreme was the verdict of the crowd ..."

The team performed at the CNE in Toronto on September 2, the Montreal Air Pageant on the 6th and 7th; Kitchener on the 20th. A final appearance was in Ottawa, before 5000 fans. Exhibition flights aside, Gobeil had experienced moments of excitement, such as his starboard tire blowing on takeoff, forcing him to land with the tire off the rim. His next sortie was on May 8, 1931, when he practiced in No.10 for 20 minutes. The team soon went to Montreal for work-ups prior to the Trans-Canada Air Pageant, which moved off on June 30. From July 1 to September 8 Gobeil performed at Hamilton, Windsor, Brandon, Grand Forks, Montana, Edmonton, North Battleford, Saskatoon, Quebec, Louisburg, Halifax and Toronto

Gobeil resumed Siskin flying on June 20, 1932—there was intense training to get the team into shape. On July 26, while practicing a loop in Siskin 23, he collided with F/L Henry W. Hewson, cutting his machine in half. Gobeil parachuted; Hewson died of injuries the next day. No blame was attached. Gobeil performed six times in Siskin 210 before Central Canada Exhibition crowds between August 22-26. He was team leader there, and at the Belleville Fair August 30-September 2. That concluded his career as an exhibition pilot. He rarely flew a Siskin thereafter. His log gives some insight into the aerobatic team. The RCAF appears to have had plans to emulate a spectacular stunt that RAF pilots had performed at Hendon. On November 18, 1932, following a 15-minute flight on Siskin 205, Gobeil wrote, "Formation with aircraft fastened together with light string. OK". Although tried, the roped-together formation was not executed before a Canadian audience. With the coming of war in 1939, Gobeil distinguished himself in a different kind of flying—fighting the Luftwaffe over France. There he made the first RCAF claims of the war—two damaged enemy aircraft. He retired as W/C Gobeil, AFC in 1956.

## No.1 Squadron

No.1 Squadron formed at Trenton with Siskins under F/L B.G. Carr-Harris in 1937. By then the Siskin was hopelessly obsolete. With a top speed of 156 mph, it barely exceeded the performance of the best WWI fighters. It carried two .303 Vickers guns, while modern European fighters like the Hurricane had much heavier armament. Nonetheless, the RCAF now could begin training fighter pilots and developing tactics, while awaiting new equipment. In August 1938 No.1, under S/L E.G. Fullerton, moved from

Gordon S. Williams Hurricane photos from Sea Island in early 1939. Based in Seattle, Gordon was an aviation fan since boyhood. At the hint of some new "action" in Vancouver he would cover it from every angle. Meanwhile, only a few RCAF photos might be taken of the occasion. Gordon kept up a lifelong correspondence with photography buddies in the US like Jim Larsen, Howard Levy, Hal G. Martin and Pete Bowers; and in Canada with Jack McNulty and Al Martin. This elite group swapped negatives. Thank goodness, for archives and private collections today are rich with top notch photos that often were missed by air force photographers (who may have little interest in their work), or later were destroyed according to regulations. Gordon spent his working years in sales and public relations at Boeing.

Hurricane 312 pranged at Sea Island, taking a wing off Ford Trimotor CF-BEP (ex-RCAF G-CYWZ). (NAC PA501521, K.M. Molson Col.)

Hurricanes arrived, but there was disappointment when their propellers were found missing (they turned up). F/L Ernie McNab now joined No.1, starting on the Delta on May 12. In 1999 Ralph C. Davis recalled his time. He mentioned how S/L Fullerton, a wartime veteran, used to rig himself upside down in a chair. Thus suspended, he felt that he was building tolerance to inverted flight (which probably was correct):

*Although I was never qualified as an instructor, I thought that a short briefing for the pilots about the Delta would be in order; but the CO was anxious to get the ordeal over. He took the co-pilot's seat, while the rank pecking order determined who got the two cabin seats. The other three pilots sat on the floor. Soon the Delta was airborne, carrying all of Canada's air defence pilots! I demonstrated a landing, then the squadron leader exchanged seats with me, took off, made a short circuit and approached to land. On touch-down he almost put the Delta on its nose. Fortunately, the pilots in the rear provided sufficient ballast to keep the tail down. The cabin filled with smoke from burnt rubber— the squadron leader condescendingly agreed that it would be better if he did not land with the brakes fully on.*

Trenton to Calgary municipal airport. Six Siskins were forwarded by rail. The squadron diary gives a sense of how slowly things happened. For September 12 the first Siskin (No.303) was "assembled, awaiting petrol and oil", No.302 was being uncrated. For the 15th: "90 gals gas and 7½ gals. of oil received from Imperial Oil... Siskin No.303 test flown by Sqn. Ldr. Fullerton." On the 24th Fullerton, F/O C.J. Fee and Sgt R.L. Davis flew formation. At the end of October No.3 Squadron, with its rickety Wapitis, joined No.1. On January 4 R.B. Bennett, former Prime Minister of Canada and Member of Parliament for Calgary West, toured No.1. On January 26-27, 1939 there was a cross-country exercise to Lethbridge.

On February 16, 1939 Fullerton and 20 other ranks reached Vancouver to accept the first RCAF Hurricanes: "Personnel arrived Vancouver and secured accommodation at the Eburne Hotel ... Arrangements were made for tools for the uncrating ... at Sea Island Airport." Hurricane 312 was uncrated on the 18th, Fullerton flying it on the 25th. Ground and air tests proceeded, but the project was set back on March 2, when Sgt R.L. Davis pranged 312 on take off: "The aircraft was totally destroyed by fire and a wing from a Ford Trimotor torn off." Davis' prang usually is described as pilot error, but this seems unfair. Nobody on No.1 Squadron was experienced on high-performance fighters. The jump in horse power alone was from the Siskin's 450 to the Hurricane's 1030. As he roared down the runway, Davis must have felt that he was on a bucking bronco. The accident opened a few eyes in Ottawa. On March 17 Sgt Ralph C. Davis arrived from Lethbridge with dual-control Delta 675, to be used at No.1 as an advanced trainer. This was a brilliant idea, if a bit late. If No.1 had had the all-metal, low wing monoplane Delta earlier, Ralph L. Davis might not have become the first in Canada to crash a Hurricane.

Much Delta flying ensued, pilots being F/Os Desloges, T.G. Fraser, K.A. Gordon and E.M. Reyno, and Sgts Briese and R.L. Davis. Briese had dual on the Delta on March 29 and 30, and April 11, soloed on the 12th, then flew the Hurricane next day. On the 14th five more crated

So much for the RCAF's original "fighter OTU"—a ropy operation that Davis could have lived without. However, he lived in hope that a flight in a Hurricane might be his reward. "The Squadron Leader did not buy that one," he reported 60 years later. On May 26 No.1 Squadron paraded in downtown Calgary, during the Royal Visit by King George VI and Queen Elizabeth. On the 29th three Hurricanes escorted the Royals, who were sailing from Vancouver to Victoria. On June 1 Fullerton delivered the first Hurricane (No.316) Vancouver-Calgary. Meanwhile, five Siskins went to High River (likely for storage) on June 3 with S/L Lewis, F/O Annis, WO2 Dave Ceiffets and Sgts Michalski and K. Birchall. For June 8, 1938 there is a sad note: "F/O T.G. Fraser killed en route to Calgary in an accident near Mission, B.C. with Hurricane 317." Three days earlier the undercarriage of No.317 had collapsed at Vancouver, while Fraser was landing.

On July 9-11 there was an inspection by G/C G.O. Johnson. On the 15th Delta 675 left for Trenton. For July 17 there is a note, "Sgt Davis, R.L. taken off flying duties and remustered from pilot to aero engine mechanic." Was this punishment for his prang, or had he requested this? On July 29 Air Marshal W.A. Bishop inspected the Calgary squadrons and watched S/L Fullerton put a Hurricane through its paces. On August 14 all personnel "living out" were ordered into barracks. For the 25th the diary notes, "All leave and passes cancelled... 7 Wapitis No.3 Bomber to East Coast." Pilots from No.1 (F/Os Gordon and Reyno, Sgt

**Delta 675 in which Sgt Davis instructed No.1's pilots. Gordon S. Williams photographed it at Vancouver in May 1939.**

Hurricane 315 at Rockcliffe in September 1939. WWII was only days old, so the RCAF had to learn much overnight about servicing and maintenance, to say nothing of safe flying and tactics regarding its new fighter. (NAC PA63512)

Briese, etc.), meanwhile, were rushing the first RCAF Harvards from Vancouver to Camp Borden.

On August 31 No.1 Squadron started the move to St. Hubert, the OC and Sgt Briese departing with Hurricanes 311 and 316. The declaration of war is mentioned in the diary for September 3, then there is an odd note for the 7th that Siskins 304, 306 and 309 had been damaged in a "Shooting fracas" at High River. This turned out to have been a prank involving Militia guards. By the 10th No.1 Squadron was at St. Hubert. On the 12th Siskin 309 is noted as returned to Calgary from High River "for instructional purposes". On September 11 No.1 was "placed on war basis and mobilization was commenced." Machine guns and oxygen equipment were installed and tested, and formation and cloud flying began. Delta 675 returned to the squadron. By September 28 strength was 5 officers and 72 airmen. Mr. Dobson of Canadian Car and Foundry, who wished to look over the Hurricane (which CCF soon would build) visited on September 28.

On October 29 nineteen airmen left for Rimouski, Moncton and Dartmouth, the advance party to service Hurricanes ferrying to the East Coast. On the 30th F/Os Gordon R. McGregor and D.R. Anderson reported for duty. On November 1 S/L E.A. McNab took over command. On the 3rd and 4th, seven Hurricanes reached Dartmouth where, by the 9th, No.1 was operational. On the 20th Anderson died in the crash of No.329. An "old man" at age 38, he was buried next day. On November 29 W/C Roy Grandy, a former WWI pilot, is noted as making his first Hurricane flight. Meanwhile, the pilots were busy training. F/O Reyno logged a typical sortie on November 20: "Diving attacks on Naval vessels in Bedford Basin for anti-aircraft defence practice." There were convoy patrols and formation practice, and all were having a go at reaching 25,000 feet with full fuel and armament. F/O John W. Kerwin is noted in the ORB of March 14 as "Crashed" in Hurricane 323. He was aloft only five minutes. Kerwin survived, fought and was shot down with No.1 in the Battle of Britain, then died ferrying a Kittyhawk in the Aleutians on July 16, 1942.

No.1 flew last in Canada on May 20, then prepared for the UK. Packing and crating began and, on the 28th, 8 officers and 86 men of 115 Squadron (Montreal) were taken on strength. On June 8 the squadron paraded in Halifax, then boarded *Steamship E.37—Duchess of Athol*, sailing off to war on June 11: "Squadron departed Halifax at 1000 hours for England under escort of 3 destroyers and 1 battleship." Stranraers accompanied the convoy until dusk.

## Trans-Canada Flying

Flying across Canada still was rare in the late 1930s, but in March 1939 Sgt Ralph C. Davis ferried Delta 675 from Rockcliffe to Vancouver. His trip illustrates the trickiness of such a venture:

*Airfields between Ottawa and Winnipeg were kept usable for wheel and ski aircraft in winter by rolling and packing the snow. This worked well in winter but, with spring break-up, a field could be used by ski planes only briefly. Thus, on March 9 we left Rockcliffe on skis to get as far as Winnipeg, where we changed to wheels. As no fuel consumption figures for the Delta (on skis) existed, I found myself approaching Winnipeg at dusk in a severe snow storm with my fuel gauge bouncing on the red "Empty". We passed over the city below the tops of some buildings. People must have wondered about the damn fool flying in those conditions.*

*Although we had a voice radio, it was unsatisfactory for radio range navigation, because a long trailing aerial had to be used. This confused the "A" and "N" aural signals when near the broadcasting station. In the weather I was unable to find the airfield, so this became my first experience with ground-controlled approach. The radio range operator offered to go out and listen for us. When he did, he gave the direction from his position on the hangar line. This did not work well at first, since our direction from the field kept changing. However, with the help of the range operator and my two crewmen, who were keeping a lookout, we spotted the field. I landed quickly. My crewman, Cpl Baxter, later reported that the gas used in refuelling was the same as the published maximum quantity for a Delta.*

*We reached Lethbridge via Regina, and waited for a good forecast from the local meteorological office... Trans-Canada Air Lines was gaining experience on this route, notorious for its winds, turbulence in cloud, and icing. Unlike today's pressurized jet flights high above the weather, our planes were unpressurized and without oxygen or de-icing equipment—we had to fly through the weather. As this was my first flight in or over mountains, it seemed sensible to await good weather. This we did—for three days. The flight was uneventful and we reached Vancouver on March 17.*

## No.3 Squadron

In June 1937 No.3 Squadron at Rockcliffe introduced the RCAF's first bomber—the Westland Wapiti, 24 of which had been acquired in 1936. Since metal monoplane bombers were available in the UK and US, one wonders how the RCAF fell for this fleet of worn-out "What-a-Pities". Nonetheless, the RCAF would learn about bomb-

Based on the D.H.9A, the Wapiti first flew in 1927. With a 475-hp Bristol Jupiter, it carried 500 pounds of bombs, had a machine gun firing forward from a fuselage side mount, and another on a Scarff ring in the rear cockpit. J.C. "Jack" Charleson, who watched the Wapitis being uncrated and assembled in Ottawa, claimed that they exuded the odour of camel dung. Aircraft 513 served on 10 (BR) Squadron, remaining in service into 1940. (CF HC8364-3)

No.3 Squadron Wapitis at Halifax airport on November 15, 1939. The war was raging in Europe, but Canada was still relying on such clunkers for home defence. (NAC PA136274)

ing with them. In October 1938 No.3 moved to Calgary. The odyssey began on the 18th. Leading the 4 Wapitis of "B" Flight was Ralph C. Davis, who recalled in 1993:

*The dismal shape of this ancient plane greatly restricted our flight. It had a cruising speed of 80-90 mph and an endurance of about three hours... we had to land on just about every airfield on the Trans-Canada Route—Emsdale, Porquis Junction, Kapuskasing, Nakina, Sioux Lookout, Winnipeg, Regina, Swift Current and Lethbridge. Our two flights moved separately, to avoid crowding at the refuelling sites. Gas, which we hand-pumped, was provided in 45-gallon drums. Since the aircraft did not have voice radios ... we communicated with hand signals. Thus we could plan only to fly in good weather. Thanks to significant west winds, we did not reach Currie Barrack, Calgary until October 24. All eight Wapitis arrived safely, floating in sedately to a small gathering of friends and well-wishers... For the majority of crews who had not crossed Canada before, the trip proved a great experience and a practical lesson in geography.*

No.3 trained at Calgary until August 26, 1939, when it began its final move. Seven Wapitis left that day on another trans-Canada epic: Nos.510, 543, 542, 512, 532, 509 and 535 with the crews of S/L A. Lewis/FSgt A.J. Bradford, F/O P.G. Baskerville/Sgt K.W. Walton, F/O C.L. Annis/Cpl M. Monson, Sgt W.J. Michalski/Sgt S.G. Cable, Sgt R.L. Davis/Sgt A.N. Roth, Sgt K. Birchall/Sgt R.F. Herbert and Sgt H.T.W. Blockley/Sgt J.H. Watts. They departed at 1345 hours, reaching Swift Current at 1900. Next day they made Regina, then Winnipeg. Soon the crews were getting frustrated, the diary lamenting: "On the subject of maintenance, the Wapiti ... has many shortcomings." Next day they left for Sioux Lookout at 0900, but weather forced a return. They tried again at 1230, reaching Sioux Lookout at 1545. Crews spent the night in sleeping bags, "as no other accommodation was available."

On August 29 the Wapitis continued, going first to Nagaming, thence, Kapuskasing to overnight after 4:15 hours. Next morning a two-hour flight was made to North Bay, then it was on to Ottawa, (4½ hours), where the aircraft were inspected. Nos.541, 538 and 544 joined No.3 at this point, crewed by P/Os Van Camp, Kenyon and Blackburn, with AC1s Simpson, Strickland and Welsh. Operating as two flights of five, No.3 departed on August 31, refuelling first at Megantic, Quebec. By late afternoon on September 1 all aircraft had reached Halifax; 2300 miles had been covered. Nos.3 and 2 Squadrons were under canvas at the civic airport, RCAF Station Dartmouth being under construction. Personnel were billeted in private homes. By this time Eastern Air Command could call upon 10 Wapitis, 4 Atlases and 7 Stranraers. No.3's role was specified in an EAC Op Order: "To be used as an Air Striking Force in co-operation with RCN, or independently against any enemy forces located within range." For flights out of sight of land, a Stranraer would accompany the Wapitis as navigation shepherd. Flying commenced on September 3, with No.3 logging 2:40 hours on five short flights.

In spite of a good start No.3 Squadron disbanded on September 5, No.2 on December 16. Most men on No.3 joined No.10 (BR) Squadron (Digbys, one of which flew No.10's first operation on June 17, 1940). On October 30 Digby No.747 destroyed U-520 with four depth charges. While the Atlases persevered in coastal artillery support, the Wapitis became static ground trainers. Unfortunately, the RCAF had no interest, let alone policy, regarding preserving historic aircraft, so no Wapitis, Siskins or Atlases were saved.

## West Coast Sentry: No.4 Squadron

Although Japan did not enter WWII until December 7, 1941, the world was wise to fear it—Japan had been brutalizing neighbours like Korea and Manchuria for years. With the help of nations like Canada and the US, which were eager to sell it iron ore, scrap metal and the latest technology, Japan maintained a huge military. In the 1930s there was debate in Canada's parliament about this, but profiteering continued to the eve of war. Initially, two units guarded the BC coast—No.4 (FB) and No. 6 (BR) Squadrons, the first with Vancouvers, Vedettes, Fairchilds and Sharks; the other with Vedettes and Sharks (both later had Stranraers and PBYs).

No.4 had formed at Jericho Beach on February 17, 1933. Its duties primarily were CGAO-oriented, but this changed in 1938, when it turned to service tasks. Diary excerpts tell of ordinary affairs at No.4, but there was occasional excitement. For February 5, 1936 they note: "Vancouver 905 dragged mooring in 60-mph wind and heavy seas. Aircraft smashed against Royal Vancouver Yacht Club Dock and sank." On April 1 Vancouver 902 arrived from Ottawa by rail. Much effort was needed to erect and test it. Meanwhile, 906 was doing most of the flying, pilots W/C Shearer, F/L F.J. Mawdesley, S/L L.F. Stevenson and FSgt H. J. Winny being involved. No.4 also had Fairchild FC-2W No.619. An airshow at Vancouver on August 1, 1936 included fly-overs by Vancouver 906, Vedette 813 and Fairchild 619. RAF Hawker Ospreys K5745 and K5748, visiting with HMS *Apollo*, also flew. For August 1936 No.4 logged 88:05 hours.

On February 8, 1937 Fairchild No.641 arrived (damaged) by rail. Repaired, it flew on March 22—a trip to Alert Bay with F/L Mawdesley and Cpls Lortie and Brown. They returned on the 25th. April was busy with a host of activities—postings in and out, VIP visits, courses (safe driving, welding, Lewis gun, etc.), medical and promotion exams. F/O H.M. Carscallen and Sgt R.I. Thomas arrived on April 23 with Deltas 667 and 669 on detachment from No.6 (GP) Squadron. On May 10 Delta 669 sank at its mooring, when a float sprang a leak. On Coronation Day (May 12) No.4 flew three Vancouvers over New Westminster, welcoming HMCS *Fraser*. The Deltas left for Prince Rupert on May 16. Mention is made on July 17 that LAC Briese was going to Trenton for flight training (his name would shine in the early days of the coming war). On various occasions No.4 co-operated in army exercises. On August 2 two Vancouvers left for Bella Bella on a West Coast meteorological survey. On December 15 S/L Mawdesley flew an injured man from Anyox to Prince Rupert in Fairchild 633. A note on December 25 states: "Twenty airmen called upon to shovel snow off hangar roof."

In April 1938 No.4 was working on a suitable bomb-loading system, and machine gun fittings for the Vancouver. On May 29 Fairchild 641 under F/L W.I. "Bull" Riddell left for Zeballos searching for missing Fairchild CF-AUX of Ginger Coote Airways. It had disappeared two days earlier, but wouldn't be found for more than two years. On July 12 Vancouver 904 was damaged taking off—holed by a submerged steel rail. Scow *M.159* retrieved 904. The same day RN Walrus amphibian K8341 of HMS *York* visited. It later departed for Prince Rupert, returning on the 30th. On the 20th S/L Wilson arrived in the sole Boeing Totem, CF-ARF. The trim little flying boat was to be weighed. On May 29, 1939 Vancouvers 902, 904 and 906 escorted the King

Vancouver 906 patrolling in September 1939 near Point Atkinson on the BC coast. The aged Vancouver, armed with one or two .303 Lewis guns and some small bombs, was the biggest RCAF West Coast threat to any encroaching Japanese. (CF via Chris Weicht)

and Queen, who were aboard ship between Vancouver and Victoria. On July 16 S/Ls Mawdesley and Jimmy Ashton delivered Stranraer 912 from the east, the first of its type for No.4. It soon entered service, searching for a drowning victim on the 26th. No.4 would re-equip with Cansos and Catalinas and finish the war after 1762 sorties and 21,677 training and operational hours. There were no enemy encounters, but 2 aircraft and 11 aircrew were lost.

## The Interwar Officer

The interwar RCAF was a small club, so paths crossed regularly. Only hindsight warns that this officer is destined for high rank, while another's promotion will be stymied. It generally proved accurate that any officer reaching squadron leader rank before the "Big Cut" (1932) was sure to rise quickly upon the restoration of normal budgets, and prewar expansion. Such men would also find themselves flying more as passengers than commanders. Prewar graduation from RAF Staff College was a sure ticket to Air Commodore rank or above. Even the last peacetime P/P/O graduates would gain rapid promotion, usually via early wartime combat, followed by executive positions at operational or staff assignments.

Clifford Mackay "Black Mike" McEwen was a distinguished WWI fighter pilot (MC, DFC and Bar, Italian Silver Medal for valour, and at least 24 kills). His wartime career is well-documented (e.g. see Ed Cosgrove, *Canada's Fighting Pilots*). As of late June 1920, McEwen was on staff as a civilian examiner. His first assignment was to reconnoitre (on the ground) the Kicking Horse Pass as a potential route for the Trans-Canada Flight. He completed this ahead of schedule. In September-October he served at Haileybury on aerial spray trials with an HS-2L, then spent that winter in Halifax, checking applications for pilots' licences and aircraft registrations. He left there in April 1921, spending that summer bush flying, commanding the Mobile Unit roaming Northern Ontario by rail. In late October he was in Ottawa for his session as a 28-day CAF officer, snappily dressed for his refresher course at Camp Borden. Then it was back to civilian dress.

S/L C.M. McEwen in RCAF ceremonial uniform. (CF PL117515)

In the summer of 1922 McEwen flew in Algonquin Park. On September 25 he force-landed on the Opeongo River with engine failure. This work finally convinced Ontario that forestry protection merited investment in aircraft. A provincial contract with Laurentide Air Service (1923) led to formation on April 1, 1924 of the Ontario Provincial Air Service (the same birthday as the RCAF). McEwen moved on, still a civilian. In May 1923 he did acceptance tests on the second CAF Viking. However, he was badly burned on July 15, 1923 when the Rockcliffe Air Station motor boat exploded on the Ottawa River. In September he resigned from his civilian appointments, donning the CAF uniform in anticipation of the force being reorganized on a more permanent basis.

Upon formation of the RCAF in 1924 F/L McEwen took command of the RCAF Depot at Victoria Island in the Ottawa River. He was posted to Camp Borden in January 1925. First as a flying instructor, then as CFI, he had a tremendous impact on the new air force. Handsome and dynamic, this war hero must have been an awesome figure to the P/P/Os passing through his hands from mid-1925 until his posting to RAF Staff College early in 1930. McEwen, who was promoted to squadron leader in January 1929, flew little, compared to pilots of later years, although this period would be the most active for peacetime flying. To the end of his service as an instructor he averaged about 100 hours yearly. McEwen's 1925-1930 tour at Camp Borden was marred by a mistake for which he made amends quickly. The incident, which illustrates service conditions of the period, was set down by W/C N.R. Anderson in a letter to AFHQ of August 9, 1926. Its contents will surprise those who traditionally describe McEwen as a spit-and-polish type:

*At the commencement of this year's Flying Training Course for Provisional Pilot Officers, all officers of the Permanent Staff were given verbal instructions that they must not on any account make "Pals" of the young Officers under instruction... Particular emphasis was laid on the fact that last year a great deal of trouble had been caused by Instructors going on "parties" with Provisional Pilot Officers to Summer Resorts and returning to Camp with them in the early hours of the morning. Instructors were warned that should similar occurrences be brought to the attention of the Officer Commanding this year, the Instructor or Instructors concerned would be required to leave the Station.*

*On the morning of August 2nd Flight Lieut. C. McEwen returned to Camp in a car together with five Provisional Pilot Officers at 0235 hours. When asked for an explanation he stated that he had been offered a "Lift" back to Camp by the Provisional Pilot Officers, whose subsequent actions and a collision with another car caused a delay which prevented them from arriving in Camp within the hours laid down.*

*The above explanation is not questioned, but the fact remains that this Instructor has been guilty of "Returning to Camp with young Officers under instruction in the early hours of the morning"... It is considered that this Officer is not putting forth sufficient effort to obey orders and Instructions and, unless he can improve in this regard, it will be necessary to recommend his retirement from the Service.*

McEwen clearly responded to the warning. On November 19, 1926 Anderson wrote to AFHQ: "It is desired to advise, please, that the above mentioned officer has been very obedient to orders and instructions since August 9th, 1926. May the matter now be considered closed, please." McEwen's attendance at RAF Staff College and the RAF School of Army Co-operation in 1930-31 indicate that he was a rising star. His subsequent career in staff and command positions proved that he was worthy of the trust placed in him. These included command of Station Trenton (1939-40) and No.1 Group, St. John's, Newfoundland (July 1941 to February 1943). In No.1 Group he showed diplomatic talents as he balanced Canadian, American and Newfoundland Commission interests. He was a forceful Base Commander in No.6 Group (March to December 1943), then commanded the group itself, where he mixed compassion and charisma in a degree probably unmatched by any officer of comparable rank in RCAF history. Air Chief Marshall Harris wanted him knighted, and wrote a furious letter, when it was explained that Canadian law forbade bestowal of this honour. McEwen was everyone's choice to command the RCAF component of "Tiger Force" (for a 1945-46 air campaign against Japan). However, he had long suffered from diabetes and the European war had exhausted him. He retired in 1946 and died in 1971—a hero of two wars and a pillar of the interwar RCAF.

## A/V/M de Niverville

Joseph Lionel Elphage Albert de Niverville was unusual—a French-Canadian who made his way in a force that was rigidly "Anglo". He was unusually competent, rising to Air Vice Marshal with scarcely a negative comment from superiors. In other ways his experiences typify the state of affairs many of his peers endured. Born in Montreal on August 31, 1897, de Niverville lived in Ottawa from 1905 onwards, becoming bilingual. In 1916 he sought to leave a stenographer position in the Civil Service to join the air services. In this he succeeded, going overseas in April 1917 with the RNAS. In England he switched to the RFC, trained as a pilot and was posted to France, where he was wounded on July 8, 1918. After a spell in hospital he took an instructor's course, then taught others to fly until demobilized in April 1919.

De Niverville was hired by the Air Board as a stenographer for a salary of $1320. He worked in the Civil Aviation Branch to 1924, then joined the RCAF. As a young flying officer and family man, he was entitled to $2550 annually (good money), including flying pay and a $200 ration allowance. He was posted to Camp Borden for refresher flying, then excelled as an instructor. As with others, he was expected to do some odd jobs (the term "job description" would have been meaningless to de Niverville and his contemporaries). In 1925-26 these included six months as Camp Borden Postmaster, being Defending Counsel at a Court Martial (October 1925), Prosecutor at another (November 1926), and President of the base Library Committee. In the winter of 1926-27 he was attached to Canadian Vickers as an aircraft inspector. He transferred to RCAF HQ in Ottawa on September 1, 1927 (just after promotion to flight lieutenant), where he became Staff Officer Training.

Wherever he went, his superiors strove to keep de Niverville for themselves. On May 5, 1931 he moved to Vancouver, his tasks being instructor, service pilot and adjutant. His commanding officer, S/L McLeod, wrote: "This officer ... is well suited as a Flying Instructor and efficient in organizing ground studies. He has a good knowledge of office administration. Since joining this unit he has made excellent progress in acquiring seaplane experience." De Niverville transferred to Ottawa in March 1932. That year he logged 98:50 hours. Although part of this was instructional, the highlight was his part in the mail flights between Ottawa and the Strait of Belle Isle (in charge of the Belle Isle base) during the Imperial Economic Conference. As an RCAF Staff College candidate, F/L de Niverville spent most of 1933 in Britain. Although promotion came slowly (squadron leader in January 1936, wing commander in April 1939) his responsibilities respecting flying decreased; those regarding administration and policy grew. He was employed almost continually in AFHQ from January 1934. Even after an Army Co-Operation Course at Camp Borden in 1935 he returned to staff duties rather than flying. Coincidental with his promotion, he became Air Staff Officer to No.4 Military District (Montreal).

F/O de Niverville, while instructing at Camp Borden in the mid-1920s. He attained Air Vice Marshall rank and commanded No.3 Training Command in 1941-43. Check out his vintage flying gear, including the Gosport speaking tube (attached to his helmet with the mouthpiece tucked into a pocket;), parachute (note the "D" ring on the left harness), sturdy gauntlets, and what looks like a rag in his right pocket. The latter would come in handy for wiping engine oil from his goggles. (CF RE18650)

The Vancouvers flown by de Niverville during the 1932 Imperial Economic Conference. They are seen at Gaspé. (NAC PA13277)

The most intense periods of de Niverville's flying career had been at Camp Borden from May to August 1925 (when he flew 98:10 hours) and June to September 1931 (102:35 hours). He returned to AFHQ on February 16, 1939, where RCAF expansion became his forte. He became Director of Air Force Manning in April 1940. In the summer of 1941 he briefly commanded No.2 SFTS, Uplands, but by September 1941 was in AFHQ as an Air Commodore. He became AOC No.3 Training Command, Montreal in November 1941 (promoted Air Vice-Marshall, November 1942). He served there for one year, then returned to Ottawa as Air Member for Training. Reorganization of AFHQ at war's end left him without a job, and he retired in February 1946. Through the war de Niverville had been Commended for Valuable Services in the Air, presented by the Czechoslovak government with the wings of their air force, awarded a CB (Companion, Order of the Bath) and appointed a Commander in the American Legion of Merit. All the time he kept the respect of his peers—"unflustered in the face of overwhelming duties" wrote one superior (January 1940).

## G/C Ernest A. McNab

Ernest Archibald McNab bridged three RCAF periods—interwar, WWII and postwar. Born in Rosthern, Saskatchewan on March 7, 1906, he was educated in Saskatoon. In the summer of 1924 he was on a prairie survey party, then was a CNR trackman. In 1925, after working for the Dominion Survey in Manitoba, he registered with the University of Saskatchewan. He studied law, then switched to engineering. He played for the Saskatoon hockey team, for the university rugby team, rowed, swam, and was counted as an all-round athlete. Motor cars were his hobby. Applying for the RCAF in 1926, he impressed F/L Harold Edwards, who made much of McNab's athletic skills, writing of him. "Very self confident. Seems the type to make a good pilot, but may or may not be responsive to discipline". McNab completed three summers of training as a P/P/O. He qualified for the coveted flying badge (August 1928) before obtaining his degree. Thus he was commissioned in the Non-Permanent Force in September 1928. He became a flying officer (still on the NPF List) in September 1929, but, to complete his studies, had to take leave of absence. Once he had his degree, he secured a permanent commission in April 1931. In the prewar period McNab would have postings to Camp Borden, Montreal, Ottawa, Trenton, Vancouver and (on RAF exchange) England.

Throughout P/P/O training McNab impressed his instructors. His final assessment was especially notable in view of coming assignments: "This officer has been an exemplary pupil throughout his entire course. Flying discipline good. He is very enthusiastic about flying and is good at aerobatics... Judgment under all conditions good." McNab's record demonstrates the type of instruction given at Camp Borden, e.g. for 1927: 19:05 hours dual, 5:25 solo (Avro 504K), 2:10 hours dual, 1:35 solo (Avro 504N), 14:20 hours solo (Avro Viper). McNab was assessed each year in such categories as taxying and handling of engine, straight and level flying, climbs and stalls, landing into wind and judging distances, turns over 45 degrees with and without engine, spinning, forced landings, taking off and landing across wind, and low flying.

Each term pupils were assessed in drill, cipline and efficiency. In the final year there was also an overall assessment of "Practical Flying", which counted for 750 points (150 on a written test, 600 from the CFI's points appraisal). McNab scored 142.5 on the written exam, 540 on the appraisal. The final 1928 flying tests might challenge the modern private pilot. McNab was expected to make 4 landings—the first 3 starting from 1,500 feet, the last from 5000—and touch down within a circle with a 25-yard diameter. His first major cross-country flight (2:50 in an Avro 504N) was on August 13, 1928 (Camp Borden-Kitchener-Hamilton, home). He had to demonstrate turns, figure-8s, stalls, side-slips and cloud flying (although the standards applied are not stated), plus 6 consecutive simulated forced landings. All this was performed under McEwen's

P/O (later G/C) E.A. McNab (right), with fellow Siskin pilots F/L David A. Harding (centre) and P/O Edwin McGowan at Camp Borden in September 1929. McNab seemed to excel everywhere, whether on prewar duties, in the Battle of Britain or postwar. (CF RE15851)

critical eye. F/O Victor J. "Shorty" Hatton examined his navigation. At Vancouver, McNab trained on Avro floatplanes, Vedettes, and Moths. He learned taxying, water handling, mooring, approaching docks and beaching. On May 6, 1929 instructor F/O C.E.F. Arthur graded him "Average", while assessing him as promising: "This officer flies accurately and with confidence and should, with practice, develop into a useful seaplane and flying boat pilot." A long Camp Borden posting (1929-31) was broken by winter study terms at university.

McNab's summers were as exciting as any young pilot could desire—flying with the Siskin aerobatic team. In 1929 he was on strength of Camp Borden only from June 14 to September 30. He flew 99:10 hours (much on Siskins). W/C G.M. Croil described him as "a good pilot... his deportment is good", although he also felt that McNab had not yet "developed as a regimental officer". In the summer of 1930 he logged 154:15 hours on the Siskin, Moth, Avro and Fairchild. For 1931 he flew 254 hours on the Trans-Canada Air Pageant. His superiors praised him for everything except administrative duties. In 1932 he participated in the air mail operation supporting the Imperial Economic Conference. He logged 152:50 hours, and was commended for this. After taking the photographic course (January 1934) McNab commanded No.6 (P) Detachment, spending that summer in northern Ontario.

As the new Non-Permanent squadrons began forming, Permanent Force officers and NCOs were assigned to help. McNab became adjutant and instructor to No.12 (Army Co-operation) Squadron (Winnipeg). On posting from there, he was praised for his tact, enthusiasm, knowledge and for leaving a legacy of high morale. He spent the summer of 1935 on photo survey Bellancas in Ontario and Manitoba. His 1935 flying time totalled 215:50 hours. The next year he served with No.15 (F) Squadron in Montreal. Before taking the post, he was re-tested for instructional skills by S/L E.G. Fullerton, who considered him a bit rusty (he had been away from training for a year). But he was competent enough to give ab initio training to his new charges. Fullerton's assessment was a small rebuke; more satisfying was McNab's promotion to flight lieutenant in April 1936. News of his selection for RAF exchange duties was exciting.

He sailed from Halifax aboard SS Alaunia on March 20, 1937, in company with F/L Frank A. Sampson and F/L R.C. Gordon. He joined No.46 Squadron at Kenley (May 5, 1937), flying Gauntlets and Gladiators, while waiting for a chance on Hurricanes. A form dated April 23, 1938 stated that he had 1,698 flying hours, 122:35 with No.46. McNab was assessed in all categories as "above average". In 1938 he added another 139:05 hours, while winning still more praise. The AOC No.12 (Fighter) Group (G/C Trafford Leigh-Mallory) rated him as: "A very good Flight Commander, reliable and painstaking... a first class fighter pilot... popular with all ranks in the squadron".

Even before sailing home, McNab was assigned to No.1 (F) Squadron, soon to be the RCAF's first Hurricane unit. There his expertise would be vital. Yet his time on type was a mere 5:00 hours to March 1, 1939. The RCAF wanted to extend his attachment, but failing this, he was instructed to return with the latest RAF training syllabi for fighter squadrons and all possible information as to Hurricane maintenance. He reached Canada in late April 1939. In September he became a squadron leader and took over No.1 Squadron. At age 39 McNab did what thousands of younger men, anticipating embarkation for war, would also do as an act of optimism and hope. On January 20, 1940 he married Barbara Huycke of Vancouver. Just before going overseas, he completed trials at Rockcliffe of a Spitfire I and a P-40.

After the Battle of Britain, now W/C McNab was acclaimed a Canadian hero. At home he did things the title demanded - addressing service clubs, speaking on the radio, appearing at Victory Bond rallies. Notwithstanding his senior rank he won command of No.118 Squadron. A peculiar aspect of his career was his failure to rise beyond group captain. A/M G.O. Johnson, reviewing McNab's record in March 1945, suggested that he had been too much of an "operations man" for his own good. There was certainly nothing wrong with him from the working view:

*During G/C McNab's tour of duty as Station Commander at Digby, he was the recipient of very satisfactory reports, both as to his ability and efficiency, and, as well, was considered to be a reliable and tactful officer, particularly in his contacts with senior RAF officials and in his dealings and cooperation with the Army authorities. As a senior Group Captain, this officer has had little opportunity during his Service career for employment on staff duties, other than a short period at Western Air Command, first in his capacity of RCAF Liaison Officer with the USAAF Forces in Seattle, and later as ASO.1 at the Command Headquarters.*

Johnson suggested more executive duties for McNab and a post "more commensurate with his present rank". He was thinking of McNab's career (but not necessarily of the service's best interests). Always an indifferent paper pusher, McNab was with an air force which would, after VJ Day, become increasingly influenced by administrative rather than operational interests. One man recognized the inequity of the situation. In January 1946, on the eve of his own retirement, A/V/M F.V. Heakes assessed Ernie McNab, whom he felt was getting a "bum rap". His reputation as a staff officer was poor, so people rashly misjudged him. Heakes described this as a "whispering campaign", fed by ignorance and jealousy. He pointed to McNab's achievements in getting No.1 Squadron moved, trained, and organized for the Battle of Britain. He cited assessments of McNab at Digby, where he had been responsible for administration, tactics and training of as many as 3600 personnel, and to his reluctance to defend himself against prejudiced rumours.

Unhappily, McNab's reputation did not improve, and the evidence is that he himself gave ample proof that he had gone as far as he could expect. Through successive postings in Northwest Air Command and elsewhere, he seemed unable to delegate work, then was swamped by the ensuing "paper blizzards". He was genial and popular, but even a sympathetic superior like A/V/M Ken Guthrie found himself embarrassed by backlogs attributable to his earnest subordinate. In 1948-49 McNab attended RCAF Staff College. Thereafter he was repeatedly commended for projects such as co-ordinating RCAF aspects of a 1951 Royal Visit and the 1953 National Air Show in Toronto. Nevertheless he had slipped on the promotion ladder somewhere between 1943-49. To his credit, however, he expressed no bitterness (at least while in the service) at his stalled career. Ernie McNab retired on October 23, 1957 and died in Murrayville, BC on January 10, 1977.

**Fairchild FC-2 No.624 in a scene from Camp Borden in the 1930s. This useful type served the RCAF into WWII. Its many uses included photo survey, anti-drug patrols, parachute training and search and rescue. (CANAV Col.)**

# Out of the Doldrums

The RCAF suffered patiently through the 1930s, faithfully conducting its civilian duties under the CGAO mandate, and doing so with its obsolete aircraft. The addition of the Atlas and Siskin gave some hope of a more military future.

By 1935, as sabres were beginning to rattle in Europe, Ottawa became more sympathetic towards its military. The RCAF, with its budget increasing yearly, could purchase new service types and increase training. These photos illustrate this period wherein the RCAF made its first serious progress in a decade.

In the early 1930s the RCAF began forming auxiliary squadrons, most members being from the local communities. No.10 (Aux) Squadron formed in Toronto in October 1932 under S/L G.S. "Geoff" O'Brian, AFC. It began with no aircraft—the first was not delivered for 2 years. In 1937 No.10 became No.110. It was flying Avro 626s and Tiger Moths with Lysanders promised. No.110 sailed for England in February 1940, where it became 400 "City of Toronto" Squadron. At Century's end 400 was flying Griffon helicopters at CFB Borden. Here No.110 parades at the CNE in May 1936. (CF RE13659)

No.111 (Aux) Squadron ready for inspection at Sea Island, Vancouver on September 3, 1937. Shown are Avro Tutor No.187, Fawn 203, three Moths and another Tutor. Such trainers lasted into the early war years, when most became ground instruction aircraft. (NAC PA133591)

RCAF Station Trenton in June 1936. Much of this infrastructure was established under a depression scheme that put unemployed men to work for 20¢ a day. Notice the hangars, then (across Highway 2) station HQ with its parade square. Trenton grew steadily—compare this view with that on page 134. (NAC PA190252)

With war around the corner, the RCAF positioned most of its service aircraft in eastern Canada. Here No.3 Squadron Wapitis wait in Ottawa after a gruelling flight from Calgary. Soon they departed for Halifax. Wapitis flew some of the RCAF's first wartime patrols. (NAC PA63515)

A replacement for the RAF's Harts and Hinds, the Fairey Battle first flew in March 1936. After losses in the Battle of France it withdrew to safer skies, primarily in training. A few were service aircraft in the RCAF (115 Squadron), but most flew in the BCATP. With a 1030-hp Merlin engine, the Battle had a top speed of 240 mph. The RCAF had 740 Battles on strength. Jack McNulty shot L4987 at Ottawa Car in 1941.

These Battles came to Canada not to fly, but to be ground trainers at the RCAF trades school at St. Thomas. (Murray Castator)

The first Canadian-built Westland Lysander flew from National Steel Car (Malton, Ontario) in August 1939. Lysanders served 110 and 112 Squadrons at Rockcliffe in 1940, then overseas (re-numbered Nos. 400 and 402, these squadrons re-equipped with Tomahawks and Hurricanes). Canada's home-based "Lizzies" mainly were target tugs, but also did army and artillery co-operation training, and air-sea rescue. Armament included a .303 Browning in each wheel housing, and in the rear cockpit; small bombs could be carried on winglets attached to the undercarriage. (NAC PA63856)

Obsolete from the start (1936), the RCAF's Blackburn Sharks dwelt mainly on the West Coast, doing stalwart duty in coastal patrol. When a pilot wrecked one after the float struts failed, his CO promised to recommend him for the Air Force Cross if he would do the same with the rest of his Sharks! This example is fully configured for combat: 4 100-lb bombs, 8 small practice bombs, a centreline practice torpedo and a Lewis gun. It looks great, but rarely (if ever) would have flown with such a load. With the crew of Herriot and Biehler, Shark 506 (shown here) was wrecked in June 1942. (David Thompson Col.)

First flown in 1931, the Grumman G-23 was built under licence in Fort William by Canadian Car and Foundry. CCF sold one each to Nicaragua and Japan, then 40 to an intermediary, who diverted them to the Communists fighting the Spanish Civil War. This violated US regulations, so the last 6 from the order were seized in Saint John, NB. In 1938 CCF failed to offload its last G-23s on the RCAF, then tried again (successfully) when war broke out. Christened "Goblins", they served No.118 (F) Squadron. One wonders if this odd deal would have happened had CCF's member of parliament not been the powerful Minister of Munitions and Supply, C.D. Howe. (CF PL5954)

The RCAF often evaluated new aircraft. In early 1940 this shiny Curtiss P-40 visited Ottawa, where F/L Ernie McNab tested it, along with Spitfire Mk.II L1090. McNab preferred the latter. Spitfires never did equip HWE squadrons, but P-40s and Hurricanes did, and admirably so. (CF PL375)

The Lockheed Hudson, which joined the RCAF in September 1939, proved valuable in desperate times. On July 31, 1942 Hudson BW625 of 113 (BR) Squadron, based at Yarmouth (pilot S/L N.E. Small) sank U-744 off Sable Island. This was Eastern Air Command's first U-boat kill. Its second came on October 30, 1942, when Torbay-based Hudson No.784 of 145 Squadron (pilot F/O E.L Robinson) sank U-658. This example (RCAF No.770) still bears the US civil registration in which it was delivered in January 1940. Within a few weeks it was lost in a crash. (CF)

To boost coastal resources, the RCAF selected the Supermarine Stranraer flying boat. Licence-built by Canadian Vickers, the first of 40 (No.907) flew in October 1938. "Strannies" spent most of their time on the West Coast. A few were lost in mishaps; the rest were sold to civil operators beginning in 1944. G/C Gord Truscott commented in 1985: "I just loved the Stranraer. It was a gentleman's airplane. It did everything right, but not too quickly." Here No.907 takes off with depth charges underwing. RCAF No.920, later CF-BXO, is the world's last Stranraer; it may be seen in the RAF museum at Hendon, England. (Canadair Archives H147)

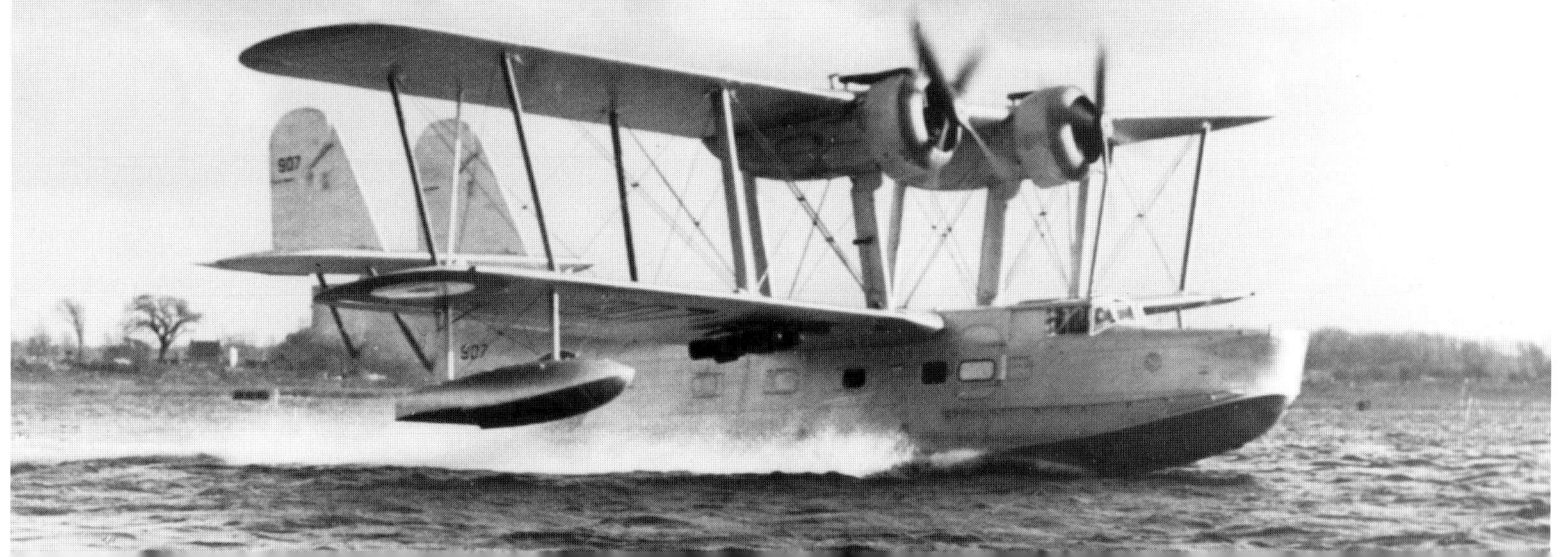

Recruiting would be the RCAF's life's blood. Here a contingent at Rockcliffe on its first parade. Then eager lads are about to take their oath after "signing up" in Ottawa. From a handful of members the day Canada declared war, the RCAF would grow to more than 200,000. (CF PL20724, NAC PA63473)

# The War on the Homefront

While the biggest wartime headlines focused overseas, Allied victory depended primarily on the North American home front—Great Britain would have foundered had it not been for the raw materials and products supplied by the United States and Canada. Rich in every resource, North America was a bastion against which the enemy could not inflict a pin-prick. While the Axis had ideas for long range aircraft to bomb North America, these such efforts failed. Their best hope was to sink the ships carrying men, equipment and supplies overseas. At this Germany initially succeeded, but North America steadily increased production. All the torpedoes of the U-boats could not stop this effort, especially once anti-submarine resources were in place.

On the other hand, from early in the war German military and urban/industrial centres were bombed. Eventually, Japan faced the same onslaught. Both nations had to adopt extreme measures, such as dispersing industry. As their labour pools shrank, slaves became more widely used, and even children were put under arms. Meanwhile, raw materials grew scarcer by the day.

In his famous June 1944 pep talk, the commander of the Allied Expeditionary Force, General Dwight D. "Ike" Eisenhower, reminded those about the embark on the invasion of France: "Our Home Fronts have given us an overwhelming superiority in weapons and munitions of war, and placed at our disposal great reserves of trained fighting men." Ike knew it, so did his men; but one wonders how Germany and Japan did not. It is as if they never had studied the simplest geography or economics.

## Canada and the BCATP

When WWII began, the RCAF was a puny organization. It numbered but 4153 men, even though it was authorized for 7259. There were 53 service aircraft, all obsolete except for 10 Hurricanes. Fortunately, this predicament would reverse almost overnight, and Canada would make great contributions in the air war, perhaps its single most important one being the British Commonwealth Air Training Plan. Originally the Empire Training Plan, it came into existence with the signing in Ottawa on

**A grand scene at No.20 EFTS as instructors and students head to their Tiger Moths. (CF PL5356)**

Constructing dozens of air training bases was one of Canada's great wartime tasks. Here a station begins to take shape. (CF PL243)

Oblique and high level views of a typical BCATP station— RCAF Station Summerside. The standard triangular runway layout is evident. (CANAV Col.)

December 17, 1939 of the "Agreement Relating to the Training of Pilots and Aircraft Crews in Canada and Their Subsequent Service". Canada, Great Britain, Australia and New Zealand were the signatories. In ensuing years Canada would graduate more than 131,000 aircrew of all trades—pilots, observers, air gunners, wireless operators, bomb aimers, flight engineers. Many books have been published covering "The Plan". One of the best is Fred Hatch's *Aerodrome of Democracy: Canada and the British Commonwealth Air Training Plan 1939-1945*. For the political background, the lists of schools, aircraft types, dates, the training anecdotes and the results one must consult all these sources.

In the early stages there was a stampede of young men to join the RCAF. Each had his sights on pilot training and flying a Spitfire. Since no training system yet existed to handle such a rush, few applicants were accepted. For the time being the RCAF stuck to its high educational standards for pilots. Most who visited recruiting centres were sent home, told that they would hear in due course. Meanwhile, aircrew training gradually increased, elementary training at first being handled by civilian flying schools such as Patterson and Hill at Barker Field in Toronto. To remedy a shortage of instructors, the RCAF took in civil pilots from the flying clubs, and even some old time bush pilots. Still short, it turned to the United States, from where many instructors soon arrived. As soon as possible, sites were chosen and surveyed for new training fields; construction began on more than 100. Elementary schools were operated under contracts to well-known flying clubs. Air observer schools also came under civil management, run by subsidiaries of companies such as Canadian Pacific Airlines. Advance training schools were either RCAF or RAF. As fast as schools came on line, eager students packed their bags and boarded trains to begin their new careers.

## Main BCATP Trainers

| Type | Wing Span | Length | Engine | Gross Weight (lbs) | Max. Speed (mph) |
|---|---|---|---|---|---|
| **EFTS** | | | | | |
| D.H.82C Tiger Moth | 29' 4" | 23' 11" | 130 hp Gipsy Major | 1825 | 109 |
| Fairchild Cornell | 36' 0" | 27' 8" | 200 hp Ranger | 2735 | 122 |
| Fleet Finch | 28' 0" | 21' 8" | 125 hp Kinner B-5 | 2100 | 113 |
| PT-17 Stearman | 32' 2" | 25' ¼" | 220 hp Continental R670 | 2720 | 124 |
| **SFTS** | | | | | |
| Fleet Fort | 36' 0" | 26' 10" | 330 hp Jacobs L-6 | 3900 | 180 |
| North American Harvard | 42' ¼" | 29' 0" | 550 hp &W R1340 | 5260 | 206 |
| North American Yale | 42' 0" | 27' 7" | 420 hp Wright R975 | 4400 | 172 |
| Airspeed Oxford | 53' 4" | 34' 6" | 355 hp A.S. Cheetah | 7500 | 169 |
| Avro Anson Mk.I | 56' 6" | 42' 3" | 340 hp A.S. Cheetah | 8000 | 188 |
| Avro Anson Mk.V | 56' 6" | 42' 3" | 450 hp P&W R985 | 9275 | 190 |
| Cessna Crane | 41' 11" | 32' 9" | 225 hp Jacobs L4MB | 5700 | 179 |
| **Other Types** | | | | | |
| Bristol Bolingbroke | 56' 4" | 42' 9" | 840 hp Bristol Mercury | 14,400 | 295 |
| D.H. Mosquito | 54' 2" | 40' 10" | 1705 hp Packard Merlin | 19,000 | 380 |
| Fairy Battle | 54' 0" | 42' 2" | 990 hp R.R. Merlin | 10,790 | 210 |
| Handley Page Hampden | 69' 2" | 53' 7" | 780 hp Bristol Pegasus | 21,000 | 265 |
| Hawker Hurricane | 40' 0" | 31' 5" | 1030 hp R.R. Merlin | 6000 | 335 |
| Lockheed Hudson | 65' 6" | 44' 4" | 1100 hp Wright R1820 | 18,500 | 284 |
| Lockheed Ventura | 65' 6" | 51' 7½" | 1850 P&W R2800 | 26,500 | 300 |
| Noorduyn Norseman | 51' 6" | 31' 8" | 570 hp P&W Wasp | 7400 | 170 |
| North American Mitchell | 67' 6" | 51' 5" | 2400 hp Wright R2600 | 24,000 | 308 |
| Westland Lysander | 50' 0" | 30' 6" | 890 hp Bristol Mercury | 5920 | 206 |

Civilian pilots of No.1 AOS at Malton in 1940 included many Americans(*). Standing are W. Hughes*, Frank Bailley, Ralph Leslie, B. McKinney*, Robert Smuck, A. Mitten*, Dan Duggan*, Art Leach, Robert LeRoy* and W. "Bill" Walker. Seated are E. Hamel*, W. "Bill" Resseguier, Henry Seiger*, W. "Babe" Woolett, C.R. "Peter" Troup, R. Mulhern*, C.A. "Duke" Schiller and W. Hilchie. Others, like Resseguier, Schiller, Troup and Woolett were seasoned Canadian bush pilots. Before long, most of the Americans had disappeared to Ferry Command. (Fred Hotson Col.)

# The Beginning of Victory— Trainers of "The Plan"

## Air Training Plan Establishments

**Elementary Flying**—Malton, Ont.; Fort William, Ont.; London, Ont.; Windsor Mills, P.Q.; High River, Alta. (Relief—Frank River, Alta.); Prince Albert, Sask.; Windsor, Ont.; Vancouver, B.C.; St. Catharines, Ont.; Hamilton, Ont.; Cap de la Madeleine, P.Q.; Goderich, Ont.; St. Eugene, Ont.; Portage la Prairie, Man.; Regina, Sask.; Edmonton, Alta.; Stanley, N.S.; Boundary Bay, B.C. (Relief—Langley, B.C.); Verdun, Man. (Relief—Hargreave, Man.); Oshawa, Ont. (Relief—Whitby, Ont.); Chatham, N.B.; Quebec City, P.Q.

**Air Navigation**—Rivers, Man.; Pennfield Ridge, N.B.

**Air Observer**—Malton, Ont.; Edmonton, Alta.; Regina, Sask.; London, Ont.; Winnipeg, Man.; Prince Albert, Sask.; Portage la Prairie, Man.; Quebec City, P.Q.; St. Johns, P.Q.; Chatham, N.B.

**Bombing and Gunnery**—Jarvis, Ont.; Mossbank, Sask.; Macdonald, Man.; Fingal, Ont.; Dafoe, Sask.; Mountain View, Ont.; Paulson, Man.; Lethbridge, Alta.; Mont Joli, P.Q.

**Service Flying**—Camp Borden, Ont. (Relief No. 1—Egenvale, Ont; Relief No. 2—Alliston, Ont.); Ottawa, Ont. (Relief No. 1—Pendleton, Ont.; Relief No. 2—Edwards, Ont.); Calgary, Alta. (Relief No. 1—Shepard, Alta.); Saskatoon, Sask. (Relief No. 1—Vanscoy; Relief No. 2—Osler); Brantford, Ont. (Relief No. 1—Burtch); Dunnville, Ont. (Relief No. 1—Welland, Ont.); Macleod, Alta. (Relief No. 1—Granun); Moncton, N.B. (Relief No. 1—Scoudouc; Relief No. 2—Salisbury); Summerside, P.E.I. (Relief No. 1—Mount Pleasant; Relief No. 2—Wellington); Dauphin, Man. (Relief No. 1—North Junction; Relief No. 2—Valley River); Yorkton, Sask. (Relief No. 1—Sturdee; Relief No. 2—Rhein); Brandon, Man. (Relief No. 1—Chater; Relief No. 2—Douglas); St. Hubert, P.Q. (Relief No. 1—St. Johns; Relief No. 2—Farnham); Aylmer, Ont. (Relief No. 1—St. Thomas); Claresholm, Alta. (Relief No. 1—Woodhouse); Hagersville, Ont. (Relief No. 1—Kohler, Ont.).

**Central Flying School**—Trenton, Ont. (Relief No. 1—Mohawk).

**Wireless**—Montreal, P.Q.; Calgary, Alta.; Winnipeg, Man.; Guelph, Ont.

**Initial Training**—Toronto, Ont.; Regina, Sask.; Victoriaville, P.Q.; Edmonton, Alta.; Belleville, Ont.; Saskatoon, Sask.

**Technical Training**—St. Thomas, Ont.

**Technical Detachments**—Toronto, Ont.; Montreal, P.Q.

**Composite Training School**—Trenton, Ont.

**Aeronautical Engineering**—Montreal, P.Q.

**Holding Unit**—Moncton, N.B.

**Air Armament School**—Mountain View, Ont.

**Depots**—Trenton, Ont.; Winnipeg, Man.; St. Johns, P.Q.; Calgary, Alta.; Toronto, Ont.; Montreal, P.Q.; Ottawa, Ont.; Brandon, Man.; Edmonton, Alta.; Quebec, P.Q.; Lachine, P.Q.; Camp Borden, Ont.; Regina, Sask.; Halifax, N.S.; Rockliffe, Ont.

The D.H.82C Tiger Moth was Canada's chief EFTS trainer, more than 1500 being on strength. No.8870 is shown at No.4 EFTS, Windsor Mills, Quebec with student pilot Regie Marsh.(via H. Halliday)

This listing of BCATP schools appears in the December 1941 issue of *Canadian Aviation*.

The Fleet Finch also did solid EFTS work, 431 serving. In this idyllic Windsor Mills scene Finches 4492 and 4490 are nearest. The former was lost on May 22, 1941. (CF PL2050)

Tiger Moth and Finch both were "Canadianized" for winter with skis and coupe tops added. Fleet Fawn 280, photographed by Jack McNulty at Camp Borden, was on strength 1938-44.

The Fairchild Cornell, built by Fleet in Fort Erie, replaced Tiger Moths and Finches at 9 schools and equipped 4 others from the outset. These were at Virden, Manitoba with No.19 EFTS. No.16630 logged 555 hours and was on strength until February 1947. (NAC PA176331)

117

The Fleet 60 Fort flew in March 1940. Failing as an advanced trainer, an area dominated by the far superior Harvard, most of the RCAF's 101 Forts became wireless trainers. (via Stephen Mouncy)

A typical wireless air gunners' class (Course 61, August 1943) at No.1 B&GS at RCAF Station Jarvis. (Roy Clarke Col.)

Hollywood star James Cagney looks over the student's position in a Fairey Battle gunnery trainer. He was in Ottawa and Trenton in 1941 for filming of the Hollywood movie *Captains of the Clouds*. (via Rob Schweyer)

The Northrop Nomad, 32 of which were in the RCAF, proved useful in drogue towing at bombing and gunnery schools. This Nomad is seen at Camp Borden. (John Buzza Col.)

With war brewing the RCAF needed an advanced trainer. The North American NA-16 Harvard Mk.I was the answer; 30 were purchased in 1939. In spite of rugged use only 6 Harvard Is were lost in crashes. The survivors finished the war as ground technical trainers. No.1321 is seen at Cartierville in experimental winter configuration. (Jack McNulty Col.)

When France surrendered in 1940, 119 of its North American NA-64s were diverted to Canada. Called Yales, they served mainly at No.1 SFTS (Camp Borden) and No.6 SFTS (Dunnville). Later, many were trainers at Nos.1 (Winnipeg), 2 (Calgary), 3 (Montreal) and 4 (Guelph) Wireless Schools. Left to rot in "back 40" settings, a few Yales have been recovered and restored to flying condition. While the Harvard had a 550-hp P&W R1340, the Yale used a 420-hp Wright R975. (Ken Smith)

The Harvard II was an SFTS stalwart. According to an August 1938 agreement, licence production was at Noorduyn in suburban Montreal. For its licence Noorduyn paid $40,000, then a royalty to North American of $1000 per airplane, reducing to $500. About 2800 were built, most for the RCAF, but others for the US which, under Lend Lease, consigned them to the RAF. Here F/L E.G. MacDonald flies 3233 from No.6 SFTS, Dunnville. Eleanor Haddons is in the back seat. (via Doug McPhail)

As late as December 1941 Avro was promoting the Anson as a service aircraft. By this time, however, its shooting days were over. (David Thompson Col.)

The 1935 Avro Anson served the RAF in bombing and coastal patrol early in WWII, then left for the training world. Canada accepted its first Anson from the UK in February 1940, then built many for the BCATP. Shown is an old Mk.I with Cheetah IX engines. Other Ansons used P&W or Jacobs power. (CF PMR74-264)

(Top) The Anson cockpit became familiar to thousands of BCATP students. RCAF mechanic Howard Anderson is in the pilot's seat of this Mk.I. (Anderson Col.)

Canada's main contribution to the Anson was the Mark V, with its US-designed, molded plywood fuselage, wooden wing, and P&W engines. The Mark V first flew at Cartierville in January 1943. Various companies built 1048, many of which survived the war as civil transports and photo survey aircraft. This spiffy No.1 AOS example, on RCAF strength till March 1954, was photographed by Jack McNulty at Malton.

Anxious for trainers Canada evaluated a 1939 project, the T-50 (US designation AT-8), that had been languishing with Cessna in Wichita. An RCAF order gave Cessna a life-saving break (according to historian Joseph P. Juptner, "Cessna Aircraft was not too far from going broke at the time"). The T-50 was the Crane in the RCAF, but Americans called it the AT-8 Bobcat or "Bamboo Bomber". Of 5400, the RCAF took 826. No.8073 is seen landing (on one engine) at Fargo, North Dakota in October 1941. It served with No.11 SFTS at Yorkton, amassing 3100+ hours. (Jack McNulty Col.)

A formation of Anson Vs flies over No.17 SFTS at Souris, Manitoba, saluting the end of the war and the closing of the station. (Mel Swift)

(David Thompson Col.)

Another BCATP twin was Britain's Airspeed Oxford. Starting in October 1942, the RCAF had 819 on strength. These Oxfords are seen at 36 SFTS, Penhold, Alberta. (Les Corness/CANAV Col.)

A pilot's view of the Oxford cockpit. (Les Corness/CANAV Col.)

The Bolingbroke, Hampden, Hudson, Mitchell, Mosquito and Ventura were important in the BCATP. The Bolingbroke served mainly in bombing and gunnery, the Hampden and Hudson at coastal OTUs, the Mitchell as a lead-in to the Liberator, the Mosquito as a specific type trainer, and the Ventura in medium bomber conversion. AJ990 and AJ991, two of 160 Hampdens built in Canada, were at Aylmer. (Eaker Col.)

Many Bolingbrokes were target tugs. As such they were yellow and black like No.9123, seen at Edmonton. (Les Corness/CANAV Col.)

No.1 (F) OTU was Canada's only BCATP fighter OTU. It took new pilots off Harvards and put them into Hurricanes, usually the day they reached Bagotville. Next stop was a refresher in the UK, then fighter operations. Here, Doug Rose of Winnipeg flies Hurricane 5461 in March 1942. After OTU he flew Kittyhawks and Hurricanes on 130 Squadron at Mont Joli, then did a Lancaster tour on 207 Squadron. Next, the armourers prepare and load rockets for a day's practice firing from Bagotville. For every aircrew member, the RCAF needed many non-flying tradesmen like the armourers. (Doug Rose and John Buzza Cols., CF PL29488)

A trades class at Central Technical School in Toronto, where many a sprog got his start before getting into RCAF uniform. (Murray Castator Col.)

"PT" (physical training) was part of daily life on every station. Here Course 60 at No.2 Wireless School works out. (CF PL12930)

Every station had its sprogs on guard duty as they awaited ITS. Here, AC2s Harry Hartsborn and Murray Cookman man a base defence post at Trenton in August 1940. (CF PL1219)

Clerical personnel were indispensable at any station. Part of the No.6 SFTS staff of telephone answerers, memo writers and general paper pushers is shown in this snapshot. (via Doug McPhail)

House cleaning was a duty that nobody could shirk. Here airmen at No.2 B&GS at Mossbank pause, while cleaning barracks the night before the station commander's weekly inspection. (NAC C74483)

The Link Trainer was a vital flight training device. Developed in the 1920s by American inventor Edwin A. Link, it began as a carnival attraction. One of Link's first big customers was the RCAF; then his little "toy" took off. Here is the Link room at RCAF Station Virden in October 1944. (NAC PA140658)

While young men from age 18 joined the RCAF, thousands of boys anxiously waited their turn. Many were in Air Cadets, which had squadrons all over Canada. Cadets learned about RCAF life, studied topics like aircraft recognition, and got the occasional familiarization flight. Shown is 173 Squadron at summer camp at Mountain View (near Trenton) in July 1943. The CO, F/O V.P. Carswell, is front and centre. His 13-17 year old cadets are (front row): Bob Weese, Earnest Ormshaw, Ron Murphy, Gerald Woods, Rex Glover, Chuck Barker and Hugh Squires. In the middle are: Bert Storey, Ken MacDonald, Doug Carswell, Gordon L. Diller, Harry Lamorre and Jack Wannamaker. Behind are: unknown, Ted Snider and Don Moore. (H.S. and Gordon Diller Col.)

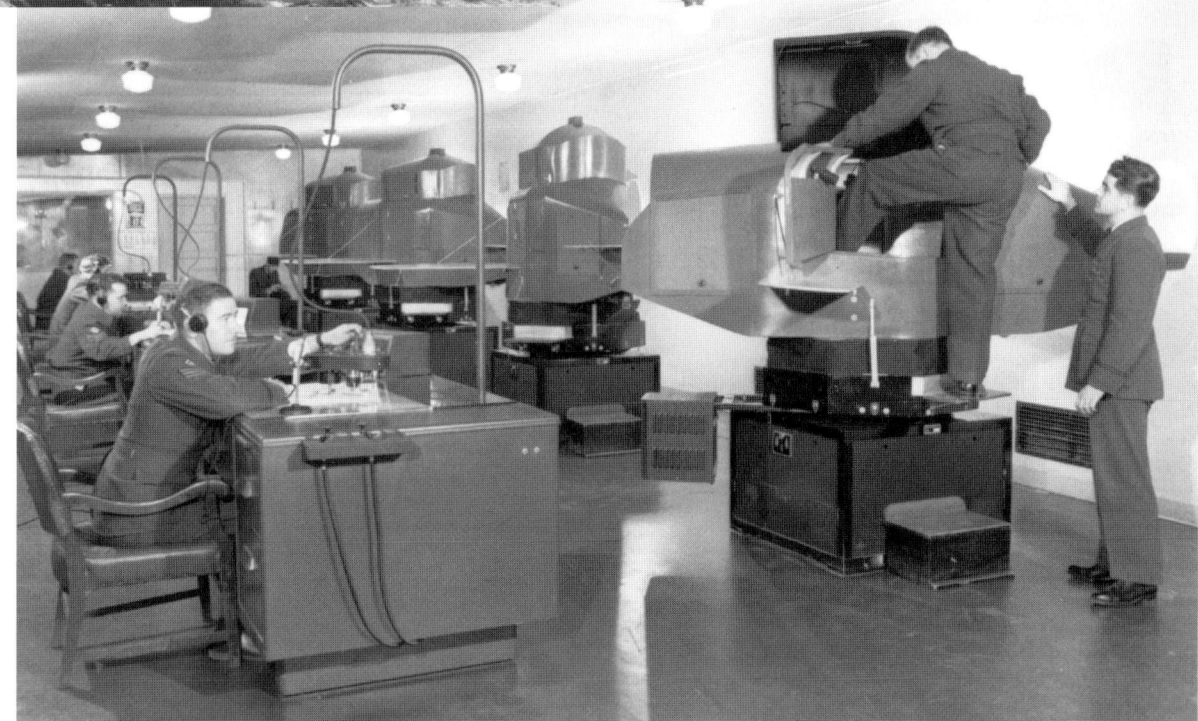

A 173 Squadron cadet who advanced to the RCAF was Gordon L. Diller, son of Harry Sylvester Diller of High River. With a Military Medal from Vimy Ridge, Harry had joined the RCAF in 1924. He had a variety of postings including 3 years on photo detachments. He retired in 1945, by when his son was a young wireless operator. Father and son are shown here. (H.S. and Gordon Diller Col.)

Senior officers often inspected a station, presented wings and addressed the students. One of the most popular such figures was Air Marshal W.A. "Billy" Bishop, VC, DSO and Bar, MC, DFC. On this occasion he was at Camp Borden on May 23, 1940. (H.S. and Gordon Diller, and Michael Vacheresse Cols.)

The greatest moment for a trainee was Wings Parade. It signalled the end of basic training and brought him to the next exciting stage—OTU. These pilot graduates are about to be winged, following their course at No.1 SFTS at Camp Borden. Proud, if apprehensive, family members always were part of such ceremonies. Then, G/C R.S. Grandy, station commander, pins the wing on a glowing student, who now would trade his LAC "prop" for sergeant stripes. (CF PL20221)

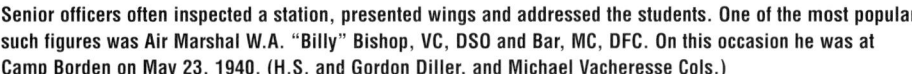

## Creating An Operational Pilot

Typical of BCATP students was a lad from Hamilton, Ontario William Mitchell "Bill" Baggs, born on November 22, 1922. After finishing high school at Westdale Collegiate in 1942, he took a job at nearby Mount Hope, where an EFTS was being set up. Helping build hangars was not for him, so he worked his way into something more interesting. Baggs was given a table in a little office from where he recorded each pilot's flying time. Here he met some of the instructors, including Ernie Weeks of nearby Waterdown. Ernie gave Baggs his first flight in a Tiger Moth.

As soon as possible, Bill Baggs enlisted in the RCAF. Stage one in his training began on August 2, 1942 at No.5 Manning Depot at Lachine in Montreal. This was a reception centre where new entries received basic RCAF indoctrination. They were issued uniforms, had their first service haircuts, were inoculated against various diseases, marched, drilled and took lectures about military life. For Baggs this phase lasted till September 15. Next came a more intensive academic period at Initial Training School, but there usually was an ITS backlog. Thus, most recruits (who had the lowly rank of AC2—Aircraftsman 2nd Class) spent a few weeks on airfield guard duty, or helping with menial jobs around the hangars and flightline (in today's air force this is known as OJT—on the job training). For Baggs this meant a stint at St. Hubert until December 24, after which he was posted to No.5 ITS at Belleville, Ontario. Here, among a variety of

A class of LACs at No.5 ITS in Belleville. Bill Baggs is in the second last row, centre. (All photos Bill Baggs Col.)

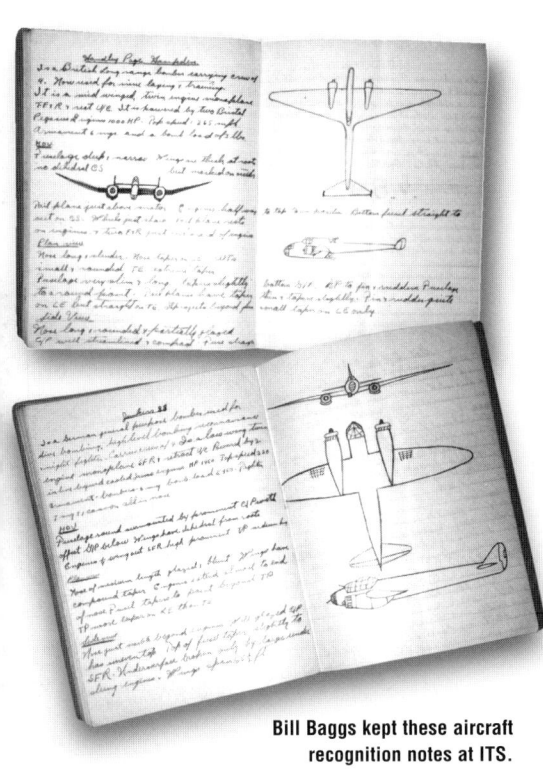

**Bill Baggs kept these aircraft recognition notes at ITS.**

activities, students had a chance to try "flying" the Link Trainer, the precursor of today's flight simulator. Introduced in 1929 by American inventor Edwin A. Link, it sometimes was referred to as a "synthetic trainer". The Link would show an instructor just who among recruits had an aptitude for flying. Anyone who could not come to grips with it was unlikely to succeed in pilot training.

ITS, which lasted for Baggs from December 29, 1942 to March 6, 1943, focused on academics—airmanship, meteorology, navigation, armament, military law, finance, aircraft recognition, etc. There was a lot more drill and an emphasis on athletics. At the end each man was assigned an aircrew trade, and left ITS wearing a white flash in his cap—the emblem of the LAC (Leading Aircraftsman).

Bill Baggs' posting was to pilot training at No.20 Elementary Flying Training School at Oshawa, east of Toronto. EFTS was where a pilot candidate's dreams could be fulfilled. Most would get through to the next stage, but some would wash out. There were many reasons for this, whether persistent air sickness, an inability to grasp the basics of flying, some minor mishap, sometimes even because a student had the bad luck to have an instructor with whom he didn't click. But for those unlucky enough to fall by the wayside, all was not lost. They could be reclassified to another aircrew trade—navigator, wireless operator, air bomber, etc.

On March 8, 1943 Baggs took his first flying lesson, going up in Tiger Moth 8990 with Sgt T.M. Davidson. He progressed quickly with the fundamental exercises, soloing on March 29 after 10:15 of dual. On April 18 he made his first solo cross-country, flying from Oshawa northwest to a point on Lake Simcoe, southwesterly to Malton, then back to Oshawa along the Lake Ontario shore. By now the flying was intensive—Bagg's flew four times and did one Link session on April 22, a typical day. EFTS instructors were competent, although some were friendlier than others. Bill Baggs recalled C.R. "Chuck" Coome, Jack Lumsden and George Slocombe as among his favourites.

On April 29 Baggs made his last flight at EFTS, then, with 70:30 hours logged, moved along in the scheme to No.6 Service Flying Training School at Dunnville, south of Hamilton on the Lake Erie shore. There he joined Course 80 on May 1, flying first in Harvard 3766 with P/O Bruton on the 4th. The Harvard cockpit reminded Baggs of a roomy, comfortable bathtub. With 550-hp the Harvard was far more airplane than the 130-hp Tiger Moth, but most students had the self-confidence to handle this. They couldn't wait for the challenges of SFTS. There were many dual and solo exercises, including formation, instrument and night flying, and air firing using a wing-mounted .303 machine gun. SFTS gave a much greater operational feeling to a young student than had EFTS. By now, wearing air force blue was fast-becoming a way of life, and the boys loved it.

**A page from Bill Baggs' log, showing Harvard time from SFTS.**

On May 15 Baggs soloed in the Harvard after nine hours. He passed his wings test on July 15 with S/L McDonnel, had his final test on the 19th with F/O Coates, and finished SFTS on August 13. He had logged 77:05 hours of dual on the Harvard, 72:30 solo, plus 25 hours in the Link. He graduated at his Wings Parade on August 20. Parents, brothers and sisters, even a few fiancées, turned out for the grand occasion, presided over by station commander

G/C Val Patriarch. The "old man" was interested in his young charges. He presented each with a copy of *Canada's Air Heritage*, a booklet honouring some of Canada's great WWI pilots. The graduates were awarded their sergeant's stripes, although a few who had excelled won commissions.

It was never guaranteed that Bill Baggs would leave Dunnville a pilot officer (which he did), for he had gone over the line at least once. On a cross-country flight he had succumbed to temptation upon spotting a train puffing along near Collingwood. He beat it up, but on pulling up, his maps flew from the cockpit. Now he met stronger-than-forecast winds, became lost and had to force-land at Jarvis, low on fuel. There he partied with some Hamilton buddies, and spent the night in the bed of a student pilot killed hours earlier. Baggs returned to Dunnville in the morning to face the wrath of G/C Patriarch, who let him off the hook after a stern lecture. Baggs was relieved to have escaped being washed out. Now the students were streamed, some to fighters, bombers, or coastal command, others to instruct. Baggs was elated when posted to No.1 (Fighter) Operational Training Unit (Course 17) at RCAF Station Bagotville, in isolated country north of Quebec City. This put him just where he wanted to be in his career. Little by little, like everyone in the BCATP, he was being groomed for operations.

No.1 (F) OTU, formed in July 1942 with Canadian-built Hurricanes, was not where most RCAF fighter pilots trained (as a rule they attended OTUs in the UK). Even so it was turning out its share of sprog pilots, and easing the burden for the RAF. Among the instructors were pilots recently returned from ops, men like F/L H.W. "Wally" McLeod, home from Malta, where he had downed a dozen enemy planes. Ironically, his posting at OTU was a sort of punishment—he had stepped on toes while overseas. His presence was awe-inspiring to Bagotville's wet-behind-the-ears students. (The Wally McLeods of Bagotville often did not stay a minute longer than required—they wanted back into the fray. McLeod returned to ops in early 1944. On September 27, 1944 he was mixing it up over Holland with a flock of Me.109s. When his squadron landed, McLeod was missing. Soon afterwards he was found dead in the wreckage of his Spitfire.)

P/O Baggs flew his first Hurricane on September 3, 1943 and was exhilarated. Training accelerated, with all the standard exercises familiar at SFTS, except for greater emphases on formation, low-level and instrument flying, aerobatics, R/T and gunnery. Daily routines grew familiar—rise early, breakfast, parade, then a half day of flying, lunch, a half day of lectures, athletics, supper, free time, and bed. There was little time for gallivanting. Besides, things were quiet in the main community of Chicoutimi, where few spoke English. Another routine was Sunday church parade, and there was the occasional funeral, when someone made a fatal mistake. Occasionally, students got a 36-hour pass. So treasured were these that on Friday afternoon Baggs would take the train all the way home to Hamilton and be back Sunday night, a 1200-mile return trip!

Everything went routinely for Baggs until October 22. While aloft that day, the weather suddenly closed in and he became hopelessly lost. All too often, this situation ended fatally, but luck, skill and cool heads prevailed. When he reported his predicament, the tower handed him over to a local experimental unit, No.12 Radar Detachment. A ground control intercept (GCI) operator, F/O A.H. MacCarthey, took over, identified Baggs on his scope and, inch by inch, worked him towards the airfield. When Baggs finally broke cloud at a few hundred feet, he was right over Bagotville and landed safely. When he stepped from the cockpit, he was met by his CO, W/C Ed Reyno, who congratulated him heartily. Better than that, he inscribed Baggs' log with a rare "green endorsement"—a good show written in green ink. Reyno noted: "As a result of instant and accurate obedience to orders given by the controller, P/O Baggs brought his aircraft safely back to base with only a few gallons of petrol left. This officer displayed considerable coolness under difficult conditions." As to the radar team, it too was recognized. In the station diary someone wrote about the incident: "12 RD has been going around with a collective grin from ear to ear ever since."

Bill Baggs pushed ahead, each flight being something new. On September 29 he coaxed a Hurricane up to 28,000 feet. He flew at OTU for the last time on October 27, 1943, finishing with 59:25 hours on Hurricanes, and a further 25:20 on Harvards. The final log book notation he received was an "Average" from his flight commander. The "fledgling eagles" now were going overseas, although they did receive two weeks of embarkation leave before this. On November 16 Baggs reported in at "Y" Depot at Halifax, boarded the liner *Mauretania* a few days later with school buddy Chuck Harrison of Hamilton, and arrived in Liverpool on December 1. He then travelled by train to Bournemouth to No.3 Personnel Reception Centre. For the next three weeks he and his friends enjoyed life in and around the pleasant seaside resort, staying at local hotels like the Bath Hill Courts. Their worse duties were morning and afternoon parades, and the odd lecture.

At last P/O Baggs was posted to further training—his orders were to report on January 25, 1944 to No.61 (Fighter) OTU at Rednal in Shropshire. This he cheerfully did on the 29th, flying first on a Miles Master. The same day he strapped into Spitfire TO-E and made his first

**G/C Patriarch's inscription in** *Canada's Air Heritage*, **as presented to Bill Baggs at Wings Parade.**

**Hurricane entries from No.1 (F) OTU.**

flight in the legendary fighter. Everything went well. Baggs was elated—he was off and running. Soon he met a number of other Canadians, including George Starkey and Ken Trumley. The intensity of training built as each strove to excel. They took their Spitfires on tail chases, low-level cross countries, cine camera and live firing exercises, and practiced formation flying, aerobatics, sector recces, and forced-landings. By March 3 they were sufficiently competent to begin more advanced work at Montford Bridge (south of Liverpool). They practiced large formations, rhubarbs, night flying and bomber affiliation. All this culminated on March 31, by when Baggs had 60:30 hours on the Spitfire (he had flown it 57 times). The course had worked out for all the students but one. F/O John W. Wright of Toronto died when he dove straight in from high level. Discussion about this led to suspicion about his oxygen system. A funeral for Wright was held in Blacon Cemetery, Chester.

Even after all the hard work at Bagotville and Rednal, P/O Baggs required more training. On April 17 he started a Hurricane course at No.3 Tactical Exercise Unit at RAF Annan (near Dumfries). The focus was on dive bombing, rocketry, and low-level flying. This involved 30.5 hours in 33 flights. On one day alone Baggs flew three rocket trips and a fourth as spotter. By now the pilots knew their postings were near. Most felt they would be going on Spitfires. Word soon came, however, that some would go to RAF Honiley (a drome about equidistant from Cambridge and Ipswich) on a Typhoon course. One was Baggs, who was anxious to try the daunting fighter-bomber. After lectures, cockpit briefings, tips from his instructor and a good look through the pilot's notes, on June 12, 1944 Baggs strapped into Typhoon XA-L, started its mighty 2400-hp Sabre engine, taxied out and took off for his 1:10-hour introduction flight. He liked it right away, writing in his pocket

After his step by step period of training Bill Baggs got on operations. He sent this snap home in October 1944 writing on the back: "Wow! What a look on my face. I guess I just had to look tough beside my formidable little war horse. What do you think of those guns, dad?"

diary: "Another one of those few days in which I've actually accomplished something. After a good check I was given my first Typhoon to fly. What power! What speed! Really enjoyed the experience and I think I'll like them very much. They sure pull one around." This was not to say that the Typhoon was a piece of cake — compared to the gentlemanly habits of the Hurricane, pilots had to compensate for the Typhoon's strong tendency to swing on takeoff. It also had a "reputation". It was said that the tail of a Typhoon could separate in flight. If that didn't get you,

there was always the Sabre with its history of failures. All things considered, pilots like Bill Baggs were not going to let rumours interfere with their fun!

After all this time in training, P/O Baggs made only 16 Typhoon flights for 20:35 hours, then was posted to operations on June 30. Such was the confidence that the RAF had in those who had come down the BCATP pipeline. His new home would be with No.164 "Argentine-British" Squadron at Hurn, near Bournemouth. On July 18 he flew in a Dakota to join 164 at B.8 (Sommervieu) and the next day logged his first op. Ironically, it was aborted by weather, so Baggs had to bring his rockets home. The trip wasn't a dead loss, however—the flak around Caen was thick, quickly teaching the sprog about a deadly foe.

Bill Baggs would have an action-packed tour (see Hugh Halliday's *Typhoon and Tempest: The Canadian Story*), surviving 92 ops. The war over, he boarded the *Louis Pasteur* and sailed for home, arriving in Quebec City on August 13, 1945. Like most veterans, he returned to civilian life, taking up a career in the paint industry. There wasn't much time for reminiscing, although there were occasional reunions. It wasn't until 50 years had passed that a group of Typhoon pilots, headed by Ed McKay, started meeting in the Toronto area. This rekindled that old RCAF camaraderie. Mostly the chitchat revolved around ops, yet it had all begun at Manning Depot, then ITS, and that first flight in a Tiger Moth or Finch. From there had come SFTS, OTU and advanced courses. All this had taken Bill Baggs two closely-orchestrated years of training. The final product off the end of the line was a superbly trained pilot; the payoff was the 12 months he would put on ops. In Baggs' case, although he finished with 147:55 Typhoon hours, those preceding 405:15 hours in training had put him on the road to fighter pilot success.

**Spitfire flying at No.61 OTU.**

**Finally, Baggs flies the Typhoon.**

## Guard Duty

Aircrew recruits were expected to put in a few weeks on guard duty as they awaited ITS. Drafts usually were large, such as the one shown here (No.31 B&GS at RCAF Station Picton in 1941), attended by AC2 W.N. "Bill" Stowe. His story appears elsewhere in the book. It is rare that the names from such a group are available, but Stowe made a good effort to get everyone's autograph before his group dispersed.

The moving thing about a picture like this is that so many did not survive—at least 13 of the 46 students. Here is a partial history of how this group fared. First in front is George Kenneth Alfred Smallwood of Toronto. A pilot, he went to 429 Squadron on Halifaxes. On April 28, 1943 he was on a Gardening trip off Norway, when shot down and killed with his 6 crewmates in Wellington JB922. Next are Jim Spicer (Toronto), Eric G. More (Toronto), Doug MacKellar (Brantford, later a Halifax pilot on 77 Squadron, postwar in the RCAF and aerospace industry) and Frederick Ward (Toronto). Over England, Ward later had a dicey experience. While flying a Tiger Moth, he was attacked by an Me.109. He managed to survive by aggressively out-turning the fighter pilot, who eventually gave up. Ward died with his No.50 Squadron crew the night of March 12, 1943, their Lancaster (ED449) going down in Holland. Wally Roberts (Oshawa) later flew Beaufighters on 177 Squadron and was lost on ops at age 23 on November 4, 1943. Harold Turner (Niagara Falls, he left the RCAF when he had trouble night flying at SFTS, but went into the RCN), then an unknown corporal (staff). Bo Bruce (Niagara Falls), unknown and Henry Lynch follow. Then comes Archibald Clarke (North Bay) who died on December 21, 1943, his 132 Squadron Spitfire not returning from a sweep over France. "Swifty" Darragh is beside Clarke, followed by Bill Stowe (Toronto), Tom Fraser (Toronto) and W. Thatcher (Hamilton). Thatcher, a navigator, was lost with his 405 Squadron Lancaster crew on January 14, 1944. Last in the row is Teddy Finch of Brantford, who would survive a tour as a pilot on 408 Squadron, and earn a DFC.

In the middle are Earl Miltemore (Brantford), Jim Hobbs (Toronto, later a bush pilot with Austin Airways), Al Puffer (Toronto, flew 2 tours on Pathfinder Force, killed after the war in a traffic accident), Bud Hannah, Sam Kinnear (he later flew Liberators in the Far East; postwar, he was a corporate pilot for the T. Eaton Co.), Harold Reeves, J. Gallimore and Scotty Detlor (Brockville). Beside Detlor is Doug Warwick (Toronto). He must have become an excellent navigator, for he ended on one of Bomber Command's exclusive squadrons—No.617, "The Dambusters". On September 16, 1943 he was killed with his crew on ops in Lancaster JA874. Next is Alex McDonald (Toronto), who went on to Spitfires with 441 Squadron. Having survived a bail-out near Caen, he was killed over Belgium on October 28, 1944. Beside McDonald is Harry Beckett (he also would survive two tours pathfinding, then take his degree at Queen's University in Kingston after the war), then Elbert L. Buell, and Floyd Wile. A navigator from Nova Scotia, Wile would die on May 17, 1943 on 617 Squadron. His Lancaster was shot down while attacking the Moehne Dam. Charlie Wattie (navigator) and Edward Miles Merritt are far right. Merritt, a navigator from St. Stephen, New Brunswick would die with his 149 Squadron crew on the night of April 15, 1943 (Stirling BK759).

In the back are John Dagmar (Madison, Wisconsin), Arthur Scothorn, Jack Westlake (who did not train for long, but returned to civilian life), Pat McCann (he would survive and also study at Queen's), W.R. Taylor, unknown, Roy Waygood, and Ralph M. Lawrence (Brantford). Lawrence became a PR Spitfire pilot on 541 Squadron, but lost his life on ops to Kiel on June 24, 1943. Then are Ben W. Warren (Niagara Falls) and Kenneth Willard Rosevear of Toronto. Rosevear was a bomb aimer on 428 Squadron. He and his crew would all perish in Holland in their Wellington on April 5th, 1943. Next is Reginald Lionel Reddy Hepburn of Brighton, Ontario. On January 8, 1943 he and his RAF navigator were killed when their Beaufighter (No.2 OTU) crashed near Bewholme, Yorkshire. (Lionel's brother F/L Donald Stuart Reddy Hepburn had died 9 months earlier on Lancaster operations with 97 Squadron.) Finally are Clarence Drake, Jim Dalgleish, Red McCartney and AC2 Stewart.

## The OTU

The vital role of the operational training unit is mentioned frequently throughout this book, examples being No.31 (Coastal) OTU at Debert (later renumbered 7 OTU) and No.1 (Fighter) OTU at Bagotville. No.31, a RAF operation with Hudsons, was the first BCATP OTU, having formed on June 3, 1941. It also was one of the last, not closing until July 1945. Few such OTUs have been featured in books, one exception being No.5 OTU at Boundary Bay, BC with its detachment at nearby Abbotsford. Both stations were near Vancouver in the Fraser River Valley. Hectic though life would be at No.5, its days would be short-lived—April 1, 1944 to October 31, 1945. On average 5 OTU showed a strength of 43 Mitchells and 38 Liberators, while its gunnery flight had 12 Kittyhawks (fighter affiliation), 4 target-tow Bolingbrokes and 1 Norseman.

The task at No.5 OTU was to train 11-man crews on the B-24 Liberator (the details are best read in Carl Vincent's *Consolidated Liberator and Boeing Fortress*). Most crews were destined for RAF Liberator squadrons in India-Burma. The instructors at No.5 were a select crowd, having flown operations or with BCATP background. A typical officer commanding was W/C D.J. Williams, DSO, DFC. After graduating in August 1941 from No.3 SFTS in Calgary, he had gone overseas, where his first tour was on Hampdens with 408 Squadron. On a raid to Kassel on August 27/28, 1942 he was attacked by a Ju.88 night fighter. As it pulled away, Williams destroyed it with his fixed nose gun. This would be his first of several kills. Williams commanded 406 Squadron (Mosquitos). His DSO was awarded for an action on July 21, 1944, when he and his navigator, C.J. Kirkpatrick, encountered two Do.217 torpedo bombers, likely en route to attack nearby RN destroyers. Williams shot down one, had an engine disabled by return fire, then pressed on to send down the other Dornier. (After the war Williams would graduate from the Empire Test Pilots School, serve on WEE Flight, be the first in the RCAF checked out on SAC's B-47, and command No.1 Wing in Marville, France. Such was the calibre of a senior OTU man.)

The CFI at No.5 OTU was S/L D.P. MacIntyre, DFC, US DFC. After graduating from No.8 SFTS in Moncton in July 1941, he joined 35 Squadron, one of the first with Halifaxes. He was shot down during a raid on the *Tirpitz* in Trondheim. He and some companions evaded, reaching Sweden, then returned to England. In September 1942 he joined a RAF Liberator squadron in the Middle East, attached to the US 9th Air Corps; it was here that MacIntyre earned his US DFC, the first Canadian so honoured. No.5 OTU's chief ground instructor was F/L A.L. Parnall, a 1941 graduate from No.1 ANS at Rivers. His first overseas assignment was navigating aircraft from the UK to Egypt in 1942, then he joined 419 Squadron on Wellingtons. Eventually he was involved in an accident wherein only he and one other man survived. Parnall later converted to the Halifax, then was posted to Boundary Bay.

Since most pilots on course at 5 OTU had no experience with a tricycle landing gear aircraft like the Liberator, all started with the Mitchell flight at Boundary Bay, formed on August 15, 1944. Once competent on tricycle twins, pilots moved to the bigger "Lib". The OTU's souvenir history describes this well-remembered machine:

*The Liberator is a very nice old gentleman's aircraft. Take-off and landing are both extremely simple, first because the aircraft sits on the ground in approximately the flying attitude and, second, because it is unwilling to groundloop or even to*

A scene at 5 OTU as Liberator trainers get set for their daily missions. (NAC PA173389)

swing, and this unwillingness increases with acceleration. It is very heavy on the controls, especially the ailerons; but with a little effort, turns up to 70 and even 80 degrees can be executed smoothly, and 60 degree bank turns can be held indefinitely without loss of height.

The stall can be violent, but it occurs well below the cruising speed ... and gives plenty of warning. Even should a stall develop into a spin, this can be corrected readily by the normal method (stick forward and bank direction of spin), but one would be unwise to try this deliberately, even at a great height. Flying on three engines is very simple; flying on two engines is possible, but it is very hard and exhausting work. The trimming devices are most effective and it is possible to fly almost hands-off, though there may be some unsteadiness in the pitching plane.

The Liberator awes and even dismays the beginner by its array of dials, knobs, pipes, cables, electric motors and red-painted valves. The main parts of this puzzle, however, fit together fairly soon, and from then on the pilot learns something new about the aircraft every time he goes near it. One would imagine that, ultimately, this process would finish. Then one would know everything ... but apparently this is impossible within the span of an ordinary lifetime.

While the pilots were getting to know their new bomber, the other trades were equally busy. Navigators had an onerous duty, for they would have to guide their crews to target over vast stretches, whether of ocean or jungle. The 5 OTU history describes a typical nav briefing:

*The WAGs, AGs, navigators and air bombers are briefed separately by their section leaders at least one half hour before main briefing. Any information which applies particularly to the trade concerned is given ... Section Commanders, Intelligence and Flying Control complete briefing pro formas which are handed to the Flight Commander before main briefing time. These ... contain specialist information which is of interest to all crew members.*

*Main briefing is held ... at least two hours before take-off time ... Aircrew are seated at tables as crews, facing a large map covering practically the entire wall... The route ... is marked by coloured tape, the outward and return legs designated by different coloured tape or by arrows. On this map also are marked the target ... defended areas (simulated), M/F stations, H/F stations, radio beacons and radio ranges.*

*The Flight Commander takes the main briefing... He first calls the Met officer to give an outline of the weather expected ... and to give the latest navigational winds. A large synoptic chart is displayed on the wall ... also a cross section of any cloud expected on the trip, complete with freezing level and temperatures for different altitudes. The briefing officer then goes through the trip ... He stresses the importance of being fully and warmly equipped ... adherence to take-off time, etc. He gives the runway in use, wind conditions and aerodrome control information. He goes through the trip from take-off, climbing, setting course, cruising and let-down to landing. ... questions may be asked of any crew members ... Captains go through the trip with the crew and work out the finer points with them ... An interrogating officer interviews all crews on return to get a complete picture of the success of the exercise, e.g. adherence to track, ETAs and time on target. Any snags encountered are noted ... engine logs are examined, number of rounds fired and bombs dropped noted ...*

As it did any large unit, trouble hit 5 OTU during its 29 courses. On October 4, 1944, for example, Liberator EW127 crashed in flames on Salt Spring Island, killing 11 men. On January 9, 1945 another 11 died in the crash at sea of KH173. On July 4, 1945 Liberators KG880 and KH107 collided on the ground at Abbotsford, killing 9 of 13 men involved. An odd case of personal tragedy at 5 OTU is that of Sgt Don Palmer Scratch. Having joined the RCAF in July 1940, he succeeded in pilot training. He was injured in the crash of 119 Squadron Bolingbroke 9064 on March 19, 1942. His CO described Scratch as a "capable, industrious pilot, shows high degree of energy and initiative..." Scratch, by now commissioned, later flew Liberators on 10 (BR) Squadron, but it was here that he seemed to start unravelling. On July 20, 1944 he took up a Liberator alone, roaring around for more than three hours and beating up Gander. He later admitted his guilt, but excused himself as being frustrated with his EAC posting. He was court marshalled and dismissed, but, astoundingly, re-enlisted in September 1944 as a sergeant pilot. The following month he was posted to 5 OTU. He seems to have progressed normally, his instructor noting of him: "a very keen average pilot ... neat in his appearance ... very quiet and generally well-liked."

But Scratch again went off the deep end. At 0454 on December 6, close to the time when he would have finished his course, he took off alone at night in Mitchell HD343. He began shooting up the airfield, continuing this for more than two hours. WAC authorized Kittyhawks to shoot him down, but only should he steer towards the US border. Day dawned and Scratch continued his antics around Boundary Bay. G/C D.A.R. Bradshaw, the station commander, later reported: "At one time during the last hour he flew the entire length of the tarmac between the lines of parked aircraft and the hangars, so low that his propeller tips could only have been inches from the ground. ... His speed practically all the time, I would estimate ... 220 to 270 mph."

Scratch was always shadowed by the Kittyhawks. Later, the accident investigator on the case, G/C F.S. Wilkins, noted: "He paid no attention to their efforts and continued his wild aerobatics." At about 1000 hours Scratch flew a short distance from the station. Suddenly the Mitchell dove straight into the ground, abruptly ending the incident and the life of a strange young man. A court of enquiry was called, but came to no firm conclusion. People could not believe that Scratch was crazy. He was, however, a drinker, and had been tippling before absconding with the Mitchell. In the end, the court felt that it was likely that Scratch had crashed either from running out of fuel, or from physical and mental exhaustion. It was revealed that shortly after 0815 on December 6 a Liberator was found in a ditch at Boundary Bay; and there was a rum bottle in the cockpit. Investigation showed that it bore a fingerprint from Scratch. A minute scribbled in the side of the chief investigator's report noted: "Now that we have more than enough pilots, I hope the policy of 'once out, stay out' in these cases will prevail."

**Mitchells and Liberators of 5 OTU at Boundary Bay in the Fraser River flood plane. (CF)**

# It Wasn't Always A Piece of Cake

Accidents plagued the BCATP at a terrible cost of both aircraft and men. Inexperience was the chief culprit, but weather and mechanical failure added to the toll. In this case Tiger Moth 5110 of No.16 EFTS (Edmonton) was lost on September 21, 1941. Unless the occupants had taken to their parachutes, they surely must have died. (Bill Dunphy Col.)

When Finch 4477 from Windsor Mills pranged in a barnyard on July 30, 1941, the locals got a close-up look! (via Tom Dietrich)

On June 14, 1943 P/O Bill Boulton and Sgt Bill Hill were taxying, when FSgt Jim Buchanan collided with them. Buchanan's prop chewed in deeply, giving Hill a close shave. (via Rob Schweyer)

On February 13, 1942 these No.1 SFTS Harvards collided at Camp Borden. No.3184 was listed as "Cat.B"—requiring repairs at a major facility. No.2647 was lightly damaged, but the following May 26 was in an even worse accident. (via Doug McPhail)

Cold weather woes. These Battles are being salvaged after "off-strip" landings. (H.S. and Gordon Diller Col.)

Ansons landed atop one another on several occasions around the Commonwealth. The location of this Canadian incident is uncertain. (Chris Hare Col.)

The funeral of someone from No.1 B&GS. Such events occurred often as the BCATP strove to meet quotas. (via Rob Schweyer)

## No.34 OTU

Air gunner George Kozoriz, a prairie boy from Trochu, Alberta, completed his ops as an air gunner with 226 Squadron (Mitchells). Long before this he had trained at No.34 OTU at Pennfield Ridge, New Brunswick. No.34, established in June 1942, was the only Ventura OTU in Canada. In 1999 Kozoriz described this period in his RCAF career:

One of 34 OTU's Venturas trundles along at Pennfield Ridge. (J.A. "Joe" Ouellette)

*After some leave at home I left for Edmonton, where I boarded the CNR transcontinental for the east coast. After five days and nights we arrived in Halifax on November 21, 1942. From there I took a bus to Pennfield Ridge, where a pine forest had been cleared to make room for the airfield. There was nothing else around, except a small lunch counter across from the station. The nearest town was the fishing village of Black's Harbour, famous for its Connor Brothers canned sardines.*

*Station routine at Pennfield Ridge was a bit of a shock. We were billeted in the standard H-huts. It was comfortable enough, considering the winter weather. However, while we could use the ante-room of the Sergeants' Mess, we had to stay out of the dining room. This is because we were under-training aircrew. Our meals were in the Airmen's Mess. This was a bit bizarre and the standard of cleanliness in our mess left a lot to be desired.*

*The food was abysmal and was the first I had eaten that was prepared by RAF kitchen staff. They did an excellent job of ruining perfectly good Canadian rations. We each were issued a set of personal cutlery. After a meal we would return our dirty dishes, which we placed in large wooden racks having heavy screened bottoms. Next, there was a long narrow trough of water where we swished our cutlery. This usually was filled with floating food debris from those who had gone before. Each man usually carried his own drying cloth. Whenever possible, I re-washed my cutlery back in barracks.*

*My first Saturday night at Pennfield Ridge arrived, but there was nothing to do. A gala, all-ranks dance was under way in the recreation hall. Students were not invited, but I decided to beat the system. Since my second uniform still had LAC props on the sleeves, I slipped it on and went to the dance as an LAC. Luckily, no one recognized me. It proved a rather dull affair, since there were too few women compared to all the men. I went back to barracks early. One day in the following week everyone was confined to barracks. There was plenty of security and we wondered what was happening. Later we heard that an RAF sergeant was to be hung that night for murder. He had stabbed the daughter of the Connor family behind a dance hall. He was tried locally, found guilty and justice was swift for his heinous crime.*

*There still was no flying for the air gunners, although observers and WAGs already had been flying for a couple of weeks. The routine for us during free time was to be ever present in the mess ante-room, where we waited to be approached by a pilot looking for his fourth crew member. I already had been crewed with a New Zealand pilot, but had yet to meet our observer and WAG.*

*One cold Sunday afternoon I wandered over to the hangar line. I was keen to make my first flight in a Ventura, which had a hydraulically-operated Boulton Paul turret with twin .303 Brownings. I couldn't wait to try out the turret, for I knew about the Browning—how to strip and re-assemble it, to recognize the three kinds of stoppages, to clear them, etc. So far, however, all my air firing had been in Fairey Battles over Lake Winnipeg, while at No.3 B&GS at MacDonald. The Battle had a rather primitive set-up, where you had to stand in the rear cockpit. The coupe-top swung up as a sort of wind break. The only safety device, as you stood in the slip stream, was a short cable that clipped on the bottom of your parachute harness. There was a post-mounted WWI Lewis machine gun, which the gunner could pivot from side to side by squeezing a hand lever. The Ventura would be pie-in-the-sky compared to the Battle.*

*I watched from the hangar line for some time and observed what, to me, was a disturbing method of starting very cold engines. They were primed repeatedly with fuel, resulting in horrendous backfires. Even with two or three tries, engines would fail to start. Nonetheless, I went to the ready room and asked the first pilot I saw if I might go take a familiarization flip. A friendly Aussie, Sgt Grant Crawford Suttie, agreed. I wasn't even wearing a flying suit, but was given someone else's parachute and the four of us trekked out to Ventura AE936 about 3:00 p.m.*

*Our engines had been warmed up, so they started immediately. I had never been in such a large airplane, the Cessna Crane being my only other twin. I climbed into the turret and got ready. Away we went on a smooth, low-level flight. Everything seemed fine for about an hour, then I started hearing some agitated conversation. P/O Tom Good, our navigator, pointed to the starboard engine. It was overheating. Now we got into a snow storm. Before long the pilot feathered the starboard propeller. All I could do was listen and hope for the best. As we headed for the nearest USAAF airfield at Presque Isle, Maine, we slowly lost altitude.*

*At last Presque Isle came into view, thanks to the good work of our WAG, Sgt Alex "Lofty" Simpson, in getting the course direction on the radio; but there was a water tower straight ahead. "Sut" banked steeply to starboard, leveled out, but didn't have a hope of getting lined up on the runway in time. Ahead were trees; within moments we were mowing tops off them. By good fortune a potato field loomed. By this time I was thinking how my parents would be receiving a telegram about my demise in northern Maine. I closed my eyes, awaiting the inevitable. The Ventura hit hard, then all was silent. We were still alive! I scrambled from the turret and groped for the door handle. We all piled out, just as fire erupted. The base emergency vehicles arrived, as did a crowd of spectators, who lined an adjacent road to watch the fireworks as AE936 sent up smoke and flames.*

*We were whisked away in a staff car to a welcome by a colonel who introduced himself simply as "Curly". That night there was a big party on base—it was the first anniversary of Pearl Harbor. We were invited to a turkey dinner—3 BCATP sergeants and a pilot officer from Canada in an American Officers' Club. What a treat this was, compared to the usual fare in our ratty Airmen's Mess! Next day we were issued American-style Irving jackets and other gear, since ours had been lost in the fire. We spent five days at Presque Isle, while a board of inquiry, chaired by an RAF squadron leader from Pennfield Ridge, investigated our accident.*

*In the end a yellow Anson flew over to collect us. Two American ground crew wore themselves out trying to start the Anson by turning the engine cranks, then the RAF squadron leader took over. He had the knack and the propellers soon were spinning. We waved farewell to our generous friends and were back at Pennfield Ridge in about an hour. The first thing we heard was that our course had moved to Yarmouth. I was concerned about my crew, but Grant Suttie put his hand on my shoulder. "You're now a member of my crew—I'm not letting you go," he said.*

*It seemed a bit odd, but we were required to submit individual memos explaining how we had lost our personal flying gear. I had lost someone else's parachute, but didn't know whose. I was reprimanded for not following the proper RAF format with my memo. "I am, Sir, Your Obedient Servant" is how I should have prefaced my signature. I finally conformed, but under duress. That, and the experience of often being referred to overseas as a "colonial", left me with a poor impression of the RAF, which endured for life.*

Trenton in 1940 looking south. This view may be compared with that on page 111, to see how much Trenton grew in a few years. Aircraft below include at least 18 Battles, some Oxfords and a Boeing 247. Some of these buildings remained in use in 2000. (CF PL1222)

Trenton at ground level showing Lockheed 10s, Harvard Is, Ansons, Oxfords, Battles, Fawns and Finches. (CF PL293)

## The Home War Establishment

The RCAF's Home War Establishment (home-based units) comprised Eastern Air Command, which focused on the U-boat war; and Western Air Command, responsible for discouraging Japanese aim in the North Pacific. Time would tell that EAC was where the action was; its squadrons would fight in dozens of actions and sink numerous U-boats. The West Coast would be quiet, save for some shooting in the Aleutians. At first the HWE had only seven squadrons with obsolete types like the Atlas, Goblin and Wapiti. This was rectified—by 1943 EAC counted 19 squadrons, WAC 18, nearly all with modern types. Meanwhile, the RCAF home front involved other activities— the Northwest Staging Route, air-sea rescue, internal ferrying of aircraft, test and development, etc. Broadly speaking, the home front also included the civilian world— industries turning out every product imaginable, farmers growing the food that kept overseas armies supplied, all the volunteer groups doing everything from knitting socks, to raising funds, to packing Red Cross packages for POWs, etc.

The groundwork for Canada's part in the war had been laid some years before hostilities began, when Ottawa began bolstering military budgets in response to the international situation. Thus did the RCAF turn from civil to service flying. Manpower restrictions were eased. Leaky old clunkers like the Vancouver now carried machine guns and bombs. Air stations were expanded and modernized. New aircraft were ordered, old factories expanded, new ones erected, e.g. National Steel Car at Malton to build Hampdens and Lysanders, Fairchild in Longueuil (Bolingbrokes), Canadian Vickers in Montreal (Deltas and Stranraers), de Havilland Canada in Toronto (Tiger Moths and Ansons), Fleet in Fort Erie (Finches). It didn't matter that these designs were not the latest. They provided the RCAF with meaningful experience. Perhaps more importantly, they gave industry the know-how that, within months, would enable it to begin mass production of advanced designs like the Harvard, Hurricane, Lancaster and Mosquito.

Defending Canada was a costly undertaking, requiring many fighter and patrol squadrons. In the northwest Bolingbroke fighters with belly packs (4 x .303 machine guns) were part of the picture. (David Thompson Col.)

## New Pre-WWII RCAF Types

| Taken on Strength | Type | Category |
|---|---|---|
| March 1936 | Fleet Fawn Mk.II | Trainer |
| March 1936 | Westland Wapiti Mk.IIA | Army co-operation |
| August 1936 | Fairchild Super 71P | Photo |
| September 1936 | Northrop Delta | General purpose |
| October 1936 | Blackburn Shark | Torpedo bomber |
| July 1938 | Grumman Goose | Transport |
| January 1938 | D.H.82 Tiger Moth | Trainer |
| November 1938 | Supermarine Stranraer | Coastal patrol |
| February 1939 | Hawker Hurricane | Fighter |
| May 1939 | Airspeed Oxford | Trainer |
| July 1939 | Harvard Mk.I | Trainer |
| August 1939 | Fairey Battle | Army co-operation |
| September 1939 | Lockheed Hudson | Coastal patrol |
| September 1939 | Westland Lysander | Army co-operation |
| October 1939 | Fleet Finch | Trainer |
| October 1939 | Lockheed 10 | Transport |

## Defining Defensive Roles

Besides training, much else had to be done on the home front. RCAF HQ pushed for fighter protection—there was near paranoia about possible enemy landings on the east coast, even rumours of a German aircraft carrier somewhere in the region. No.1 Squadron stood guard with Hurricanes at Dartmouth. No.118 also was there with "frontline" Grumman Goblins. These tough, manoeuvrable and hopelessly obsolete fighters were replaced by Kittyhawks, then No.118 moved to the Alaskan theatre. Other squadrons took shape, typical being No.125 at Torbay (Hurricanes). These even did over-water patrols, carrying underwing depth charges, lest a U-boat be sighted (none ever was). In time there were many Hurricane and Kittyhawk squadrons on both coasts. Some argued that there were far too many HWE fighter squadrons, to the detriment of other types.

Canadian-built Hurricanes equipped HWE squadrons from Tofino on remote Pacific Ocean shores to Torbay, facing the Atlantic. BW850/BV-T served No.126 "Flying Lancers" Squadron. Formed in April 1942, its first COs were Battle of Britain veterans F/L A.M. "Jeep" Yuile and S/L Hartland Molson. Operating from Dartmouth, Bagotville and Gander, No.126 flew 1919 sorties, and 2596 operational/12,238 training hours. Accidents claimed 9 aircraft and 6 pilots. (CF REA253-48)

The Curtiss P-40 Kittyhawk played a key role in the HWE. This one of 132 Squadron is running up at Patricia Bay, BC. (NAC PA197485)

A large Japanese force invaded the Aleutians in 1942, threatening Alaska and BC. A US-Canada force, including RCAF Kittyhawks and Bolingbrokes, rushed to the region and the Japanese were crushed. Flying from Umnak Island, the Kittyhawks of 14 Squadron flew 190 sorties against the enemy at Kiska. They usually carried 300-pound general purpose and 20-pound fragmentation bombs. On September 25, 1942 S/L Kenneth A. Boomer of 111 Squadron shot down a Zero near Kiska, the only RCAF kill in this theatre. Boomer also made a concerted attack on a Japanese submarine during his tour. (On October 22, 1944 Boomer was lost over Germany in his 418 Squadron Mosquito.)

There is no confirmed case of a Japanese naval vessel being spotted off BC or the Alaska panhandle. Some attacks were reported on alleged submarines, but these usually turned out to be whales or deadheads. One case rated a banner headline in the Toronto *Evening Telegram* of Saturday, December 26, 1942: "Toronto Boy Tells How He Bombed and Sank Jap Submarine"

Just as 118 Squadron was settling at Annette Island, this headline flashed across Canada. A Japanese submarine had shelled the Estevan Point light house, causing no damage. This would have been the only time in WWII that Canada was fired upon. (Globe and Mail)

roared the paper in an article by former Toronto mayor and WWI pilot Bert Wemp, DFC. According to Wemp, he was greeted at an RCAF base "somewhere in Alaska" by the CO, S/L Gordon Diamond, who began, "Let me introduce you to one of our lads who sank a Jap sub." The keen young fellow was P/O W.E. Thomas of Toronto. Wemp quoted him: "The submarine was surfaced when we sighted it. We put our nose down and went for it... one 500-pounder hit right on top of the craft as she went under for a crash dive ... The navy patrol came along later and finished the job with depth charges. There was nothing left but wreckage." This nutty report smells like pure wartime propaganda. The very best that might have happened is that Thomas thought he had hit a submarine and his story was not doubted. If so, the wreckage in the sea was not much more than fins and blubber from a local whale. (As to Thomas, in August 1945 he died of natural causes while hospitalized in England.) In retrospect the war on the West Coast proved more like a training exercise, but the RCAF could take no chances—better to be safe than sorry.

Art Yuile, Buck Newsome, Allan Studholme, Bernie Petley, Frank Grant, Bill Stowe and Pete Wilson—seven of those on the Dartmouth—Annette Island odyssey. (Bill Stowe Col.)

## Kittyhawks on Annette Island

No.18 Squadron was an auxiliary bomber squadron formed at Montreal with Moths in September 1934. Flying did not commence until May 1936. In November 1937 No.18 became No.118 Squadron. When war broke out, it was assigned to coastal artillery co-operation at Saint John, NB (Atlas and Lysander). The first patrol took off on November 11, 1939, but was cut short when Atlas 404 suffered engine trouble. After logging 1023 hours, the squadron disbanded on September 27, 1940. On December 13, 1940 it re-formed at Rockcliffe under F/L E.W. Beardmore, a Battle of Britain pilot. Its first aircraft were antiquated Goblins that Canadian Car and Foundry had been trying to pawn off on the RCAF for years. In July 1941 No.118 moved to Dartmouth to help defend Atlantic Canada. Its prospects improved in November when the first Kittyhawks were delivered. By this time W/C Ernie McNab was CO.

In June 1942 S/L Yuile led 118 on a trans-Canada flight to its new base at Annette Island, a US base near Ketchikan, a refuelling stop for aircraft ferrying to and from Alaska. There 118 took up its duties in the defence of North America. Also at Annette Island was 115 (BR) Squadron with Bolingbrokes, keeping watch for Japanese naval movements. No. 118 set up in tents four miles from the main US base. The men dubbed their home "Camp Tokyo", or "Father Nesbitt's Boys Camp" (named for the well-liked W/C Deane Nesbitt, the RCAF's first station commander at Annette Island). The operation was supported from Prince Rupert by the RCAF vessel *Sekani*. One airman recalled: "The 10-hour trip to Prince Rupert was rough and uncomfortable. Many meals were contributed to the fishes from the rail of the Sick Annie." Social life for the Canadians varied. There was a decent mess hall, and the airmen built a canteen from scrap lumber. In summer there was outdoor activity—

No.118 Squadron worked up to operational status at Rockcliffe and Dartmouth in 1940-41, first with Goblins, then Hurricanes and Kittyhawks. With strong leadership from Battle of Britain COs F/L E.W. Beardmore, W/C E.A. McNab, S/L H. deM Molson and S/L A.M. Yuile it was ready for its posting later to Alaska. Here Goblin 337/RE-G sits at Dartmouth. Note that wheel chocks are in place and a pan is under the engine to catch oil drips. "RE" was 118's squadron code, "G" represented the particular aircraft. The code changed to "VW" with the Kittyhawk. (Les Corness/CANAV Col.)

No.118 Squadron ready to depart St. Hubert for Alaska on June 6, 1942. VW-V shows the RAF serial AK809, but really was RCAF 1035. (Bill Stowe Col.)

## Annette Island Pilot

William "Bill" North Stowe served with 118 Squadron from Dartmouth to June 1943. Born in Edmonton on November 2, 1922, he returned to his parents' home in Yorkshire as a child, then the family re-emigrated, this time to Toronto. There Mr. Stowe worked in the printing trade. Young Bill was fascinated by airplanes, and sometimes would cycle from his home in the Yonge-Lawrence neighbourhood to Barker Field on Dufferin Street to watch the activity. As a teenager he developed a love for speed, getting into motorcycling. When his chance came, he joined the RCAF.

May 1941 found the 18-year old beginning his career at No.1 Manning Depot at the CNE grounds in Toronto. Next he was posted to guard duty at RCAF Station Picton. In August he was on the first course at No.5 ITS, recently opened in the School for the Deaf in Belleville. The following month he joined Course 39 at No.1 EFTS at Malton. There instructor George Slocombe took him on his first flight (Tiger Moth 5114) on September 26. The course involved a half day of class work, a half day of flying. Stowe soloed in Tiger Moth 4381 on October 20. Some of his lessons were with the CFI, E.L. "Les" Baxter (later well-known in Ferry Command and as a postwar commercial photographer). On November 16 Stowe was ready for his cross-country. In the morning Mr. Holden took him Malton-Kitchener-Brantford-Malton. After lunch Stowe took off solo, this time flying Malton-Brantford-Mount Hope-Malton. A few days later he graduated, his log showing 29:30 hours of dual, 31:05 solo.

LAC Stowe now was posted to No.2 SFTS in Ottawa, where he flew initially on November 26 in Harvard 3040 with P/O Airey. He joined "C" Flight under P/O Bob Hyndman, who later would fly Spitfires, then become one of the RCAF's leading war artists. Sgt Bob Pentland gave Stowe most of his instruction. He soloed in Harvard 2516 on November 29. In 1999 Stowe remembered how hard everyone worked and how little time there was for relaxation. Everyone's greatest fear was washing out. Stowe succeeded, making his last flight on March 6, 1942, finishing with 60:20 hours dual and 65:35 hours solo on the Harvard. No time was wasted, the RCAF sending Stowe to 118 (F) Squadron (Kittyhawks) at Dartmouth. He flew there first on March 17 in Harvard 2618 with WO Parsons.

Since the RCAF as yet had no fighter OTU, 118 did its own conversion of sprog pilots, the first being Stowe, Ivens, Frostad and Baxter. First there were a few hours in the Harvard, then came the jump to the Hurricane. After some ground school, studying the pilot handling notes, and getting genned up from the experienced pilots, a sprog would take up his first Hurricane. So it was that on March 21 away Bill Stowe went aloft in Hurricane BV835.

Some Hurricanes belonged to the RN's Merchant Ship Fighter Unit at Dartmouth. These aircraft served aboard freighters as defence against the Luftwaffe, which was sinking vessels around

fishing, hiking, swimming, boating. The Canadians formed an amateur theatre group, put on sing-alongs and had bingo nights. There were visiting entertainers, and even tours to the local salmon cannery. A host of other activities from horse shoes to baseball and volleyball tournaments, ping pong, table hockey, band practice and photography helped keep everyone occupied in off-hours. One thing missing, however, was women, the only ones being in Ketchikan, where the US Coast Guard seemed to have first choice. The RCAF published a base newspaper—the *Cannette*. It was filled with gossip, news of postings, gen about enemy aircraft, weekly movie listings, etc. A typical item appeared in the July 10, 1943 issue:

*"Blackout of 1943", that smoothly executed variety show presented by the Air Force Troupe of talented entertainers, more than lived up to its advance billings. Each and every one of the hundreds of airmen and soldiers who viewed the performance, voted it by far the best show to hit Annette Island. The ladies of the cast proved to be more than mere novelties—certainly any members of the opposite sex would have little difficulty entertaining the boys at Annette. But these girls were gifted with oodles of talent. Their singing and dancing would charm the most hard-hearted audience ... Most certainly they delighted the lonely and oft times bored personnel at Annette ... the troupe included numerous members of the sterner sex with talent to burn—smart comedy, music, songs and skits all presented in a fast and smooth-moving manner with every evidence of expert professional direction.*

The RCAF was well treated in Ketchikan, although the rough, four-hour boat trip challenged the toughest of airmen. Winter also was rough, with long nights, snow and dampness that seeped in everywhere. Activities moved indoors, especially to the movie theatre run by the Americans. Use of the US base library increased, but so did the drinking. There was considerable flying, although the enemy never visited the region. Several Kittyhawks were wrecked in accidents. F/L Arthur Jarred, age 27 from Lansing, Michigan and P/O George Baxter of Victoria were 118's only Alaska fatalities. On March 28, 1943 Jarred crashed while rolling a Kittyhawk with a 300-pound, centre-line bomb. This was a foolish stunt, for rolling was difficult in such a configuration.

In August 1943 No.118 Squadron departed for Sea Island. On October 14, 1943 most personnel went on embarkation leave before going overseas, where they became 438 Squadron on Hurricanes, then Typhoons. On November 1 they boarded the *Mauretania*, destination Greenock. When they docked on November 16 they were greeted by an RCAF band. Very exciting times lay ahead for the squadron.

Some of 118's 28 technical men who travelled west by train. The porter and one of the men's wives got into this snapshot, taken while passing through Winnipeg. (Murray Castator Col.)

The groundcrew had plenty to do servicing engines, arming, fuelling and putting wrecks back together. Here a crew give a Kittyhawk a good check. Those who flew this fighter always felt safe behind its superb Allison engine. (Murray Castator Col.)

VW-K after pranging off the runway at Annette. The Kittyhawk was no Spitfire, but had far greater range—more than 2 hours. It packed six .50 cal. machine guns, plus stores. (Murray Castator Col.)

This Kittyhawk had to be foamed after catching fire. (Bill Stowe Col.)

Rigged for 8 practice bombs, Annette Island Harvards 2618 and 3289 helped pilots keep their skills sharp. (Murray Castator Col.)

No.115 (BR) Squadron did anti-sub and shipping patrols. Equipped with a belly pack with four .303 machine guns, its Bolingbrokes also were fighters. (Murray Castator Col.)

Bolingbroke 9044 threw a propeller out to sea on February 11, 1943, then broke its back landing. (Murray Castator Col.)

This 115 Squadron Bolingbroke was not so lucky. It burned on the runway at Annette. (Murray Castator Col.)

No.115 Norseman 695 at Annette. Originally CF-AZA, it was the second of its type, having flown in April 1936. It served with Mackenzie Air Service, was impressed into the RCAF in February 1940, but was scrapped late in the war. (Murray Castator Col.)

No.115 aircrew at Annette Island with the boss, S/L Ralph Ashman, in the middle with the cigar. (Bill Stowe Col.)

Men from Annette Island make the boat trip to Ketchikan for the 1942 Memorial Day parade; then they come marching by, looking quite smart. (All, Murray Castator Col.)

Fishermen of 118—Sgt Burmaster, Cpl Castator and FSgt Pedley have made a nice haul.

Although the fellows complained about life at Annette, there was always something to do. If not work, then something like pay parade, as shown in this snap.

Talent night in the rec hall. An RCAF accordion player is at work, as other US and Canadian amateurs wait their turn. Then, an impromptu visit by Bob Hope (striped tie) and Gerry Colonna (mustache). Canadians swarm around.

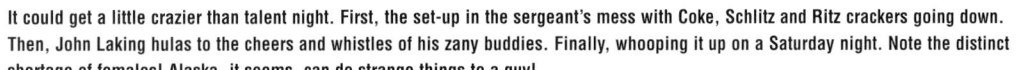

It could get a little crazier than talent night. First, the set-up in the sergeant's mess with Coke, Schlitz and Ritz crackers going down. Then, John Laking hulas to the cheers and whistles of his zany buddies. Finally, whooping it up on a Saturday night. Note the distinct shortage of females! Alaska, it seems, can do strange things to a guy!

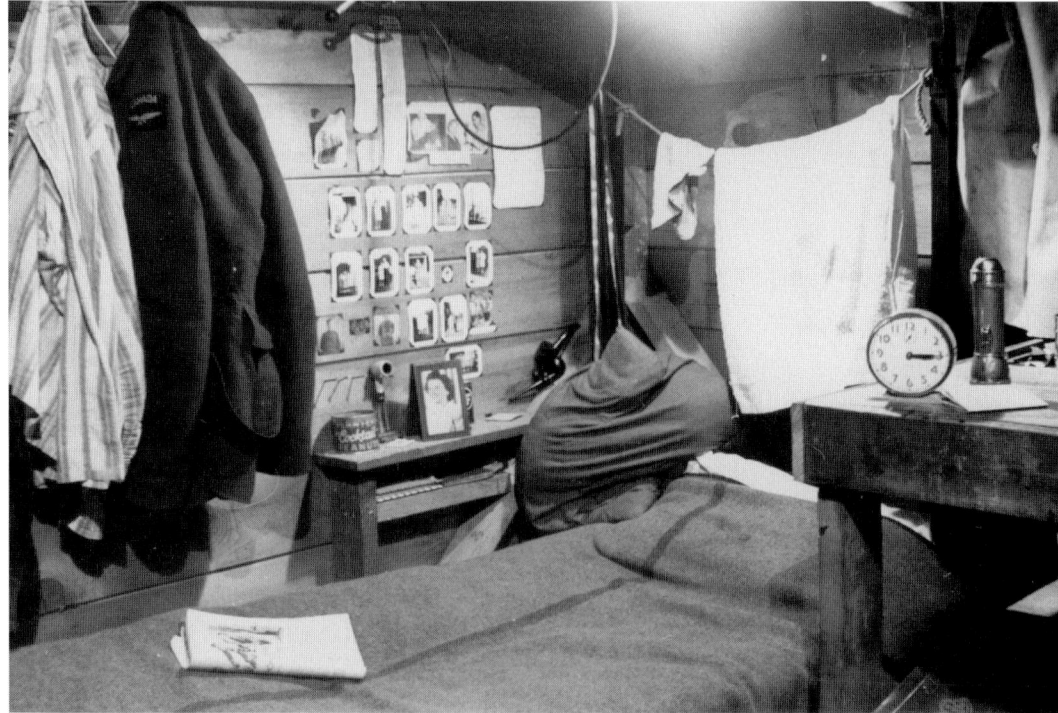

Murray's corner in the tent he shared. Wherever the fellows went, they tried to have some sort of a cozy little spot.

Murray Castator displays a standard practice bomb. His T-shirt sports the Bryant Press logo—that's where he worked in Toronto before and after the war. Many years later Bryant Press helped put CANAV Books on the map by printing its first titles.

The RCAF first aid tent and infirmary. The walkway was typical, Annette Island being largely muskeg.

Although things were quiet on Annette Island, there was the possibility of an enemy landing—everyone had to know what to do in that event. These 118 men were training in hand-to-hand fighting.

The layout of the 118's aircraft serviceability board was W/C Nesbitt's idea, based on the style he was familiar with during the Battle of Britain.

RCAF launch M-291 was stationed at Annette Island. (Bill Stowe Col.)

Besides RCAF types, many other aircraft were based at Annette, or regularly passed through. Most were going to or from Alaska, or ferrying to Siberia. In the opening photo Martin B-26s are nearest, then 115 Bolingbrokes, a C-47, a PanAm Lodestar and P-39s, wings covered against the weather.

C-47 41-38717 on the tarmac to carry F/O Jarred to his funeral in Michigan. Note the guard of honour, also the US Navy Venturas.

This early Douglas C-54 drew a lot of attention when it stopped into Annette Island.

Interesting amphibian—Douglas Dolphin NC26 of the US Civil Aeronautics Administration.

A B-24 with anti-surface vessel (ASV) radar (note antennae underwing).

RCAF Canso No.9752 at Annette. It was escorting No.14 Squadron's Kittyhawks to Umnak Island in March 1943.

Lockheed P-38 41-2069 taxis at Annette.

# Some of the Pilots of 118 Squadron

W/C A.D. Nesbitt, DFC ran the RCAF show on Annette Island from June to October of 1942. On his previous posting he was CO of 401 Squadron at RAF Digby, Lincs. May to July 1944 he commanded No.144 Wing (441, 442, 443 Squadrons—Spitfires) and January to September 1945 he commanded No.143 Wing (438, 439, 440 Squadrons—Typhoons). (Murray Castator Col.)

Frank G. Grant was 118's CO from February 28 to July 28, 1943, then went overseas 438's first CO. He survived the war with a DFC. (Bill Stowe Col.)

Sgt Arthur C. Brooker in winter flying gear. He later joined 438. On November 30, 1944 he and fellow pilot Bill Beattie were in a Dakota returning to Eindhoven from leave in England. Passing over Dunkirk, the "Dak" was blasted by flak and crash-landed. Luckily, the fellows escaped. (Bill Stowe Col.)

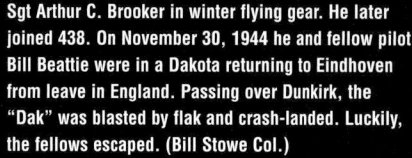

Big and little. Don Connell and Jack Beirnes at Annette Island. (Bill Stowe Col.)

At age 27 F/O Art "Flash Gordon" Jarred was one of 118's older pilots. One day he tried rolling a Kittyhawk carrying a 300-lb bomb. This no-no cost Jarred his life. (Murray Castator Col.)

Bill Stowe and Pete Wilson take in Alaska's scenery. (Bill Stowe Col.)

George Baxter on 118's Hurricane Flight at Dartmouth. The aircraft is BW835 named "Gary Madore" to honour P/O Garfield Madore of Saskatoon, shot down in his 242 Squadron Hurricane, as he covered the Dunkirk evacuation in May 1940. As to George, he was killed in a freak accident while flying Kittyhawk AL210 from Annette Island on October 12, 1942. His seat dinghy somehow inflated, making it impossible for him to control his aircraft. (Bill Stowe Col.)

Bill Stowe of Toronto and Jack Beirnes of Tofield, Alberta in an Annette Island gag shot. Bill went on to fly Spitfires on 41 Squadron. Jack moved to 438, which he twice commanded. On June 1, 1945 this competent pilot was in a Victory flypast over Copenhagen, when he somehow crashed. He is buried in Aabenraa Cemetery in Jutland, Denmark. (Bill Stowe Col.)

Competent and well-liked, Pete Wilson of North Vancouver was a popular 118 pilot. On New Year's Day 1945 he was caught trying to take off at Eindhoven. German fighters riddle his Typhoon. His mates hauled him from the cockpit, but he died within minutes.(Bill Stowe Col.)

The great Art Yuile at the controls of a motor boat off Annette Island. (Bill Stow Col.)

Albert "Buck" Newsome of Prescott, Ontario poses with Kittyhawk "Marie". An original with 438 Squadron, he was hit by flak and killed near Borken, Germany on October 7, 1944. (Bill Stowe Col.)

After his 118 Squadron tour P/O Herb Ivens flew Hurricanes and Thunderbolts in Burma with 261 and 146 Squadrons. He was shot down in December 1944, captured, and abused by the Japanese. After the war he studied law, then travelled to Japan. He learned the language with the purpose of determining why he had been so mistreated by his fellow human beings. Apparently he even met some of those responsible for his plight as a POW. He passed away about 1980 on the West Coast. Ivens would have been disheartened in 1999 when Japan re-adopted its wartime flag and anthem, two of humanities great terror symbols. (Murray Castator Col.)

Ross Reid of Toronto while in Alaska. He also would have a full career with 438 on the Typhoon. (Bill Stowe Col.)

Alan Studholme of Toronto later joined 401 Squadron, was shot down over Holland, and spent much of the war as a POW. This photo is accompanied by the newspaper report of his predicament. This fine gentleman passed away in Unionville, Ontario in June 1999, a victim of Alzheimer's Disease. (Murray Castator Col.)

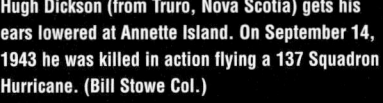

Hugh Dickson (from Truro, Nova Scotia) gets his ears lowered at Annette Island. On September 14, 1943 he was killed in action flying a 137 Squadron Hurricane. (Bill Stowe Col.)

the UK. A Hurricane was mounted on a launch rail on a ship's bow (usually one such ship per convoy). When needed, the fighter would take off, assisted by rocket boosters. After a patrol it would return to the convoy, where the pilot would ditch or parachute, expecting to be rescued. This was not a pilot's dream posting, yet men did volunteer. Occasionally, an MSFU Hurricane needed servicing. As an MSFS ship neared Halifax, its fighter would fly to Dartmouth. Now, until needed again by the MSFU, 118 Squadron could use it for training. F/L Freddie Turner (RAF) was OC in charge of the MSFU at Dartmouth.

Bill Stowe flew his first Kittyhawk (RE-Y) on April 18, 1942. Training now encompassed such skills as formation flying, air-to-ground and drogue firing and searchlight co-operation. Eventually he made his first dawn and dusk patrols as a member of F/O A.E. Studholme's "B" Flight. Now the squadron got news that it was posted west. Although pilots had an idea about what was going on, the airmen were in the dark. Even after they entrained, they didn't know their destination. The pilots were briefed to fly en masse across Canada to a "Y" Wing at Annette Island. They began their marathon trip on June 6, Bill Stowe taking VW-W as shown in his log:

| Date | Routing | Duration |
|---|---|---|
| 6 | Dartmouth-Pennfield Ridge | 1:00 |
|   | Pennfield Ridge-St. Hubert | 1:45 |
| 7 | St. Hubert-Rockcliffe | 0:30 |
|   | Rockcliffe, test flight | 0:30 |
|   | Rockcliffe-North Bay | 1:45 |
| 8 | North Bay-Porquis Junction | 1:00 |
|   | Porquis Junction-Armstrong | 2:15 |
| 8 | Armstrong-Winnipeg | 2:05 |
| 9 | Winnipeg-Regina | 1:50 |
|   | Regina-Lethbridge | 2:00 |
| 10 | Lethbridge-Edmonton | 1:35 |
| 12 | Edmonton, test flight | 0:35 |
| 21 | Edmonton-Prince George | 2:05 |
|    | Prince George-Annette Island | 2:15 |
| Total flying | | 21:10 hours |

On June 25 Stowe made his first flight (in VW-W) from the rough gravel strip at Annette. On July 9 there was a "sub search" (2:25 hours). On such patrols the Kittyhawks usually carried a 300-lb high explosive bomb on the centreline. On the 28th there was a scramble after some unidentified vessel. The summer weather proved ideal, so flying was constant. The pilots got to love their Kittyhawks. "We were bloody lucky to have those kites," recalled Stowe. They trained ceaselessly, especially at formation flying, gunnery, dive bombing, and dog fighting. Two or three sorties a day was typical. Stowe logged 24:15 hours in July. A typical log book entry is from December 17: "Attacks on Kingfisher. Shoot up field. Escort Goose. 1:05." This good fun was had by Stowe in company with his friend, Jack Beirnes. Now the weather turned—it was cold and damp in the tents and shacks. Flying hours fell, but having leaders like W/C Deane Nesbitt and S/L Art Yuile kept up squadron morale.

Winter weather resulted in lost airplanes. Some were found, others not. A note from Stowe's log on New Year's Day, 1943 reads, "Lockheed lost in mountains." For the 14th: "Search for B-24." On February 16 Stowe flew VW-W to intercept a Stranraer near Sandspit in the Queen Charlotte Islands. He then escorted it to Alliford Bay. Next day he was pleased to note, "Bombing. 300 lb demolition. Direct hit." On the 19th he flew VW-W: "Escort 3 Bolingbrokes. Base-Rose Point-Prince Rupert, return. Interception." Next day he noted that Bernie Petley had belly landed.

June 5-7 Stowe took Kittyhawk 1082 Annette Island-Juneau-Yukatat-Elmendorf-Kodiak Island, delivering it for another squadron. F/O Wallace of 115 Squadron flew him home in Bolingbroke 9122. A heart-breaker came on June 8: "Dawn patrol, F/O Petley. Sub search. Freddie Currie, Dowling and Stapleford, plus 3 USO hostesses crash off Metlakatla in Norseman." Currie of 115 Squadron often flew the Norseman on general trips. This day he got into a blizzard, tried landing, but didn't make it. On June 25 Bill Stowe left Annette Island. An American DC-3 flew him to Seattle, then he headed home by rail to Toronto on leave. In September he was in Halifax to board SS *Strathnaver*. The ship sailed, but had mechanical trouble and had to put into St. John's. On September 29 Stowe left Torbay for Dartmouth in Dakota 650, then received orders to entrain for New York. On October 8 he boarded the *Queen Mary* for Greenock and a new adventure flying Spitfires.

## HWE Coastal Patrol

If HWE fighters seemed superfluous, it was long-range patrol planes that the RCAF needed. At first there were only Vancouvers in the west, Stranraers in the east. Although they flew many early patrols, they were pitiful as weapons of war. This lead Air Commodore G.O. Johnson in September 1939 to warn that the Vancouvers "should be kept within easy reach" of repair facilities, and that to send them too far afield "is to invite disaster." Even so, a look at No.4 (BR) Squadron shows the important work being done. The squadron mobilized on September 10, its first operation being flown on the 12th by FSgt J.W. McNee and crew in Vancouver 906. The same day S/L Mawdesley and P/O Fred J. Ewart patrolled in Stranraer 912. Hereafter, there were almost daily patrols with 902, 906 and 912. The pace accelerated, diary comments suggesting a war footing—gas masks, increased gunnery practice, etc.

On October 4 Vancouver 902 and "Stranny" 915 escorted *Empress of Japan*, *Empress of Canada* and *Empress of Asia* in days following. This became routine work. On October 21 S/L Mawdesley took over No.4. On November 3 he flew 915 on a gruelling patrol to centres from Esquimalt to Bella Bella to Prince Rupert, logging 17:50 hours. On the 12th he transported army officers to Esquimalt, including MGen A.G.L. McNaughton and BGen C.R. Pearkes. Much training was done in navigation, seamanship, mooring and handling on the water. At the beginning of 1940 squadron aircraft included Fairchild 633, Vancouvers 902 and 903, Vedette 816 and Strannies 912 and 915. Eight pilots were on strength. The ORB for this period shows Mawdesley leading most of the operational patrols. On December 14, 1940 he flew 912 three times: on a weather recce (0:30 hours), on a passenger trip to Esquimalt (1:00), and on a local recce (0:35). A patrol in 915 on December 23 listed Air Commodore A.E. Godfrey as captain, Mawdesley as co-pilot. Godfrey, always excited about operations, flew another patrol on the 26th.

On New Years Day 1940 Vancouver 903 operated Bella Bella-Alliford Bay under F/L C.M.G. "Con" Farrell, while F/O S.R. McMillan flew Fairchild 633 to and from Esquimalt. Both men were famous bush pilots. On January 7 S/L Mawdesley flew 11 passengers to Alliford Bay, then returned with another 11, having been away 9 hours. On the 10th a note is made that F/O Vanhee was taking instruction in 633 from F/O McMillan (Vanhee, a prewar bush pilot, appears to have made his first operational flight on the 14th in 915, as the CO's co-pilot.

Vedette 816 was added to No.4's roster on January 24, Mawdesley flying it that day and finding it "entirely satisfactory". The next day

**No.4 (BR) Stranraers, one on beaching gear (previous page), the other about to dock. These sturdy "boats" filled a gap until the PBY arrived. (CF UC114, Gordon S. Williams)**

902 began its 200-hour overhaul (completed on the 30th), thus explaining the arrival of 816. F/O C.C. "Chuck" Austin, well-known in Northern Ontario with Austin Airways, first appears in the ORB on January 28, taking instruction in 916 from F/L J.E. Jellison. Even this late in the game the Vancouvers were on armed patrols, e.g. January 4 F/L Con Farrell flew 903 Bella Bella to Alert. A 7:10-hour reconnaissance of January 13 in 903 was under FSgt J.W. "Johnny" McNee with F/O J.K. "Jack" Herriot (another prewar bush pilot), Cpl John A. Biehler, and AC1 W.H. Adrian as crew. They went up Georgia Strait from Jericho Beach, to Alert Bay, Bella Bella and Alliford Bay. They overnighted on a barge at Bella Bella. "We damn near met our Waterloo in Alert Bay," Biehler mentioned in 1999. "It wasn't easy mooring a Vancouver, and with a 12-foot tide running, we came close to losing a wingtip." The next day 903 returned to Vancouver without incident. (On another trip Biehler was with McNee when the latter decided to take his Fairchild under the Lion's Gate Bridge.)

Pilot proficiency was stressed at No.4. On February 11, 1940 F/O McMillan and F/O Vanhee were aloft in 902, the ORB noting: "Dual instruction for F/O Vanhee: taking off, circuits, alighting, leaving and approaching buoys." All flying on February 25 is listed as training, 633, 816, 902, 912 and 915 logging 20:05 hours on 16 flights. On March 17, 1940 Vedette 809 appears for the first time, flying 3 times that day. Vedette 812 appeared on March 19. On April 2, 3 and 4 Vancouvers 902 and 903 flew a number of gunnery trips. A 2:30-hour training flight on the 17th by 903 under F/O P.J. Grant and F/O J.R. Wight appears to have been the RCAF's last Vancouver flight. At the beginning of May, 1940 No.4 began moving to new quarters at Ucluelet on Vancouver Island's remote Pacific coast.

Now some new equipment appeared—Sharks 501, 504, 506, 545 and 546, their first use being on May 3, 1940. Some action occurred on May 19. Two Stranraers were despatched to search for 506 with F/L J.K. Herriot and Cpl Biehler, and passenger F/L Williams. Herriot had taken two passengers to Ucluelet, to help set up direction-finding towers. Coming home, he failed to switch tanks; his Tiger engine quit over Georgia Strait and he alighted on a swell, damaging a float strut. Biehler recalled: "I climbed up on the wing, took off my shirt and held it up on a paddle as a signal. Soon a boat came over and towed us to Nanaimo." Shark 506 returned to Vancouver for repairs aboard an RCN tender. (If Herriot had had a radio, he could have reported his situation and saved the trouble of a search, but few RCAF aircraft then had radios.) The Shark would serve at No.4 (BR) to January 1942. Thereafter, the squadron carried out its duties mainly with the PBY-5 Canso and Catalina. The days of the prewar "pretend air force" types were by now over.

**Another important coastal type in 1940 was the Douglas Digby. A Digby made the third HWE U-Boat kill, when No.747 of 10 (BR) Squadron (F/L D.F. Raynes, skipper) sank U-520 off Newfoundland. (NAC PA136273)**

**The RCAF ordered the long-range Consolidated PBY-5 in December 1939. Soon there were deals with Boeing in Vancouver and Canadian Vickers in Montreal to build the amphibian (Canso) and flying boat (Catalina). Here is Catalina 9706 at Dartmouth. (Ross Lennox Col.)**

## Coastal Forces Mature

Clearly, more modern types were needed in the anti-submarine campaign, especially in Atlantic Canada and Newfoundland, where U-boats abounded, even venturing up the St. Lawrence. One of Canada's earliest designated anti-submarine units was No.11 (BR) Squadron, the first in the RCAF with the Lockheed Hudson. The Hudson, a civil airliner modified for war, was makeshift, but would serve well until America's war machine geared up. If the Hudson was makeshift, so was the Douglas Digby (USAAC B-18), a patrol plane based on the DC-3. Nevertheless, these types proved valuable, sinking a few U-boats, keeping many others on edge. Gradually, PBY-5s, Liberators and Venturas took over. By war's end there were so many RCAF maritime patrol squadrons that, some argue, the Home War Establishment had far too much overkill.

## A Month in the Life: September 1943

The Monthly Review of RCAF Operations, North America for September 1943 provides a sense of what was happening operationally on the home front. Covered first was the valuable work done by the Liberators of No.10 (BR) Squadron (EAC). The Liberator had entered Coastal

EAC Liberator GR.V No.591/K in white and gray camouflage. TOS in April 1943, it served to war's end with No.10 (BR) Squadron. The Liberator, 149 of which were on RCAF strength, was 66' 4" long, had a wing span of 100', was powered by P&W R1830s of 1200 hp, had a normal fuel capacity of 1890 gal. and a gross takeoff weight of 60,000 lb. The nose bulge housed ASV radar. With the Liberator the RCAF helped close the mid-Atlantic gap that had stymied efforts to track and kill U-Boats. (CF REA318-484)

Command service in mid-1941, giving the RAF the range to patrol in mid-Atlantic waters. Now the RCAF, making do with less capable types, also sought the Liberator. Neither Britain nor the US would give up any, until continuing losses to U-boats brought a change of attitude—in early 1943 the RAF allotted Canada 15 Liberators. The RCAF was ready for this step. Many of its crews already were experienced in over-water patrols, knowing the worst of Atlantic weather, the airfields, communications routines, etc. No.10 (BR) Squadron received the first Liberators. In April it sent crews to Dorval, where George Lothian and J.L. "Lindy" Rood, top TCA captains, began converting pilots. In early May they went to Gander for follow-up work. Here they wore RCAF battle dress, but with TCA caps. Many had no idea who they were or what unit they represented. Everyone was saluting them, so the TCA men started saluting back. Neither Lothian nor Rood had an RCAF serial number. Thus, on occasions like signing on and off base, they quoted their laundry numbers!

By May 27, 1943 No.10 (BR) had 15 Liberators; 35 more were pried from the Americans, allowing No.11 (BR) Squadron at Dartmouth to equip, beginning in July. Now came a flood of Liberators, 148 being delivered by war's end. At first No.10 (BR) had trouble getting operational, since it had few spares. It robbed parts from some aircraft, but the solution lay with US Liberator units at Gander. The ever-generous Americans provided the needed spares and support, so that No.10 (BR) could get on with its work. The squadron's first attack came on June 28, when the crew of P/O R.R. Stevenson damaged U-420.

In mid-September an increase of intercepted wireless traffic in the North Atlantic tipped the Allies to an imminent U-boat campaign. Eight U-boats were thought to be gathering, with others expected. They were converging on two juicy targets—convoy ONS.18, which had sailed from Milford Haven on September 12 (27 merchant ships, 12 escorts) and ON.202, following from Liverpool (42 merchantmen, 6 escorts).

On September 19, 1943, Liberator "A" of 10 (BR) Squadron (skipper: F/L R.F. Fisher) was en route Reykjavik-Gander after escorting HMS *Renown* (carrying Winston Churchill). "A" spotted a surfaced U-boat and attacked with six 250-pound depth charges (DCs). Fisher now set course for Gander, landing after nearly 13 hours. Results from his attack were unclear, the monthly report noting: "Some seconds later a large oily patch appeared, but there was no other evidence of damage to the submarine." In fact, U-341, which had sailed on patrol from France on August 31, had been destroyed, making it the first RCAF Liberator kill. (F/L Fisher, age 22 from Armstrong, BC would die along with 23 others in the crash of a 10 (BR) Liberator near St. Donat, Quebec on October 20, 1943.)

Liberators "L" (WO J. Billings) and "X" (F/L J.R. Martin), escorting convoys ONS.18 and ON.202 on September 22, each made U-boat attacks. "L" dropped DCs alongside U-270, but was hit by return fire and had to withdraw. U-270 did not crash-dive. This was a new German tactic—stay on the surface and fight it out, using dual and quadruple 20-mm anti-aircraft guns. This vicious weaponry would spell doom for many a Coastal Command crew. "X" attacked U-377 about 20 miles from the convoy. Heavy fire drove it off after it dropped four 250-lb DCs, then the U-boat, pouring forth ack-ack, dove. The Liberator returned to drop two Z-type homing torpedoes. Now "X" attacked U-402, also on the surface, but only with .50 machine gun fire. U-402 slipped into a fog bank. (U-270 would be destroyed by a 461 (RAAF) Squadron Sunderland on August 12, 1944; U-377 and U-402 survived the war.) On September 23 Liberator "P" of 10 (BR) under S/L J.F. Green attacked a surfaced U-boat, but the enemy escaped after a brief shoot-out. Later, "P" got a radar contact (U-422) a few miles off the convoy. The monthly report recorded:

*This time the submarine was taken by surprise and a depth charge attack was completed before any AA fire was opened. Two 250-lb. depth charges were dropped and the upheaval entirely obliterated the U-boat... The U-boat made off on the surface towards a fog bank, maintaining heavy AA fire as it went, and submerged in a normal dive some 27 minutes later as it entered a fog patch. The aircraft continued to the point of disappearance and dropped two 600-lb. depth charges. (These really were acoustic torpedoes, perhaps referred to as DCs for security reasons.)*

On this patrol "P" had an irregular crewman, A/V/M A.E. Godfrey. He manned the port waist gun with considerable effect. The ever-enthusiastic Godfrey thus became the most senior RCAF officer in combat. In a third 10 (BR) attack of September 23 "Y" (F/L R.R. Ingrams) caught a surfaced U-boat, ran in, but its DC's fell short. The U-boat dove, "Y" dropped an acoustic torpedo, but no results were noted. So went a month with 10 (BR) Squadron, its busiest of the war. Hundreds of hours of routine patrol work were flown, some furious combats were fought, and one U-boat was sunk. There also was tragedy (not mentioned in AFHQ's monthly report). On September 4 Liberator "D" crashed into Gander Lake moments after taking off. The four-man crew perished.

The September 1943 report notes certain accidents, one at Gander on August 6. Ventura 2169 failed to climb after a pre-dawn take-off in duff weather. It crashed, killing the crew, about a half-mile along track, having gained no height. AFHQ would not automatically blame the pilot, calling for "a little conjecture" as to why he did not climb: "It's a man-sized job to take a fully loaded, fast, twin-engine aircraft into the air and go 'on the panel' immediately the wheels are clear. Even the airlines insist that two pilots must share the task of lifting the undercart, adjusting throttles, pitch, mixture, checking the dials, etc., plus the proper handling of the controls." The commentary finished with a plea for greater attention to instrument flying.

The September report covered many other topics. One focused on a study by three RCN medical officers. They had treated 150 survivors of the Atlantic, who had spent from 30 hours to 22 days adrift. All had suffered "immersion foot", a condition of numb, then, swollen feet, resulting from sitting in a lifeboat or dinghy with one's feet in cold water. The doctors learned that the worse way to treat immersion foot was to apply warmth. Their survivors were treated with cold (ice, or frigid air from a fan), allowing circulation to resume gradually over days and weeks. Few suffered amputations, whereas nearly all the survivors of another sinking lost their feet to amputation. They had been treated with warmth, a process which induced unbearable pain, rapid infection and shock. The RCAF was interested in this report, since more of its men would end in the sea. If properly treated after rescue, they could return to flying.

Also noted was some of the work of the Aircraft Detection Corps, a civil organization that watched for enemy movements on land, sea or air. There was a large force of ADC observers: Eastern Canada—12,680, Newfoundland and Labrador—2195, Central Canada—8902, Western Canada —2061. In September they made 20,197 reports, more than 99% of which involved ordinary aircraft movements. Otherwise there were 15 U-boat reports, 13 of aircraft in distress, 8 of mines and wreckage, and 7 of suspicious vessels. A note was included that the Department of Transport was providing 85 sets of binoculars to Atlantic region light stations, so that staff could watch for U-boats. In one case, the September report noted how the captain of SS *John A. McDougald*, an ADC officer, rescued two RCAF men from Lake Ontario.

The value of using and maintaining one's oxygen mask was also covered in September, along with a description of the workings of the RCAF's C2-type mask. Various HWE statistics also were offered. For Western Air Command, for example, it was noted that weather severely had hampered operations. Tofino completed only 33% of its bomber reconnaissance patrols. Of BC's six air stations, the best record was Prince Rupert's with 68%. Aircraft serviceability in BC showed: Canso (13 on strength) 51.5%, Catalina (4) 51.5%, Stranraer (10) 38.3%, Ventura (45) 64.7%, Bolingbroke (16) 54% and Shark (3) 55.5%. Other WAC statistics included:

| Type | Number of Patrols | Hours Flown | Average Duration (Hrs) |
|---|---|---|---|
| Canso | 79 | 586 | 7.4 |
| Catalina | 20 | 134 | 6.7 |
| Stranraer | 11 | 51 | 4.9 |
| Bolingbroke | 50 | 153 | 3.1 |
| Ventura | 16 | 26 | 1.6 |
| Shark | 9 | 23 | 2.6 |

## September 1943 U-Boat Sightings in EAC Waters

| Date | Position | Aircraft | Unit | Duty |
|---|---|---|---|---|
| 4 | 45 31N 59 31W | Ventura "N" | 113 Sqn | Escort convoy SC-141 |
| 8 | 46 59N 47 45W | Canso "A" | 5 Sqn | Escort HX-255 |
| 19 | 58 50N 25 30W | Liberator "A" | 10 Sqn | Sweep ONS-18 |
| 22 | 53 57N 40 32W | Liberator "L" | 10 Sqn | Close cover ONS 18/ON-202 |
| 22 | 53 17N 37 29W | Liberator "X" | 10 Sqn | Close cover ONS 18/ON-202 |
| 22 | 53 24N 37 28W | Liberator "X" | 10 Sqn | Close cover ONS 18/ON-202 |
| 22 | 53 58N 40 38W | Liberator "N" | 10 Sqn | Close cover ONS 18/ON-202 |
| 22 | 50 20N 50 00W | Catalina | RAFTC | Transport |
| 23 | 53 13N 40 48W | Liberator "P" | 10 Sqn | Close cover ONS 18/ON-202 |
| 23 | 52 45N 43 08W | Liberator "P" | 10 Sqn | Close cover ONS 18/ON-202 |
| 23 | 52 36N 42 23W | Liberator "Y" | 10 Sqn | Close cover ONS 18/ON-202 |

September's report commented on an Anson which lost an engine off the BC coast. Unable to maintain height, the pilot ditched. He was knocked out, but three others were not badly hurt. The Anson floated for 45 minutes, giving time for the crew to get the pilot into the dinghy. Their predicament, however, went unnoticed, since their distress call had not been heard (the Anson was almost on the water before an SOS was sent). By good fortune another Anson flew by and was attracted by a flare. Soon other aircraft were involved and the HSL *Takuli* was despatched from Ucluelet. It picked up the stranded airmen less than four hours after they had ditched.

In an odd West Coast incident of September 2, 1943 the BC Provincial Police reported that 3 men had stolen goods and money at Tofino, before fleeing by sea. A Bolingbroke spotted their boat off Ucluelet. It dropped float flares and police soon arrested the culprits. Another topic this month covered the introduction of rocket projectiles (RPs) for EAC Hudsons. These would come with either 25-pound armour-piercing (AP), or high explosive heads. The AP head was considered ideal against U-boats. A standard load would be four RPs on rails under each wing.

EAC generally worked in the Northwest Atlantic north of 40°N and west of 32°W. For September it totalled 885 sorties and 6081 hours. On account of weather, flying was less than in August. Besides normal EAC operations, there were contributions from peripheral units. Hudsons on training from 31 OTU were credited with 162 anti-submarine sorties, and Lysanders from No.4 (Coastal Artillery Co-operation) Flight made 11 harbour entrance patrols at Sydney. Ansons from No.1 GRS at Summerside and No.31 GRS at Charlottetown, which flew some 1700 training sorties over the Gulf of St. Lawrence, were always on the lookout for U-boats.

For this period EAC showed a formidable total of anti-submarine aircraft: Gander: 13 Liberators; Botwood: 12 Cansos/Catalinas; Torbay: 15 Cansos, 15 Venturas; Gaspé: 5 Cansos; Mont Joli and Chatham: 15 Hudsons; Sydney and North Sydney: 10 Cansos/Catalinas, 12 Venturas; Dartmouth: 8 Digbys, 15 Hudsons; Yarmouth: 25 Cansos. Along with USAAC, USN and RAF anti-submarine aircraft and surface craft (destroyers, corvettes, etc.) covering the Atlantic,

they made nervous wrecks of U-boat crews. For September there were 10 EAC U-boat sightings resulting in 7 attacks. Seven eastbound and 8 westbound convoys completed passages through the EAC zone, each spending 7.9 days within it. Of EAC patrols despatched to meet specific convoys, 86% succeeded, bad weather usually accounting for convoys "not met".

The worst losses in September concerned ONS.18 and ON.202. The onslaught against them cost 6 freighters and 3 naval escorts. Among these was HMCS *St. Croix* with only 80 men surviving. Their fortune would be short-lived—they were saved by HMS *Itchen*, which itself was torpedoed. Only 2 of those aboard survived. For their part, air and surface escorts claimed 1 U-boat sunk, 1 probably sunk, 2 badly damaged, 4 damaged and 2 probably damaged. Escort Group B3 signalled No.10 (BR) Squadron: "All who took part in the recent convoy operation would like to express their thanks for the magnificent air cover which was given over the convoy in extremely difficult weather." The senior naval officer aboard HMS *Keppel*, escorting ONS.18 and ON.202, praised the RCAF:

*The dense fog which prevailed ... made flying near the convoy very unpleasant. At times I wondered if the aircrews were serving any useful purpose in risking their lives to come so far, only to find nil visibility... This doubt was well answered when, the instant the fog lifted, three Liberators were not "on the way" or "expected in two hours", but actually flying around the convoy and giving it invaluable protection. After living under a blanket of fog for so long, it was very nice to come into the open air and find it filled with Liberators.* (On September 22 HMS *Keppel* rammed and sank a U-boat.)

## EAC Hours Flown by Type, September 1943

| Liberator | 965 |
|---|---|
| Canso / Catalina | 2,187 |
| Digby | 493 |
| Ventura | 900 |
| Hudson | 1,536 |
| Total | 6,081 |

## EAC Hours Flown by BR Squadron, September 1943

| Squadron | Aircraft | Station | Hours |
|---|---|---|---|
| 10 | Liberator | Gander | 965 |
| 5 | Canso | Torbay | 608 |
| 145 | Ventura | Torbay | 208 |
| 116 | Canso | Botwood | 213 |
| 119 | Hudson | Mont Joli-Chatham | 1161 |
| 117 | Canso/Catalina | Gaspé | 420 |
| 117 | Canso Catalina | North Sydney | 217 |
| 113 | Ventura | Sydney | 692 |
| 161 | Digby | Dartmouth | 493 |
| 11 | Hudson | Dartmouth | 375 |
| 162 | Canso | Yarmouth | 328 |
| 160 | Canso | Yarmouth | 401 |

## EAC Operational Flying, September 1943

|  | Sorties per Aircraft | Average Duration |
|---|---|---|
| Liberator | 5.6 | 13.0 |
| Canso / Catalina | 3.3 | 9.8 |
| Digby | 8.1 | 7.6 |
| Ventura | 6.7 | 5.0 |
| Hudson | 11.5 | 4.5 |

GR.V No.600/N waiting at Edmonton with a Hudson and Lancaster. This Liberator was with 10 (BR) Squadron from June 1943. Postwar it went to NWAC, then was sold for scrap at pennies a pound. (Les Corness/CANAV Col.)

## Gander: The Heart of EAC

Gander, Canada's most historic airport, once was nothing but a swamp around Mile 213 of the Newfoundland Railway. Then it was selected as a site in a British scheme to link the Empire by air. Botwood Lake, not far to the east, would be a flying boat base, complementing the airport. The first survey crews appeared in the woods and bogs around Mile 213 in 1935. Slashers cleared trees and construction men put up the first buildings. Local rock was crushed for runways. Meanwhile, the Irish Free State, also part of the airways scheme, established a flying boat harbour at Foynes on the River Shannon. By the summer of 1937 Botwood and Foynes were ready, and radio and weather reporting facilities were in place. In July inaugural experimental flights were made—a Pan American Sikorsky S.42 operated Botwood-Foynes, while Imperial Airways' Short S.23 flying boat *Caledonia* crossed in the opposite direction. Other flights ensued. Meanwhile, Germany was touting the airship as the solution to intercontinental travel. Its pride, the *Hindenburg*, completed many trans-Atlantic flights, especially to South America.

Finally, Gander was ready. Newfoundland's only airport, it was grander than anything in Canada. On January 11, 1938 Fox Moth VO-ADE, flown by Douglas Fraser, made the first landing. Gander, however, would remain quiet—only the war would bring it to prominence. Things finally began to happen in February 1940, when two RCAF Hudsons arrived with a survey party, to study Gander as a western terminus in ferrying aircraft to Britain. A quick decision was taken. Soon infrastructure was being put in place and trial deliveries were begun under an arrangement with Canadian Pacific Airlines. F/O Lawrence L. "Slim" Jones, who already had pioneered on the Atlantic, delivering Catalinas to the RAF, took 4 Hudsons from Gander to Prestwick between August 5 and September 21, 1940, demonstrating the feasibility of the concept. On November 10 Donald C.T. Bennett of Imperial Airways led the first mass flight—7 Hudsons from Gander to Prestwick. All arrived safely. Henceforth, Gander would thrive as the chief northern hemisphere departure point for aircraft ferrying overseas.

While ferrying got into gear, Gander simultaneously became a military establishment, mainly for anti-submarine work. In April 1940 it came under RCAF control, defended by Canadian Army units. No.10 (BR) Squadron sent a detachment of Digbys from Dartmouth on June 16. A USAAC B-18 arrived from Florida with a US anti-submarine advance party. On May 5, 1941 Gander opened as RCAF Station Newfoundland Airport under station commander G/C A. Lewis. Facilities were still being prepared, everyone making do in rough and ready conditions. Weather curtailed flying through May. Crews were anxious to get airborne with news of the sinking of HMS *Hood* on May 24. An ORB note the next day stated, "all CPR [ATFERO] aircraft ordered to stand by to co-operate with air force." Besides the RCAF, by month's end there were about 480 US airmen at Gander with 6 B-18s and 2 B-24s of the 21st Reconnaissance Squadron. These flew many patrols, sometimes with RCAF observers.

On June 22 the first loss was reported in the station diary. S/L Ralph Ashman had to ditch Digby 752; his crew was saved by a fishing schooner. Many VIPs are listed week by week, passing through on trans-Atlantic trips, notables like Lord Beaverbrook, the Duke of Kent, A/C/M Sir Frederick Bowhill, Toronto industrialist J.P Bickell, Prince Bernhard of the Netherlands, A/V/M L.S. Breadner (RCAF Chief of the Air Staff). It seems amazing how a Liberator often would land at Gander full of VIPs. On July 24, 1941, for example, one arrived with Hon. G.G. "Chubby" Power (Minister of National Defence), A/V/M Breadner, Air Commodore Edwards, G/C Nairn, S/L McKell, Sir Gerald Campbell and other British and US officials. The loss of such an aircraft would wipe out much irreplaceable expertise. Although this would happen from time to time, it was a gamble taken in those gloomy days.

To July 1, 1941 the Gander diary lists 103 Hudsons, 22 Liberators and 21 Flying Fortresses passing eastward with ATFERO. This day 8 Hudsons departed for the UK. The pace quickened, the July 8 entry noting: "Officers have been forced to double up and double-tier beds have been placed in many of the rooms." Sometimes crews had to sleep in tents. For the July 16, 19 Hudsons departed, all reaching Prestwick. The same day Digby 755, flown by FSgt Raynes and fully armed, made an emergency landing in Maine with engine trouble in bad weather. This was unprecedented—the US was neutral; but the red tape was cut and 755 was released. On July 21 a 116 (BR) detachment is noted as operating at Botwood with 2 Catalinas under S/L Carpenter.

On July 26, 1941 Digby 742 of 10 (BR), returning from ops, crashed in bad weather near Gander. F/L M. Tomsett, age 27 from Regina, and his 5 crew died. They were buried in Gander's new cemetery, some of the earliest among the 101 airmen who would end there. For the month 213 Hudsons passed through Gander with ATFERO. August 8 was a record day for Hudson departures—24. To this time the Victoria Rifles of Canada had been defending Gander, but they were replaced by the Lincoln and Welland Regiment on August 9. On August 31 the 21st Reconnaissance Squadron left for the US, replaced by the 41st with B-17s.

In spite of overall success ATFERO suffered losses. Of 21 Hudsons despatched from Gander on September 27, two were lost. First, F/O Harold W. Oldham (age 34 from Woodstock, New Brunswick), flying AM940 with Sgt W.R. Lance (age 22 from Hamilton, Ontario) and Mr. Cyril H. Small, sent an SOS about 3 hours after departure. A search found nothing. Then, F/L Louis R. Dubuc (age 23 from Lacolle, Quebec) with Sgt Frederick J. Goodwin (RAF) and Mr. Samuel R. Kenny set in at Dublin en route to Prestwick. Departing thence, they crashed fatally. At the end of September 1941 RCAF strength at Gander totalled 35 officers, 387 airmen, 6 nursing sisters, plus 152 civilians.

On September 3 Digby 748 crashed (without injury), while taking off with F/L R.A. Butts at the helm. The same day a US PBY brought in 2 stretcher cases from Greenland. They were taken to Gander hospital, then rushed on to the US.

On October 11 F/L Butt had another prang, this time wrecking Digby 754 after landing downwind. Another ATFERO Hudson disappeared in the Atlantic murk this day—AM951 with Capt William J. Guy, William A. Herron and Clinton L. Larder, all civilians. On September 18 a Catalina from 116 (BR) set out to search for reported U-boats. It returned 23 hours later with nothing to report. Such were the gruelling patrols flown from Gander. On the 25th Digby 740 under S/L C.L. Annis attacked a U-boat, placing two 600-lb bombs near the target, but there were no explosions. The diary explained: "The fuses had been removed without the pilot's knowledge." The airman, who had relieved the one who had fused the bombs, had decided to safe the bombs, then nobody had checked before the patrol departed!

Henceforth, many flying days were washed out. A typical entry is that of December 27, 1941: "Low overcast, mild with light rain... turning to light snowfall during the evening. Wind NW 20 mph. No patrols or local flying." On the 29th Digby 744 failed to return after a patrol. Next day 4 RCAF station aircraft, 2 B-17s and 2 Goose of the USAAF, and 4 ATFERO Hudsons searched without success. This continued next day, augmented by 2 US Norsemen, but the diary had to report, "Nothing sighted, next of kin have been notified." On January 2, 1942 F/O Maltby successfully ditched burning Digby 738 in Locker Bay. It was towed ashore, but never again flew. On the 15th, No.10 (BR) Digbys and US aircraft searched for survivors from a torpedoed ship; 2 rafts were sighted and a USN destroyer picked up 5 survivors.

On January 19, 1942 Digby 756 under F/L Young attacked a U-boat 40 miles east of Newfoundland. It dropped 6 depth charges (no results seen). On the 22nd F/L Williams in Digby 740 depth charged a crash-diving U-boat. The same day a Hudson, under Robert W. Whitmore (American civilian), turned back for Newfoundland from 400 miles out, then all was silent. On the 27th an 11 (BR) Hudson from Torbay (four crew under F/L Alan S. Pilcher) disappeared. The next day it was sighted a few miles from Grand Falls. A dog team was despatched from Botwood. On February 2 the crew was saved, two by a Fox Moth flown by station commander G/C R.H. Foss, the others by ground rescuers. On February 7 the crew was back on squadron. (Like many others toiling in the Home War Establishment, Pilcher eventually won an overseas posting. Nearly all such men might have stayed in the relative safety of Canada, but most pined for "the real war". Pilcher joined 544 Squadron. On December 2, 1943 he was killed flying a Mosquito in England.) For the rest of the war there would be few uneventful days at Gander, except when the weather made flying impossible.

### Christmas Flap in EAC

In the fall of 1943 W/C Ralph C. Davis became acting station commander at Dartmouth, then Canada's largest air base. Being called in before G/C Basil Hobbs for the "one-minute hand-over" (as Davis called it) came as a surprise. The groupie's chief instruction was: "Be sure my wife gets a staff car whenever she wants it." He then disappeared for points unknown. Davis was startled, so consulted the adjutant, who also had no idea what was going on. Davis asked him "to use his influence to ensure that staff cars were used in accordance with the book."

Later in the year the pilots of 127 Squadron, a local Hurricane unit, were posted overseas and their aircraft hangared, pending arrival of new pilots. This was a quiet time for EAC, the U-boats seeming to be too far out for Nova Scotia's

Lockheed PV-1 2207/Q of 145 Squadron. An outgrowth of the Lodestar, this type served the US Navy and RAF in large numbers. The RCAF had 286, the main squadrons being Nos.113 (Sydney and Torbay), 115 (Patricia Bay and Tofino), 122 (Patricia Bay, Port Hardy), 145 (Torbay, Dartmouth, Yarmouth), and 149 (Annette Island, Terrace). The Ventura OTU was No. 34 at Pennfield Ridge. In peacetime several ex-RCAF Venturas served Spartan Air Services of Ottawa on jobs around the world. Others were converted into luxurious corporate aircraft in Texas by Dee Howard. (NAC PA175615)

Bolingbrokes, Cansos, Hudsons and Venturas. (This is not to say that things were slack. BGen E.A. Hale (Ret.) was CO in Dartmouth of 161 Squadron. He recalled in 1999 that the board in operations sometimes showed as many as 100 U-boats being plotted. Naturally, the flying was a challenge, Hale commenting: "We flew when the sea gulls walked.")

On Christmas Day 1943 most of those at Dartmouth were ready for a festive time. Turkey dinners were prepared and the senior NCOs were welcomed in the officer's mess early in the afternoon. From there all proceeded to the airman's mess for dinner. Davis describes one of the activities: "The Senior Warrant Officer brought the youngest airman to me. We quickly exchanged jackets, as was the custom, and off he went to finish his beer and be the centre of interest among his buddies." As dinner was about to begin, Davis was summoned to the phone—acting AOC EAC, Air Commodore Walter Orr was on the line. The word was sobering: "Two enemy aircraft carriers reported off the coast of Yarmouth. Arm the Hurricanes and get them airborne as soon as possible. Alert the two BR squadrons. Re-arm the Venturas with bombs." The alert had originated with US intelligence, although no one had seen a German carrier, other than *Graf Zeppelin*, which had not yet been commissioned. But this was no time for talk, especially since *Graf Zeppelin* apparently had disappeared from its usual port. Now a call went to 145 Squadron at Dartmouth to download the DCs on 7 Venturas, replacing them with 300-pound HE bombs; 113 Squadron at Sydney was to do the same with 5 Venturas. No.145 Venturas 2165, 2171 and 2208, once armed, left for Yarmouth. No.34 OTU (Venturas) was not alerted.

Davis returned to the mess for an impromptu COs conference. There were no qualified Hurricane pilots on base, but S/L Ed Hale agreed to organize something. Davis later called this as "one of the most unorthodox fighter operations in the history of the RCAF." Pilots with Hurricane time, armourers, and mechanics familiar with Merlin engines, were to report to Hale. All others were directed to their usual posts. Hale got 7 Hurricanes armed and fuelled. Four pilots with Hurricane time (no matter how little) volunteered to fly. Then came word from EAC that Davis remain on the ground. S/Ls Hale and Henderson, and F/L Robinson took off. They flew out to sea, tested their guns and commenced searching. They returned in due course, refuelled and went on standby. Meanwhile, the station was on high alert, including its defence force, rusty though it was.

At 1830 hours EAC cancelled the alert. What had been behind this flap? According to Ralph Davis, "The report of carriers proved a hoax and no explanation was ever provided. Fortunately, no one was injured and no damage was caused during this hectic afternoon." As to the makeshift Hurricane patrol, what did the fellows think? In 1999 Hale reminisced: "We did it with very little—all we really had was the urge. Our chief emotion was being terrified that we might run into the enemy."

EAC suffered its shared of men killed. Unlike the RCAF overseas, those killed in Canada often lie in their home communities. One such was Sgt Frederick Riess, born on March 19, 1920. Having trained at No.18 EFTS and No.15 SFTS, he was posted to 119 Squadron at Yarmouth (EAC). On December 11, 1941 he and Sgts Earl G. Bawtinheimer of Hamilton and William C. Whitman of Laurencetown, Nova Scotia died in the crash of Bolingbroke 9053, while landing at Yarmouth. Riess' remains were returned for burial in Woodland Cemetery in Hamilton. Bawtinheimer is buried in Grove Cemetery in nearby Dundas, Whitman in Fairview Cemetery in Laurencetown. (Larry Milberry)

## Air Transport: The West

Air transport was important in the HWE, the RCAF using various types to carry senior staff and government officials to daily meetings; to haul material on construction projects such as to Goose Bay and along the Northwest Staging Route; to fly passengers, mail and supplies to remote bases; and on air-sea rescue and mercy flights. In the west there was 165 Squadron at Sea Island (Vancouver) with Lodestars and Dakotas (detachments at Edmonton, Rivers and Winnipeg). Its main role was to support the NWSR, along which moved aircraft from the US to the USSR. (The NWSR also was a vital link in the defence of North America, mainly in Alaska and the Aleutians. An excellent book for those pursing this story is *Warplanes to Alaska*.) A 165 Squadron detachment at Rivers helped train Canada's first airborne troops. Supplementing 165 Squadron were smaller units such as 122 and 166 Squadrons in Vancouver with a hodgepodge of types from Norseman to Shark, Anson, Goose and Hudson. Meanwhile, various station flights existed. With light twins, even Harvards, they served useful transport roles.

One of Canada's leading transport pilots was Wesley H. McIntosh. A Winnipegger, he had learned to fly in the 1930s, served in the RCNVR, then switched to the RCAF as soon as war broke out. He made his first RCAF flight at Camp Borden on October 19, 1939, going up in a Fleet with F/O Ab Hiltz. He then had a distinguished career instructing at various BCATP stations. Hundreds of students passed through his hands. After a stint at Ferry Command in late 1942, during which he delivered a Boston to Prestwick, McIntosh returned to instructing at Trenton. One day he asked W/C J.G. "Joe" Stevenson about getting into transport. Stevenson wrote a supporting letter to AFHQ, emphasizing McIntosh's proficiency at instrument flying.

In April 1943 McIntosh moved into transport. By this time his log showed 3200 flying hours. He began at No.12 (Communications) Squadron at Rockcliffe, then, in July joined 165 Squadron in Edmonton. His work there involved skeds, plus ad hoc trips that came up on most days. His first trip was with F/O Ritzel on July 22, going Edmonton-Fort Nelson-Whitehorse in Dakota 654 (7:05 hours). Business boomed in August, especially with 12 trips between Whitehorse and Snag in Dakota 656, supporting airfield construction. For the month McIntosh logged 160:40 hours (today's airline pilot may log half that in a busy month).

September flourished with as many as six return Dakota trips daily from Fort St. John to Beaton River (airfield construction) with P/O Clare Sweetman and Sgt Iverson. Destinations served in October were Beaton River, Edmonton, Fort Nelson, Fort St. John, Grande Prairie, Snag, Vancouver, Watson Lake and Whitehorse for 69:40 hours. On October 23 McIntosh was posted to 168 (HT) Squadron at Rockcliffe to begin a new stage in transport—long-range operations.

## The East

In eastern Canada there were three key transport squadrons. With everything from the Fairchild 71 to the Goose and Dakota, 12 (Comm) Squadron at Rockcliffe specialized in VIP duties—transporting Members of Parliament, Air Marshals, etc. according to their bidding. One such flight for Wess McIntosh was on July 2, 1943, carrying Prince Bernhard of the Netherlands from Rockcliffe to New York (return) in Lockheed 10 No.7654. The squadron had a famous sked— the Blueberry Run—which operated Ottawa-Montreal-Moncton-Halifax with Lockheed 10s. Besides such serious business, No.12 often would fly VIPs to isolated hideaways in the Laurentians, where they could fish, hunt and relax. Stories abound of certain "big wigs" who were regular passengers. Pilots would talk about old so-and-so, who every Friday would board a Lockheed in Ottawa to fly to Montreal. There he would spend the afternoon with a girlfriend, before returning to Ottawa, all at the King's expense! Also at Rockcliffe was 168 (HT) Squadron, a specialized air mail operation serving the troops overseas with Dakotas, Fortresses and Liberators.

With Lodestars and Dakotas, 164 Squadron was similar to 165. Besides skeds, it supported Goose Bay, carrying hundreds of tons of material there from Montreal, Mont Joli and Moncton. Also important in transport were Nos.124 and 170 Squadrons, which ferried aircraft internally. Many details of air transport in the HWE are documented in such books as *Consolidated Liberator and Boeing Fortress, Air Transport in Canada, RCAF Squadrons and Aircraft, Sixty Years,* and *The Royal Canadian Air Force at War 1939-1945*.

# Home Front Transports 1939-45

The HWE operated a variety of transports. The Norseman, first flown in 1935 in Montreal, at first was slow to catch on—the RCAF bought none before the war. Suddenly sales boomed, hundreds going to the US military. The RCAF took 99 for transport, photo, search and rescue, and wireless training. No.791 "Maggie" of No.12 (Comm) Squadron is shown on photography around James Bay in the summer of 1944. (Rae Reid)

Lockheed twins did yeoman HWE service, especially the Model 10A. Acquired one by one from Canadian and US civil sources, they totalled 15. No.1529, seen at Edmonton, had been CF-AZY. Delivered to Canadian Airways in 1937, this 10-passenger beauty was the first of its type in Canada. The speedy L.10A cruised at about 195 mph over a range of 500 miles. The RCAF also had 10 smaller Lockheed 12As. (Les Corness/CANAV Col.)

The larger Lockheed 14 first flew in July 1937. When war broke out, Lockheed offered a bomber version—the Hudson. The RCAF took 247, nearly all for coastal work. Some, however, were used in high-speed transport and rescue. This example was doing a practice lifeboat drop at Quiddi Viddi Lake, Newfoundland. (Ken Smith)

The Lockheed Lodestar entered airline service in 1939. The RCAF had 18, others served CPA and TCA mainly supporting the war effort. Les Corness photographed this CPA Lodestar at Edmonton. Because of the vital war work it was doing on the Alaska Highway and Canol Pipeline, CPA was supplied with Lodestars from a US military order. (Les Corness/CANAV Col.)

## Lockheed Transports in the RCAF

| Type | Length | Span | Empty Wt. | Gross Wt. | Max. Speed |
|---|---|---|---|---|---|
| Model 10A | 38' 7" | 55' 0" | 6,450 lb | 10,500 lb | 210 mph |
| Model 12A | 36' 4" | 49' 6" | 5,960 lb | 8,650 lb | 225 mph |
| Model 14 | 44' 4" | 65' 6" | 11,000 lb | 15,650 lb | 260 mph |
| Model 18 | 49' 10" | 65' 6" | 11,630 lb | 18,500 lb | 250 mph |

Another important RCAF twin was the Beech 18. This modern type first came to Canada in 1937 for Starratt Airways of Hudson (in prewar times commercial operators usually were ahead of the RCAF in technology). No.1381 is seen in the spiffy markings of No.12 (Comm.) Squadron. (CF PL24264)

Able to haul 3½ tons of cargo or 30 paratroops at 160 mph over long distances, the Douglas C-47 Dakota was the most valuable transport of the war. Most RCAF "Daks" served overseas with 168, 435, 436 and 437 Squadrons. While their losses were nothing like those on bomber and fighter squadrons, Canada's Dak's did not get off lightly. 435 lost 7 aircraft, 25 aircrew and 6 passengers; 436—5, 9 and 0; 437—14, 16 and 0; totals—26, 50 and 6. No.168 focused on overseas mail and passengers, operating from the UK down to Gibraltar, through the Mediterranean and North Africa as far as Cairo. Here Dakota 974 "Gravel Gertie" offloads at Edmonton at war's end. It served to 1964, when it was struck off charge and sold to Austin Airways of Toronto, becoming CF-AAC. In July 1970 it crashed at Val d'Or, Quebec. (Les Corness/CANAV Col.)

Dakota 664 "Northern Witch" in Edmonton in 1945. HWE Dakotas did outstanding work with Nos.12, 164, 165 and 168 Squadrons. No.165 primarily supported the Northwest Staging Route and Alaska Highway. No.164's labours were in the east, where a big job was logistics for Goose Bay. Most of the RCAF's 169 Dakotas served postwar, the last few into the 1990s. (Les Corness/CANAV Col.)

A 168 Squadron B-17 air mail scene at Prestwick. The RCAF's six B-17s proved to be serviceable mailplanes. In the case of 9204, however, it had a short career. Following an undercarriage collapse on September 17, 1944, it was scrapped. (CF PL22544)

B.17 No.9204 taxying at Rockcliffe on December 15, 1943. It later was damaged beyond repair after the landing gear collapsed. (CF RE3577)

One version of the 168 Squadron patch.

## RCAF B-17 Flying Fortresses

| Serial | TOS | SOS | Notes | Gross Wt. | Max Speed |
|---|---|---|---|---|---|
| 9202 | 6-12-43 | 19-2-46 | First example delivered (Rockcliffe 4-21-43), crashed near Muenster, Germany 4-11-45 killing the five crew | | |
| 9203 | 6-12-43 | 7-6-45 | Lost at sea 15-12-44 | | |
| 9204 | 9-12-43 | 11-10-44 | Flew first MAILCAN 15-12-43, disabled by landing gear collapse at Rockcliffe 17-9-44, scrapped | | |
| 9205 | 15-12-43 | 27-12-46 | Survived mid-air collision with Wellington over Bay of Biscay 23-1-44, sold in US after the war | | |
| 9206 | 21-12-43 | 7-6-46 | Sold in US after the war | | |
| 9207 | 2-2-44 | 3-5-44 | Crashed on takeoff from Prestwick 2-4-44 killing the 5 crew | | |

### Civil Ferry Pilots

The Atlantic Ferry Organization (ATFERO) and RAF Ferry Command story is one of aviation's great ones. Several histories and memoirs cover the topic, one of the best being Carl A. Christie's *Ocean Bridge*. Such books must be consulted for the details. ATFERO resulted from a July 1940 agreement between Britain's Ministry of Aircraft Production and the CPR, whereby the latter would ferry aircraft from North America to the UK. Ottawa would co-operate, so long as Canada's defence priorities were not compromised. ATFERO was given facilities at St. Hubert, DOT radio and weather reporting resources were made available, etc. Those guiding ATFERO were seasoned trans-Atlantic men including D.C.T. "Don" Bennett and G.J. "Taffy" Powell, former Imperial Airways flying boat captains. The first big job would be to ferry some Hudson bombers, desperately needed by the RAF. On November 10, 1941 Bennett successfully led a formation of 7 Hudsons from Gander to Aldergrove, Northern Ireland, thus inaugurating ATFERO operations.

In May 1941 President Roosevelt proposed that the RAF take over the ferry operation between Canada and the UK. Ottawa and the CPR did not agree fully, but went along with the change. Thus did RAF Ferry Command come into being in July 1941. The RAF organization was small; ATFERO staff remained intact. The flow of aircraft increased. In spite of losses more than 11,000 aircraft would be delivered, using North or South Atlantic routes.

At first, ferry crews were mainly American and Canadian civilians; but as the war progressed, more air force personnel became involved. Among the many Canadian civilians was Fred Hotson of Fergus, Ontario. He had been an avid aviation fan since boyhood days. In 1938 he was working at de Havilland Canada, building Tiger Moths. As soon as the war began he became a staff pilot on Ansons at No.1 AOS at Malton, then at No.9 AOS at St. Jean, Quebec. Although the AOS was a good environment for someone like Hotson, pilots always had an eye for new opportunities.

A great Ferry Command scene at Dorval on August 24, 1942. The occasion was the arrival in Canada of the first Lancaster—R5727. Beside it is an RAF Coastal Command Fortress II. Examples of the Canso, A-20, Hudson and Ventura, B-25 and B-26 are in view—most awaiting delivery. The Lancaster had been flown across the Atlantic to Dorval by the American celebrity pilot Clyde Pangborn. After being shown off around North America, it went to National Steel Car in March 1943 for conversion as the first Lancaster mailplane. (CF PL11070)

The best was a slot in ATFERO. AOS management, however, was not thrilled at losing a pilot, so did not make it easy for a fellow to leave. Fred Hotson bided his time, eventually becoming assistant operations manager at St. Jean. Finally, the day came when he left for ATFERO. He immediately started ground school at Dorval. Celestial navigation was emphasized, but lectures also covered the B-25. Hotson recalled in 1999: "We began circuits on Mitchells 20 days into the program. Next day I teamed with my friend Don Murray in picking one up in Elizabeth City and ferrying it to Dorval." A few days later the pair set off to deliver a B-25: "We deposited FW271 at Prestwick on April 10, then promptly returned to Dorval in an American C-54." The return flight, made on the 11th, totalled 19:00 hours routing Prestwick-Meeks, Iceland-Stevenville, Newfoundland-Albany, NY. Henceforth, Hotson usually would get home by air, but three times he returned by sea, sailing aboard the *Queen Mary*, *Louis Pasteur* and *Mauretania*.

After several trans-Atlantic deliveries, Fred Hotson transferred to 45 Group RAF Ferry Command. His duties now involved passenger and cargo flights between Dorval and Bermuda, in the Caribbean and across the Atlantic. Following the war, he flew in the bush, had a long career in corporate aviation, then finished his flying days in the Twin Otter program at DHC. Always enthusiastic about aviation, Hotson spent many years as national president of the Canadian Aviation Historical Society, authored articles and books, and was inducted into Canada's Aviation Hall of Fame.

### ATFERO delivery—Mitchell FW271**
March-April 1944

Crew: Don Murray, Fred Hotson (pilots), T. Roberts (flight engineer), G. Roberts (radio operator)

| Date | Route | Duration |
|---|---|---|
| March 25 | Dorval-Mont Joli | 1:55 |
| 27 | Mont Joli-Mingan | 1:30 |
| 28 | Mingan-Goose Bay | 1:45 |
| 30 | Goose Bay-Bluie West One | 5:15* |
| April 8 | Goose Bay-Bluie West One | 3:40 |
| 9 | BW1-Meeks Field (Iceland) | 5:30 |
| 10 | Meeks Field-Prestwick | 5:35 |

*return to base (weather),
**226 Squadron, dbr, u/c collapsed landing Oldenburg April 17, 1945.

### ATFERO delivery—Liberator EW184
April 1944

Crew: E.G. Carlisle, Don Murray, Fred Hotson, Al King, Mulaney and Monroe

| Date | Route | Duration |
|---|---|---|
| 20 | Dorval-Boston- Bermuda | 6:20 |
|  | Bermuda-Lagens (Azores) | 12:25 |
| 22 | Lagens-Rabat | 8:40 |
| 24 | Rabat-Castel Benito-Cairo | 13:00 |

### ATFERO delivery—Liberator KH337
November 1944

Crew: S.E. Alston, Fred Hotson, A.D. Sutton, F/O W.R. Perry, FSgt R. Lewis

| Date | Route | Duration |
|---|---|---|
| 5 | Dorval-Bermuda | 5:40 |
| 6 | Bermuda-Lagens | 11:40 |
| 8 | Lagens-Prestwick | 10:15 |

### Fred Hotson's ATFERO Deliveries

| Type | Serial | Destination | Misc. |
|---|---|---|---|
| B-24 | EW184 | Cairo | 99 Sqn, SOC 11-4-46 |
|  | KH155 | Algiers | 178 Sqn, SOC 17-3-47 |
|  | KH290 | Algiers | 86 and 466 Sqns, SOC 11-11-47 |
|  | KH191 | Prestwick | 8 Sqn, SOC 11-4-46 |
|  | KH337 | Prestwick | 53 and 59 Sqns, SOC 7-7-48 |
|  | KE271 | Algiers | Not delivered |
|  | KK347 | Algiers | 40 Sqn, crashed Foggia Main 15-4-45 |
|  | KL383 | Rabat | 37 Sqn, SOC 14-3-46 |
|  | KL494 | Algiers | 206 Sqn, SOC 7-7-48 |
|  | KL599 | Algiers | ME, SOC 14-3-46 |
|  | KL640 | Algiers | ME, SOC 11-11-47 |
| B-25 | FW271 | Prestwick | SOC 8-2-47 |
| C-47 | KG670 | Prestwick |  |
|  | ?? | Prestwick |  |

# Ferry Command Scenes

Another Dorval panorama, this a snapshot through someone's office window. Awaiting the green light for Prestwick are A-20s, B-25s, Lancasters, Liberators and a Mosquito. (Halliday/CANAV Col.)

While Gander and Dorval were the chief Ferry Command bases, Goose Bay also was vital. Building Goose Bay was a great undertaking, with everything from tarpaper and nails to heavy machinery coming in during summer by sea, in winter by air. This shows Goose Bay in May 1943, when it still was a congested mud hole. (NAC PA141356)

Winston Churchill frequently travelled in Liberator AL504 "Commando", seen at Dorval. It was named by Taffy Powell, head of RAF Ferry Command, who was aboard for its first trip—London-Ottawa non-stop on September 7, 1944. "Commando" disappeared off the Azores in March 1945. (Halliday/CANAV Col.)

Gander from high level in August 1944. Dozens of bombers await ferrying to the UK, there to join the US 8th Air Force in the daylight onslaught against Germany. (Wheeler Col.)

Ferry Command woes. Boston Mk. IIIA BZ295 slithered on its belly at Gander on December 9, 1942. Then, USAAF B-24 No.44-42503 on its nose at Gander in 1944 (the "Lib" suffered from a weak nose gear). Hudson FK794 had trouble at Goose Bay on June 15, 1943 (it was scrapped). (Halliday/CANAV Col.)

The 2400-mile Northwest Staging Route rivalled Ferry Command in numbers of aircraft and dangers en route. Beginning at factories in the US, aircraft funnelled northward to Great Falls, Montana, then traversed Canada towards Alaska. There they might remain as part of the bastion against Japan, or push on to Siberia if Lend-Lease aircraft for the USSR. This Edmonton scene shows RAF and US A-20s, and a US Mustang and Mitchell. (Les Corness/CANAV Col.)

An Edmonton scene with Soviet C-47s, P-38s (likely for the defence of Alaska) and a US C-47. Considering only those for the USSR, nearly 8000 aircraft followed the Northwest Staging Route, the most numerous being 5015 Bell P-39 and P-63 fighters and 1363 Douglas A-20 bombers. (Les Corness/CANAV Col.)

Early in the war B-17 No.41-9205 landed at Whitehorse, while ferrying to the US for overhaul. Minor repairs were made, then the captain invited a few people for a short test flight. But soon after takeoff he had to ditch in Bennett Lake. George Simmons and others rushed to help in the only small boat available. All 17 aboard got out, but the water was frigid. Rescuers reached the pilot, who directed them to another crewman. A few minutes later Simmons got back to the pilot, but he was down beyond reach in the crystal clear water. Another fellow looked like he was a strong swimmer, but suddenly disappeared. In minutes 11 lives were lost. Later, the B-17 was salvaged for used as a fire fighting training aid at Whitehorse airport. (Rae R. Farrell Col.)

## The RCAF Women's Division

From the outbreak of war women lobbied for the chance to be in the armed forces. Members of Parliament and senior officers received letters and telephone calls. An early supporter of their cause was Air Marshal "Billy" Bishop, seconded by G/C F.S. McGill and A/C G.O. Johnson. Nevertheless, the Chief of Air Staff, A/V/M Croil, rejected the idea on December 16, 1939. Meanwhile, in November 1939 South Africa had organized its Women's Auxiliary Defence Corps. New Zealand formed a female component of its air force in February 1941, Australia followed in March 1941.

At first there was little need for RCAF women, until full mobilization had been achieved. By the summer of 1940, however, with the BCATP launched and the situation overseas looking grim, the prospect of women resurfaced. A model existed—the Women's' Auxiliary Air Force in Britain—and planning for a Canadian equivalent began in July 1940. It was proposed that even if women were recruited, they should be "immobile"—attached to specific large bases near their homes for domestic duties such as cooking, and not subject to posting elsewhere. Still, the Air Council found reasons to delay even this limited concept.

Unwittingly, the British forced the RCAF's hand. Civil and military mobilization had gone much further in Britain, and in February 1941 the Air Ministry informed the Canadian government that they proposed to introduce WAAFs into RAF schools. The Air Ministry also suggested that it might seek to recruit Canadian women directly into the WAAF—or form a CWAAF under RAF auspices. Meanwhile, other agencies were forcing hard questions upon the services (similar discussions were going on within the army and navy). The Canadian Red Cross, for example, reported that its volunteer drivers often were used to transport military personnel. Did this put them in violation of the Geneva Convention?

By June 1941 RCAF HQ was bowing to the advice of men like A/C Johnson and the pressure of events. The need for women was becoming obvious; it also was clear that the concept of "immobile" personnel was too restrictive. Britain was asked to lend two WAAF officers. Meanwhile, a Training Depot was to be established and a cadre of officers and non-commissioned officers trained, on which the larger organization would rise. An Order in Council was passed on July 2, 1941. The army took similar steps on August 13; the navy held off until 1942.

The first Canadian woman commissioned in the new organization, the Canadian Women's Auxiliary Air Force, was Flight Officer (later Wing Officer) Kathleen Walker, widow of an RCAF officer. At the time she was director of the Red Cross Motor Transport Corps in Ottawa. She initially received Service Number C6358, but almost immediately this was changed to V30001. V30002 was Section Officer Jean Flatt Davey (commissioned August 18, 1941); V30003 was a ceremonial appointment—Princess Alice,

The RCAF Women's Division, in which thousands served, fulfilled many vital duties. Here a WD does routine daily maintenance on a BCATP Anson. Then, on a more formal note, HRH Princess Royal inspects WDs at No.6 RCAF Group HQ in England. She is speaking to Cpl H.I. Thompson. The inspection party includes A/V/M C.M. McEwen (AOC 6 Group), Flight Officer H.M. Kendall (beside the princess), W/C S.A. Terroux (OC HQ, behind McEwen), Squadron Officer K.L. Ball and Wing Officer S.S. Dowson (RAF WAAF officer). (CF)

Countess of Athlone (wife of the Governor General) became Honorary Air Commandant. Airwomen began their service careers with a number prefixed with a "W"; if they were commissioned, they received a "V" number. Thus, on December 1, 1941, 58 recent graduates from the new female Manning Depot became officers. W301101 AW2 Wilhelmina Walker (better known as Willa Walker) became V30005 Assistant Section Officer Walker, W0050 AC2 Sylvia Isabel Evans became V30006 Assistant Section Officer Evans, and so forth. The organization now became the RCAF (Women's Division) on February 3, 1942, and its members were thereafter universally known as "WDs".

The credentials possessed by some of these first recruits were impressive. Willa Walker had served two years as a secretary in the Canadian embassy in Washington, giving her a background in politics and management. In 1939 she had married David Walker, then an army officer. She rose rapidly in the service, and for much of the war outranked her husband. Although her work centred at AFHQ, she crossed the Atlantic three times. In June 1944 she became a Member, Order of the British Empire (MBE). Yet, when hostilities ended she closed the book on military life, as did most of those who served. Lesser ranks were, nevertheless, impressive. A male instructor found himself in front of a WD class that included three MA degree holders, 15 with BAs, and more than a few school teachers. BCATP pilot training stations were the first to receive WDs, beginning with No.2 SFTS at Uplands in January 1942.

Next came the Bombing and Gunnery Schools, followed by isolated stations such as Gander and Prince Rupert. WDs went to Joint Staff appointments in Washington and to the RCAF Requirements Detachment at Dayton, Ohio. Approximately 600 served in the United States.

WDs came from a variety of backgrounds and places. There were British evacuees and European refugees, and women from British colonies throughout the western hemisphere. Recruiting was not easy. Many parents opposed their daughters going off to serve. More important was competition from industry, which offered better pay and more freedom (though less adventure). The RCAF waged publicity campaigns using stories in the press, booklets, radio announcements, WD Precision Drill Squads, and inviting whole families to attend ceremonies and "At Home" days on stations. Probably the most pressing incentive for a woman to join, however, was the presence of a brother in the force.

By September 1941—before even the first 150 recruits had been trained—RCAF Overseas HQ was inquiring about CWAAF postings. Priority was to be given to home establishments, however, and the first CWAAF overseas draft did not sail until September 1942. Even so, the majority of WDs would not see foreign service. As of October 31, 1944 RCAF Overseas HQ reported the following WDs in the UK:

|  | Officers | Other Ranks |
|---|---|---|
| RCAF Overseas HQ | 42 | 812 |
| No.3 Personnel Reception Centre | 11 | 5 |
| Repatriation Depot, Warrington | 2 | 15 |
| No.6 Group Headquarters | 18 | 258 |
| Station Linton | 11 | 181 |
| Station Leeming | 7 | - |
| Station Middleton St. George | 6 | - |
| Station Topcliffe | 7 | - |
| Station Digby | 1 | - |

The original WD trades were basically "women's work" tasks, e.g. replacing men in cooking, messing, clerical, and telephone work. Smaller numbers were dental assistants, transport drivers and parachute riggers. They soon expanded into other jobs—photographers, teleprinter operators, meteorologists, airframe and aero-engine mechanics, and wireless operators (ground). More significant was their growing presence in the operations rooms and cipher centres filled with secrets. Unhappily, they began with a pay scale only two-thirds that of male equivalents. In 1943 this rose to four-fifths.

Those compiling historical reports of the women overseas did the WDs a disservice. Accounts sent back to Ottawa stressed athletic contests and social events, particularly weddings. One finds numerous references to trivial matters, such as one dated October 13, 1943: "Corporal Burcham, LAW McBride and LAW Douglas debate against three airmen, 'Resolved, That Service Life Makes Women Undesirably Independent'—satisfactorily disapproved."

As to overseas life, a posting to London was the best. WDs generally lived in apartments, 2 or 3 per unit. Outside the capital it was different. Describing conditions at No.6 Group HQ in the fall of 1943, one writer recorded: "Accommodation at No.6 Group is unlike anything in Canada. The airwomen sleep in tin Nissen huts for 14-16 airwomen... heated by coal fires similar to Quebec heaters, in the middle of the room. Each hut has an adjoining toilet, but the main ablutions are all in a separate building, and in some cases the airwomen have quite a walk." The overseas experience included more than cathedral tours and damp quarters, though one must search diligently in the records for insights. A report, covering WD overseas activities in February 1944, illustrates that Air Raid Precautions had been allowed to run down since May 1941 (when the Luftwaffe shifted most of its bombing to the eastern front). A renewed "Little Blitz", commencing in January 1944, revived concern for ARP measures:

*Owing to the recent air raids in London, it was felt that airwomen at this Headquarters should have lectures on Air Raid Precautions, bomb disposal, etc. An excellent lecture was given to the airwomen at RCAF Overseas Headquarters, Lincolns Inn Fields, on this subject. They were told how to behave in air raids, which precautions to take such as having warm clothes ready and other necessary items such as a bag containing a torch, chocolates, socks, hankies, a book, thermos of tea or water ... if they had to spend the night ... down in a shelter... The airwomen were urged to go down into shelters or even to the basement of their homes and the foolhardiness of staying on upper floors was pointed out... In the recent raids two airwomen from this Headquarters were injured, also two male officers. Several people were bombed out, losing everything they owned...*

That there was a more serious side to overseas life, even off duty, was demonstrated by another activity. Hospital visits were not part of the job; they were nevertheless part of many women's routines. No.10 General Hospital, Watford, was a Canadian centre. From the summer of 1943 onwards groups of 15 or more WDs dropped by on weekends to provide such cheer as they could. In August 1943 LAW E.M. Hassett, RCAF Overseas HQ, organized the first visit of Canadian airwomen to the services wards of the Queen Victoria Hospital (Plastic Surgery), East Grinstead. Three airwomen made the first visit. Every week thereafter a minimum of five airwomen travelled by rail at their own expense to this institution. The convalescents looked forward to these calls, and those taking part received many letters from staff and patients.

Not all WDs overseas joined in Canada. From May 1943 onwards the force recruited limited numbers already overseas. Some were Canadians who had been living in Britain for years; others were married to RCAF men. There was some additional paperwork; the Ministry of Labour and National Service had to give consent if the lady was already employed in war-related industry. They were trained at a WAAF Depot to which RCAF WD instructors had been assigned. A few Canadian-born WAAFs also applied for transfer to the RCAF. In all 17,038 women served as WDs; 612 being officers. Peak strength was in January 1944 (597 officers, 14,959 other ranks). This was considerably more than the other services—the army recruited about 12,000 CWACs, the RCN some 6,000 WRENS. Total WD fatalities numbered 28, including those who died of natural causes; three died in air crashes. In spite of the Women's Division motto, "We serve that men may fly", it is doubtful that they enabled more than a handful of males to remuster to aircrew. They did, however, free many tradesmen for overseas service, thus promoting top-to-bottom "Canadianization" in RCAF squadrons abroad.

WDs garnered several awards—1 OBE, 14 MBEs, 39 BEMs and 50 Mentions in Despatches. Examples of these honours illustrate the work performed and the value placed upon it. Wing Officer Winnifred May Taylor was made a Member, Order of the British Empire in June 1946. Born on October 30, 1909 in Montreal, she was educated in Toronto, worked for Canada Wire and Cable, then Lever Brothers, where she headed the order department. Taylor enlisted in October 1941. She became the first woman CO of an RCAF unit, and became Senior Staff Officer, WD, in the autumn of 1944. After the war she worked at Canada Customs, but was brought back into the RCAF briefly in 1954 for consultative duties. Taylor died in September 1972. The citation to her MBE reads:

*Wing Officer Taylor was among the first Royal Canadian Air Force Women's Division officers appointed. Her work from the very first was outstanding and as a result she was later assigned to increasingly responsible positions, first as Officer Commanding Training Wing, No.6 Manning Depot (WD), Toronto, the Commanding Officer No.6 Manning Depot (WD), after which she became Commanding Officer of No.7 Manning Depot (WD), Rockcliffe. In these important appointments she displayed outstanding administrative ability, coupled with loyalty and devotion to duty of a high order. By her example, initiative and leadership she created a very high standard of initial training of airwomen, the great majority of whom passed through her hands. This officer is now the Senior Women's Division Staff Officer, and through her character and personality has continued to exercise a tremendous influence for good among the airwomen. It is considered that the splendid Service attitude and discipline of thousands of airwomen are in large measure a result of Wing Officer Taylor's example and effort.*

Cpl Ruth Taverner, a Newfoundlander, enlisted in the RCAF in June 1943 and was posted to RCAF Overseas HQ. In June 1946 she was awarded a BEM, the citation for which reads:

*This airwoman has given over two years outstanding service in this Directorate in her capacity as Clerk in charge of Motor Transport Records. This has been extremely difficult as Royal Canadian Air Force Motor Transport has been supplied at various times by units all over this country and the Continent. This airwoman's never failing application to duty has ensured accurate records being available, without which efficient Motor Transport operations would have been impossible and losses would have been great.*

Sgt Dorothy Helen Belyea of Saint John, New Brunswick had been a bookkeeper before enlisting in November 1941. She made her mark in Canada, performing administrative duties at No.1 General Reconnaissance School. Sergeant Belyea was awarded a BEM in June 1945, receiving it formally on parade six months later:

*This non-commissioned officer has set a splendid example. She has carried the entire weight of the voluminous work of the General Reconnaissance School Orderly Room, the efficiency for which she alone is responsible. Understaffed, she has worked long hours without complaint to ensure the continuity of trainee records and trainee reports. The trainee reports in particular have a direct bearing on the war effort for the promptness of their despatch and meticulous accuracy in their preparation provide the only source from which General Reconnaissance graduates are selected for employment in the branch of Coastal Command for which they are most suited. This Women's Division Sergeant has made a valuable contribution to the war effort.*

LAW Doris Edith Armitage of Dunrobin, Ontario also received the BEM:

*This airwoman has, since being on strength of this unit, proven herself to be an exceptionally dependable Driver Transport who has at all times given her services to the Royal Canadian Air Force beyond normal demands. In addition, on the night of February 3rd, 1946, her quick thinking and fortitude during a fire in the unit Motor Transport Section was instrumental in preventing possible fatal injuries to a comrade and reducing property loss to a minimum. Her devotion to duty and exceptional judgment merits recognition.*

Although most WDs were released in 1945, some remained until December 1946. The RCAF was thus the last of the services to dispense with their duties. Five years later, women would again be wearing air force blue. Several books cover the topic of the WDs. These include *Props on Her Sleeve: The Wartime Letters of a Canadian Airwoman* by Mary Hawkins Buch, *Time Remembered: A Woman's Story of World War II* by Gwen Evans Pilkington, and *The Memory of All That: Canadian Women Remember World War II* by Ruth Latta.

## Motor Transport WD

In August 1999 Rosalie Woodland looked back on her days as a WD. A Brantford, Ontario girl, she had a particular interest in military life—her father, LCol A.B. Cutcliffe, DSO, MiD, had been Assistant Director of Veterinary Services for the Canadian Corps in France during WWI. After the war he earned his living as a food inspector for the city of Brantford. Rosalie's parents sent her to Havergal girls' school in Toronto, but her father was disabled by a stroke in 1936. Now she switched to Day's Business College, then worked as an office girl at the Bank of Commerce. She volunteered as a driver for the Red Cross, something more to her liking. When the war started, her father tried to re-enlist, so that he might help in any capacity, but was turned down on account of his disability. This convinced Rosalie that it was her turn to carry the torch.

*I applied to join the RCAF as soon as they would take the girls, and was sworn in at Hamilton on January 22, 1942. I would serve until discharged on June 28, 1945, working all the time as an MT driver. Initially, I was posted on February 7 to the WD Manning Depot (formerly Havergal College) on Jarvis Street in Toronto. The course included telephone operators, parachute packers, cooks, MT drivers and GDs (general duty). It lasted a month, at which time our squadron, which included six MT drivers, was posted to No.14 SFTS at Aylmer. The day after we arrived, we reported to our sections. This took nerve, believe it or not. All the MT chaps were lined up waiting for us. We weren't sure how we'd be received, but they welcomed us. These were some of the lads that the Station Commander, G/C G.C. Irwin, had described to us the day before as the "Gentlemen of Aylmer". They lived up to his expectations.*

*Each of us was given a vehicle to look after and drive. I became the Station Commander's driver. This usually meant sitting in the MT Section waiting to be called. G/C Irwin was a*

**Just before LAW Rosalie Cutcliffe left for overseas, the boys in the MT Section at Aylmer insisted on getting their picture with her atop a bowser. (All photos, Rosalie Woodland Col.)**

treasure of a CO, but had his own car, even his own little plane for longer trips. So I mostly just sat and waited. Then fate took a hand—I came down with the mumps and spent 10 days in hospital at the Technical Training School at St. Thomas. When I returned to duty, I learned that G/C Irwin had taken to driving everywhere on his own, so I was without a car. I was made spare driver. One day there was a stake truck run, but the usual driver had been posted away. The MT sergeant put me in charge and I became the regular stake truck driver. That was heaven for me, driving every day (usually accompanied by an airman from the Supply Section), and often getting off the station. Then one day I was put on a fuel tender. After a few minutes of instruction I was on my own at No.2 Hangar. A female gas truck driver—that was a shocker to most of the guys! It wasn't long, however, before I was accepted by the groundcrew. They were a great bunch. I drove that truck every day, except when flying was cancelled due to weather, or when I was off on a 48-hour pass. Everything went normally, except on the day that I and another WD driver each had an accident. The tarmac was ice-covered with ruts where trucks had been driving. In both cases, our trucks got stuck in the ruts. No matter how we tried to stop, we each bumped into a Harvard. We were paraded before the CO who asked us just whose side we were on! Then he let us off the hook.

On another occasion we were refuelling a visiting Tiger Moth. To do this you had to get up to the filler cap on the top wing. The corporal was yelling at us to get on with it, but the refueller was being careful not to let the gas come gushing out, which it would do if the person controlling the fuel flow from the truck wasn't careful. The corporal, however, thought we were wasting time. He clambered up on the Tiger Moth, grabbed the nozzle and yelled, "Let her go, full blast!" The gas went all over the place, soaking the impatient corporal! Another highlight for me at Aylmer was getting two flights in the Harvard. That made the whole posting worthwhile, if nothing else happened.

In November 1943 I was posted overseas, sailing aboard the Mauretania on November 25. It was a cold, rough crossing. We got in to Liverpool, where the train was standing by at dockside to take us to Bournemouth. One day we were paraded and told that half of us were going to Linton-on-Ouse, half to 6 Group HQ. Being crazy about airplanes, I was happy to get Linton, but next day I was switched to 6 Group. This was disappointing, but I soon realized how lucky I was—6 Group was a wonderful posting! There I was given a little Hillman to drive— quite a change from Aylmer.

Our main task was driving officers to all the 6 Group stations. Each morning after breakfast we would go to the MT Section and get our first "658"—the form that told us which vehicle we had, which passenger, at what time, the mileage, etc. If it was cold, whoever was on night duty would have to run the engines up every so often, since there was no antifreeze available. Sometimes our days were gruelling, but we didn't complain. On June 5, 1944, for example, I was on a trip to our farthest station, Middleton St. George. When I got back late in the afternoon, there was an important dispatch to take to Linton, which I delivered. It was a bit strange that day—everything seemed oddly quiet. Then came the night and never before had we seen or heard more aircraft. Next day we learned the reason—it was D-Day. From that morning till I went off duty, I put in 52 hours! By this time we realized why all leaves had been cancelled in the weeks before, and why nobody was allowed to travel within 20 miles of the coast, unless on essential duty.

Some of my senior passengers at 6 Group included A/V/M "Black Mike" McEwen (Commander, 6 Group), A/C Roy Slemon, G/C Johnny Fauquier, W/C Hutton (Senior Medical Officer), and G/C McCarthy (Senior RC Padre overseas). One day I had a trip to Bomber Command HQ at High Wycombe, not far from London. There the security was so tight that I was not allowed on station. My passenger got off at the gate and I drove away. There are so many great anecdotes about Allerton Hall. Usually our routine was to enter the castle and go straight to our passenger's office—to let him know we were standing by. Then a new Winco took over the castle, declaring that we "scruffy" MT people would no longer be allowed in, but could phone to our passengers from the MT compound. This did not go well with Slemon, who soon put the Winco straight—MT drivers would return to the usual practice.

One day my task was to drive A/V/M "Black Mike" McEwen and his adjutant to a function in Leeds. I tried to find out exactly where to go, but neither the despatcher nor the "adj" were sure. I drove along High Street and all around the centre of Leeds, but to no avail. Black Mike was getting irritable. At last I spotted an airman and asked him for directions. He was going to the very function and I soon was at the door. I took umbrage, however, when Black Mike grumbled, "Make sure I never get her for a driver again!" He even thought that I should be put on charge! Ironically, I did have him for a passenger again— I took him in our big Desoto to Dishforth, but he didn't seem to remember me, thank goodness!

One day I felt vindicated about how Black Mike had treated me. A bomber had piled up and there was fear of an explosion. A second aircraft was about to follow, but its crew, except for the pilot and engineer, was ordered to run for it. The bomber exploded, but a worse explosion was feared—there was a cookie in the bomb bay. At that moment a car came by. The driver yelled for the crew to jump in. Off they sped, but the car made a sudden turn, causing the pilot to yell: "You dumb son of a bitch. You're driving into the bomb dump!" The driver muttered some apology and got out of there in a hurry, but the pilot was still ticked off, warning him to keep on the perimeter track. Once they reached safety, the two aircrew suddenly noticed the "scrambled egg" on the driver's hat—it was the AOC himself— A/V/M McEwen.

One day a young Mosquito pilot, Eric Smith, showed up at 6 Group, to visit his cousin. I chanced to meet him and his little group. Eric was a character. Since he wanted to have some fun with the airmen, he donned an LAC's uniform, and got away with it. More than 50 years later I heard from Eric again. Life stayed mostly routine in the MT Section, but there was the occasional excitement. One day upon entering Linton-on-Ouse, I was told that I had driven over an unexploded bomb in the nearby village! The bomb was supposed to have barriers around, but someone had failed to put them up. Another time I was rushing two officers from RCAF Overseas HQ in London to York, where they had a train to catch. Along the way I had a wheel come off my Hillman. I was saved when another of our vehicles came along, took on my passengers, and got them to the train just in time. Looking back on it all, life in the RCAF was thoroughly enjoyable and many good friends were made. Even though there was a war going on, those were the happiest days of my life. I wouldn't have missed them for anything.

Just before Rosalie Cutcliffe had arrived at Aylmer, a certain LAC Arthur Woodland, MT Driver, was posted to the UK from there. When she was later posted overseas, one of the boys at Aylmer asked her to say hello for him to "Woody" from Aylmer, should she ever bump into him. That was OK by Rosalie, except she knew what the odds were of ever crossing paths with him. Then one day she was waiting in her Hillman for a passenger. An airman came by. "Must be nice to be a WAAF, sitting around with nothing to do," he cracked. Rosalie put him straight pretty quickly, but he really got her steamed by calling her "Freckles" before he made off.

Later, Rosalie asked around MT if anyone knew a Woody. "He comes in once in a while," somebody said. One day he did show up and Rosalie met him—it was the same smart aleck wisecracker—Arthur Woodland from Aylmer. He was driving at Topcliffe, but eventually came down to the MT Section at Allerton Hall. He and Rosalie started chumming around. The upshot came in February 1945, when they were married in the little church beside Allerton Park. Rosalie sailed for New York on the Île de France on May 25, 1945 for New York. Arthur followed, and they settled in his hometown of Orangeville, Ontario. He and his father did home renovations for a while, he worked at Avro Aircraft in Malton, then became county land registrar. The marriage lasted 53 years, till Arthur passed away in 1998.

# Rosalie's Album

Rosalie in 1941, while driving in Brantford for the Red Cross; then behind the wheel of a fuel tender at Aylmer.

Ruth MacGregor ready for a flip in the Harvard.

WDs Rosalie Cutcliffe, Cherrie Bakewell (of Pennsylvania) and Ruth MacGregor.

WD Lilias Vanbuskirk, an Ottawa girl at Aylmer.

There was plenty of good fun on any BCATP station, as with this station dance at Aylmer.

RCAF Station St. Thomas, where Rosalie spent time recovering from the mumps. Then, two shots that she took there of ground training aircraft: Delta A143 and Battle A99.

Harvards refuel at Aylmer. No.3003 was in a mid-air collision with 2554 on July 19, 1943, both being lost. Then, two others that tangled. Although badly damaged, they were repaired.

Rosalie with her Hillman at 6 Group HQ in May 1944.

Some of the drivers in the MT compound. Their passengers would come out the door at the left. The drivers are Cherrie Bakewell, Jimmy Hulme, Mickey Dixon, Harold Freestone and Margaret "Zeke" Prouse.

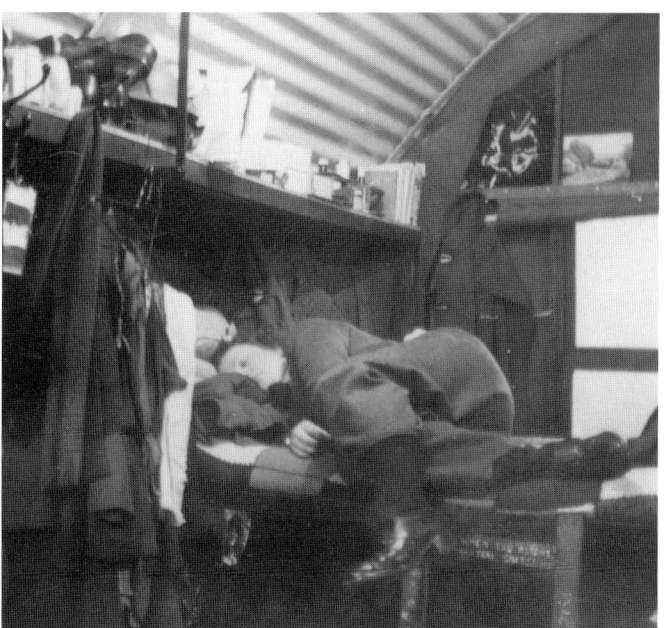

Airmen and airwomen at Allerton Hall were quartered in Nissen huts on the castle grounds. Rosalie's group called their hut "The Jeeps". Here Cpl Dorothy "Clem" Clements catches 40 winks in a typical Nissen hut scene.

The 6 Group MT Section enjoys a special evening. In the back row centre (by the curtains) is Rosalie. On her left is her future husband, Arthur Woodland.

Arthur Woodland and Frank Afelski, RCAF drivers at 6 Group. Truck 26924 may have been RAF property, but note how the MT Section has emblazoned a maple leaf on one fender.

LAW Rosalie and LAC Arthur Woodland with LAW "Mac" Hardie at a WD's reunion in Ottawa in June 1985.

The "Ruhr Express", the first Canadian-built Lancaster Mk.X. It's shown during the roll-out ceremony at Malton on August 6, 1943. (NFB WRM3502)

## Canada's Aircraft Industry in WWII

Through the 1930s Canada's aircraft industry comprised a few small firms building outdated types. These usually had metal-framed fuselages and wooden wings, covered in plywood or fabric. In 1935 plant space totalled about 500,000 sq. ft. About 4000 employees were producing perhaps 40 aircraft a year. There was additional work rebuilding worn and damaged machines. It was a branch plant industry, e.g. de Havilland Canada being owned in the UK, Canadian Pratt & Whitney in the US. Then in 1935 the industry began a gradual transition. The RCAF ordered the high-performance Northrop Delta. Twenty were built by Canadian Vickers in Montreal, test pilot A.S. Schneider flying the first from the St. Lawrence River in August 1936. The Delta, being Canada's first all-metal aircraft, gave industry a start with a new technology. (Industry still would need its tried-and-true skills in wood and fabric once war began, whether building prosaic Tiger Moths and Ansons, or 400-mph Mosquito "Wooden Wonders".)

There seemed to be no master plan in 1939 for Canada to mass produce aircraft. Preparations went no further that the building of a few Deltas and Stranraers, and setting up the Canadian Associated Aircraft consortium to build 160 Hampden bombers. Great Britain seemed confident about handling its own needs, should hostilities erupt. Besides, it looked askance at competition from the Dominions, which still were viewed as underlings. Then came Dunkirk, and Germany's U-boat stranglehold. Suddenly, Canada became a vital, even appreciated, source of supply for "Mother England".

Companies like de Havilland and Fleet at first struggled to provide aircraft for the BCATP. Meanwhile, much war material on the way from the US and Canada to Britain, and some aircraft and components coming from the UK to Canada, ended at the bottom of the Atlantic. The situation worsened—the U-boat seemed invincible. Britain had to admit that it would need more help than imagined. Thus was a meeting held in Ottawa on June 20, 1940 between the Department of Munitions and Supply (headed by C.D. Howe) and the Department of National Defence (headed by J.L. Ralston, whose predecessor, Norman Rogers, had died in a plane crash a few days earlier). Plans were discussed about building operational aircraft in Canada. Months passed with only a few modest orders. These would not even total the 500+ aircraft that the RCAF had planned before the war.

At last things started to happen. Hundreds of contractors and sub-contractors went to work producing large quantities of airframes, engines, instruments, propellers, tires, etc. Behind the effort were a few top men from industry and finance, all under Howe. As the wheels began to turn, there remained some catching up to do; most machine tools, raw materials and skilled labour had been tapped off by industries that had been going full blast since 1939—steel mills, shipyards, arsenals, vehicle and tank plants, etc. The Aircraft Production Board in Ottawa, comprising production, purchasing and repair/overhaul divisions, also had to find workers, then train most from scratch (30,000 women were included).

At first the program was disheartening. One aircraft type, the Martin B-26, to be built at Malton by National Steel Car, had to be abandonned after much preparation. NSC also was considered for production of Short Stirling bombers, but this died when the Stirling's limited performance came to light. Other projects foundered. Then the APB, in consultation with Britain's Ministry of Aircraft Production, settled on a number of types showing long-term promise. The value of DND contracts awarded by the Defence Projects Construction Branch illustrates the importance of aviation in Canada's overall war effort. These statistics, published by Ottawa in 1950, make the point: Air Force—$250,388,303; Army—$100,045,938; Navy—$74,357,385.

# Industry on a Wartime Footing

De Havilland Canada built more than 1000 Mosquitos at its Downsview, Ontario factory. First flight there was on September 23, 1942. Here, "Mossies" come down the line. Most would go overseas with Ferry Command. (DHC/Hotson Col.)

The repair and overhaul bay at Noorduyn's Cartierville factory. Nearest is California-built Harvard I No.1336, then a Noorduyn-built Harvard II, a Yale (with a Fleet beside it), and various types, even a Bolingbroke. (H.S. and Gordon Diller Col.)

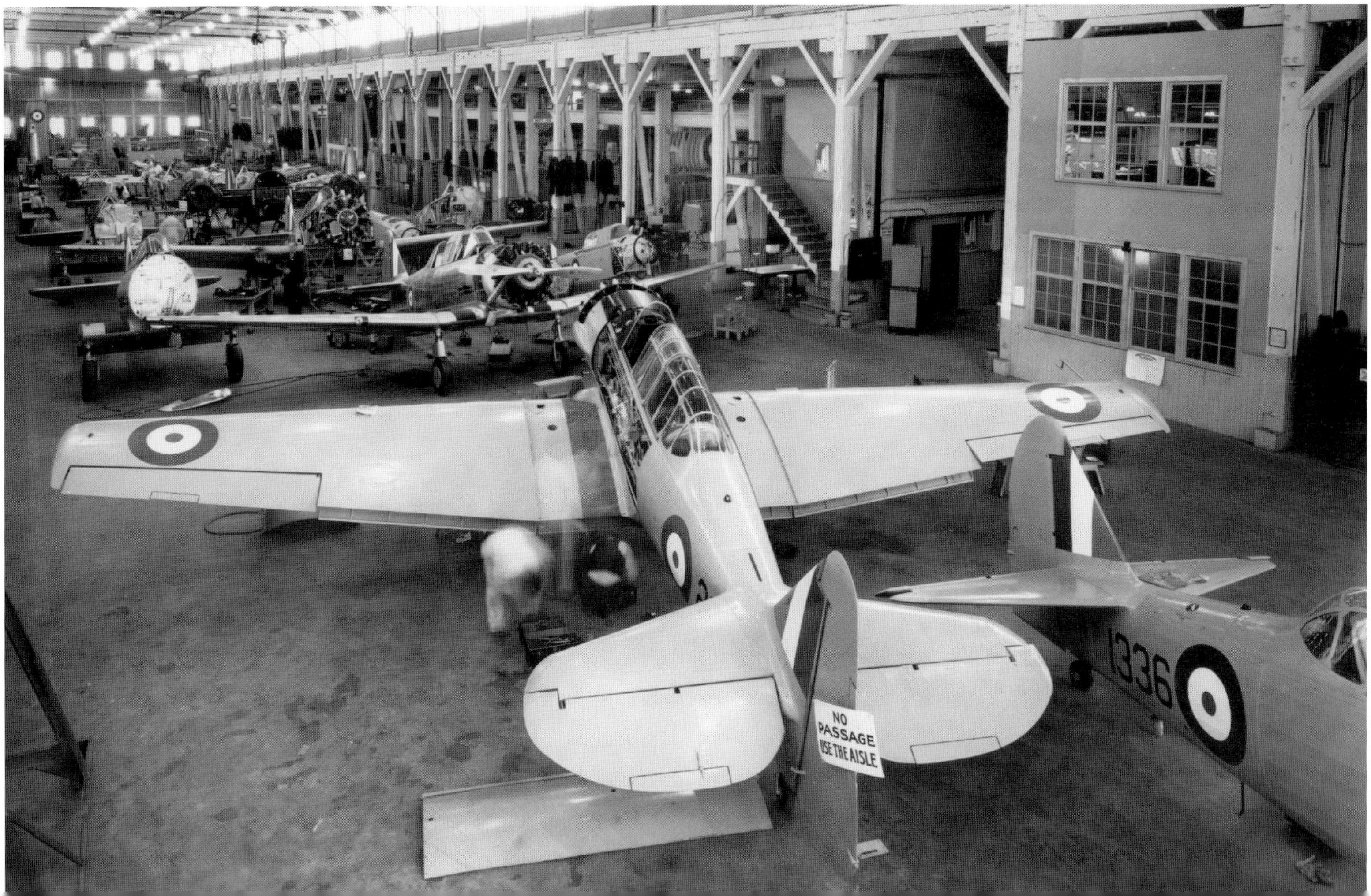

## Canada's Aircraft Manufacturers and Principal Types, 1939-45

**Boeing Canada:**
Vancouver: Blackburn Shark,
Consolidated PBY-5

**Canadian Car and Foundry,
Fort William, Ontario,
Montreal, Quebec,
Amherst, Nova Scotia:**
Grumman Goblin,
Hawker Hurricane,
Curtiss Helldiver

**Canadian Vickers (later known as Canadair):
Montreal, Quebec**
Northrop Delta II,
Supermarine Stranraer,
Consolidated PBY-5A

**de Havilland Canada, Downsview, Ontario:**
D.H.82C Tiger Moth,
Avro Anson,
D.H.98 Mosquito

**Fairchild, Longueuil, Quebec:**
Bristol Bolingbroke,
Handley Page Hampden,
Curtiss Helldiver

**Federal Aircraft Ltd., Montreal:**
Avro Anson

**Fleet Aircraft, Fort Erie, Ontario:**
Fleet Finch,
Fleet 50 Freighter,
Fairchild Cornell,
Handley Page Hampden

**MacDonald Brothers, Winnipeg:**
Avro Anson

**National Steel Car (later known as
Victory Aircraft Co.), Malton, Ontario:**
Westland Lysander,
Handley Page Hampden,
Avro Anson,
Avro Lancaster,
Avro York

**Noorduyn Aircraft, St. Laurent, Quebec:**
Noorduyn Norseman,
North American Harvard

**Ottawa Car:**
Avro Anson

### Aircraft Production Branch, 1939-45

| | |
|---|---|
| Value of Production | $850,000,000 |
| Purchase orders | 41,000 |
| Aircraft produced | 16,418 |
| Employment—1940 | 17,000 |
| Peak employment—August 1944 | 116,000 |
| Employment in overhaul and repair | 18,000 |
| Aircraft overhauled | 9,000 |
| Engines overhauled | 30,000 |
| Aircraft exported to USA | 5,096 |
| Aircraft exported elsewhere | 4,854 |

Anson Is at MacDonald Brothers in Winnipeg. Among other work going on, their windows were being "boarded up", to reduce winter draftiness. (NFB WRM1407)

PBY-5A Cansos on the line at the new Canadian Vickers plant at Cartierville. Besides RCAF orders, CV (later re-formed as Canadair) also built PBYs for the United States. (Canadair Archives)

## Canada's Key Aircraft Manufacturing Projects 1939-45

| Type | Chief Contractor | Location | No. Built* |
|---|---|---|---|
| Avro Anson | Canadian Car and Foundry | Fort William | 641 |
| | D.H. Canada | Toronto | 375 |
| | Federal Aircraft | Montreal | 3** |
| | MacDonald Brothers | Winnipeg | 1,067 |
| | National Steel Car | Malton | 736 |
| | Ottawa Car & Aircraft | Ottawa | 60 |
| Avro Lancaster X | Victory Aircraft | Malton | 422 |
| Avro Lancaster T/PP | Victory Aircraft | Malton | 8 |
| Bristol Bolingbroke | Fairchild | Longueuil | 626 |
| Consolidated PBY-5 | Boeing Canada | Vancouver | 352 |
| | Canadian Vickers | Montreal | 312 |
| | Canadair | Montreal | 57 |
| Curtiss Helldiver | Canadian Car and Foundry | Fort William | 835 |
| | Fairchild | Longueuil | 300 |
| D.H.82C Tiger Moth | D.H. Canada | Toronto | 1,520 |
| D.H.98 Mosquito | D.H. Canada | Toronto | 1,033 |
| Fairchild Cornell | Fleet | Fort Erie | 1,642 |
| Fleet Finch | Fleet | Fort Erie | 437 |
| Handley Page Hampden | National Steel Car | Malton | 80 |
| | Canadian Associated Aircraft | St. Hubert | 80 |
| Hawker Hurricane | Canadian Car and Foundry | Fort William | 1,451 |
| Noorduyn Norseman | Noorduyn | Cartierville | 815 |
| N.A.A. Harvard | Noorduyn | Cartierville | 2,798 |
| Westland Lysander | National Steel Car | Malton | 225 |

*Data drawn mainly from the work of K.M Molson in Canadian Aircraft since 1909.
**Federal functioned mainly in administration and subcontracting.

# Canadian Car and Foundry: The Fort William Scene

Canadian Car and Foundry entered aviation with a licence-built Grumman G-23. This work gave CCF experience with a relatively sophisticated design (metal construction, retractable undercarriage, etc.), and in marketing a high-risk product. This G-23 was nearing completion; from the main plant it would be towed to Bishop's Field for flight testing. (All, CANAV Col.)

CCF also built the speedy FBD-1 fighter (see *Sixty Years, RCAF at War*, etc. for various photos). Designed by Michael Gregor, it flew at Fort William in December 1938. By this time air forces were equipping with advanced monoplanes, so CCF received no orders. The sole FBD-1 was lost in a hangar fire at St. Hubert in 1945. CCF used a full-scale mock-up to prove FBD-1 design concepts.

CCF manufactured the Hawker Hurricane for the Royal Navy, USSR and RCAF. V.J. "Shorty" Hatton flew the first at Fort William on January 10, 1940. No.5434, which served from August 1942 at No.1 (F) OTU, is seen at Cartierville with old time CV pilot Fred King at the controls.

The type of machinery needed in a modern wartime plant; office staff at work, with RCAF technical officer S/L F. Hems at his door; and fabric shop workers covering Curtiss Helldiver dive bomber components. S/L Hems was an artillery veteran who emigrated to Toronto from the UK soon after WWI. One day he spotted a newspaper notice from Col. Joy of Camp Borden—Joy was looking for machinists. Hems jumped at the chance and joined the CAF. He worked steadily through the interwar doldrums, then WWII, retiring as a wing commander. He then joined Avro Canada, staying till the demise of the Arrow. He passed away in 1985.

One of the greatest wartime manufacturing projects in Canada featured the Helldiver, built by CCF and Fairchild. In this CCF scene from late 1943 Helldivers come down the line.

A "Stranraer"—largest aircraft ever built in Canada

*Leading the Flight of Progress!*

**CANADIAN VICKERS LIMITED**

PIONEERS IN AIRCRAFT DESIGN AND MANUFACTURE IN CANADA

## MALTON —
### A GREAT CANADIAN ACHIEVEMENT

*Skilled workers throughout Malton plant, realize the importance of their duties. They are determined that our airmen shall get the best, lots of it, and fast.*

The creation of a great airplane plant — such as Malton, Ontario — is the creation of an immensely valuable Canadian asset. In this plant, rigid control and constant checking, plus the willing eagerness of the workers, assures the utmost in dependability, safety and performance.

In war, as in peace, National Steel Car provides safe, fast, comfortable and economical TRANSPORT — with railroad cars, automobile and bus bodies, and planes.

**NATIONAL STEEL CAR**
*Corporation Limited*

HAMILTON   MONTREAL   MALTON

BUILDERS OF TRANSPORT FOR CANADA

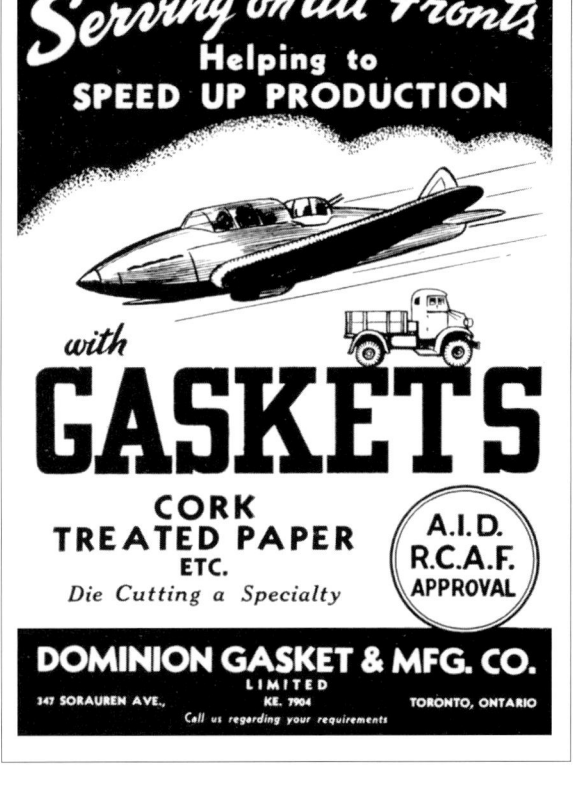

*Serving on All Fronts*
Helping to
SPEED UP PRODUCTION
with
**GASKETS**
CORK
TREATED PAPER
ETC.
Die Cutting a Specialty

A.I.D. R.C.A.F. APPROVAL

**DOMINION GASKET & MFG. CO. LIMITED**
347 SORAUREN AVE.,   KE. 7904   TORONTO, ONTARIO
*Call us regarding your requirements*

**SHERWIN-WILLIAMS AIRCRAFT FINISHES**

Approved by:
The British Air Ministry (Canadian Associated Aircraft Limited); the Department of National Defence (Royal Canadian Air Force); the Department of Transport (Civil Aviation Division).

Made by
**THE SHERWIN-WILLIAMS CO.** of Canada, Limited

A Coast-to-Coast Service!

HALIFAX, SAINT JOHN, CHARLOTTETOWN, QUEBEC, ROUYN, MONTREAL, OTTAWA, KINGSTON, TORONTO, HAMILTON, BRANTFORD, KITCHENER, STRATFORD, WINDSOR, FORT WILLIAM, WINNIPEG, REGINA, SASKATOON, CALGARY, EDMONTON, VANCOUVER.

*We can help you KEEP EM' FLYING*

With five offices and distributing warehouses located at key points across the Dominion, Railway & Power Engineering Corporation Limited is in an unrivalled position to assure aircraft manufacturers of a constant smooth-flowing supply of essential components and accessories.

From anti-icer equipment to wheels; from propeller nose-spinners to tail lights, Railway & Power Engineering Corporation distributes nearly every basic item of equipment and material used by the aviation industry.

Check the list on your left. You may find the answer there to some of your production problems.

**Our Wide line Includes...**
Aircraft Hardware, Anti-Icer Equipment, Navigation Equipment, Bar Stock, Bolts, Bomb Control Devices, Carbon Bushings and Bearings, Engine Materials, Electrical and Control Cables, Flares and Signals, Flightex Fabric, Fuel Pump Drives, Hydraulic Equipment, Instruments, Interphone Systems, Lights, Mooring Equipment, Morganite Brushes, Propellers, Propeller Spinners, Radio, Radio Ignition Shielding, Rivets, Screws, Dzus Fasteners, Shock Cord, Solenoids, Tachometer and Remote Control Drives, Terminals, Tools, Parker Tube Fittings and Valves, Tubing, Washers, Wheels, Windshield Material, Lord Bonded Rubber Shear Type Mountings.

EXCLUSIVE DISTRIBUTORS IN CANADA FOR AIR ASSOCIATES, INC. BENDIX, N.J.

**RAILWAY & POWER ENGINEERING**
CORPORATION LIMITED

TORONTO   HAMILTON   MONTREAL   WINNIPEG   VANCOUVER

# A 100% CANADIAN PRODUCT DISTINGUISHED BY QUALITY

Atlas Steels Limited is proving to you right now the benefit of a Canadian plant for the manufacture of tool steels, stainless steels and mining drill steels. This domestic source of supply helps the war effort by keeping money in Canada and saving foreign exchange. It further assures that Canadian industry will get its requirements of fine steels at reasonable prices, regardless of what develops in other countries.

ATLAS FINE STEELS ARE STOCKED BY DISTRIBUTORS ACROSS CANADA AND AT BRANCH OFFICES LISTED BELOW

| Winnipeg | Hamilton | Toronto |
|---|---|---|
| 237 Garry St. | 42 James St. N. | 401 Fleet St. W. |

Swastika, Ontario — Montreal, 1744 William St.

DISTRIBUTORS:
C. H. HENZE COMPANY, WINDSOR, ONT.
WM. STAIRS, SON & MORROW, LTD., HALIFAX, N.S.
MARSHALL-WELLS B. C. LIMITED, VANCOUVER

**ATLAS STEELS LIMITED — WELLAND, ONTARIO**

---

## Fighting On the HOME FRONT

To assure a continuous flow of war supplies, fire protection must be carefully planned. Pyrene and C-O-Two Extinguishers are "on duty" in munition plants, defense industries, R.C.A.F. hangars and crash trucks, and on transportation and fighting units on land, sea and in the air. Moreover, Pyrene is at your service with fire protection engineering to help you defend against fire hazards and provide maximum safety.

*Priorities enable you to secure reasonably prompt deliveries.*

**Pyrene Manufacturing Company** OF CANADA, LIMITED — Toronto

**C-O-TWO FIRE EQUIPMENT** OF CANADA, LIMITED — Toronto

---

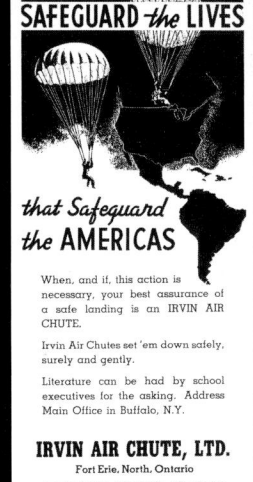

## SAFEGUARD the LIVES that Safeguard the AMERICAS

When, and if, this action is necessary, your best assurance of a safe landing is an IRVIN AIR CHUTE.

Irvin Air Chutes set 'em down safely, surely and gently.

Literature can be had by school executives for the asking. Address Main Office in Buffalo, N.Y.

**IRVIN AIR CHUTE, LTD.**
Fort Erie, North, Ontario

Factories at Buffalo, N.Y.; Glendale, Calif.; Fort Erie, Canada; Letchworth, England; Stockholm, Sweden and other countries.

**IRVIN Air Chutes**
SERVING OVER 40 GOVERNMENT AIR FORCES

---

# WE'VE NEVER BUILT A PLANE
### PRODUCED A TANK
### OR LAUNCHED A BATTLESHIP

## BUT We Have Made Thousands of Parts For Them!

That's right. We've never manufactured a completed war machine in our lives. But we have efficiently, quickly and economically turned out many thousands of "AGS", standard aircraft parts, and radio contact parts; screw machine parts for naval and air-craft instruments, and special machine parts of many kinds.

For years the Coleman Company has specialized in work on high-precision gasoline pressure products.

Work that requires skill and accuracy to the ultimate degree. Today, the Coleman plant is set-up with machines, assembly lines, and backed by the experience of capable executives and craftsmen to perform a real service on "sub-contracts" for you.

That unusual angle in specifications, that particularly tough precision problem may be just the one we are better equipped to handle than anyone else! Let us figure it on it for you. Write, wire or phone today.

**The Coleman Lamp and Stove Co. Ltd.** Toronto, Canada

---

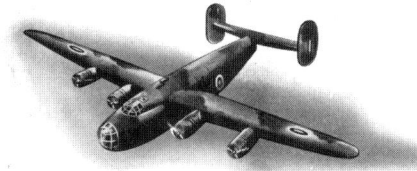

## FULFILLING OUR PLEDGE: *To Keep them Flying*

ENGINE overhaul is a little realized, but vitally important operation. Below is a view of a section of the Canadian Pratt & Whitney plant at Longueuil where almost twice the number of engines originally planned for is being handled. And steps are now being taken again to double this capacity.

Every engine received is stripped down to its smallest part. Each part is cleaned, inspected, tested, replaced if necessary, then the engine reassembled and run for hours on the test block before reshipping. A 1200 h.p. Pratt & Whitney Twin Wasp Airplane Engine has 8600 precision made parts.

About 25% of the engine overhauls for the R.C.A.F. since the beginning of the war have been entrusted to Canadian Pratt & Whitney.

## Canadian Pratt & Whitney
AIRCRAFT COMPANY LIMITED • LONGUEUIL, P.Q.

PRATT & WHITNEY ENGINES — HAMILTON STANDARD PROPELLERS — De HAVILLAND PROPELLERS — PESCO ACCESSORIES

---

## A Modern Plant for Modern Aircraft

In our modern aircraft plant at Malton, Ont., we have the latest in production machinery and a fully trained staff. Facilities are available for assembling and repairing aircraft and for quantity production of airplanes and parts.

We are also equipped to produce steel and light alloy aircraft forgings.

**NATIONAL STEEL CAR CORPORATION LIMITED**
Aircraft Division
MONTREAL — HAMILTON — MALTON

A new Lancaster Mk.X is assembled at Malton, No.3 engine being the centre of attention. (NFB WRM3239)

## The Lancaster

The plum of Canadian wartime aviation contracts was the Avro Lancaster. Soon to be hailed as Britain's grandest heavy bomber, it had evolved from the twin-engine Manchester (first flight, July 1939), which had failed in service. The solution was to re-design the Manchester with four engines, in which form it flew in January 1941. The go-ahead for Canada came that December. This plan would assure Bomber Command an alternate supply, lest enemy air raids curtail deliveries at home. A new government entity, Victory Aircraft Co. of Malton, would build the Lancaster Mk.X, supported by Canadian and US subcontractors, e.g. hundreds of Merlin engines would be supplied from the US by Packard. Victory would operate in the expropriated National Steel Car factory. The plant expanded suddenly, as modern machines and tools poured in. Thousands of workers were recruited; components and jigs materialized. The first Lancaster, KB700 "Ruhr Express", rolled out in the summer of 1943. First flight was on August 1 and KB700 soon was overseas. On the whole, Mark Xs proved compatible with British-built Lancasters, although there were discrepancies requiring modifications at British maintenance units. In March 1944 Lancaster Mk.Xs began reaching the RCAF in Yorkshire, 419 Squadron being first.

Victory Aircraft completed 422 Lancasters, plus eight civilian versions for the Canadian Government Trans-Atlantic Air Service, a high-speed mail and priority passengers operation. The first batch of Lancasters for the RCAF were serials KB700 to KB999, delivered between September 1943 and June 1945. A second batch was FM100 to FM229, the first of which flew the Atlantic in March 1945. Early KB Lancaster Xs saw fierce combat, although subsequent examples generally arrived too late for ops. None from the FM series saw combat.

### "KB" Series Lancaster Xs on Ops

| Serial | Squadrons | Fate |
|---|---|---|
| KB700 | 405, 419 | Destroyed by fire (DBF) Middleton St. George 2-1-45 |
| KB701 | 419 | Crashed (Cr.) on night training 2-1-45 |
| KB704 | 419, 408 | Cr. Middleton St. George 11-5-44 |
| KB705 | 428, 1666 HCU | Struck off charge (SOC) postwar |
| KB706 | 419 | Missing Aachen 24/25-5-44 |
| KB707 | 419 | Cr. landing (lndg.) Middleton St. George 19-9-44 |
| KB708 | 419 | Cr. Boscombe Down 26-8-44 |
| KB709 | 428 | Missing Stettin 29/30-8-44 |
| KB710 | 419 | Missing Louvain 12/13-5-44 |
| KB711 | 419 | Missing St. Ghislain 1/2-5-44 |
| KB712 | 419 | Missing Cologne 28-10-44 |
| KB713 | 419 | Missing Louvain 12/13-5-44 |
| KB714 | 419 | Missing Cambrai 12/13-6-44 |
| KB715 | 419 | Missing Lohausen 24-12-44 |
| KB716 | 419 | Cr. Middleton St. George 7-5-44 |
| KB717 | 419 | Missing Dortmund 22/23-5-44 |
| KB718 | 419 | Missing Villeneuve St. George 4/5-7-44 |
| KB719 | 419 | Missing Stuttgart 24/25-7-44 |
| KB720 | 419, 1664, 1666 HCU | SOC postwar |
| KB721 | A&AEE, 419 | SOC postwar |
| KB722 | 419 | Cr. lndg., hit by friendly fire St. Quentin 6-1-45 |
| KB723 | 419 | Missing Villeneuve St. George 4/5-7-44 |
| KB724 | 419 | Cr. on takeoff 27-8-44 |
| KB725 | 428 | Cr. County Durham 3-2-45 |

| Serial | Squadron(s) | Fate |
|---|---|---|
| KB726 | 419 | Missing Cambrai 12/13-6-44 |
| KB727 | 419 | Missing Villeneuve St. George 4/5-7-44 |
| KB728 | 419 | Missing Sterkrade 16/17-6-44 |
| KB731 | 419 | Missing Cambrai 12/13-6-44 |
| KB732 | 419 | SOC postwar |
| KB733 | 419 | SOC postwar |
| KB734 | 419 | Cr. near Zeist 16/17-6-44 |
| KB735 | 419 | Cr. on landing East Moor 18-9-44 |
| KB736 | 419, 1660 HCU | SOC postwar |
| KB737 | 428 | Missing Essen 25-10-44 |
| KB738 | 419 | Missing Opladen 27/28-10-44 |
| KB739 | 428 | SOC postwar |
| KB740 | 428 | Cr. lndg. Woodbridge 25-7-44 |
| KB741 | 428, 434 | Missing Chemnitz 14/15-2-45 |
| KB742 | 428 | Cr. Middleton St. George 4-11-44 |
| KB743 | 428 | Missing Bremen 18/19-8-44 |
| KB744 | 428 | SOC postwar |
| KB745 | 419 | Cr. Northumberland 4-10-44 |
| KB746 | 419 | SOC postwar |
| KB747 | 428 | SOC postwar |
| KB748 | 419 | SOC postwar |
| KB749 | 428 | Missing Soesterberg 15-8-44 |
| KB750 | 419 | Missing Wiesbaden 2/3-2-45 |
| KB751 | 428 | Missing Stettin 16/17-8-44 |
| KB752 | 419 | Missing Homburg 7/8-4-45 |
| KB753 | 419 | Missing Buer 29/30-12-44 |
| KB754 | 419 | Missing Bochum 9/10-10-44 |
| KB755 | 419 | Missing Caen 7/8-8-44 |
| KB756 | 428 | Missing Villeneuve St. George 4/5-7-44 |
| KB757 | 428 | SOC postwar |
| KB758 | 428 | Missing Brunswick 12/13-8-44 |
| KB759 | 428 | Missing 28/29-7-44 |
| KB760 | 428 | SOC postwar |
| KB761 | 419 | Missing Hamburg 31-3-45 |
| KB762 | 419 | Severe damage, Middleton St. George 23-4-45 |
| KB763 | 428 | Cr. County Durham 28-1-45 |
| KB764 | 428 | Ditched Azores en route Canada 4-6-45 |
| KB765 | 419 | Missing Buer 29/30-12-44 |
| KB766 | 428 | Icing, cr. Beauvais-Tillé 2-12-44 |
| KB767 | 419 | Cr. Manston 1-11-44 |
| KB768 | 419 | Mid-air collison, cr. Warwickshire 16-11-44 |
| KB769 | 419 | Missing Merseberg 14/15-1-45 |
| KB770 | 428 | Missing Stuttgart 28/29-1-45 |
| KB771 | 428 | SOC postwar |
| KB772 | 419 | SOC postwar |
| KB773 | 428, 431 | SOC postwar |
| KB774 | 419, 431 | SOC postwar |
| KB775 | 419 | Missing Russelsheim 25/26-8-44 |
| KB776 | 419 | Missing Essen 23/24-10-44 |
| KB777 | 428 | Missing Hildesheim 22-3-45 |
| KB778 | 428 | Icing, cr. 5-3-45 |
| KB779 | 419 | Missing Osnabruck 6/7-12-44 |
| KB780 | 428 | Missing Duisberg 14-10-44 |
| KB781 | 428 | SOC postwar |
| KB782 | 428 | Missing Dusseldorf 2/3-11-44 |
| KB783 | A&AEE, 419 | SOC postwar |
| KB784 | 428 | Missing Kiel 13/14-4-45 |
| KB785 | 419 | Fire, cr. County Durham 24-11-44 |
| KB786 | 419 | Missing Hemmingstedt 20/21-3-45 |
| KB787 | 419 | Mid-air with Lancaster, cr. Ardennes 4/5-2-45 |
| KB788 | 419 | DBR by flak 30-11/1-12-44 |
| KB789 | 428 | SOC postwar |
| KB790 | | In UK but not on squadron, scrapped |
| KB791 | 428 | SOC postwar |
| KB792 | 428 | Missing Wiesbaden 2/3-2-45 |
| KB793 | 428 | Fire, cr. County Durham 13-1-45 |
| KB794 | 428 | SOC postwar |
| KB795 | 428 | Cr. Middleton St. George 7-4-45 |
| KB796 | 419 | SOC postwar |
| KB797 | 419 | Missing Desau 7/8-3-45 |
| KB798 | 428 | Missing Opladen 27/28-12-44 |
| KB799 | 419 | Missing Merseberg 14/15-1-45 |
| KB800 | 419 | Missing Duisberg 14-10-44 |
| KB801 | 428 | SOC postwar |
| KB802 | 419, 431 | SOC postwar |
| KB803 | 428, 431 | Engine failed, cr. Yorkshire 26-1-45 |
| KB804 | 419 | Missing Dortmund 20/21-2-45 |
| KB805 | A&AEE | SOC postwar |
| KB806 | 428, 431 | Missing Merseberg 14/15-1-45 |
| KB807 | 419, 431 | SOC postwar |
| KB808 | 428, 431 | Missing Hildesheim 22-3-45 |
| KB809 | 419 | Missing Dortmund 20/21-2-45 |
| KB810 | 431 | SOC postwar |
| KB811 | 419, 431 | SOC postwar |
| KB812 | 431 | SOC postwar |
| KB813 | 428 | Cr. Bedfordshire 25-10-44 |
| KB814 | 428, 434 | Missing Hagen 15/16-3-45 |
| KB815 | 419, 431 | Missing Hagen 16-3-45 |
| KB816 | 431, 434, 428 | Cr. Church Broughton 14-4-45 |
| KB817 | 419, 431 | Missing Oberhausen 1/2-11-44 |
| KB818 | 431 | Cr. Ford 7-2-45 |
| KB819 | 431 | SOC postwar |
| KB820 | 428 | SOC postwar |
| KB821 | 431 | Missing Hanau 6/7-1-45 |
| KB822 | 428, 431 | Mid-air with KB831, cr. at sea 25-4-45 |
| KB823 | 428, 431 | SOC postwar |
| KB824 | 419, 434 | SOC postwar |
| KB825 | 428, 434 | SOC postwar |
| KB826 | 428, 434 | SOC postwar |
| KB827 | 431 | SOC postwar |
| KB830 | 419, 434 | SOC postwar |
| KB831 | 419, 431 | Mid-air with KB822, cr. at sea 25-4-45 |
| KB832 | 419, 434 | Cr. Croft 22-3-45 |
| KB833 | 419, 434 | SOC postwar |
| KB834 | 434 | Missing Essen 11-3-45 |
| KB835 | 431, 434 | Missing 15/16-3-45 |
| KB836 | 431, 434 | SOC postwar |
| KB837 | 431 | SOC postwar |
| KB838 | 428 | SOC postwar |
| KB839 | 431, 419 | SOC postwar |
| KB840 | 428, 434 | SOC postwar |
| KB841 | 419 | SOC postwar |
| KB842 | 428, 434 | Battle damage, cr. lndg. 5/6-3-45 |
| KB843 | 434, 428 | SOC postwar |
| KB845 | 419 | Cr. Bedfordshire 5/6-3-45 |
| KB846 | 434, 428 | Missing Hagen 16-3-45 |
| KB847 | 431 | SOC postwar |
| KB848 | 428 | SOC postwar |
| KB849 | 431, 434 | SOC postwar |
| KB850 | 419 | Missing Zeitz 16/17-1-45 |
| KB851 | 428, 419 | SOC postwar |
| KB852 | 431, 434 | SOC postwar |
| KB853 | 434, 431 | Missing Essen 11-3-45 |
| KB854 | 419 | SOC postwar |
| KB855 | 419, 428 | Cr. Middleton St. George 20-2-45 |
| KB856 | 431 | SOC postwar |
| KB857 | 419 | SOC postwar |

| | | |
|---|---|---|
| KB858 | 431 | Missing Chemnitz 5/6-3-45 |
| KB859 | 431 | Missing Hamburg 31-3-45 |
| KB860 | 419 | SOC postwar |
| KB861 | 431 | SOC postwar |
| KB862 | 434 | SOC postwar |
| KB863 | 434 | SOC postwar |
| KB864 | 428 | SOC postwar |
| KB865 | 419 | SOC postwar |
| KB866 | 419 | Missing Kiel 3/4-4-45 |
| KB867 | 428 | SOC postwar |
| KB868 | 431 | SOC postwar |
| KB869 | 419 | Missing Hamburg 31-3-45 |
| KB870 | 419 | Missing Hagen 15/16-3-45 |
| KB871 | 419 | SOC postwar |
| KB872 | 431 | SOC postwar |
| KB873 | 434 | SOC postwar |
| KB874 | 431 | Cr. lndg. Manston 25-3-45 |
| KB911 | 434 | Missing Hamburg 31-3-45 |

When KB700 arrived on squadron, Bomber Command was at a peak. It was sending hundreds of crews out every night, so long as weather permitted, to plaster German cities. The aim was simple—lay waste everything, industrial or residential. Bomber Command made no bones about this. But the cost was high; there were times when Bomber Command was almost crippled by losses of men and machines. Enemy defences were formidable; their effectiveness is seen in the carnage wreaked upon the first batch of Lancasters, from KB706, lost the night of May 24/25, 1944. KB911 would be the final one lost to enemy action, being destroyed on March 31, 1945. Other than the 70 aircraft lost on operations from 175 listed here, 39 others were lost in accidents. Nearly all the others returned to Canada. Naturally, combat losses dropped in the waning months of the war, by when the Luftwaffe had been broken.

After May 1945 surviving Lancaster Xs returned to Canada in anticipation of action in the Pacific. When the US brought that war to a halt with the atomic bomb, the RCAF mothballed most of its Lancasters. Many were sold for scrap. Others, converted for aerial photo work, played a huge part in the postwar mapping of the Canadian Arctic. When the West and the Soviet Union began to menace each other, a new arms race began. The RCAF re-activated its Lancasters for coastal patrol. The last did not leave service till April 1964.

## Experimental Flying: The Impact of A/V/M Stedman

Since 1920 the CAF/RCAF had a great interest in research and development. This came largely from the efforts of A/V/M Ernest W. Stedman, the first Director of Technical Services in the air force. The small organization which he founded went from Test and Development, to Winter Experimental Establishment, and Central Experimental and Proving Establishment, to today's Aerospace Engineering and Test Establishment.

An Englishman, Stedman had started in aeronautics before WWI. He worked in design, modification and evaluation on projects from a flying boat for the Admiralty, to Handley Page heavy bombers. He was on the crew making the first Handley Page O-400 raid, bombing a rail yard near Metz. At war's end he was on the RAF's biggest bomber, the Handley Page V.1500. After the war Stedman joined a crew planning to fly the Atlantic from Newfoundland in a modified V.1500. With 1700 gallons of fuel, Stedman calculated that the 1600-mile flight was feasible, cruising at 65 mph. This would be the first airplane flight across the Atlantic. The V.1500 arrived by sea in St. John's in May 1919, then went by rail to Harbour Grace for assembly and test flying. Meanwhile, Alcock and Brown arrived from the UK, assembled a Vickers Vimy bomber, and completed the first non-stop Atlantic airplane flight. Thus pre-empted, the V.1500 sponsors opted instead for a publicity flight to New York.

On July 4, 1919 the V.1500 left Harbour Grace, flying southwesterly through the night. But engine trouble forced a landing near Parrsboro, Nova Scotia. There was considerable damage, but Handley Page decided on repairs, with Stedman in charge. Spares were shipped to Parrsboro, and the V.1500 flew again on October 2. A week later the great plane was in Long Island, where Handley Page demonstrated its commercial potential. For several days joy rides were flown, each with about 30 passengers. Next, some freight was loaded for Chicago, but,

A/V/M E.W. Stedman, CB, OBE—the father of RCAF aeronautical research. (CF RE2397-1)

en route, there was another landing mishap. This time, Handley Page gave up on its North American endeavour.

Stedman decided to settle in Canada and open a sheet metal shop. Instead, in October 1920 he accepted a position as director in the Canadian Air Board's Technical Branch. His work involved making a number of HS-2L and F.3 flying boats airworthy, modifying them for photography, etc. Much R&D into ski and float flying, and cold weather engine operations (a perpetual theme in Canadian aviation) awaited. In 1926 Stedman's department conducted winter trials at High River with two Siskins. He was especially interested in the use of metal framing in Siskin construction. He kept the RCAF abreast of anything new, whether airframe/engine design and modification, equipment and tools, or fuels and lubricants. In 1928, for example, he toured major British manufacturers. One of his comments is a touch melancholic: "A number of good flying boats was seen, e.g. Short's Calcutta and Singapore II, and the Supermarine Southampton; but they were all too large or, perhaps I should say, too expensive for our purposes."

Something important came from Stedman's involvement with the US Fairchild and Pratt & Whitney companies. In 1927 the RCAF purchased some Wright-powered Fairchild FC-2s, then became the first operator to combine the FC-2 and the new P&W engine. Stedman also laid the foundation for RCAF accident investigation and aircraft salvage. The example in Chapter 2 of the Viking structural failure shows how advanced were his methods. He worked closely with the National Research Council in Ottawa, this leading to the construction of a float testing basin, an engine and fuel testing lab, and an aircraft instrument lab. In 1935 he led the move to the Northrop Delta.

In 1937 Stedman, with his key men, A/V/M Alan Ferrier and G/C Oliver Adams, established the first RCAF technical detachments. These placed RCAF representatives in plants such as de Havilland and Canadian Vickers to inspect work being done. At the same time he and some other leading aeronautical minds, like John H. Parkin of the NRC, were promoting a national aviation museum. Exhibits were obtained from important figures like Mrs. Fairchild, J.A.D. McCurdy and W.R. Turnbull. This year also heralded the first trans-Atlantic flying boat proving flights. Having planned such a flight years earlier, Stedman paid close attention to this activity. In the late 1930s he collaborated with Tommy Siers of Canadian Airways in developing a system of oil dilution that allowed easier cold weather engine starts. As Stedman explained in his memoir, "... oil was diluted with gasoline before the engine was shut down so that, when the engine was started again, it would be lubricated with ... oil of low viscosity ... after running for a short time, the gasoline would be evaporated and the oil restored to its original viscosity."

Just before WWII Stedman was part of a mission to the US making arrangements to purchase urgently needed service types—the Catalina, Digby and Harvard. Meanwhile, the NRC was expanding, and in July 1940 opened a new aerodynamics building near Rockcliffe. Included was a horizontal wind tunnel and the first vertical spinning tunnel in Canada. Other investigations were into transport gliders, helicopters and robot weapons. On a more immediate level, tests with models were made to discredit claims that the Cornell trainer, important to the BCATP, may have design flaws. Studies into Harvard ground looping resulted in improvements to the tail wheel steering mechanism. In the NRC engine lab many projects were undertaken. One investigated cylinder barrels for the Cheetah engine. These could not be obtained in the UK, so the NRC designed a substitute, which the Ford Motor Co. then produced.

Stedman also oversaw research into aeromedical affairs. Pilot blackout, an anti-G suit, and night vision were some of the topics. Dr. Frederick Banting and Dr. W.R. Franks were the chief men involved in the anti-G suit, which was developed in Toronto. In April 1942 Stedman visited Wright Field in Ohio to observe the Sikorsky XR-4 helicopter. Greatly impressed, he recommended that six XR-6s be ordered for rescue operations in remoter parts of Canada. His requisition went forward, then was blocked by the Treasury Board, which protested the cost (which was excessive, since Sikorsky had built in R&D charges). Stedman and Parkin visited Wright Field again in October 1943, where Colonel Cooper of the USAAF flew Stedman in a YR-4. Stedman wrote:

*After demonstrating the possibilities of the machine, we returned to our starting point, where the pilot asked me to pick out the exact spot where I would like him to land. I pointed to a small light patch of grass and he placed the machine ... over it and about three feet in the air... Shortly afterwards I had an opportunity to visit the Sikorsky factory where I saw the details of construction and discussed the subject with Igor Sikorsky... The thing that interested me most at the factory was the flight area, which was about the size of a very small car-parking lot, surrounded by high wire fencing and with factory chimneys nearby... this area was extremely satisfactory for the operation of their three experimental helicopters.*

A/V/Ms Stedman and Lloyd Breadner also visited the Bell company in Buffalo, where Larry Bell demonstrated his flying helicopter model. Stedman, showing his usual foresight, noted years later: "The recent development of the helicopter has been particularly interesting as confirmation of my early conviction that it was invaluable for rescue and similar work in inaccessible areas in the north ... the helicopter freed us from the bondage of the runway." During the war he also studied such matters as ground control approach, jet-assist takeoff, propeller manufacture and the gas turbine engine. In May 1942 he visited Frank Whittle at Power Jets Ltd. to learn about the jet engine and observe the Gloster E.28, the first Allied jet: "It was watching this aeroplane fly that convinced me that here was something of the greatest interest to Canada." In October 1942 Stedman recommended to the Air Board in Ottawa that Canada do its own gas turbine studies. That month a committee was set up to study the possibilities; in January 1943 some Canadians, including Kenneth F. Tupper and Paul B. Dilworth of John Parkin's staff, went to England. Out of this visit the NRC established a cold weather test facility in Winnipeg, where several gas turbines were evaluated. Further, a Crown company was formed to pursue engine development. Founded in July 1944, this was Turbo Research Ltd., headed by H.J. Carmichael. Stedman was a board member. Meanwhile, he continued to watch developments by visiting UK and US companies involved in jet engine/aircraft design. In the spring of 1945 Turbo Research began design of an axial flow compressor (compared to the centrifugal flow type such as the D.H. Goblin in the Vampire). About this time Stedman retired from the RCAF, but continued watching the gas turbine from the Chinook through to the Orenda.

During the war Stedman followed many aircraft projects. In October 1942 he and John Parkin discussed the XB-35 flying wing with John K. Northrop; "This tailless aeroplane had been of interest to us because we also had built the NRC tailless glider, but by now we were convinced that this idea had been given a good trial and should be abandoned." This is perhaps the only major concept where Stedman's judgment was off. In modern times the tailless design was resurrected, most notably with the B-2 flying wing, the direct descendent of the XB-35. After the war Stedman remained close to aviation. In 1946 he was a UN observer at the 1946 Bikini A-bomb tests. He also served as Air Advisor in the Defence Research Board, an organization set up in 1947 under Dr. O.M. Solandt. A/V/M Stedman, the father of research and development in the RCAF, passed away in 1957.

## Wartime Test and Development

RCAF Test and Development Flight (T&D), formed in the 1930s, was the logical result of the pioneer aeronautical engineering and test flying done by pioneers like Stedman, Grandy and Tudhope. Such men encouraged new designs like the Vedette, then test flew them. As the RCAF added types like the Delta, Lysander and Stranraer, T&D had much new work. Each had to be evaluated, mods devised, installed and proven, performance and fuel consumption data gathered, etc. Often, mods related to Canadian conditions—floats for the Delta, skis and a cockpit hood for the Tiger Moth, etc. Projects also involved fitting new engines such as Pratt & Whitneys in the Bolingbroke, and a Wright in a Battle. T&D conducted such work in conjunction with industry tech reps.

With WWII there was greater urgency at T&D. Some types had to be brought to service (i.e. combat) standards. Sometimes efforts seemed fruitless, e.g. converting antiquated Vancouvers into fighting machines. Nonetheless, such work was essential, until modern designs arrived. Sometimes it was expeditious to make mods in the field, as with the Vancouvers, which may not have endured a long flight from Jericho Beach to Rockcliffe. In such cases, T&D pilots would go on the road. The unit's daily diary shows a variety of projects. What especially stands out is the pace that T&D had to maintain. F/O Colpitts flew Delta 670 on armament trials in the first week of September 1939. On the 11th F/L Gordon Truscott tested Oxford 1503, then left on a TCA flight for Vancouver to do an acceptance flight on Shark 516. The next day F/L Wray took Goose 917 to Longueuil to test Fairchild Super 71 No.665, an outdated aerial photo plane (outdated, but, under the conditions, the RCAF needed it). Then Wray brought 665 to Ottawa for more testing. On September 13 he visited Victory Aircraft to test new Lysanders. On the 16th Truscott test flew Harvard 1324, recently delivered from North American. Meanwhile, several T&D pilots were familiarizing themselves on Fairey Battle 2185, and on the first RCAF Hudson. A Wapiti is also mentioned as being at T&D. On September 26 F/L Wray was at Malton accepting Lysanders with civil pilot Leigh Capreol. The diary noted a Malton flight with Lysander 421 on October 13 as a test of a "carbon monoxide detector in cockpit."

F/L Truscott and FSgt McManus brought Stranraer 915 from Montreal to Ottawa for trials on September 27, while F/O Colpitts was in Fort Erie testing Fleet trainers. On October 14 Colpitts and Sgt Lamoureux ferried Atlas 407 from Trenton to Ottawa. Three days later Wray left by rail for Fort William to discuss with Canadian Car and Foundry the idea of converting a Goblin to dual controls. Truscott, still on temporary duty in Vancouver, declared Shark 517 "suitable for acceptance" on October 11 and set off next day by air for Ottawa. On October 31 he and Cpl Greenway flew Avro 626 at Uplands to determine why its propeller was vibrating.

On November 2, 1939 S/L Wray left Ottawa by car to test the first Fairchild-built Bolingbroke (No.702) at Longueuil. This he did next day with company chief pilot Red Lymburner. On subsequent flights in 702 they evaluated power settings, handling, speed and altitude. Also on the 2nd, F/Ls Truscott and Colpitts drove to Canadian Vickers in Montreal to accept Stranraer 916. On the 7th they tested an auto pilot installed in the same plane. On the 11th Colpitts made a famil flight in Beech 18 CF-BMU, while F/L Truscott and FSgt Miscampbell set off to deliver 916 to Halifax via Quebec City. On the 15th S/L Wray declared Bolingbroke 702 fit for RCAF acceptance. Soon it was in Ottawa, where other T&D pilots could check out on it.

In December 1939 F/L Colpitts went again to Fort Erie to test trainers, often with Fleet chief pilot T.F. "Tommy" Williams. Colpitts conducted climb, service ceiling, cockpit heating, sliding hood, metal airscrew, airspeed and aerobatic tests on aircraft 1006 and 1007. He also flew solo in front and back cockpits, to evaluate trim. Before Christmas F/L Truscott went to Malton to accept more Lysanders. At the end of January 1940 he became OC of T&D Flight.

For February 27 F/L Colpitts flew Fairchild 71B 634 for 1:05 hours so that 23 test parachutes could be dropped. The first T&D Anson flight appears in the entry for April 15, F/L Colpitts flying N9914. Thereafter this machine was busy on flight tests and training at Rockcliffe. Fairey Battle 1319 was involved in April with target tow trials. On May 9, 1940 Spitfire L1090 joined T&D, although it does not appear as flying with the unit and was struck off strength on June 25. For June 1940 T&D logged 15:45 hours with its "aircraft on charge": Battles 1315, 1634 and 1649, Shark 502, Delta 684, Ansons 6002 and 6006, and Bolingbroke 706. New types that appear on test projects in the second half of 1940 include the Nomad and Yale. A number of civilian pilots took famil flights in the Yale in late September, as indicated by diary entries stating, "Flying test, civilian pilot". In the fall of 1940 Goblin 342 was being flown to gather performance figures. Handley Page Harrow 794 turned up on December 7, flown by S/L Truscott. Dr. Solandt of the NRC appears several times towards the end of 1940, riding as a passenger in Harvards and Yales.

The strength of T&D for January 1, 1941 is noted as 12 officers and 61 airmen. W/C T.R. Louden of the University of Toronto was the OC in this period. Night photography flights were made this month with Hudson T9385. Spin tests were done with Fawn 4675, sometimes flown by Dr. Green of the NRC. The Cessna T-50 Crane appears for the first time in a series of engine, airframe and undercarriage tests. In February intensive trials were conducted with Digby 751, including bombing. On May 3 the Anson Vidal plastic fuselage arrived at T&D. On June 6 Anson 6008 was tested with Jacobs engines. Later in the month Digby 751 was tested with ASV (anti-surface vessel) radar equipment. S/L Briggs and Dr. Green conducted spin trials in a Harvard, and a Menasco Moth was tested with a radio installation for training students.

The Vidal made its first flight with F/O King on July 21. T&D was anxious to test it for performance and weathering. Also in July the Bolingbroke made comparative flights with Bristol Mercury and P&W Wasp engines (25 flights). On January 31, 1941 F/O De Pret ferried the first Canadian-built Harvard (No. 3034) from Montreal to Ottawa. Much routine flying was recorded in the following months, all safely carried out until August 1, 1941. That day Battle 1315 crashed across the Ottawa River from Rockcliffe, but the crew of F/O Culliton and Cpl Arnold survived. Battle 1628 arrived from Mountain View a week later. A cloth wing cover for the Norseman was tested, as was a Fairchild F.24 camera in a Crane, and bomb racks in the Digby. Tragedy finally visited T&D Flight. On September 19, 1941 S/L F.E.R. Briggs and F/L W. Richards, conducting dive tests in Crane 7919, died when it crashed near Ottawa. That month the first Canadian-built Anson, No.7069, was delivered to T&D. The flight would go on to complete hundreds of projects before war's end, dealing with such sophisticated mods as skis for the Hurricane, Harvard and Ventura.

## Winter Experimental & Training Flight

On October 1, 1943 Winter Experimental and Training Flight came into being under the AOC, No.1 Training Command. Its job was to compile operating experiences for RCAF aircraft; develop mods, lubricants, ground equipment, and clothing; establish techniques for cold weather servicing and maintenance; and teach the art of cold weather operations. For a base WETF selected Kapuskasing, Ontario at 49°25N latitude. "Kap" already was a DOT and TCA station, and a regular stop for 124 Ferry Squadron. It had a hangar and 3 runways (2 paved, the longest 3000- by 150-feet).

W/C H.J. Phillips was the original WETF OC. The unit's first aircraft was Lancaster Mk.III EE182, loaned on September 29, 1943 from RAF Ferry Command. It went first to Victory Aircraft at Malton for mods, then was delivered to Kap on December 9. Lancaster flying began on January 6, one of the first projects being cold weather propeller feathering. On the 16th de-icing research Hudson BW456 was taken on strength (first flight, January 19, 1944.) On the 25th a rep from the US arrived to study the effectiveness of Goodrich de-ice equipment being evaluated with BW456. Another winter project involved Ventura AE860, used to test streamlined NRC skis. In this period A/V/M Stedman visited the NACA engine lab in Cleveland, Ohio. Scientists there were studying icing, but were not getting the funds to complete their work. In his memoir Stedman noted: "The official in question had just flown across to England and had sent word back that, when flying the Atlantic at night, he had heard the ice, detached from the propellers, hitting the sides of the fuselage. He was now prepared to give his full support to their de-icing program."

On February 29 EE182 started gun firing trials. At this time WETF had 10 officers and 52 airmen. Visits from industry reps were commonplace, such people evaluating the performance of their many products—from spark plugs to lubricants. On March 27 WETF began a move to Gimli, Manitoba. There, it was hoped, colder conditions might be found. The Lanc carried 17 men to Gimli, the Ventura 7, the Hudson 4. Equipment went by rail. On April 9 Spitfire Mk.VIII JG480 was assembled at No.8 RD in Winnipeg. On the 21st reps from Compact-O-Hangar of Chicago visited Gimli to supervise erection of a new type of portable hangar to be evaluated in cold conditions. In late April the Lanc flew to St. Petersburg, Florida. In May 1944 the Hudson was out looking for ice, a strange pastime under normal conditions. On September 25 Mosquito B Mk.XXV KB428 was delivered to Gimli from DHC. F/L S.O. Partridge was busy flying the Spitfire which, in October, was fitted with special wing covers in Winnipeg (the Mossie also had wing covers fitted). DHC reps J.F. Neal (later C.N. Forrest) supported the Mossie. On October 19 Battle R4052 arrived at Gimli from No.3 B&GS for fuel consumption tests.

Late in October BW456 was noted as "detached" to Rockcliffe. The NRC, meanwhile, was studying the icing problem. The WETF Lanc was used on propeller de-icing. On December 21 Hudson BW654 was allotted to WETF. In January 1945 the Hudson and Battle had their engines fitted with NRC-designed priming jets to improve cold weather starting. There were fuel flow and oil dilution tests with these types. In January 1945 F/L Don Laubman, a well-known fighter ace in Canada between combat tours, was posted to WETF.

**WETF's Lancaster EE182 sits in sub-Arctic weather at Kapuskasing. Engine tents and heaters are being used to prepare for flight. (NAC PA196937)**

The NRC and CCF developed streamlined skis for the Hurricane and Harvard. The NRC also co-operated with the US in ski trials with types like the P-39. On March 8, 1944 this Hurricane was lost in an accident at Bagotville. (CANAV Col.)

At month's end unit establishment was 93 personnel; 30:20 hours were flown. Flare drops and dinghy release tests were made with the Lanc that month. The Huter engine pre-heater was evaluated in this period. Laubman did cold weather firing with the Spitfire on February 18 and 24. Hours logged in February totalled 62:40.

In March the Lanc was bombing at Shilo, Manitoba. Since the 29th was the final day of the BCATP, S/L Partridge and F/L Griffin flew displays with the Lanc and Spitfire. By April WETF was winding down (only 15:45 flying hours for the month). On April 12 the Lanc left for Eglin Field, Florida, via Ottawa and Washington. At Eglin a Lanc bombing-up demonstration impressed all viewers. While there Partridge qualified as captain on the B-29 and B-32, and others in his crew took instructions. On April 26 he took the Lanc to an airshow at Patuxent River, Maryland, then flew home. On June 19 the Lanc left Gimli for the UK. On July 19 Battle R4052 ferried to No.5 SEHU at Neepawa, Manitoba for disposal. The Lanc returned on July 20. In August the Mossie and Spitfire were despatched for Victory airshows around country. On September 22 word came of a move to Edmonton. On the 30th an organization order was received changing WETF to the RCAF Winter Experimental Establishment. The last WETF flying appears to have occurred on September 30 with Liberator 600. Now the unit faced a new era of cold weather testing, where the emphases would be on turbine power and new weaponry.

## Other Specialties

Many trades receive little notice in the host of books about the RCAF. Yet each was indispensable, whether a member was a cook, driver, clerk, firefighter, medical aid, armourer, radar technician, etc. Aerial photography was such a trade. It had its start in the Canadian military in 1920, when the Air Board appointed Lt E.R. Owen, a WWI RNAS and RAF veteran, as Air Photographic Inspector. The department grew yearly. In 1925, for example, 235 aerial 7 x 9 inch rolls were processed, each with 115 negatives. More than 50,000 prints were made, all by hand. In 1935 the organization moved to a new building at RCAF Station Rockcliffe; 15 airmen were employed under S/L E.R. Owen.

In January 1943 Org. Order No.107 described the formation at Rockcliffe of No.1 RCAF Photographic Centre. This was a joint RCAF-NRC unit doing photographic research "for Canadian requirements." Its flying component was the Photographic Reconnaissance Flight. In March 1944 the unit became No.1 Photographic Establishment under No.7 Wing. Its building, the famous "White House", was shared with the Geographical Section General Staff until 1945. Then a plan went ahead to map Canada as part of a grand scheme to assure North American defence. Now the White House was turned over to the RCAF. In April 1947 the operation came under No.22 (P) Wing. In 1954 the name RCAF Photographic Establishment was adopted.

Wartime priorities at No.1 PE had included training trades required to staff all stations in Canada, and operational squadrons overseas—PRU units like 400, 414 and 430 squadrons. In March 1944 work at Rockcliffe included flying practice survey lines using Ansons 11633, 11534 and 11533. Pilots included S/L "Squirt" Wiseman, F/L Gubb DFC, P/Os Pook and Snasdell-Taylor and WO2 Fraser. Hurricane 5625 and Sea Hurricane BW884 also were used, as were Mitchells 891 and 892, and Spitfire Vs R7143, X4492 and X4555. When weather allowed, there was flying training, airborne photo assignments (e.g., evaluating new, overhauled or modified equipment, or film) and test flying. The Ansons often took air photo students on famil rides. On the 19th S/L Wiseman is noted as flying 891 to Oklahoma City and St. Louis to pick up a camera mount, returning on the 23rd. At the end of April 1944 strength at No.1 PE was 299.

Along with Nos. 7 (P) and 13 (P) Squadrons, Rockcliffe would send out detachments in the spring to photograph various regions. Most days these would ship film to Rockcliffe for processing, keeping the dark rooms busy till the detachments returned in the fall. George Craig was typical of the pilots doing such photo work. He had joined the RCAF in December 1940, his tours being as a staff pilot on RAF training units on the Isle of Man; and Halifax ops on 424 Squadron. Afterwards, he was a flight commander at No.6 Bombing and Gunnery School at Mountain View, then was posted to Rockcliffe. Like many RCAF pilots, in 1946 he was looking forward to returning to civil life. His plan was to go into diamond drilling with his brother in Northern Ontario. One day he accompanied his friend Gil Wass from Rockcliffe, to Montreal. Wass had an interview with F.T. Jenkins, who was looking for pilots for CPA's Survey Division. Before long, Wass, Craig and others from Rockcliffe were flying for CPA on aerial photo contracts from Newfoundland to Venezuela. This postwar "brain drain" from the Photo Establishment was an on-going problem. Once trained by the RCAF, photo techs, after a season or two of field work, usually could find a job on Civie Street, earning better money. Most of the top men at Spartan and Kenting, Canada's leading aero photo companies in the 1950s-60, learned their trade at Rockcliffe.

Rockcliffe in 1938 showing the first of the new hangars at the foot of the bluff; the Officers' Mess, quarters and water tower atop, and the Photographic Unit's boxy little "White House" below. (CF)

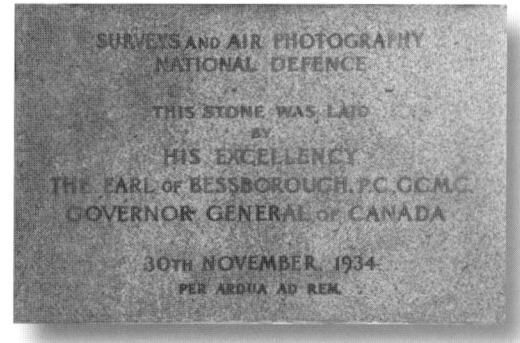

The White House (corner stone inset) in a view from December 1989. It was demolished in the early 1990s, as the DND downsized and closed bases. (Larry Milberry)

## The Firefighter

Melvin Swift, born on July 26, 1920, grew up in Toronto's Riverdale area. In 1937 he finished his schooling at Danforth Technical School, then had various jobs in printing and machine shops. When he decided to join the RCAF in August 1941, he was earning $19 a week, a decent wage for the time. Although Swift was hoping for the radio trade in the RCAF (radio was his hobby), there were no such positions at the time, so, in 1941 he joined on general duties. The same day that he was sworn in he was on a train for Brandon, the farthest he had ever been from home. There he and his mates did basic training at No.2 Manning Depot, living in a tent city at the overcrowded station.

Swift's first posting was to Calgary, still on general duty. Finally there was word of positions as fire fighters and he put in his name. Soon he was on course at Trenton. Upon graduation he was posted again to Calgary, this time to No.2 Wireless School at nearby Shepherd airfield. In time he moved to Souris, Manitoba, an Anson school. Fortunately, there was little call at the station for the fire department. One day, however, there was an accident. While several Ansons were aloft, Souris closed in with bad weather. A Harvard went up to find as many Ansons as possible, to point them toward Virden, which was open. One crew missed this warning, tried landing, but crashed fatally at the Souris station gate.

Mel Swift remained at Souris until it closed in late 1945. There then were various offers of postings. The best seemed to be RCAF Station Watson Lake on the Northwest Staging Route. He took this opportunity, serving there as a fireman into early 1946. He then left the service, first for the Toronto Fire Department. In 1949 he returned to Calgary to pursue various interests in brewing, fire fighting, oil and, finally, stationery sales.

The firefighters at No.17 SFTS at Souris in 1944. Back row: Jack Strickland, Art Hope, Mel Swift, Red Foggerty, Lyle Thompson, J.C. Richards and Bill Makas. Front row: first 2 unknown, LAC Klem, LAC Taylor, E.R. Bowick, unknown, Bert Hicken. (Swift Col.)

With accidents an every day reality, each station needed a fire brigade. This scene involved fire fighters cleaning up the mess at Dorval after Hudson FK789 swung on takeoff on June 10, 1943. (CANAV Col.)

Firefighters at No.6 SFTS, RCAF Station Aylmer. (Rosalie Woodland)

## EAC Meteorological Flight, Yarmouth

Many other "unsung" units performed vital, if unexciting, wartime duties. Typical were the meteorological flights, so important in helping the RCAF produce accurate weather forecasts. Such forecasts often were a matter of life and death for aviators, so as much information as possible had to be gathered on an hourly basis. One such operation was the Eastern Air Command Meteorological Flight, formed at Yarmouth on September 3, 1943. The officer commanding, F/L R.D. Renwick, reported in on October 12. On October 31 he ferried Canso 9760 down from Scoudouc; 9796 arrived two weeks later. There was much practice flying as the organization built up. In this regard Harvard 3330 proved handy. Staff totalled 14 by the end of November 1943.

The ORB gives a detailed report of each day's weather, e.g., for December 15, 1943 it notes: "Snow ceiling was 500 to 1000 feet in the first quarter. 2500 overcast for the rest of the day. Visibility 1-3 miles in snow, otherwise 3 miles. Wind was south at 10 shifting to north at 10-15, decreasing to 5-10 in the last quarter." Eventually, long-range met flights went out every day, weather permitting. For July 1945 there were 31, six of which recovered in Bermuda, the flight's main alternate. That month 360:20 hours were logged by its 4 Cansos, 1 Harvard and 1 Anson.

The EAC Meteorological Flight disbanded on August 12, 1945, now becoming the "Liberator Met Flight", equipped with aircraft 3732, 3734, 3735 and 3742. Famil flights commenced to Bermuda; the first met flight was operated (aircraft 3742) on August 10 under F/L J.E. Lynes. On August 13 duties were officially taken over from the Canso flight. But the war was over, so the need for met data dwindled. All the rumours around Yarmouth now focused on the unit standing down. The final patrol was made on September 16 by F/L M.M. Hay in 3735. Due to an oil leak, he had to return to base. By month's end the four Liberators were ferried to Mont Joli for disposal. There they were sold for scrap, the total flying time for all four machines being a mere 1217:50 hours. Soon the met flight's 32 staff had been demobilized and the unit faded into history. For their solid work in this often drab, though sometimes dangerous work, a number of met flight men received the Air Force Cross.

## No.1 Port Transit Unit, RCAF, Pier 21, Halifax

Canada posed a real challenge for those in charge of arranging transportation for thousands of RCAF personnel. Huge resources were consumed in such work. Unfortunately, there are no books devoted to the work performed, and the success of it all by war's end. Day by day from September 1939 men and women had to be transported to one training school or another, to this flying station or that. Some had postings to remote radio sites on isolated BC or Labrador shores. A member could be despatched from one coast to another, usually travelling by train, but sometimes by road, water or air. He could reach a posting in BC, having travelled from Atlantic Canada, then, be turned around and sent back east, his posting having suddenly changed! In each case, somebody had to make the arrangements.

Of all this to-ing and fro-ing, what could have been more daunting than sending RCAF personnel in their thousands out upon the waves to their overseas postings? Many remember such trips, but few know anything of the organization that made it all work smoothly. To begin, the RCAF established a "Y" Depot at Halifax to receive personnel and transfer them from one mode of transport to the next. Sometimes this meant housing thousands of airmen for days, using barracks and billets, whether drill halls, or Salvation Army hostels. If heading overseas, personnel usually arrived at Y Depot by rail, then awaited embarkation. If returning from overseas, they would disembark, then entrain for destinations throughout Canada. Trains came in and out of the Halifax dockyards by the hour. Managing such traffic must have been akin to conducting an orchestra.

In the great port the movements of ships and the disposition of those aboard became the task of No.1 Port Transit Unit, RCAF, Pier 21, Halifax. No.1 PTU had formed on December 1, 1941 under S/L M. Jones. The first entry in the ORB reads: "Stood by awaiting arrival of

**Canadian airmen embark at Halifax for overseas in 1941. (CF PL5320)**

*Queen Elizabeth* (83,600 tons) is processed at No.1 Port Transit Unit during a wartime visit. Then, *Louis Pasteur* (29,260 tons) and *Niew Amsterdam*, (36,670 tons) two other important liners that carried thousands of Canadians on the Atlantic. Of all the Allied trans-Atlantic troop ships, *Empress of Scotland* carried the most passengers—292,000, sailing 712,000 miles. (CF PL37233, Castator Col.)

SS *Pasteur* which is expected to dock with troops either on Saturday, November 29th or Sunday, November 30th." Hereafter, the ORB, kept meticulously by some top-notch diarist, tells the story of the RCAF in Halifax harbour. No.1 PTU not only covered Halifax, but other regional ports like Saint John and Sydney. The affairs of distant places like Port aux Basques, Newfoundland, also are noted, so long as they involve RCAF personnel or material. Certain aircraft movements, such as scheduled flights to and from Torbay, also were included in No.1 PTU's mandate (airmen might reach a port for embarkation by flying in from elsewhere).

For December 2 the ORB notes: "SS *Pasteur* docked and staff prepared to disembark ship's 2651 [RCAF] passengers and baggage." The *Pasteur*, a prewar liner converted for trooping, often carried RCAF personnel. Each time such a great ship arrived, it posed a mammoth task for No.1 PTU (usually about 135 strong) and all other

Masters aboard RCAF Marine Section supply vessels that often used No.1 Port Transit Unit included F/O A.K. Sonnichsen of Lunenburg, Nova Scotia (MV *Beaver*), and (shown here) F/L John Howell of Metegham, Nova Scotia (MV *Eskimo*). Both were mentioned in despatches. (CF PL24425)

establishments involved in receiving, then turning the ship around. At least a day before a ship docked senior port and RCAF officials would convene. How many are aboard? Who are they? Where are they going and how are they going to be furthered towards destination? What of their baggage? What of clearing out the ship and making it ready to embark the next overseas drafts? Disinfecting with DDT will be a must. What are the arrangements for provisioning, water, bedding, ship's drills, even entertaining the troops? What will be the weather? So many matters had to be attended. All this went on day by day, as thousands of ship movements were successfully logged. Herewith are a few 1944 entries from the No.1 PTU ORB (at Halifax unless noted):

**February 14:** "SS *Aborjon* arrived from the United Kingdom with ammunition, general cargo, four Swordfish and three Mosquito aircraft."

**February 12:** "SS *Île de France* arrived and docked 1700 hours. Prisoners of war debarked during evening."

**February 19:** "Forwarding signal to Roycanairf giving name of ship of E806. SS *Erria* completed discharging cargo. SS *Aborjon* unloading air force equipment from United Kingdom."

**February 23:** "Halifax experiencing tremendous snow storm and all work in harbour tied up. A meeting was held ... relative to the arrival of hospital ship SS *Lady Nelson*."

**February 27 (Saint John):** "Supervising discharging of 11 Swordfish aircraft ex SS *Riverview Park*. One case damaged while being unloaded."

**March 1:** (North Sydney): "SS *William H. Gray* sailed this date with export shipments for United Kingdom"

**March 5:** "Owing to fire which broke out in Army delousing machine a good part of Pier 21 was destroyed. Trains were delayed ... instead of embarkation of SS *Andes* commencing at 0800 hours, embarkation took place at 1400 hours."

**March 6:** "Number embarked [*Andes*] 680 officers, 1192 sergeants, 707 other ranks. Total 2579 personnel."

**March 10:** "Two Seamew aircraft unloaded on SS *Sydland* and MV *Westmoor* destined for United Kingdom."

**March 13:** "SS *Fort Townsend* arrived from Newfoundland W654. Personnel were disembarked and arrangements made for unloading and forwarding of one damaged aircraft to Canadian Car and Foundry Co., Amherst."

**March 14:** "Arrangements made for loading two Seamew aircraft on SS *Riverview Park*."

**March 21:** "Three Seamew aircraft loaded on SS *Uranienberg* and two Seamew aircraft loaded on SS *Pacific Enterprise* for United Kingdom."

**March 22:** "Advanced party embarked on E847 *Niew Amsterdam*. Commanding Officer W/C Jones, attended a debarkation meeting in Lt. Col. Gwynne's office at 1000 hours."

**March 24:** "Advance party of 63 Officers and 120 sergeants embarked on E847. E822 SS *Andes* arrived from United Kingdom. First troop train from No.1 "Y" depot, Lachine, PQ, arrived on time at 0800 hours and embarkation proceeded very satisfactorily on E847 SS *Niew Amsterdam*."

**March 25:** "SS *Niew Amsterdam* sailed at 0800 with 912 officers, 2399 sergeants and 3063 other ranks... Grand total 6374."

**March 24:** "Security guard of ... 245 sergeants being embarked on SS *Andes* ... "

**March 30:** "SS *Andes* E822 embarked with 835 officers, 1728 sergeants and 838 other ranks totalling 3401."

**April 1:** "SS *Pasteur* E817 arrived and arrangements made to embark advance party of security guards [10 officers, 31 sergeants, 179 other ranks]. Discussion held with F/L G.H. Brown of the RAF Film Unit relative to taking debarkation pictures of RAF trainees arriving in Canada. Cunard Steamship Lines contacted and berthing arrangements made for E817."

**April 1 (North Sydney):** "Advised by Newfoundland Railway Co. that SS *Kyle* stuck in ice. Approximately 100 Service personnel on hand exclusive of American personnel for embarkation to Newfoundland."

**April 4:** "SS *Pasteur* departed for United Kingdom with 360 officers, 1152 sergeants and 1085 other ranks."

**April 6 (North Sydney):** "Advised by Navy that SS *Kyle* has arrived in Louisberg."

**April 15:** " The MV *Beaver* completed loading and cleared for Seven Islands. The SS *Manchester Trader* with air force equipment arrived from the United Kingdom."

**April 22 (Saint John):** "The case ex. SS *Argus Hill* picked up for fumigation."

**April 22 (St. John's):** "Two aircraft arrived and were despatched today ... S/L Hoyt and F/Lt Milliken of the 164th Squadron at Moncton are staying over night."

**April 28:** "F/L F. Baxter embarked 54 officers, 180 sergeants (air gunners) and troop deck details on the *Île de France* and arranged berthing cards for the rest of the drafts."

**April 29:** "*Empress of Scotland* arrived at 1300 hours. Personnel embarked on the *Île de France* (E848). Ship's final inspection took place at 2300 hours. Baggage from the *Empress of Scotland* discharged and forwarded to destinations... Rations for 32RD received at 0930 hours and delivered to Newfoundland Railway freight shed. The SS *Kyle* arrived at 1530 hours and 20 service personnel debarked."

**May 1 (North Sydney):** "Twelve aero engines from No.5 Equipment Depot, Moncton, were transferred to the Newfoundland Railway for loading on the SS *John Cabot*. Examined men's dormitory in the K of C hostel and had a discussion with Mr. McGillivary, supervisor of the hostel, regarding the inadequacy of fire escapes."

**May 3:** "The embarkation of the balance of the air force draft took place today, numbers as follows: 700 officers, 1764 sergeants and 1242 other ranks. Total loaded on the *Empress of Scotland* (E842) 3706. A/V/M G.O. Johnston visited and inspected the *Empress of Scotland*. Final inspection at 2030 hours."

**May 11:** "Two Mosquito aircraft arrived from the United Kingdom. Arrangements made for discharge and loading on freight cars."

## No.3 PRC ... Bournemouth

For most RCAF members going overseas, the first step after disembarking (usually in Greenock or Liverpool) was to gather one's kit, then catch the train for Bournemouth, where all would report in at No.3 (RCAF) Personnel Reception Centre for their postings. Initially, arrivals would get billeting instructions, often to one of Bournemouth's resort hotels. These, like the ocean liners that had carried them from Canada, had had their roomy suites subdivided and re-subdivided, until thousands more than ever intended could be squeezed in. Once settled, airmen or airwomen would be assigned duties (few were posted immediately). This usually involved parades and courses, although, for those destined to spend some weeks in Bournemouth, generous leave was common. Thus, a new arrival often had plenty of time to get familiar with the new environment, hitchhiking or travelling by train around the country. Inevitably, a posting would come through. The easy life of Bournemouth soon was a memory, replaced by the seriousness of OTU. Staffing at No.3 PRC was considerable. For December 1944, when the centre was in one of its busiest periods, personnel numbered 158 officers, 1749 other ranks. Among these were 706 WAAF. At No.3 PRC in October 1943 those processed included:

### Posted In

| Officers | | Airmen | |
|---|---|---|---|
| Pilots | 65 | Pilots | 565 |
| Observers | 280 | Observers | 570 |
| Air Gunners | 3 | Air Gunners | 166 |
| | | WOp/AG | 86 |
| | | u/t* Observer | 48 |
| | | u/t Bomb Aimer | 64 |
| Bomb Aimers | 3 | | |
| Air Bomber | 1 | | |
| Administrative | 13 | | |
| Medical | 6 | | |
| Ground Trades | 72 | Airmen Ground Trades | 574 |
| Totals | 443 | | 2081 |

*u/t—untrained

### Posted Out

| Officers | | Airmen and NCOs | |
|---|---|---|---|
| Pilots | 209 | Pilots | 926 |
| Observers | 298 | Observers | 381 |
| Air Gunners | 4 | Air Gunners | 118 |
| Air Bombers | 5 | Air Bombers | 64 |
| WOp/AGs | 5 | WOp/AGs | 41 |
| | | u/t Observers | 139 |
| | | u/t Pilots | 1 |
| | | u/t WOpAGs | 19 |
| Administrative | 11 | | |
| Medical | 3 | | |
| Ground Trades | 74 | | 536 |
| Totals | 609 | | 2225 |

As the war progressed more commissioned aircrew were created. For June 1944, for example, No.3 PRC pilot intake was 295 officers and 42 NCOs. In certain trades, airmen remained the majority, with officers growing in number. For the same month there was an intake of 65 officer WOpAGs, compared to 334 NCOs. Also, by this time there were progressively more flight engineers, 107 arriving in June 1944 (all NCOs). A great challenge at PRC was keeping all those passing through occupied. Thus were various schemes always afoot. One is described in the unit diary of December 11, 1944—the Aircrew Canvas Camp at Tarrant Keynston:

*This scheme, originated by C/Capt Hutchison, while the Unit was stationed at Innsworth, was started near Painswick in the heart of the beautiful Cotswold countryside. About 2000 Aircrew Officers and NCOs have attended the camp for 7 to 14 days, where all ranks did their own work, cooking, cleaning and all fatigues. Activities ... included PT, games, cross country runs, fieldcraft exercises, rabbit shooting, fishing, harvesting, fruit picking, lumbering, fencing and ditching, discussion groups and a variety of games, such as volley ball, soft ball, golf, tennis and swimming, all run on an Inter-Section competitive basis.*

Life at Bournemouth was not always so jolly, nor so drab, as many have related. It also could be deadly. The Germans knew what went on in this famous seaside hide-away and knew where men were billeted. Thus did they conduct a number of hit-and-run raids, trying especially to bomb the hotels. Their timing was no accident, hitting the hotel strip one day at noon, another day at tea time. Many died in these indiscriminate raids. Three RCAF were lost on June 6, 1942, when the Anglo Swiss Hotel was hit by an Me.109: P/O Russel Norman Bailey, a 21-year old WOp from Winnipeg; P/O Jacob Alexander Epp, a 25-year old WAG from Manitou (Manitoba); and P/O James William Morgan, a 30-year old WAG from Port Arthur, Ontario. It was only mild consolation that the offending Jerry was despatched on the spot by two Spitfires.

One of the strangest Spitfire case studies is that of ER824. (via H.J. Russell)

## Spitfires in Canada

While hundreds of Canadians flew Spitfires overseas, throughout the war very few "Spits" appeared in Canadian skies. There were some for trials at Test and Development Flight (especially in cold weather research), others for aerial photo work at Rockcliffe. Perhaps the most unusual case, however, is that of Spitfire Vb ER824. This is one of those stories best told by the man who knew it, Herbert J. Russell. During his tour at Torbay, ER824 was cast ashore at St. John's in strange circumstances:

*In December 1942 I was the Engineering Officer at RCAF Station Torbay, where three squadrons were based, one each flying Hudsons, Hurricanes and Lysanders. On December 26 I received a telephone call from a naval officer at nearby St. John's, asking if I would examine the deck cargo of aircraft on a damaged merchant vessel that had just arrived in the harbour. The ship, Empire Kingsley of British register, had sailed from the UK with a cargo consisting mainly of Spitfire Vb aircraft in crates. En route to Takoradi in West Africa, Empire Kingsley had met severe weather, that had severely damaged some of its Spitfires. On Christmas Eve it limped in to St. John's, its nearest port.*

*To repair the ship, it was necessary to remove most of the deck cargo. The Navy wanted my opinion of the damaged Spitfires, and whether there was any point in reloading them afterwards. The alternative was to consign them to the scrap heap. Naturally, it was urgent that the ship resume its voyage and deliver its cargo to the RAF in Takoradi.*

*Examination showed that three Spitfires were extensively damaged. My opinion was that there was little point in sending them on to Africa. If it would help, I offered to have the wrecked aircraft transported to Torbay for disposal. This idea was accepted. The next day I returned to the ship with a crew of NCOs and airmen, and suitable trucks for moving the aircraft. For some reason the stevedores assigned to help with the job did not turn up. After a lengthy wait, we took matters into our own hands and, by operating the ship's winches, the aircraft were soon on our trucks. We heard later that, when the stevedores learned of this apparent breach of labour law by the RCAF, they threatened a general strike; the Navy talked them out of this.*

*At Torbay the three damaged Spitfires were uncrated and stored in the back of a hangar. On examining them in detail, it occurred to me that there might be enough parts to make one good aircraft. A number of NCOs and airmen volunteered their services toward this end. There were two good engines, one fuselage, one propeller and a pair of wings that seemed repairable. A major project was to rebuild a badly damaged tailplane.*

*None of the crew had ever worked on a Spitfire, and we had no drawings or manuals for assistance. Thus, the work went slowly and there was a good deal of trial and error. All the restoration work was done in the men's free time. When assembly was complete, all systems were carefully checked, the aircraft was placed on jacks, and the landing gear was cycled. Finally, "ER824" was rolled out. We were greatly satisfied when the engine started on the first try. All along the Station Commander, who had been aware of our efforts, had neither approved nor disapproved of our project.*

*S/L R.W. "Bob" Norris, Commanding Officer of 125 Squadron (stationed at Torbay with Hurricanes), had flown Spitfires in the Battle of Britain. He agreed to test fly our aircraft. On March 16, 1943 ER824, wearing its desert camouflage and with its desert air scoop, took off from snowbound Torbay. S/L Norris found that the Spitfire performed well and reported no problems after landing.*

*ER824 was flown later by F/O L.M. "Lal" Parsons, a native Newfoundlander on 125 Squadron. His log book shows a one-hour flight on April 19, 1943. This was his first Spitfire flight. He later told me that ER824 had been one of 750 Mk.IV Spitfires ordered from Castle Bromwich Aircraft factory on August 23. 1941, but that it was built as a Mk.Vb. It had been one of the "Presentation Spitfires" donated by Kabala Province in Nigeria.*

*Soon after Parson's flight in ER824, I was posted to a new station, so have no firsthand knowledge of what happened to ER824 in the subsequent months. It appears, however, that Eastern Air Command heard of its existence. Apparently, the aircraft was ordered to be shipped to Halifax, from where it was returned to the UK. Lal Parsons later heard that a former 125 Squadron pilot one day had spotted a Spitfire at a RAF station. Painted on the fuselage was the name "Miss Torbay". Was this our old ER824? The last known details in ER824's life are that it served on 442 Squadron, then on No.12 (Communications) Group. It was sold for scrap in July 1948.*

*Although a minor detail in relation to the momentous events of 1943, the restoration of ER824 was an achievement for those involved. In the end, it is another little part in the great saga of a truly remarkable aircraft—the Spitfire.*

Other RCAF Spitfires at home included A166. Originally Mk.IIB P8332 it served on 222 Squadron at RAF Coltishall, then came to Canada in May 1942. Here it sits dilapidated in Edmonton after its career as a ground training aid. Luckily, someone had the sense to save A166. It may be seen today in the National Aeronautical Museum. (Les Corness/CANAV Col.)

X4492 made its first flight at Farnborough on September 17, 1940. For the next 2 years it was a developmental aircraft, e.g. it was the prototype Mk.FVI. After a tour on 140 Squadron at RAF Benson it came to Canada in February 1943 as a Mk.V. Based at Rockcliffe with No.9 (T) Group, it mainly operated on photo work. X4492 remained on RCAF strength till September 1947, then disappeared, its fate unknown. (CF PL20225)

R7143 started as a Mk.I, its first flight being on February 20, 1941. After developmental work at Farnborough, it served on 140 Squadron. In Canada it flew as a Mk.V at Rockcliffe. (Gordon S. Williams)

Mk.VIII JG480 was on strength with No.2 Training Command (Winnipeg) while serving WEFT at Gimli. (Mike Comar Col.)

## The Cost

From September 1939 to May 1945 hundreds of RCAF personnel lost their lives on the homefront, whether in flying accidents, mishaps on the ground, or natural causes. These men are still remembered by their families and may be read about in numerous documents, articles and books. There also are many commemorative plaques and monuments around the country. This plaque is the facing of a cairn at the Brantford, Ontario airport. It recognizes the BCATP, No.5 SFTS and those who gave their lives there:

**F/O Macdonald Joseph Andrews**, age 34 from Cornwall, Ontario killed in the crash of Anson T6261 near Burtch on March 23, 1941.

**Sgt Ernest Alman Coleman**, age 22 from Charlestown, Massachusetts, killed in the crash of Anson 6835 near Brantford on May 8, 1942.

**LAC Kenneth Frederick Davis**, age 27 from Des Moines, Iowa, killed in the collision of Ansons 6154 and 6190 while landing at Brantford on August 27, 1941.

**LAC Glen D. Faris** of Palo Alto, age 19 from California, killed along with Sgt Coleman.

**S/L Charles Duncan Bremner Green, DFC (WWI)**, age 44 of Toronto, died of natural causes in a Toronto hospital.

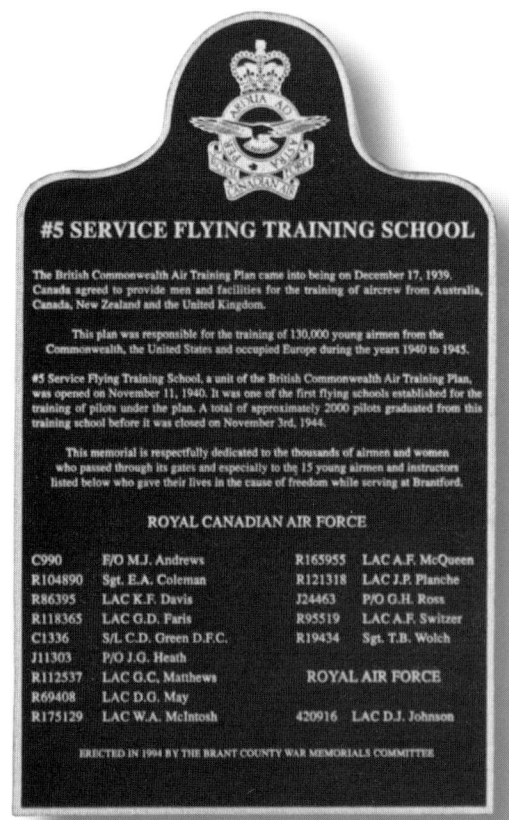

**P/O James Gilmer Heath**, age 24 from Richmond, Virginia, killed when Anson 7070 struck trees while low flying near Lynedoch, Ontario. LAC D.J. Johnson (RAF) died in the same accident.)

**LAC D.J. Johnson**, RAF. No details available.

**LAC Grant Carmichael Matthews**, age 20 from Pembroke, Ontario, killed in the crash of Anson 6118 near Thornton, Ontario on December 7, 1943.

**LAC Douglas Gordon May**, age 32 from Toronto, killed the night of August 26, 1941 at Brantford, when struck by a whirling Anson propeller.

**LAC William Albert McIntosh**, age 28 from Toronto, killed in the crash of Anson 11415 near Glenmorris, Ontario on June 29, 1943.

**LAC Donald Gordon McQueen**, age 26 from Hamilton, Ontario, killed along with LAC McIntosh.

**LAC John Paul Planche**, age 19 from Calgary, killed along with Sgt Coleman.

**P/O George Hugh Ross** of Vancouver, killed along with LAC McIntosh.

**LAC Alexander Frederick Switzer**, age 27 from Pembroke, Ontario, killed along with LAC Davis.

**Sgt Theodore Benjamin Wolch**, age 23 from Toronto, killed when Anson 8488 collided with another Anson, crashing near St. George, Ontario on June 22, 1942.

**Fleet Forts flying over Manitoba prairies from N0.3 Wireless School at Winnipeg. (Harry Mosher Col.)**

# ✦ THE DAY FIGHTER GAME ✦

RCAF fighter pilots excelled from the Battle of France to D-Day, and into The Occupation. They scored against countless enemy air, ground and water targets. Of fighters they flew, the commonest was the Spitfire. Then came the Hurricane, followed by a long list— Kittyhawk, Mustang, Thunderbolt, Whirlwind, Beaufighter, Mosquito, etc. Canadians served from the home front and Alaska to Great Britain, through Europe, in Malta, North and West Africa, the Middle East, Ceylon, India-Burma, Singapore, Hong Kong, even icy Murmansk. Here, 443 "Hornet" Squadron Spitfire Mk.XVIEs sit at Eindhoven, Holland. 2I-W carries the name "Klondike". This squadron completed 5850 operations, logged 12,763 hours (training and ops), shot down 42 enemy aircraft (e/a), with a further 31 probables and damaged, and plastered 1112 road transport targets, 25 locomotives and 8 tanks. Operations cost 9 pilots dead plus 6 POW. (CF PL43158)

F/O Larry Seath of 400 Squadron was an early RCAF pilot on Mustangs. Born in Saskatoon, he grew up in St. Lambert, Quebec. After enlisting, he trained at No.21 EFTS (Chatham, NB) and No.9 SFTS (Summerside) before going overseas. On December 27, 1942 he shot up 7 locomotives over Brittany, flying through a tree on one of his passes. On October 24, 1943 he claimed an Hs.126 and damaged an He.111. Later he flew P.R. Mk.XI Spitfires, then had leave in Canada. Returning overseas, he took ill, people at first thinking he was sea sick, when what he had was appendicitis. His liner was bound for LeHavre but, so serious was Seath's case, that it put in at Southampton, letting him off almost as it sailed by the jetty! Once recuperated, Seath flew on 412 Squadron during the occupation, then came home to a career in insurance. In 2000 he was enjoying retirement in Vankleek Hill, Ontario.(CF)

With war on the horizon in the late 1930s many Canadians began thinking of adventures ahead. With spirits on the rise following the Depression, what could be more enticing to a young man than visions of soaring among the clouds in a speedy fighter! Soon dozens of Canadians had crossed the Atlantic to enlist in the RAF. Once the shooting began, recruiting offices were beset by those anxious to join the RCAF. Nearly every applicant wanted to be a pilot, a trade for which there was little need at first. Thus, many a recruit had his wings clipped before getting into uniform— he would be a wireless operator, an air gunner, observer, radar tech, fitter or rigger, an armourer, driver, cook, firefighter, clerk, etc. Inevitably, the need for pilots grew, but there remained a problem—after all those exciting newsreels about the Battle of Britain and Malta, the dream of most hopefuls was to fly Spitfires. But the RCAF needed other pilots at least as urgently—instructors to teach on Tiger Moths at Oshawa, someone to fly a Norseman in Prince Rupert, or a Lodestar in Winnipeg. It also needed pilots for Hudsons in Iceland, Halifaxes in Yorkshire, Sunderlands in Scotland. Who would get to fly those coveted Spitfires?

Sometimes a new pilot, chest puffed out showing off his wings, got onto fighters. Perhaps he had excelled at SFTS, in which case he sometimes could request his posting. Otherwise, a CFI, flight commander or station commander might write a letter of recommendation, knowing that a certain graduate not only wanted fighters, but showed the aptitude. Further along the line, a postings officer, reviewing a fellow's records, might get a sense that he would do well on fighters—perhaps he had good overall marks and had scored well in gunnery.

The fighter pilot's war would be one of quick and furious action. A sortie sometimes lasted only 15 or 20 minutes, but a man could fly 3, 4 or 5 such in a day, operating from forward strips sometimes in sight of enemy positions. Many would lose their lives at this, others would suffer everything from disfiguring burns to wounded minds. After the war many biographies would be published, graphically detailing the Canadian fighter pilot's war. *Black Crosses off My Wingtip, Blue Skies, Lucky 13,* and *Woody: A Fighter Pilot's Album* are good examples. By profiling a few individuals, this chapter adds to the vast body of published material. To understand the story, however, the reader must delve far more deeply than a single book.

## Canadians in the RAF

In the mid-1930s Britain was combing the dominions for recruits as it vied in the military build-up with Germany. Advertisements seeking aircrew appeared in Canadian publications; these brought many enquiries. Hopefuls might work their way aboard cattle boats or tramp freighters to the UK, where they could enlist. Others signed on directly with British agents in Canada and were given passage. In time hundreds of Canadians won their Wings at RAF schools (they became known as "CAN/RAF" men), and went on to fly on every sort of operation. Many would earn gongs and rise in the ranks. Typical was fighter pilot Vernon C. Woodward, who fought in North Africa, Greece and Crete, shooting down more than 20 Axis aircraft, then staying in the postwar RAF to command squadrons of Hunters and Canberras before returning to his native British Columbia. Another leader was W/C Lawrence "Slim" Jones, who earned a DFC on Sunderlands at the outset of war, then became a key figure in Ferry Command.

Thousands of other Canadians would serve in the RAF, but in RCAF uniform. Most air trades from early BCATP classes ended on RAF squadrons, some of which were as much as half RCAF. Canadian fighter pilots in the RAF would fly in the battles of France, Norway, Britain, in North and West Africa, Ceylon, India-Burma, and Singapore. Some of the first hot action for CAN/RAF fighter pilots was in France and Norway in 1940. One of those involved was Allen Laird Edy. Born in Winnipeg in 1916, he joined the RAF in 1939. By that October he was on 613 Squadron, flying Hector army co-operation biplanes. In May 25 he was in action around Calais, bombing and dropping supplies to beleaguered British troops in the area. For this Edy earned a DFC, the citation for which noted: "On 25th May 1940, Pilot Officer Edy was a member of a formation of aircraft detailed to carry out a dive-bombing attack on a heavy battery near Calais. He pressed home his attack in the face of severe anti-aircraft fire with the utmost courage... Edy has shown a complete disregard of personal danger and has set a fine example by his keenness and magnificent spirit."

On September 9, 1940 Edy joined 602 Squadron on Spitfires. On the 15th he knocked down a Do.17. On November 12 he was shot down near Folkestone. He may have been injured, since he was not back on ops until December 26. In the New Year, he moved to 315 Squadron (Polish, Hurricanes); then, in June was with 457 Squadron (RAAF, Spitfires). Much excitement followed, including a deadstick landing on October 24. On December 5, 1941 Edy took off in a Spitfire; 15 minutes later his plane was down close to base. He had bailed out, but was found dead. He was buried at Adreas.

W/C Howard Peter "Cowboy" Blatchford, DFC, MiD was born in Edmonton on February 25, 1912. He joined the RAF in 1936, serving on various squadrons. On October 17, 1939 he was on patrol east of Whitby with 41 Squadron, when he shared an He.111 with FSgt Shipman and Sgt Harris. After they beat up on the Heinkel, it ditched in the sea, two of the crew climbing out on the wing. This was one of the first kills of WWII involving a Canadian. By mid-May Blatchford was flying Photo Development Unit Spitfires from Heston. Targets (naval facilities, aerodromes, etc.) were anywhere from Rotterdam to Brest to Abbeville to Wilhelmshaven. This carried into September 1940, when Blatchford joined 17 Squadron, then 257 (which he would command in mid-1941). On October 29, 1940 the Luftwaffe clobbered 257 at North Weald, 2 Hurricanes being lost and a pilot killed. Blatchford, airborne at the time, took a pounding from an Me.109.

About this time some Italian units appeared over the UK. On November 11, flying Hurricane V6962, Blatchford shot down a BR.20 bomber, sending it, as he put it, "like snowflakes falling" into the sea. Getting carried away when out of ammunition, he deliberately chewed up the top wing of a CR.42 with his prop. In this action 257 was credited with 8 BR.20s and 2 CR.42s destroyed—a grim day for the Italians. Elsewhere that day Blatchford destroyed an Me.109 and a Ju.87, and had 4 Me.110s damaged. By this time his score was 12 destroyed, plus 9 probable or damaged. (A "probable" indicated an aircraft so severely damaged that it was likely to have gone down; but nobody reported seeing it crash.) On November 24 he was awarded the DFC. On February 5, 1941 he led 257 on its first offensive sortie, a sweep over the Pas de Calais. Many ops were flown in this period (including at night). On the 28th Blatchford shot down an errant barrage balloon; on the night of May 11 (in Hurricane Z3183), an He.111. On the latter he used all his 368 20mm rounds. On June 5 he made an emergency landing, when his engine quit, but he was uninjured. On June 13 he was on a search for a downed Whitley crew, found the crew in a dinghy, alerted ASR and waited till he saw the crew rescued 45 miles off Yarmouth.

On September 8 Blatchford became a wing commander and was posted to Digby. On April 4, 1943 he was up with the Coltishall Wing. His combat report gives a good description of a Ventura escort to Holland. He felt that the presence of some 30 Fw.190s, which gave the bombers a hard time, likely was a trap, for the Venturas had been to Holland 5 times in succession, always coming out the same way.

*Coltishall Wing set course with 24 bombers ... at 1830 hours for Rotterdam. We arrived ... with the bombers at 12,000 feet and ourselves at 15,000 feet. Flak was relatively heavy and two appeared to be severely damaged. The Venturas turned left after bombing and started out at The Hague and the Hook of Holland. The time was now about 1917 hours. The two Venturas hit by flak were losing speed rapidly. Off to the right of the Hook there was a large flat layer of cloud. The main body of aircraft was about 4 miles ahead of me at 10,000 feet. I decided to sit with my section over the Venturas for a couple of minutes, or at least long enough, if possible, to give them a chance to get into the cloud ... I called up the Wing and informed them of this and about this time the FWs started coming up. The Poles were near us and R/T conversation was almost impossible.*

*To the left and slightly below and behind me I saw several aircraft ... overtaking me. These aircraft apparently had no difficulty overtaking the main body... They were FWs for, although I did* not identify them myself, I heard 118 Squadron warning the Wing that they were coming up. At this time a fight started with the main body and I tried to work myself into a position to head off some of the FWs coming up, but I did not have the speed to do it. About now I saw an aircraft being fired at below and to the left of me and it crashed into the sea. I could not identify it, nor the machine that shot it down, owing to the glint of the sun on the water. An FW flew underneath me about 4000 feet below toward the Dutch coast... I pulled the plug, used full throttle and dived towards the FW... He started pulling away from me and I opened fire at extreme range but saw no results. I do not agree with long range shooting, but in this instance I consider I had a reasonable excuse ... because one of the bombers was not far to the left of the FW ... this particular bomber was attacked by an FW at sea level and my No.2 drove him off.

I turned left and started to climb into the sun again... about 7000 feet ... I saw 4 aircraft in line astern above me coming up on the right. I had a good look at them, but could not get a proper silhouette, though I could plainly see duck egg blue spinners and yellow leading edges to the wings. These made me think momentarily that they were friendly. Within a matter of seconds they attacked and so did 4 more which were coming up directly behind me. I now saw they also had yellow cowlings. The time would be about 1930 and from now until 1950 I fought off 8 aircraft single handed. I do not know what happened to my No.3. I did not see him go down. My No.2 returned to base having fired.

The German aircraft were obviously manned by experienced personnel. They made numerous attacks trying to come directly up my tail, and each time I would turn and take my attackers head on. When I did this some of them would nip around behind me again, so I must have, in the course of 20 minutes, carried out from 10 to 15 head-on attacks. On numerous occasions I did not fire myself as I obviously would not have hit anything anyway. At the beginning of the fight I was hit in the right wing by machine gun bullets and in the tail plane by cannon... Of special interest was the attack carried out by their leader... The leader started to fire in a slight dive. As I took him head on he rolled on to his back and continued firing at me with considerable accuracy... When we were almost close to colliding he pulled back his stick and slipped underneath me. The other 3 then attacked me one after another. They gradually dropped off in pairs until at 1950 hours there were only 2 left... I crossed the coast at 2000 hours just south of Great Yarmouth and landed at Ludham at 2010 hours.

On May 3, 1943 Blatchford was leading the Coltishall Wing on a Ramrod—escorting 11 Venturas to Amsterdam. Ten bombers were lost in this disastrous effort, and so was Blatchford, who crashed at sea 40 miles off the English coast. His final score was 1½ He.111s, ½ Do.17, 1½ BR.20s, 1 Me.109, and 1 Fw.190 destroyed; 2 CR.42s, 1 Ju.88, 2 Fw.190s probables.

F/O Mark H. Brown (born on October 9, 1911 in Portage-la-Prairie, Manitoba) finished with 18½ kills, perhaps his earliest being with No.1 Squadron (RAF, Hurricanes) on November 23, 1939. That day he and S/L "Bull" Halahan shot down a Do.17 over France. On March 2, 1940 Brown and Sgt F.J. Soper of No.1 Squadron attacked a Do.17. Brown suddenly had his propeller fly off; he force-landed at Nancy, France. The next day he badly mauled an He.111, forcing it down in a field with 2 crew dead. Brown gave some rudimentary advice about attacking German bombers: "Astern attack from slightly above... for best results. Do not approach before gaining the same height. This applies to He.111 (not to Do.17)." On April 20 Brown and 3 other Hurricanes fought 9 Me.109s with uncertain results. He knew that the 109 had a speed edge on the Hurricane, but commented: "Although we were greatly outnumbered, the enemy seemed to want to evade combat." On May 1 Brown test flew a captured Me.109E at Amiens, fighting with a Hurricane. The RAF machine out-fought the 109 at low level, but was no match higher. On May 4 Brown ferried the 109 to Boscome Down for evaluation by the RAE. He was awarded the DFC in July 1940, the citation noting: "Flight Lieutenant Brown has shown courage of the highest order, and has led many flights with great success and determination when consistently outnumbered." In May 1941 he received a Bar to his DFC, with this remark in the accompanying citation: "His splendid leadership and dauntless spirit have been largely instrumental in maintaining a high standard of efficiency throughout the squadron." W/C M.H. Brown, DFC and Bar died over Sicily on November 21, 1941, when shot down in his 249 Squadron Hurricane.

The short but furious air war over Norway involved CAN/RAF types. On May 25, 1940 P/O Phillip H. Purdy of St. Stephen, New Brunswick, and Sgt Kitchener of 263 Squadron (Gladiators) encountered a Ju.90 around Harstad, Norway, shooting it into a fjord. The next day Purdy was in other action at Harstad, where Do.17s were bombing shipping. After making a quick attack on one, then sending another down trailing smoke, he was set upon by another Do.17, so broke away. Other 263 action in Norway involved F/L Alvin T. Williams of Toronto and Sgt G.S. Milligan. On May 26, 1940 they departed Bardu Foss. At 1030 hours they encountered a Ju.88 bombing Skoanland. Firing from 400 yards astern, Williams set the e/a aflame. He pressed his attack until both wings came off the Junkers. Canadians died in such early actions. F/L Williams and (by then) F/O Purdy, DFC, MiD were lost off Norway, when the RN carrier HMS *Glorious* was sunk by the *Scharnhorst* on June 8, 1940. Many books tell of CAN/RAF fighter pilots. Recommended is the excellent Grub Street series, e.g. *Fledgling Eagles (France and Norway 1939 -40)*, *Malta: The Spitfire Year 1942*, *Dust Clouds in the Middle East*, and *Bloody Shambles* (Singapore).

### The RCAF Overseas—No.1 Squadron

Canada's first squadron overseas was No. 110, having sailed on February 16,1940. In England it trained on army co-operation Lysanders, but saw no action until re-equipped with Tomahawks a year later. No.112 Squadron and No.1 Squadron followed. No.1, profiled earlier, had converted to Hurricanes in Vancouver and Calgary. In November 1939 it moved to Dartmouth from where it embarked after a year of training. At first No.1, equipped in the UK with Hurricanes, stood by; but as the Battle of Britain developed, it was transferred to Middle Wallop on June 21. It moved to Croydon on July 4, from where its first flying was done by S/L E.A. "Ernie" McNab, F/L Gordon R. McGregor and F/O C.E. Briese on July 27. No.1 moved again—to Northolt on August 17. Meanwhile, McGregor and Briese attended the RAF Air Fighting Course at Northolt. This provided the latest gen about the Battle of Britain, tactics, etc. In these days McNab and McGregor had a chance to fly with 111 Squadron. On August 15 McNab became the

**No.1 Squadron gained renown in the Battle of Britain. Here its Hurricanes are scrambled at Croydon. (Nesbitt Col.)**

first pilot on an RCAF fighter squadron to fly operationally, scrambling with 111 and shooting down a Do.215. Later that day the Luftwaffe beat up No.1 at Northolt, initiating it into the shooting war.

At Northolt, No.1 formed a wing with 229 and 303 squadrons. For August 1940 the Canadians flew 685:05 hours. (mostly training). Their first noteworthy "do" was on August 26, when 10 Hurricanes scrambled against 25-30 enemy aircraft (e/a). In the mêlée F/O McGregor tallied a Do.215, No.1's first kill. But F/O Robert Leslie Edwards, age 44 from Cobourg, Ontario was lost, becoming the first WWII combat casualty in an RCAF squadron. No.1 finally got into the thick of battle on September 1, its Operational Record Book (ORB, or daily diary) noting:

paper of what had gone on. Individual reports were typed and placed into the official record. Countless such narratives are available for study in archives. Others, including those of Canadians on fighter squadrons in North Africa, seem to have disappeared.

*Ten Hurricanes No.1 Squadron took off from Northolt 1400 hours 1-9-40. Combat near Biggin Hill at 1415 hours. Orders were given to intercept Raid 29 and F/L McGregor, after two vectors, sighted about 20 enemy bombers with escort above. He was about 3000 feet below and climbing. Gave orders for a head-on attack. After forming line astern by sections, he led his section echeloned to port, delivered his attack about 10 degrees off enemy port bow and below, seeing*

*before going out of sight. He is of the opinion that this e/a, from its manoeuvres, was about to crash.*

*F/O Reyno was forced to make a beam attack with deflection, as the enemy were now turning, and made his second attack from astern. He observed an e/a break away from the formation and his No.3 F/O Christmas on second attack fired all his remaining ammunition at the damaged bomber, which has been claimed as damaged by the section.*

*Yellow section, F/O Briese, who had been acting as rear guard, as the enemy swung to the south, made his head-on attack. His No.2 F/O Peterson hit starboard engine of e/a, and also port engine with second burst. It fell out of formation, but he was unable to observe any further, and is positive that the machine is badly damaged.*

*It is reported by an independent witness on the ground that he observed with his glasses Yellow 3 F/O Kerwin engage and shoot down an Me.110 in flames before being shot down himself and baling out. This witness picked up F/O Kerwin, who was burned on the legs, arms and face. The intelligence officer No.1 Canadian visited this pilot, but he was still unconscious and, until his report has been gotten, no claim will be made for this Me.110 destroyed.*

Pilots of No.1 Squadron during the Battle of Britain. Standing are F/L W.R. Pollock, F/Os C.W. Trevana, C.E. Briese, Paul B. Pitcher, Peter W. Lochnan, F.L. E.M. "Ed" Reyno, F/O S.T. Blaiklock (IO), R.W. "Bob" Norris, A.M. "Art" Yuile and Capt W.D. Rankin (MO). In front are F/Os Otto J. Peterson, W.P. "Bill" Sprenger, S/L E.A. McNab, F/O. Eric W. Beardmore, F/O Deane Nesbitt and F/O B.E. "Bev" Christmas. No.1's CO, S/L E.A. "Ernie" McNab, was not a young man when he fought his first battle over England—he already was 33. He trained his men hard at Croydon and Northolt. His first combat was with 111 Squadron on August 15, 1940, when he attacked a Do.215: "I followed him down, firing. His engines began to smoke and he crashed in some marshy ground just west of Westgate-on-Sea." Two days later McNab led No.1 into battle. The squadron scored kills, but took its own losses. In an August 26, 1940 combat with Do.17s F/O Robert L. Edwards of No.1 became the first RCAF pilot of WWII to die in combat on a Canadian squadron. Peterson was shot down and killed on September 27. Sprenger died in a crash on November 26, Lochnan in May 1941. Several No.1 Squadron men would rise in the ranks. Beardmore had 118 Squadron (Goblins) at Rockcliffe, then No.4 (Cansos) at Ucluelet. Reyno was CO in Canada of 115 (Bolingbrokes), 135 (Kittyhawks) and was station commander at Bagotville. Postwar he attained air rank. In 1955-57 G/C Christmas commanded No.4 Wing at Baden-Soellingen, West Germany and was the last Battle of Britain pilot in active Canadian Forces service. (CF PL907)

*There was only one scramble during the day. A raid consisting of about 20-30 bombers with a large fighter screen was intercepted at 1415 hours over Biggin Hill at 18,000 feet... one Do.215 was destroyed and one damaged by F/L McGregor and three others damaged by F/Os Peterson, B.E. Christmas and J.W. Kerwin.*

F/O Kerwin also damaged an Me.110, but himself was shot down. He bailed out with face and hand burns. All other aircraft returned to Northolt, 2 being badly damaged. The details of this action are given in a report, prepared by No.1's intelligence officer (IO), F/O S.T. Blaiklock. His report was based on pilot combat reports—standard narratives given to an IO after landing. In compiling such a document, he had time only for the essentials, although he tried confirming these with at least one other pilot. By day's end he would have a fair picture on

*tracer entering e/a, which changed course, pulled up violently, and appeared damaged. Attack did not develop quite as expected, as squadron was climbing to the bombers' height to close range.*

*Red 2 F/O Pitcher followed and did not break away until he had expended all his ammunition (with no result that he could observe at the moment) and returned to base. Red 3 F/O Yuile had his aircraft heavily hit after firing a long burst, caught on fire, but was able to bale out and returned safely to base after coming down near West Malling.*

*F/L McGregor was able to make a second attack on the A.A. [attacking aircraft], which appeared to have broken formation and was turning toward the coast. This was a close, small-deflection, quarter attack, and he observed one bomber break inwards from formation and make several turns of a steep, left-hand, diving turn*

It was not always easy to down a bomber with machine guns. No.1's diarist complained, "These babies seem pretty tough ... every time we fill them full of lead, but cannot claim anything much more than damaged." A note on the 3rd states that No.1 was unable to intercept enemy aircraft at 30,000 feet. Later the ORB noted that the Luftwaffe was being persistent about staying at or above 20,000 feet, and that the Hurricane groaned when fighting above 18,000. On September 4 No.1 intercepted 12-15 Me.110s at 15,000 feet near East Grinstead: "The enemy formed a protective circle, but our height advantage enabled us to position ourselves before attacking, with the result that 2 enemy aircraft were destroyed by F/Os Smither and Nesbitt, 1 probably destroyed by F/O B.D. Russel, and 5 others damaged."

On September 7 there was a Luftwaffe raid, but No.1 did not engage any of the 400 e/a, since it was tasked for airfield defence. However, S/L McNab, aloft by himself, became entangled with some Me.109s, shooting down one. On the 9th F/O W.M. Miller was shot down: "He baled out but was found to be wounded in the leg and to have sustained burns." Also on the 9th this comment appears in the IO's report: "F/O Peterson, who was No.2 in second section, fired at e/a who was making an attack on him and this Me.109 completely broke up in mid air, one of the pieces hitting his aircraft and breaking about 8 inches of a propeller blade and smashing windscreen. He was temporarily blinded, received a few cuts on his face, but regained his vision partially at about 4000 feet and was able to bring his aircraft back to base. Action on September 11 included F/L McGregor and F/O Molson each bagging an He.111, and F/O Art Yuille (who was in on one of the 111s) claiming a Ju.52. The combat report states that Yuille's action took place south of London, some 30 miles inland. It is unusual for a big, slow Ju.52 to be over England at this point:

*We came cross a lone Ju.52 with vertical light blue and white stripes on the square rudder and fin. Two other F/Fs were attacking from astern, when we [Yuille and his Hurricane VK-W] got into position and waited our turn. I fired one long burst of about four seconds and broke away left. There was some very scattered return fire at me. I climbed and, turning, came at him in a quarter head-on attack from above, breaking away again. I turned and made one more astern attack and fired one burst of five seconds, observing black smoke coming from between the fuselage and the starboard engine. The starboard undercarriage fell off and down. Immediately after I had reversed from my breakaway, I observed several parachutes opening and floating down and the e/a half-rolled and dove straight for the ground, where it exploded near Tunbridge Wells.*

Nesbitt shot down an Me.109 on September 15, but then was hit. A witty diarist recorded: "F/O Nesbitt had to bale out, injuring his head and tail, both of which are in hospital for observation." F/O R. Smither, a 30-year-old from London, Ontario, was lost in this action. Also on the 15th F/O Yuille was wounded, but returned to base. F/O Lochnan, taking off late to try joining the squadron, came across a lone He.111. This he attacked, then was joined by 2 Hurricanes and 3 Spitfires. Among them they forced the enemy into West Malling, where, the IO's report notes that Lochnan landed beside the e/a and helped the crew out. The RAF getting its hands on a German bomber and crew would have been cause for great celebration.

On the 17th Briese survived a crash-landing. Next day F/O E.W. Beardmore was hit by an Me.109 and had to bail out, coming down on an island near Sheerness: "He was taken to the Navy Hospital, where a cut on his mouth was stitched up. He returned to base by road in the evening, then was sent to No.5 Canadian Hospital, Taplow, for further examination." Beardmore had been the lone pilot on No.1 singled out earlier by S/L McNab as being weak: "His written test is so far below the average, that he will require special attention..." No.1's diarist, meanwhile, complained about red tape, having heard that "Scarves and sweaters will not be tolerated in future in place of collars and ties." He anticipated that pilots soon would have to be inspected as being "neat and tidy" before scrambling!

The diary of the 19th notes: "Three subsections were scrambled during the morning, F/Os Pitcher and Briese being vectored far out over the channel. F/Os Beardmore, Nesbitt, Desloges, Little and Millar, all wounded, are progressing favourably. F/L Corbett visited the squadron on his way to convalesce in Wales. F/O Smither was buried today..." On the 21st Northolt conducted its first wing scramble, this being humourously described by the diarist: "When the smoke cleared away and the spectators opened their eyes again, 36 Hurricanes were to be seen, all airborne and in very good disorder... This first attempt at operating the three squadrons as a wing ended without disaster to ourselves, or the enemy." As to September 23, little happened: "Some sissy Me.109s were around and above us, but they refused to play ball." Next day some bombs fell near a local pub, The Orchard. The ORB moaned that the pilots from No.1 who were drinking there might have been "trampled to death in the rush for cover, that is, those who were able to locomate."

On September 25 Air Marshal W.A. Bishop visited Northolt, the King on the following day. On the 27th No.1 listed three engagements. The section of McNab, Brown and Christmas shot down a Ju.88, an Me.110, and had a probable. For the Ju.88 McNab reported: "He broke away and dropped his bombs, turning towards the south coast and I followed in with section.

No.1 Squadron had a straight-forward idea for a unit badge. With a Canada goose and a maple leaf, it could not be mistaken as anything but Canadian. (Nesbitt Col.)

Two parachutes came out, but man in first fell away from his, the second was badly ripped. The aircraft continued to fly south, so I closed and gave 2-second burst. The aircraft dove straight down and burst into flames at Lingfield." There were other claims, but F/Os P.W. Lochnan and William P. Sprenger were so badly shot up that they had to seek haven at other 'dromes. Lochnan reported about helping get a Ju.88: "I fired again as he was coming towards me and going down. He turned and crashed at Gatwick aerodrome. I landed there and left my aircraft to be repaired and returned to base by Magister." F/O Otto John Peterson (age 24 from Lloydminster, Saskatchewan) was shot down and killed on the 27th. He is noted in one source as "killed while descending in his parachute." F/O Brown gave a vivid description of a kill made the same day:

*Closed in behind e/a but saw another Hurricane make an attack... E/a was apparently not damaged by first attack and I experienced extreme gunfire ... held firing button for 275 rounds per gun. Pieces were flying off and one engine smoking as I broke and climbed to make second attack. E/a burst into flames and dove straight for the ground. Saw e/a crash 5 miles east of Tunbridge Wells. No person jumped out of e/a. This enemy casualty is confirmed by F/L Kent of 303 Squadron.*

A September 28 wing operation seemed to bring great confusion on the R/T and with the ground controllers. No.1 was "unable to find the enemy raid." However, noted the ORB, "A great distance was patrolled at a great speed, as the Poles were leading, and they never operate at less than full boost." On September 30 the diarist again was descriptive: "Most patrols today consisted of wing formations going up from the station, but each time there was an unwelcomed addition of yellow-nosed Me.109s above and behind, in the sun, in fact, all over the place. They occasionally came down to join us and in the first battle, about nine in the morning—before our pilots were really awake—they shot three of our aircraft full of holes, but with no damage to any of our pilots."

For October 1 the diarist wrote: "F/Os Nesbitt and Beardmore were back on squadron today for the first time since their parachute jumps. There is a decided shortage of pilots at the moment as the CO, F/L Reyno and F/Os Briese, E. P. Brown and R.W. Norris are in hospital with colds. At midnight a stick of bombs dropped very close to the south boundary of the aerodrome, but no damage was done." F/O Hyde returned to squadron the following day. On October 4 S/L McNab was awarded the DFC, this prompting the squadron to get into a party mood at day's end. On October 2 F/O Molson had to bail out after being wounded, but No.1 had its revenge, downing 3 Me.109s and damaging 3 other e/a. On the 6th a lone Jerry dropped a 1000-pound bomb on Northolt, wrecking 2 Hurricanes and killing a pilot from 303 Squadron and an airman. Later the diarist complained that the station defence wasn't able to get a shot at him as "either no one had told them that the raider was coming, or their guns were not loaded, or the corporal wasn't there to give the order to fire, or something. Words stronger than 'poor show' are required here ... it was a goddam bloody awful balls-up ... apologies to the padre and any others who object."

On October 7 No.1's ORB notes: "F/L Nesbitt got hosed pretty badly and came down at Biggin Hill..." On the 9th No.1 was relieved by 615 Squadron and sent to Prestwick near Ayr in Scotland. The Hurricanes left immediately, followed by a road convoy. On the 13th, No.1 began operations. Now things got quiet, with convoy patrols, dawn and dusk patrols, and a lot of training. This was a chance to wind down after the madness of September. The diary for October 13 mentioned: "Leave is being inaugurated, the first since our last leave years ago last June ... it's a grand and glorious feeling here at Prestwick and the war seems very far away." On the 18th the diarist commented: "Periodically Ops over at Turnhouse send through a signal 'to be on our toes'... It is believed that this is merely done to remind us that we are still operational, and that the war still goes on, which is somewhat hard to believe in this part of the world." In Ayr the Canadians enjoyed pub life, golf, squash, ice skating and grouse shooting. There was a course of Indian pilots on station, who were well-liked by all. Some boasted of their multiple wives, one claiming to have six! About this, No.1's ORB noted, "These are looked on with envy by some of the more virile pilots and with awe by the older ones."

A close look at squadron records may explain No.1's move from its hot spot at Northolt to the quiet of Ayr. On October 7, 1940 squadron medical officer, Capt R.J. Nodwell, RCAMC, had written to RCAF HQ in London about problems. This is a perceptive commentary by an MO, whose thinking was ahead of his time.

*On joining the squadron 30-9-40 it was noted that there was a marked change in the general reactions of the pilots as compared to three weeks previously. There is a definite air of constant tension and they are unable to relax as they are practically on constant call. The pilots go to work with forced enthusiasm and appear to be suffering from strain and general tiredness. They have been working long, hard hours with not as much as 24 hours off over a period of 2 or 3 weeks in numerous cases... None of the pilots have had leave since arrival in England and the most time off at any time has been 24 hours.*

*This constant strain and overwork is showing its effects on most of the pilots, and on some it is marked. They tire very easily, and recovery is slower. Acute reactions in the air are thereby affected. There is now a general tendency to eat irregularly or to have a sandwich in place of a hot meal. The pilots are becoming run down and infections, which would otherwise be minimal, are becoming more severe. There is a general state of becoming stale. Needless casualties are bound to occur as a result of these conditions, if continued. It is considered that personnel engaged in active flying should have at least 24 hours off once a week in which to get a good sleep; a 48-hour leave regularly every 2 weeks; and a 2-week leave every 2 months. It is recommended that the squadron as a whole be given respite from their strenuous duties, to allow for recuperation, and that definite leave periods be enforced to provide proper relaxation and rest.*

Meanwhile, a night out on November 1, 1940 led to a humorous entry: "The dance was a great success, the free refreshments possibly having something to do with it. Everybody enjoyed it except F/O Russel, who was grossly exploited. It appears that as the taxi drove off, having been paid in advance to take his guest home, a brother officer stepped in on the other side and had the pleasure of escorting the lady home at F/O Russel's expense. A very nice piece of timing, or two-timing, as the aggrieved party asserts."

In time some No.1 personnel were posted home, Desloges, Kerwin, Millar, Pollock and Rankin being the first. On November 11 some replacements, pilot Kenneth A. Boomer included, reached Ayr. A diary entry for November 21 reads: "Winds of gale force and the soggy condition of the field limited flying today ... The station showed a movie on the methods employed by the Germans to extract information from prisoners." On the 22nd F/O Yuille had a practice dog fight with an FAA Grumman Martlet of 804 Squadron, flown by F/L Reid. The diary noted, "one-sided in favour of the Grumman." On the 24th crates of spares arrived from Ottawa, but the engineering officer, F/O Briggs, was hard-pressed to understand why he had received, among other things, spares for Lysanders!

On November 26 No.1 reported sad news. Having survived the Battle of Britain, F/O Sprenger, a 29-year-old from Montreal, got into bad weather and crashed in Scotland. He was buried next day "in a very pretty cemetery in the sight of Loch Lomond." The diarist could only write a simple "Great guy, Bill." Another new pilot about this time was F/O Fumerton, who would go on to fame on night fighters. Sir Harry Lauder visited Ayr on December 5 to entertain. Winds hitting 100 mph put the kibosh on flying the next day. Also aggravating was a lack of mail, but this was explained in a note from HQ—the ships carrying all the mail sent November 17-23 had been sunk by the Germans!

For December 1 the diarist commented about No.1's aircraft: "Our Hurricanes are gradually getting their new war paint on, the spinners being ducks egg blue, a band of the same colour painted around the end of the fuselage, and one wing painted black." On December 8, 15 aircraft and support personnel from No.1 moved to RAF Castletown where, "accommodation ... is strained to the utmost with the addition of half of our squadron." The diarist wrote further: "Thurso [the nearby town] is a small village with little entertainment, few women and the coldest hotel rooms." At this time of year the sun set about 1530, not rising till 0915.

Practice flying began on December 13. Soon the squadron was thinking of the coming season, a note of the 22nd reading, "F/O Hyde searched the surrounding country and finally located the only Christmas tree in the district." Christmas dinner was served on the 24th, the diary for the 25th reading, "Christmas Day was routine with the usual practice flying. The Grumman squadron celebrated by shooting down a Ju.88 to the north of our station. A Magister aircraft was attached to our strength." One Christmas Day comment, omitted from No.1's ORB, was found in the diarist's rough notes 60 years later: "During the night we had a local Lady Godiva, without her horse, wandering around our rooms at the hotel, to the embarrassment of our pilots. She was just a little too plastered to be of any use to anybody."

On January 29 baths were arranged for the airmen at a location 16 miles distant. This was the first time the men had enjoyed a bath in 21 days. The same day Nesbitt, Boomer and Norris had a long over-water patrol, the weather forcing them down to 50 feet. Lost for some time, they finally made landfall near Lossiemouth. But Boomer and Norris ran out of fuel and crash-landed: "F/O Boomer hit obstruction poles [planted to prevent German landings] ... F/O Norris somehow landed between these poles in heavy mud with wheels down and no damage." Later the diarist remarked that the army engineering officer was a bit annoyed: "He seemed to think that F/O Norris had no right to land in between his poles without writing off both his aeroplane and himself."

Overall, No.1's "rest" proved of little value, due mainly to conditions at Castletown. In a December 13, 1940 memo to No.1's OC, S/L P.B. Pitcher, Capt Nodwell summarized some of his worries:

*The airmen are accommodated in iron-sided huts which are drafty and cold. The floor along the sides remains uncovered with linoleum, and a strong, cold draft rises at the head of the beds, through the numerous cracks... The huts are very inadequately heated by two small coal stoves ... the airmen have been sleeping with their clothes on to keep warm ... The airmen's mess hall is in a filthy condition. The floor is covered by a mucky layer ... The food is served in a sloppy, unappetizing manner, and the waiters and kitchen staff do not appear cleanly... The latrines are extremely filthy... There is no evidence that any chemicals have been poured into the toilet buckets... There are no wash basins and personnel are required to scoop up water with their hands ... to wash their hands and face.*

On January 3, 1941 there was a lecture about the horrors of poison gas, much feared at this time. The ORB remarked: "The places on one's person, that a certain kind of gas attacks first, were closely noted by the pilots." Next day F/O Brown returned from Turnhouse to report that chances of getting a Harvard for No.1 were slim, but that a Whitley unit at Dyce had offered transport (when they were heading south) for Canadians going on leave. January 7 was a red-letter day: "We got our first confirmed during the morning, F/O Lochnan shooting down a sitting rabbit from about 5 paces." A note from January 9 indicates that tables finally arrived for the airmen's mess. Now the men could eat in normal comfort. Through this period, colds, flu, even measles plagued No.1. A daily issue of rum ($\frac{1}{64}$-gallon per man, per issue) helped ease such problems. During this period No.1 Squadron had the use of a Hornet Moth, Tiger Moth, Magister and, on occasion, an Anson or Harrow to get around on business or pleasure. On February 3, long-standing No.1 pilots George G. Hyde and Peter W. Lochnan were posted to No.2 and

No.110 squadrons. En route in his Hurricane on the 6th, Hyde had a forced landing, but survived. (On May 17, 1941 Hyde, age 27 from Montreal, died in a Spitfire, doing aerobatics for a Victory Bond event. Five days later Lochnan, age 27 from Ottawa, died in a Tomahawk in duff weather.)

On February 7, 1941 No.1 Squadron moved to Driffield, its 20 Hurricanes refuelling en route at Drem and Leuchars. There it replaced 213 Squadron, which moved to North Africa. Flying commenced on February 16, but with a surprise: "F/O 'Handlepuss' Hanbury, on the first scramble, rushed out and leaped into his comet-like aircraft." As it happened, no sooner had he lifted off and tucked in his wheels, than Hanbury came down with a thump. In honour of this snafu, he was presented with "a book suitably engraved with a hundred coloured illustrations on the theory of flight." The next day the ORB warned about the evils of Hull: "We should like to warn one and all concerning a certain little item which, if contracted, means a loss of pay." Hull, it seems, led the nation in "this charming game"—venereal disease. On February 18 F/L Joseph B. Reynolds, age 20 from Pembroke, Ontario was killed in a flying accident: "We regret to report that the squadron lost a very swell guy this morning... It is just one of those things that will never be known." There was an interesting scramble on the 21st, the target turning out to be "a very snooty Blenheim which would not give the letters of the day." F/O Wallace put a burst across its nose and the Blenheim identified itself. A typical day's flying from Driffield, as recorded by No.1 on the standard Operations Record Form for February 23, 1941, is listed below:

On February 28, 1941 No.1 Squadron again was on the move, this time to Digby, taking 18 Hurricanes on the 35 minute flight. On March 1 No.1 became 401 Squadron. A summary of its efforts from August 17, 1940 showed that its busiest day had been August 26, when 51 sorties were flown for 27 hours. Overall statistics show 1694 sorties, including 1569 hours on ops. In battle it suffered 3 pilots killed, 10 wounded, while tallying 30 kills, 8 probables and 34 damaged. The top scorers were S/L McNab with 4½-1-3, and F/L McGregor with 4-3-5. Three DFCs, the first of the war to RCAF members, were awarded—to McNab, McGregor and Russel. This trio's cumulative score was 11½-6-11. For a squadron that had started less than 4 years earlier with a few clapped-out Siskins, this was praiseworthy. Later (as 401) the squadron would amass 195-35-106 e/a at the cost (in battle) of 61 aircraft, 34 pilots killed and 18 POW.

| Hurricane | Pilot | Duty | Time Up/Down | Remarks |
| --- | --- | --- | --- | --- |
| V6853 | F/O Weir | local recce | 1000/1125 | |
| V7605 | F/O Elliott | local recce | 1025/1130 | |
| V7004 | F/O Hanbury | local flying | 1120/1220 | |
| V7185 | F/O Boomer | local flying | 1105/1155 | |
| V6869 | F/O Fumerton | local flying | 1110/1200 | |
| V7605 | F/O Neal | scramble | 1340/1500 | patrol convoy |
| V7004 | F/O Napier | scramble | 1340/1500 | patrol convoy |
| P3767 | F/L Johnstone | local recce | 1500/1520 | |
| V7605 | F/O Boomer | local flying | 1530/1605 | |
| V6558 | F/O Fumerton | local flying | 1530/1605 | |
| V6869 | F/L Morrison | scramble | 1750/1830 | dusk patrol |
| V7185 | F/O Mitchell | scramble | 1750/1830 | dusk patrol |

Gas attacks were much feared by Great Britain early in the war. Here No.1 Squadron practices in full anti-gas apparel. (Nesbitt Col.)

Views from distant Prestwick, where No.1 was on rest after the furious weeks of August and September 1940. (Nesbitt Col.)

No.110 "City of Toronto" Squadron starting for overseas in Ottawa on February 13, 1940, planning to enter the Battle of France, but that ended in May. Thus, while No.1 fought in the Battle of Britain, No.110 was in abeyance, training on Lysanders for a year at Old Sarum and Odiham. Late in 1941 it began operations as 400 Squadron, flying Tomahawks, then Mustangs in the tactical recce role. (NAC PA63842)

No.112 Squadron of Winnipeg embarked on June 9, 1940 for the UK, where it trained on Lysanders under RAF Army Co-operation Command. In December it became 2 Squadron, then 402. Operations commenced from Digby in March 1941. Duties included hit-and-run sweeps across the Channel, pilots looking for ground targets or Jerry aircraft. Here Hurricane BE485/AE-W of 402 carries 250-lb bombs. (CF PL6898)

Pilots G.D. Robertson and K.A. "Ken" Boomer flew with 402 early on. Robertson was the first to score, damaging a Ju.88 on June 26, 1941. Later, while CO of 411, he downed an Fw.190 on D-Day + 1, two others by month's end. Boomer later commanded 111 (Kittyhawks) in Alaska, where he was the only Canadian to score against the Japanese in combat (a Zero destroyed on September 25, 1942). He died on ops with 418 Squadron in October 1944. (CF PL4666, '4990)

In the 1980s this Canadian-built Hurricane came to the Canadian Warplane Heritage. Here it taxis at the CWH on June 19, 1992. In No.1 markings (YO-A) it appeared at many airshows, but was lost in a hangar fire at Hamilton on February 15, 1993. (Larry Milberry)

CAN/RAF men also fought in the Battle of Britain, one being P/O Alfred Keith "Skeats" Ogilvie of Ottawa. In a year of action with 609 Squadron he scored 6-3-3. Shot down in July 1941, he spent months in hospital, then sat out the rest of war a POW. During the Great Escape from Stalag Luft III, he was one of 76 men to get away through a tunnel. Recaptured, he was lucky to survive—the Gestapo gathered up most of the 76, then executed 50. When WWII ended, Germany would release 13,022 RAF, RCAF, RAAF, RNZAF and SAAF aircrew. (NMC 73-160080)

## A Fighter Pilot's Career

Arthur Jewett was born at Aroostook Junction, New Brunswick on July 24, 1923, son of a CPR man. The family lived on a small subsistence farm. Arthur helped with the chores and delivered milk in the neighbourhood. When he was 8 years old, he saw his first airplane. Before the war he finished high school, then studied at Mount Allison University in Sackville, New Brunswick. In 1942 he joined the RCAF. His career started at No.5 Manning Depot in Lachine, a Montreal suburb. Next he went to Camp Borden on tarmac duty—washing and helping refuel aircraft. ITS at Toronto followed, then EFTS at Goderich. Jewett's first flight was in Tiger Moth 1224 on March 31, 1943 with Sgt Ribble. He soloed on April 23 and finished his course on May 15 with a flight in Tiger Moth 255, a prewar relic. His time totalled 56:50 hours.

Jewett was posted to 2 SFTS at Uplands, where he started flying Harvards on May 19. His "50-hour" test came on June 21. Training was intense as shown by his log for the week July 19-23: 7:20 hours of dual instruction, 4:55 day solo and 3:00 night solo. There were 15 flights, including 5 on the 22nd. Jewett finished SFTS on August 3, 1943 having 77:15 hours dual (day) instruction on the Harvard, 66:00 hours solo (day), 7:20 dual (night) and 15:35 solo (night). The BCATP was doing its job well with Arthur Jewett. Now he was posted to Course 18 at No.1 (F) OTU in Bagotville. There he started with a familiarization flight in Harvard FE385 with F/O Miron on September 21, 1943. On the 25th he flew the Hurricane for the first time and immediately was engrossed in the high-pressure OTU world—day and night, formation and low level flying, dog fighting, cross-countries, etc. For October 1943 alone he flew the Hurricane for 21:30 hours, the Harvard for 11:55.

On November 10 Jewett noted in his log that a pupil had died in a Hurricane ("Probably blacked out"). OTU sped by, Jewett finishing with 49:25 hours on Hurricanes, 20:25 on Harvards. He was posted overseas, sailing from New York aboard the *Queen Mary*. With his draft were several others from Course 18—Neil Burns, Jake Copeland, Jimmy Flood and Alex McIntosh. As Jewett recalled, "We enjoyed the trip, drinking all the way to Greenock, then took the train down to Bournemouth." There he was pleased to meet a friend from Aroostook—Charles Gaines of 519 Squadron, who was flying weather recce. Gaines would die in August 1944 in Ventura JS518. Apparently, some control locks had not been removed before takeoff. From now on death would touch Arthur Jewett frequently.

From Bournemouth P/O Jewett was posted to 57 OTU at Eshott, flying first on March 3, 1944 in a Miles Master. The next day he took up his first Spitfire (PW-K). It was a 15-minute flight about which he noted, "I'm glad there weren't any pictures." On the 14th there was tragedy as F/O James M. Flood and Sgt Long collided in formation. Long spun in and died, Flood force-landed OK. (From Hearst, Ontario, Flood would be killed in action in August 1944 near Dieppe, flying Spitfire ML308 with 421 Squadron.

Arthur Jewett with Ronald G. Lake and Guy E. Mott in 1998. Mott, who tallied 5½-0-⅓ had a close call on August 11, 1944, when shot down by flak. Concealed by a French farmer, he soon was liberated by American troops. One of Lake's combats included damaging an Me.262 jet. (Larry Milberry)

One report states that he was shot down by an Fw.190, while other talk was that it was an RAF fighter.) OTU was busy—in March Jewett logged 24:40 hours on Spitfires. April brought more misery. One day a USAAF P-47 flew through the Eshott circuit, colliding with a Spitfire flown by a Norwegian. Both planes crashed fatally, Jewett commenting in his log, "Ruined my dinner." Another student on OTU, FSgt Jewell, was killed when he collided with power lines. When he returned from a flight that day, Jewett found that his belongings had been removed from his room and his bed prepared for a new occupant. It seems that someone in the orderly room had confused him with the unfortunate Jewell and wasn't wasting any time with formalities!

Jewett finished at Eshott on May 11 with a flight in Spitfire JZ-H. Leaving OTU with 69:50 on type, he travelled to No.1 Tactical Evaluation Unit at Tealing, Dundee for advanced training on the Spitfire I, II and V, including with the gyro gun sight. He left after 18:25 additional Spitfire hours, the CO inscribing in his log, "This pilot is gyro trained." At last Jewett was ready for operations and was posted to 441 Squadron, joining it soon after D-Day at St-Croix-sur-Mer (B.3), which he reached by Dakota. He learned that he would replace a pilot who had made a wheels-up landing and been turfed off 441 by W/C J.E. "Johnny" Johnson. The winco called in Jewett for a quick introduction. Jewett recalled in 1998 that Johnson was "very brief and very sharp." He simply warned him to be careful to do just what he was told.

Jewett made his first flight with 441 in Spitfire 9G-Y. Now all the daily action on 441 came hot and heavy for the sprog pilot. He notes for July 17: "6 Fw.190s sighted. /L Brown 2 destroyed, F/O Kimball 1 destroyed." On August 13 Jewett was on patrol as No.2 to F/L W.W. "Bill" Brown. He went down after some German trucks. Brown, a seasoned warrior from Edmonton, followed. Seconds later he was dead—they had been lured into a flak trap. On August 18 the Allies were embroiled on the ground and in the air around Falaise. That day Jewett flew 3 ops in 9G-R and 9G-Y, adding 3:55 hours. He noted in his log that the action involved 2000 enemy MT, 144 Wing accounting for 400 destroyed. His numbers were close—*The RCAF Overseas: The Fifth Year*, one of the best narratives of the Canadian air war 1939-45, gives an excellent picture of this action as it involved various wings:

*Late in July the Americans had broken through between St. Lo and Coutances and, sweeping southward towards the Loire, fanned out westward into Brittany and eastward towards the Seine, cutting in behind the German Seventh Army. On August 13 they entered Argentan. Meanwhile, the British and Canadian forces had launched an attack on the 7th, driving southward from Caen towards Falaise across the line of retreat of von Kluge's army. By the 15th the Canadian Army was within a mile of Falaise; on the 17th the town was captured. The American forces pushing up from Argentan were closing the narrow gap and trapping the Nazi army in a steel-ringed pocket. As the entrapped Nazis sought to escape through the narrow neck between Falaise and Argentan, Allied air strength cracked down with all its might.*

*In the late afternoon and evening of the 17th sorties by the fighter-reconnaissance squadrons reported hundreds of vehicles, tanks and staff cars streaming eastward from Falaise. Large convoys of 100 to 300 vehicles were on the move, running the gauntlet of Spits and Tiffies in their desperate effort to get out of the Falaise trap. When darkness finally intervened the MacBrien-Johnson wing had amassed a total of 196 flaming, smoking and damaged vehicles and three battered tanks. One sortie in particular, carried out by Wood's and McLeod's pilots around Trun, had wrought destruction in masses of over 400 vehicles; 29 were set on fire, 40 began to smoke and 36 others were damaged. The Wolves, leading with 105 MT and two tanks, lost three pilots to the intense barrage of light flak. F/O R. Weber returned ten days later; F/O M.L. Garland also subsequently rejoined his unit, but F/O H.V. Boyle is still missing. In the meantime, the Tiffies were pranging barges and boats on the lower Seine; four barges were sunk, another was set ablaze, at least two more and a dredge were damaged, while a river boat was damaged and probably destroyed by a direct hit from a 1000-pound bomb.*

*At dawn on the 18th the work was continued and the destruction reached unprecedented heights. From the first light to nightfall the Spits flew an almost continuous armed reconnaissance over the road running eastward from Falaise and Argentan to the Seine. At midday reconnaissance Mustangs reported all roads leading into Vimoutiers from the south and southwest jammed with vehicles. At 1500 hours Dean Dover and F/L C.W. Fox of the Falcons, on a special mission to spot MT, located 1000-1500 vehicles jammed bumper to bumper in a large wooded area near Argentan. When night interrupted the destruction, the roads were lined with blazing, smoking, shattered trucks, tanks and cars, the wreckage of an army in full flight. The Spitfire wing accounted for almost 1200 vehicles... Every serviceable aircraft was put into the air on these operations. No attempt was made to fly large formations; two, four or six aircraft would go out together, dive and fire until their ammunition was gone and then return for more... Our losses were only three pilots missing and two of them returned within a week. Halcrow was forced to bale out near Vimoutiers and was captured. He persuaded his captors, however, "that they didn't have a chance, so they let him go on the understanding he would arrange to have Allied troops come back and take the Germans into custody." Hugh Trainer of the Rams was also brought down by flak behind the lines, but evaded capture and returned to his unit. F/O C.E. Fairfield's aircraft was seen to crash after a direct hit from an anti-aircraft shell...*

Arthur Jewett was part of this exhilarating scene. There were endless patrols—3 or 4 a day. On September 14 he noted that F/L R.G. Sim had not come home from strafing. Local civilians helped him and he returned on October 29. He later joined 443 Squadron and assisted in shooting down a Ju.88 on May 3, 1945. After a patrol near Nijmegen on September 30, 1944 Jewett noted in his log: "Sighted Hun jet. F/L Lake probably damaged it with a short burst." This could have been an Me.262 or an Ar.234, both types being active against bridges in the area. For September, Jewett logged 23:00 hours, 19:30 on ops.

In the fall of 1944, 441 Squadron was near Antwerp. One evening Art Jewett and J.A. "Mac" McIntosh went on the town—they were not on the ops list for the next day. After a raucous time they got a taxi back to the house that had been commandeered for 441. Their driver, uneasy about his passengers, let them off short of destination. The boys started home. Inexplicably, Jewett and McIntosh began shooting out store windows as they passed, firing with .38 revolvers. Reaching home, they stumbled noisily in. Pilots who had to be up at sunrise did not appreciate this and words were exchanged. Jewett continued shooting, putting bullets into the ceiling. The miscreants at last were subdued and a kangaroo court was convened. F/L J.C. "Jake" Copeland was suggested that the offending duo be taken outside and shot! But S/L Kelly Walker, the CO, sent everyone to bed. The fuss blew over, but Jewett and McIntosh suffered one punishment—hereafter they could only carry sidearms when flying; upon landing, they had to turn them in.

Besides armed recces 441 flew bomber escorts, one being on November 16. The target was Duren, being softened up ahead of a US Army push. It was attacked that day by 485 Lancasters and 13 Mosquitos, 3 Lancs being shot down (USAAF bombers also struck Duren). In *The Bomber Command War Diaries* Martin Middlebrook described the day's work: "The RAF raids were all carried out in easy bombing conditions and the three towns were virtually destroyed." Poor Duren suffered 3127 civilians killed. Neither was this a great day for Arthur Jewett, who was on his 52nd sortie. Over Duren at 20,000 feet, his engine died. He glided as far as possible towards Allied positions, then bailed out, coming down at Frelenberg, Germany. The Americans had taken this area and he was back with 441 on November 20, having learned a little about the ground war by going on a few patrols with the Americans. He was back in the air on the 25th, doing an air test.

On January 1, 1945 Jewett again had engine failure. He was over the North Sea, but managed to glide to safety. For this good show W/C R.A. Barton, OC of RAF Skaebrae in the Shetlands (where 441 Squadron was training in fighter affiliation), inscribed a green endorsement in Jewett's log: "When flying Spitfire MK926 at RAF Sumburgh on 1 January 1945, F/O Jewett force-landed without further damage after an

No.441 re-equipped with Mustang IIIs late in the war. Here are several "Silver Foxes" in this era: F/O D.C. "Don" Gildner, F/L D.H. Kimball, DFC, F/L A.A. Smith, F/O G.D. Morrison, F/L H.E. Derraugh, S/L R.H. "Kelly" Walker (the CO), F/O G.E. Heasman and F/O J.A. McIntosh. Several of these had scored in aerial combat. Heasman shot down an Me.109 on September 18, 1944. Kimball got two Me.109s, Derraugh one on the 25th, while helping to stave off a Luftwaffe push against the Rhine bridge at Nijmegen. (CF PL45116)

June and Jake Copeland with Art Jewett during the Canadian Fighter Pilots Association convention at the Chateau Laurier in Ottawa in 1998. (Larry Milberry)

engine failure, thereby displaying sound judgment and good airmanship." On this deployment 441 lost WO J.E. "Boe" Bohemier of St. Anne, Manitoba on January 23, 1945 and F/L Bill Martin of Toronto on March 10, so this was a rough time. Bohemier, age 22, went down in Spitfire MK585 off Lerwick. Martin disappeared in ML216 while on a GCI exercise near Ronaldsay in the Orkneys. On April 11, 1945 Arthur Jewett completed his 55th and final sortie, flying Spitfire 9G-N on a 2:50-hour trip: "Escort Lancs to Bayreuthe. Uneventful. Nice bombing." (The escort was provided for 100 Halifaxes and 14 Lancasters attacking a rail target.)

On April 12, 1945, 441 Squadron moved to Hawkinge, where it remained till war's end. On May 12 it took part in a 2½-hour Victory Sweep to the Channel Islands. After returning to base the pilots were reprimanded for being too enthusiastic in their beat-ups, supposedly causing damage to farmers' greenhouses. Jewett made his last Spitfire flight (9G-N) on May 31, 1945. Now 441 began converting to the Mustang III. He flew this type from Digby on June 1, but didn't fall in love with it, recalling in 1998, "I didn't care for the Mustang. It seemed heavy and awkward compared to the Spitfire. It did, however, have great range."

Even though the war in Europe was over, there still were casualties. On June 18 Jewett was on a flight from Bognor to Portsmouth, Southampton and Bournemouth when he spotted a parachute in the sea. Later he heard that it was from F/O Vernon F. "Junior" McClung, a 25-year-old 442 Squadron pilot from Stoney Creek, Ontario. He had been flying KH694 from Digby when a wing panel flew off. The aircraft broke up and dove into Portland Bay, Dorset. A log entry of July 24, 1945 reads, "Cross-country. My birthday. McCabe killed. Spun into a junk yard. Convenient." F/O Edward J. McCabe, age 21 of Toronto, had been low-flying in Mustang KH569 when, for unknown reasons, he dove right into an auto wrecker's near Hull.

By this time many RCAF aircrew had become hard and cynical. They had flown feverishly on so many ops, killed perhaps hundreds of the enemy, been shot down, bailed out, crash landed, been wounded, recovered and gone back on ops. Many were drinkers and, in general, seemed oblivious to the misery in which they existed. It was a whirlwind existence, one that young men neither deserved nor enjoyed. They pined to be home with their families. Thus it was a happy day for Arthur Jewett when he made his last flight, taking a Mustang from Digby to Molesworth, and recording in his log, "441 Squadron officially disbanded today. C'est la guerre." He had flown the Spitfire for 254:45 hours, the Mustang for 37:15. His grand total was 682:45 hours.

Once home, Jewett returned to university in Sackville. There, in 1948, he met and married Joyce, a secretary. Life was not smooth sailing, however, for Jewett suffered psychologically and drank heavily. He quit school, but got involved with Alcoholics Anonymous. He worked in construction, then joined New Brunswick Power, remaining there until 1987. By 1998 he was a widower living in Fredericton, busy with his children and grandchildren, and in community affairs, especially charity fund-raising. He still enjoyed his great pastime of fly fishing, having used only one fly in his whole life, the Royal Coachman dry fly. Besides fishing, Jewett played piano (a lifelong hobby) and did some skiing (he specified "downhill ... never cross country"). As to booze, in August 1998 he reported, "On September 30, 1998 it will have been 44 years since I have had one ounce of C2H5OH."

In May 1997 Arthur Jewett was visiting a daughter in Cold Lake, where she was teaching school. Word got around that there was an old Spitfire pilot on base. This reached the ear of LCol Dave Burt, CO of 441 Squadron. Jewett was invited to visit his old squadron. LCol Burt had a sense of history. He appreciated the sacrifice that young men like Art Jewett had made for Canada more than a half-century earlier. It only seemed fitting to the CO that the old timer should see what the "Silver Foxes" were doing in 1997. The best way was to strap Arthur into a CF-18 Hornet and take him flying. That is what happened on June 3, 1997, when Jewett spent 1:10 hours aloft in Hornet 188916 with LCol Burt as his tour guide.

A pleased-looking Arthur Jewett following his CF-18 ride with LCol Dave Burt, CO of 441 Squadron. (Jewett Col.)

# Larry Spurr

Lawrence Esmond Spurr was born in Middleton, Nova Scotia on June 15, 1923. He grew up on the family farm with his parents (Charles, a fruit inspector, and Lola), two sisters and brother. As a boy Larry enjoyed sports, photography and music. After finishing high school, he worked as a bank clerk, then joined the RCAF on June 22, 1942. A bank manager, school principal, insurance agent and clergymen were his character references, as noted on Spurr's RCAF Attestation Paper. He completed the preliminaries—Manning Depot, ITS, etc., then reached No.12 EFTS at Goderich. There, on February 25, 1943, he made his first flight with Sgt Hank O'Mara of Toronto in Tiger Moth 3936. Spurr moved along quickly, making his last flight on April 15, by when he had logged 31:25 hours of dual, 27:15 hours solo and 2:00 hours of night flying. Years later he and O'Mara would renew their friendship in the Sabre world, but this time Spurr would be the instructor.

In 1942 there was an urgent call for aircrew. Replacements were needed in Bomber Command, where losses were horrendous. On May 12-13, 1943, for example, the RAF's 572-strong raid on Duisberg cost 10 Lancasters, 10 Stirlings, 9 Halifaxes and 10 Wellingtons. The November 1943-January 1944 Berlin campaign cost 384 RAF heavy bombers, or some 2500 aircrew, more than three times Fighter Command's losses of the Battle of Britain. In view of its own losses, the formation of new squadrons, and preparations for the forthcoming cross-channel invasion, Fighter Command also needed men, so student pilots were being hurried from EFTS to SFTS. Larry Spurr found himself at No.6 SFTS, RCAF Station Dunnville, less than two weeks after leaving Goderich. Soon he befriended Bill Baggs from Hamilton. They started flying on May 5. Their paths would cross in more ways than one in the coming years. F/O Byrnes took Spurr on his first and last flights at SFTS, the latter being on August 12. In a few days station commander G/C Val Patriarch wished the course well at the Wings Parade. Now a proud bunch of fresh pilots, each flashing sergeant stripes in place of LAC propellers, and with log books fattened by about 170 hours of Harvard time, left Dunnville for futures no one could predict.

As for Spurr and Baggs, they were off in a few days to Course 17 at No.1 (F) OTU at Bagotville. On August 31 they flew the Harvard, Spurr with F/L Brady, Baggs with F/O Corbett. This gave the boys back the feel of the Harvard, familiarized them with the local area, and let their instructors see what kind of students they had in the front seat. Now began the job of creating a batch of new Hurricane pilots. All went well, the lads piling up the hours on a host of exercises, and having some fun along the way. On September 6 Spurr and Baggs flew together in Harvard FE391, supposedly doing instrument training. But fine conditions tempted—they spent an hour seeing who could fly the lowest over Lac St-Jean. Luckily, they didn't get caught.

**Larry and Nann Spurr on the occasion of him being awarded the United States DFC for a 1952 tour flying Sabres in Korea. (Nann Baggs Col.)**

Course 17 wound up at the end of October. This time it was W/C E.M. Reyno saying the good-byes. Larry Spurr departed with a further 81:05 hours in his log, 59:10 on Hurricanes. On November 24, 1943 he embarked at Halifax. From Bournemouth he went first to No.3 EFTS at Shellingford, where he made four Tiger Moth flights on January 1-2, 1944. Both he and Baggs turned up at No.61 OTU at Rednal, Spurr taking up Spitfire "S" on his first flying day. Now began a busy period where the two OTU flights alternately flew in the morning (usually two trips per student), then did ground school in the afternoon.

As pilots got the feel of the Spitfire, confidence grew. Spurr noted a flight in "HX-P" where he exceeded 300 mph, probably for the first time. His last flight was on March 6, by when he had 58:10 Spitfire hours. Now he attended No.3 TEU for advance flying on Hurricanes (April 16-June 8); then was posted to RAF Honiley, where he made three Typhoon flights June 11-13. But Spurr suddenly was sent to 41 Squadron at West Malling (Spitfire XIIs). He was thrown into ops, as this was the busy time after D-Day. He first flew on June 24. His log shows scrambles, convoy patrols and Divers—V-1 buzz bomb patrols. For July he logged 27:10 hours. On August 4 he noted, "Fired at a couple of buzz bombs." He next attended No.83 GSU for a week of advanced Spitfire training, then joined 416 "City of Oshawa" Squadron at B.26 (Illiers l'Eveque, France) under S/L John F. McElroy, a Malta ace. His first flight was in DN-K on August 24. Hereafter his log is crammed with 2 or 3 entries a day, mainly low level shoot-ups against MT. For August 26 he flew DN-T and noted, "Got 2 flamers (destroyed), 1 smoker, 1 damaged." On October 2, while at Grave, Holland, he went after a German jet: "Fired everything but the dinghy at jet. No effect."

For October 29: "Chased PRU 190 up to 28,000 feet." The pace was wild, but everyone kept up until tour-expired, injured or dead.

For November 21 Spurr jotted down that Al Collins and Al Fleming had collided. "Very bad luck," he noted. F/O Allan W. Collins, flying MK837, was from Lethbridge. Age 25 and a new man on squadron, he did not survive the incident over Holland. For December 8 Spurr recorded: "Lost Bill Simpson shot down by 109s. I had a long dog fight with 8 109s for 10 minutes alone." This was tense stuff, but, happily, it turned out that Simpson was a POW. Christmas 1944 proved exciting. A few days earlier 416 had re-equipped with Spitfire XVIs, giving up its tired LFIXcs. It joined the back-and-forth fray of the Battle of the Bulge. Every available fighter was thrown in. On December 24 there was a hot welcome for 416. F/L Phillip, F/O Picard, F/O J.R. "Ron" Beasley, P/O Spurr and Sgt J.G.M. Patus all took flak. Phillip and Picard got home badly damaged. Patus went down, but survived. As to Spurr, his entry for the day is a beauty: "My most exciting doo of the war. Nearly shot by the Americans after I was shot down. Walked through the lines. Had Christmas dinner at Field Marshal Monty's HQ." After being hit by flak in SM335 at about 1430 hours near Malmedy, Belgium, Spurr had crash landed, but used his radio to report being safe. As he approached the lines, some US troops, mistaking him (in his blue battle dress) for a German, were ready to fire, till he could explain himself. Finally, Spurr had the good fortune to meet up with Montgomery's HQ. Beasley's luck was just the opposite—the 24-year-old from Ottawa died the day before Christmas. Meanwhile, Spurr's father was telegrammed that his son was missing. Mr. Spurr would have been overjoyed some days later on Christmas morning to hear from the RCAF Casualty Officer, "I am pleased to advise that your son, Pilot Officer Lawrence Esmond Spurr, is now reported safe... I join with your son and the members of your family in your joy in your son's safety."

In an even more ironic case 21-year-old F/O Alexander G. "Sandy" Borland of 416 was lost. RCAF Overseas notes of this: "During the morning F/O Sandy Borland of the Oshawa Squadron went out on a sortie but failed to return and was reported missing." Larry Spurr knew what had happened. In his log he notes that Borland was shot out of the sky by a US fighter. There was talk about such cases involving Luftwaffe pilots flying captured US fighters, but the historic record points to USAAF 9th Air Force Thunderbolts as the usual culprits.

On January 1, 1945 Spurr was with 416 at Evère, when the Luftwaffe pounced at 0940. He noted: "Aerodrome strafed by 50-75 109s and 190s. Lost most of our kites. I was strapped in the cockpit when they started. We were about to take off. 20 minutes of hell." Three taxiing Spits (F/L Nault, P/O Ken Williams, WO Lou Jean) were clobbered. The rest of the squadron could not taxi around their shot-up Spits, since the ground was soggy. F/L D.W.A. "Dave" Harling,

DFC of Montreal did get airborne, only to be shot down and killed. After tallying the results of the German raid, No.127 (RCAF) Wing at Evère was smaller by 11 Spitfires, with 12 more damaged. Two groundcrew were dead, 8 wounded. Even so, this was the Luftwaffe's last gasp. That day it lost hundreds of aircraft and pilots. While it had no replacements, the Allies did—within a day or two 127 Wing was back in action, Larry Spurr commenting for January 5, "Today we got some trucks and some bods. Nice to watch them run."

On January 28, while on a weather recce, Spurr had mechanical trouble and made an emergency landing at Eindhoven. He noted for February 14, "Narrow escape, engine blew up, crashed in a field. Not a scratch." Something also must have gone wrong on an armed recce a week later: "A bit shaky." Then, on the 24th, "Engine cut out. Forced landing on aerodrome. OK." Finally he had some more good luck, noting the next day, "Good Do. Fired at 2 Me.262s, got 1 damaged." His combat report states: "I half-rolled and attacked it from about 8 o'clock and gave it about a 4-second burst of cannon and .5 machine gun at about 500 yds. There were Fw.190s circling above us, so I could not finish the attack. Enemy aircraft pulled away. No one else saw the attack."

Meanwhile, casualties continued, with F/O W.F. Bridgman of London, Ontario killed near St. Vith, Belgium on January 13. F/O J.J.M. Menard force-landed after engine failure on February 25. He became a POW. March brought more of the same, Spurr's log for the month showing 47:50 hours. He got through safely, but not F/L Neil Russell. On March 17, 1945, the day he was to finish his tour, he was shot down by American flak. Russell escaped this snafu by bailing out. On March 31 a USAAF Mustang shot down two 416 Spits, F/Os V.W. Mullen and S.A.R. "Sam" Round ending as POWs. F/L F.G. Picard pursued and damaged the Mustang, but it got away. Years later Mullen recalled having watched the P-51 break cloud and climb for the six Spitfires. It milled around for a while in clear skies then, unbelievably, attacked. Mullen saw bullet holes stitch across his fuselage, then his belly tank exploded. The cockpit fill with flames, but his flying kit, especially his gauntlets, helmet and mask, shielded him till he fought loose and jumped.

By misfortune Mullen landed midst a German flak emplacement, so immediately was taken POW. The Germans proved friendly and Mullen was put on the road to Osnabrück with two guards. At one point a sentry at a bridge suggested that he be shot right there, but one of his escorts objected. Elsewhere, irate villagers called for Mullen's head. At last he was processed in Osnabrück and jailed. There he had the company of RCAF Typhoon pilot Frank Johnson. Soon they were going cross-country, a 6-day journey. At one stage the POWs were paraded through Hamburg. Mullen recalled, "The city seemed crushed, still burning and smoking in places, smelling of death. A huge rubbish dump." While being interrogated in there, he met another RCAF fighter pilot, Bob McCracken (411) of Lakefield, Ontario. Mullen's excursion ended on April 6 when he entered Stalag Luft I POW camp. There his stay would be brief—on April 30 the camp awoke to a strange situation. The guards were gone, fled westward into the hands of the British—before they could be captured by the Russians! On May 13 Mullen and some 40 compatriots flew aboard a USAAF B-17 to RAF Ford. On June 1, 1945 he sailed into Halifax harbour aboard *Louis Pasteur*, anxious to begin a new life at home. As to the P-51 pilot who had spoiled Mullen's war, he ended before a court martial and was reduced in rank. Mullen summarized his view of the man: "His aircraft recognition was not up to our standards."

Pilots of 416 Squadron's at Petit Brogel in March 1945. The six across the back are Chris Preston, G.M. "Gord" Hill, W.D. "Wally" Hill, K.J. "Ken" Williams, Jack Leyland and Keith Scott. Low on the wing are S.H. "Steve" Straub, Chuck Darrow, Larry Spurr, B.E. "Bert" Parry, C.W. "Cliff" Haines and N.M. "Mac" McGregor. Standing in front are L.J.R. "Lou" Jean, S.A. "Sam" Round, W.L. "Mac" McCallum, Walter Norman Douglas, J. F.G.H. "Pic" Picard, L.P. "Len" Comerford, Neil G. Russell, W.I. "Rocky" Gordon, S/L J.D. "Jake" Mitchell (the CO), G.A. "Gord" Cameron and Vernon W. "Moon" Mullen. Crouching in front are W.G.D. "Bill" Roddie and Sgt Brechnel, a ground crew member. On New Year's Day 1945 416 was taxying just as 40 enemy fighters pounced. F/L Nault and F/Os Williams and Jean were shot up on the spot. F/L D.W.A. "Dave" Harling got airborne, but quickly was shot down. On March 31, 1945 Round and Mullen were shot down by an over-anxious USAAF Mustang pilot. Picard got off a damaging squirt at this dolt. Happily, the Canucks escaped death. Had Douglas known the fate that awaited him on May 14, 1945, he surely would have chosen death in combat. While he was in quarters at B.154 Soltau, Germany, a squadron mate, fooling with a shotgun, blew off Douglas' face. The 24-year old from Haileybury, Ontario lies in Beckington War Cemetery. While many of those on squadron at this time were relatively new men, some had greater experience. Russell, for example, had flown a tour on Kittyhawks in North Africa. (Nann Baggs Col.)

For April 16 Spurr noted, "Got some staff cars and busses around Lunenburg." On May 3 he was leading six Spits when he noticed flak. Closing, he identified a Do.217. He and F/O Rex Tapley attacked. Of this he later noted: "Do.217 destroyed with Tapley. Great show, crashed in flames." In his Combat Report he noted:

*I was flying Bulldog Green Leader when I saw flak at 11 o'clock about four miles away, so immediately turned towards it and saw an a/c weaving. About 400 yds from the a/c I identified it as a Hun and closed in about 200 yds from it. The a/c broke away from me and into Green Three who was ahead and above. I dropped in line astern and it half-rolled. I gave it about a 4-second burst on the way down, seeing strikes on the starboard side. I was overtaking it, so I had to break off the attack. Green Three then attacked, started the port engine flaming. He broke off his attack about 400 ft from the deck and I made another attack with strikes on both sides and the a/c crashed into a field. Cine gun and gyro sight used.*

Fighting in Europe officially ended on May 5. Now ops declined everywhere. For May 7 Larry Spurr flew Spitfire "F" from B.154 (Soltau, Germany). His flight of 4 headed towards Copenhagen on a VIP escort. Such an escort would have been prudent, considering possible rogue Luftwaffe units, or Soviet fighter patrols, whose purposes could not be predicted. When he got home, Spurr's log annotation was, "Dive bomb on 4 ships. 2 MT damaged. What a spot, beautiful country." For war's end Spurr's log showed 211:25 hours on ops, with total flying by May 30 at 560:50. It had be a bittersweet war for 416. Although it had 75 German aircraft destroyed, plus probables and damaged, not to forget hundreds of ground targets, "City of Oshawa" had lost 42 aircraft, 19 pilots killed, and 13 POW. In the back of his log book, Spurr listed those whom he knew, 20 in all. Some had died ignominiously, the saddest case being F/O Alexander Graham Scott, a 19-year-old from Montreal. On April 19 he reported being in trouble and crash landed. There it was his misfortune to fall in among one of history's worst bunch of cut throats. Spurr's comment about Scott's end is a chilling "Shot by the SS."

Now 416 Squadron settled in for a long occupation. There was still lots of flying (July 1945 till disbandment in March 1946) this being from Uterson, Germany. Larry Spurr was much involved, but rotten luck kept dogging him. On a flight of June 20, 1945 to Amersfort, Holland with his friend Chuck Darrow there was a dicey do. They were in a Luftwaffe Bu.181 trainer (RAF letters DN-X). As they landed beside a motor transport park, a vehicle drove in front, forcing Spurr to ground loop. "Crash landed in car park. A bit shaky," he noted. On November 9 he reported: "A.K. Price killed today flying No.3 in my section. Just off end of runway." Appropriately, there was a memorial flypast—10 Spitfires Spurr noting in his log, "Perfect formation cross over his grave. Red hot guy."

On January 12, 1946 Spurr heard that he had received a Mention in Dispatches. This was near the end of his Spitfire days, his last flight being in "H" on March 20, 1945 going from Manston to Swindon. He had a grand total of 663:20 hours. The squadron already had disbanded (March 11); Spurr was one of the last of the 416 D-Day pilots remaining. Soon he was back in Canada, wondering, like most vets, what the future held. He collected the $343.96 War Service Gratuity payment (after deductions). This was for 1422 days of service (859 overseas). In April 1946 he must have enquired about re-enlisting in the RCAF, for a letter of April 18, 1946 from S/L J.F. Mitchell (RCAF HQ) mentions, "An opportunity exists for you to be re-engaged in the ranks for training in a ground trade." Although many would revert to the non-flying ranks in this period, the offer had no interest to Spurr, whose love was flying. He tried school, signing up at Acadia University as a cadet in the Canadian Officers' Training Corps, Royal Canadian Artillery. But neither was the student's life for him, and he left Acadia after two years. Someone once said that the best part of this interlude had been that he and some of his buddies learned to make a decent moonshine in their spare time. Larry Spurr was again thinking about air force life.

## Alexander G. Scott

Alexander Graham Scott (J42479) was born on April 21, 1925 in Iroquois Falls, Ontario. The family moved to Montreal, settling at 4385 Western Ave. Scott was known as a good student, having finished Grade 11 with honours at Westmount High School. He also was an avid sportsman and Air Cadet. After high school he worked in 1942-43 with Bell Telephone; but was

**The temporary grave of Alexander Graham Scott showing the wooden cross and propeller blades. (J.D. "Danny" Browne)**

anxious to get into the RCAF. He was barely 18 when his father, Arthur A. Scott wrote on February 18, 1943: "My son ... has my permission to join the Royal Canadian Air Force." Scott joined on February 26. On his application he was assessed by F/L B. Leclerc as: "Clean cut. Speaks clearly and fluently. Lots of confidence. Hopes to be a fighter pilot... unusually good material."

After his indoctrination to air force life, Scott took his primary training in August-September 1943 at No.11 EFTS (Cap-de-la-Madeleine, 74:00 hours on Finches). His OC noted, "No outstanding faults ... a smooth instrument flier." Advanced flying was at No.13 SFTS (St. Hubert, 150:25 hours on Harvards). Scott went overseas in April 1944. After No.3 PRU at Bournemouth, he attended No.7 (P) AFU (Hurricanes) at Peterborough, No.16 SFTS at Newton Notts, No.61 OTU (Spitfires) at Rednal and No.83 GSU at Dunsfold. He received average reports, but his potential always was recognized. His AFU assessment noted "Definitely suitable". Scott went briefly to 416 Squadron on March 24, 1945, then joined 421 on April 19, going on ops the same day.

An inexperienced replacement, Scott now was with the toughest of battle-tested fighter pilots. In August 1998 his OC, an American in the RCAF, S/L J.D. "Danny" Browne, DFC, recalled this era: "Canadian fighter pilots were the cutting edge at this stage of the war. They were over the beaches on D-Day, and 441 Squadron moved into Normandy just nine days later [the first RCAF squadron to reach France, it took up quarters under canvas at a quickly constructed air base—B.3 Ste-Croix-sur-Mer]. The Canadians distinguished themselves all along. Fighter bombers like the Typhoon and Spitfire were the ascendant power in the skies of Normandy. They assured absolute air superiority and made life hell for the Germans on the ground, terrorizing them to the point that they would abandon their tanks on the field as soon as we turned up. They knew that with every sunrise we would be there." Scott took off from B.114 Diepholz at 1330 in Spitfire SM242. The 8-plane mission was an armed recce of the Haganow-Lubeck-Hamburg-Neumunster area. But things ended tragically, 421's OC immediately submitting a Circumstantial Report, which included this narrative by F/O J.V. Marsden:

*F/O Zobel decided to strafe the marshalling yard at Sterley (west of Echwal Lake) and attacked it roughly out of the sun with F/O Scott (his No.2), following him down about 600-900 yards behind. There was quite a bit of light flak (40 mm) and F/O Zobel's fire hit some goods wagons probably loaded with ammunition, because there was a big explosion and lots of smoke. I saw F/O Zobel and then F/O Scott pull up, just clearing the smoke of the explosion. However, F/O Scott's plane was trailing white smoke (it turned out to be glycol). About a minute passed during which I lost sight of Scott due to cloud. Then he called on the R/T saying that he* was hot in the radiators and his glycol temperature was rising. He also said he was west of Sterley, losing height and would have to crash land, being at 2000 feet and crossing the Elbe Trave Canal. That is all he said, so we approximate his landing position as in the vicinity of Schwarzenhek.

A document of April 21, 1945 reported Scott missing, his final words quoted as: "Radiator temperature off clock. Can smell oil. Am going to crash land." In August 1998 the CO recalled: "I noticed a slight wisp of glycol coming from Scott's radiator. As we were operating so low, I warned him not to try bailing out—the Spitfire was not an easy fighter to exit quickly. 'Lay it down by those woods,' I suggested, 'and we'll draw away any attention from the Germans. Hide in the woods. The line is moving so fast that you'll be safe in a day or two.' Scott force-landed perfectly." Most of Marsden's report was included in an April 25 letter from S/L Browne to Scott's father. As was procedure, until a missing man's situation was confirmed, there was hope, Browne stating: "There is a possibility that your son is a prisoner of war, in which case you will either hear from him direct, or through RCAF Headquarters, which will receive advice from the International Red Cross Society... May I offer the sincerest sympathy from myself and all members of the squadron." Soon, however, it was clear that Scott was dead, and a letter (by either S/L Browne or W/C J.F. Edwards—both their names appear on a draft) was sent to his father, mentioning: "In the short time he spent with us, Scott certainly showed himself to be a keen, aggressive type ... He would have made a first-class fighter pilot." Scott's uncle in the UK LCol H.H. Hemming, also was informed of the death.

In a letter of May 20, 1945 to the DND in Ottawa, Mr. Scott quoted from S/L Browne's report, then asked what was being done to locate his son. He asked, "Has his squadron tried to see any evidence from the air?" (On June 11 the padre wrote, "The squadron commander flew over the area where he was last seen flying, saw the wreckage of an aircraft which he recognized as a Spitfire..." Mr. Scott finished, "Now that the war is over, the continued silence is more than usually oppressive." The father wanted some facts. Meanwhile, there came the letters of condolence, one on June 29, 1945 from A/V/M Hugh L. Campbell:

*It is with deep regret that I must confirm our recent telegram informing that your son, Flying Officer Alexander Graham Scott, previously reported missing, now is reported killed on Active Service... A burial report and photographs of your son's grave are being forwarded to these Headquarters... May the same spirit which prompted your son to offer his life give you courage.*

Scott had gone down near Woltersdorf, Germany. Perhaps he was hit by flak, perhaps his glycol tank was damaged when flying through the train explosion. An F/L Mulligan visited Scott's grave (probably within a day or two—dates in the files are ambiguous). Mulligan identified the site as map reference 921573, Sheet L5 Lauenburg, and submitted a report (undated) noting:

*The place of the crash was a meadow, just behind and belonging to a farm owned by farmer Herr Muegge. The aircraft was in flames and exploded on impact. The pilot was found lying near the wreckage. Members of the Wehrmacht who were stationed in the village took charge of the situation and buried the pilot in a wood near the scene of the crash.*

*The following inscription was carved on a tree beside the grave. First, there was a cross carved and under that the date "19-IV-1945", then, under that, a large W... There was also a propeller by the grave with markings R.A. 10046/RP F... I visited the burial place, but could not find the propeller. The cross had been removed, but I saw the carving on the tree.*

Mulligan found a small piece of wreckage, possibly part of a fuel tank. He recorded the information off a data plate attached to this item. Then he noted, "This is all the information I could gather, as the villagers seemed very reluctant to talk, and all claimed that they knew nothing." We aren't certain about the completeness of Mulligan's report, or whether it was edited to be read by Scott's next-of-kin. The fact remains that Scott's death was not accidental (another report notes that his death was the result of action involving an enemy aircraft, which 421 Squadron knew was not the case). The circumstances were so unusual that, once some details came to light, Scott's death was investigated by No.1 Canadian War Crimes Investigation Unit, which was informed of the case by HQ BAOR. A report (undated) by an F/O Gilbert notes:

*A Polish farm worker saw an aircraft crash on or about 19 Apr 45 at the village of Woltersdorf. The Pole saw the pilot wave to some French DPs nearby and he [Pole] was then chased away from the aircraft by the Germans, including a farmer, Theodor Mugge. The German soldiers then dragged the pilot from the aircraft "like a dog to the wood." The Pole saw the soldiers remove everything from the pockets of the pilot and gave the contents to Mugge. The Pole says he was hidden behind some bushes 10 meters from where the pilot was lying and heard a pistol shot.*

Closer to the truth, it seems that the shot was fired while Scott still was in his cockpit (this was explained by S/L Browne in 1998). The perpetrators of this execution were identified in this report as Eric Beginnen, an SS officer cadet, and Mugge. On October 26, 1945 Arthur A. Scott wrote to the Minister of National Defence: "I wish to thank you ... for the beautiful memorial cross and the message of consolation from the King and the Queen over the loss of our son, F/O A. Graham Scott, J42479. These will be cherished as long as we shall be granted the privilege of honouring his memory." Mr. Scott's wife, Clarissa, added her postscript: "I would like to add, too, my sincere appreciation for your wonderful appreciation of the valour of our lovely boy. We, like you, cherish his memory, for he gave all he had for all of us."

Scott was exhumed on January 1, 1946. That month his father applied for Graham's War Service Gratuity (this was submitted in April 1946—$362.55). He asked in this letter: "Will you please inform me when I may expect to receive from overseas the personal property of my late son ... I believe that he would have been entitled to wear some medals, had he survived the war. Will it be possible for me to obtain any such medals ... Finally, if permissible, I should like to receive his log book." The log was forwarded in February through the offices of G/C T.K. McDougall, RCAF HQ in Ottawa. Mr. Scott also was informed that he would receive the 1939-45 Star, France and Germany Star, and Canadian Volunteer Service Medal and Clasp. In February 1946 Mr. Scott wrote to the Director of Estates, RCAF in Ottawa thanking him for four pieces of baggage returned "intact and in good order." This had all been inventoried, items including clothing, toiletries, notebooks, a Falcon camera, flashlight, pipe, basin, keys, mail and playing cards. Scott's bank balance was noted as $670.98, a cheque for this amount being sent to his father that month.

In August 1946 Mrs. Buchanan, Graham's aunt in the UK, visited the RCAF Casualty Liaison Office in London to learn more of her nephew's death. She requested "that the war crime aspect of this file be withheld from the mother of F/O Scott", and that letters written at the time by the padre, A.J. Jackson, be considered adequate—that Scott had been found in the cockpit, then was removed by some Polish forced-labourers and buried. The padre noted that S/L Browne had visited the grave and that flowers had been placed, and "a simple service was conducted."

In July 1947 Mr. Scott enquired further: "I understand that the bodies of fallen airmen which occupy temporary graves in Germany will be removed and buried in special cemeteries elsewhere... I should like to know how matters are progressing with reference to the reburial of his body." The reply was that Mr. Scott would hear from the Commonwealth War Graves Commission, but that it was overloaded and could not help immediately regarding individual casualties. Meanwhile, routine procedures were completed. In February 1952 Scott's parents received a photograph of their son's grave in Ohlsdorf Cemetery near Hamburg, Germany, sent by W/C W.B. Gunn, RCAF HQ, on behalf of the CWGC.

In 1998 S/L Dan Browne lamented: "It was an awful thing that an innocent kid like Scott, with everything to look forward to, had to die in this meaningless way. In contrast we were hardened veterans. Sometimes we even viewed ourselves as savages. So, it hurt deeply when we lost Scott." Sadly, F/O Scott was not the only executed Canadian airman. A serious researcher could find enough cases to write a book. That of Gerald L. "Deke" Passmore is yet to be told. A squadron mate of F/O Buck Jenvey (see below), he was shot down on February 8, 1945, crash-landed, was taken to a POW facility, but was never seen again. The annals of other commands reveal their own black details, as with F/O Roy E. Carter, a 23-year old navigator from Burketon, Ontario. On June 17, 1944 he bailed out of his 431 Squadron Halifax on night ops. Landing in Holland, he was hidden by civilians in Tilburg, but was captured on July 8. Along with two other airmen he was executed by the Germans.

### F/O Jenvey, DFC

F/O Donald E. "Buck" Jenvey, born on January 9, 1921, grew up in Ingersoll, Ontario. Early in WWII he is noted as a radio technician and ordinary seaman in the RCNVR, but transferred to the RCAF in August 1941. He attended No.10 EFTS (Mount Hope, Tiger Moths), earned his wings at No.14 SFTS (Aylmer, Harvards), then instructed for a year. July 1943 found him on course at No.1 (F) OTU, a posting for which he likely had been manoeuvring. In late 1943 he was at No.59 OTU (Milfield), then progressed to No.57 OTU (Eschott) and No.3 TEU (Aston Down) where he converted to the Typhoon. On June 30, 1944 he joined 440 Squadron.

On December 27, 1944, following a dive bombing sortie near St. Vith, Jenvey tangled with some Me.109s, destroying one; but on the 29th, while bombing a rail target in Holland, he was hit by flak. In 1999 Bill Clifford recalled: "Buck made a nice belly landing, got out of his aircraft, waved to us and headed for the woods. Besides carrying the usual escape kit with its silk map, local currency, vitamin and Benzedrine tablets, morphine, photos (for making phony ID cards) and so on, we knew that Buck took extra precautions about safety, so were sure he would survive."

But the 24-year old from Ingersoll, Ontario would not make it. In April 1945 Jenvey's wife received a letter from Pieter Scholten of Delistraat 54 in Enschede, Holland. He mentioned their plight under the Nazis, and referred to his own family as "your near relatives". As Jenvey's "helpers", while he evaded, the Scholtens felt moved to write to Mrs. Jenvey—they had agreed to put down the details should anything befall their guest. The story begins when Jenvey was hit by flak. At first he found his canopy jammed, so had to belly land near Bentheim (in Germany near the Dutch border). He escaped his burning Typhoon, lay low until after dark, then headed towards Holland. Once he was fired at by sentries. Entering Holland, he soon was under the care of the underground, staying first with a family in Oldenzaal, but they did not speak English: "Finally the people where he was staying got in touch with the underground in our town and he was brought on bicycle by a girl to our house [a ride of about 15 miles to the south]". With the Scholtens, Jenvey's spirits rose. All the while he pined to get back to Eindhoven, but that was more than 100 miles southwest by road.

On March 21 Jenvey was photographed, so that false papers could be prepared. These were delivered next day—he now was a deaf and dumb man named Hermens Meere. A new contact entered the picture, an agent in Hengelo called K. Huska. On March 22 Jenvey set out for Hengelo (about 10 miles northwestward) with someone called Onderweegs. Next, Mr. Jacobs took over as escort, but they soon returned to the Onderweegs household in Enschede with bicycle troubles. Two German police soon arrived, but Mr. Onderweegs delayed their visit long enough for Jenvey to slip back to the Scholtens. It was a terrifying night, Mr. Scholten mentioning: "Perhaps you know that the whole family is shot down as a rule by the Germans, if it is discovered that you have been hiding soldiers, especially pilots." Jenvey and the Scholtens discussed their fear that someone in Hengelo could be a traitor, but he concluded otherwise. On March 24 arrangements were made for Jacobs to deliver two bicycles. Jenvey again set off: "I must admit that with some fearful feeling we saw his start on the dangerous voyage."

On April 10 Mr. Onderweegs appeared at the Scholtens' to deliver a package containing some of Jenvey's clothing (to this point the Scholtens still had not known Jenvey's name). Mr. Scholtens now visited the town police station where he was told that Jenvey was dead. The story went that he had been apprehended and returned by car to Enschede. There he wrestled with his escorts. Suddenly a man whom Scholtens described as "the civil person next to the driver" stepped from the car and shot Jenvey, "so that your husband and our best friend was finished at once ... that was on the 25th of March".

Enschede was liberated on April 1. On April 12 the Onderweegs and Scholtens witnessed the exhumation of Jenvey at which time several of his belongings, sewn into his clothing, were recovered. Mr. Scholtens described the gravesite as "a quiet corner with simple wooden crosses". He explained how the British had allowed him to write his letter and promised to enclose it along with Jenvey's few possessions. He finished by inviting Mrs. Jenvey and her son David to visit any time and that theirs would be a place of honour at the Scholtens. It soon was learned that Huska and Jacobs, double agents working for the Gestapo, were the chief culprits in Jenvey's demise.

Donald E. "Buck" Jenvey and friends at 440 Squadron. In the rear are John Villiers, Jenvey, Harry Hardy, Art Simard and Percy H. Kearse. Then, Gerald L. "Deke" Passmore, Bill Clifford and Nelson L. "Chuck" Gordon. On December 29, 1944 Jenvey was shot down, evaded, then was captured and executed. On December 25, 1944 Hardy was hit by flak, bailed out, then got back to Eindhoven for Christmas dinner. Kearse was killed in Holland on January 21, 1945. Passmore was hit on February 8, 1945, crash-landed, was taken prisoner, but never seen again. Gordon died on February 1, 1945, while instructing on Typhoons at 56 OTU. (Clifford Col.)

As much as Canadians do not wish the subject mentioned, it cannot be ignored that Luftwaffe members also were killed by the Allies, e.g. in lieu of being taken prisoner. Examples are extremely rare, but RCAF pilots have talked of strafing downed Luftwaffe aircraft to make sure that those aboard were finished off (usually this was done to assure destruction of the aircraft, not to kill the crew). Some openly have admitted to firing on German aircrew suspended in their parachutes. Another story is recounted on page 390 of *Royal Canadian Air Force at War 1939-45*—while one Canadian pilot would not shoot down an Italian aircraft in Red Cross markings, another efficiently did the job.

## The MSFU

With Luftwaffe maritime patrol bombers (usually Fw.200 Condors, He.111s or Ju.88s) sinking many ships in early convoy days, UK high command was desperate. After all, from August 1, 1940 to the following February 9 the Luftwaffe sank 85 Allied ships. With no fighter cover so far from land, the solution proved ingenious—the Merchant Ship Fighter Unit. This saw a lone Hurricane placed on a ship's bow on a rocket-propelled trolley at the end of a 70-foot launch rail. These were the "CAMs"—catapult-armed merchant ships. A convoy usually would have one CAM ship. Should an intruder appear, the Hurricane would start up then, at full RPM and with a boost from 13 small rockets, blast off the rail to pursue the enemy. After its sortie the Hurricane returned; the pilot ditched or parachuted to await rescue. Amazingly, pilots volunteered for the MSFU. They trained at Speke, near Liverpool, doing a few practice take-offs, then awaited their postings.

The first Condor fell to an MSFU Hurricane (sometimes called a "Hurricat") on August 3, 1941. Hereafter, the Luftwaffe showed a lot more respect for convoys, generally sticking to their fringes, or looking for lone targets. The availability of Beaufighters in Coastal Command also helped in this battle against the Luftwaffe. On September 17, 1943, for example, 8 Beaufighters of 235 Squadron shot down an Fw.200 over the Bay of Biscay, while losing one of their own. A few minutes later, 235 met 3 Ju.88s, despatching 2 and damaging the other. Maritime intrusion no longer was a sure thing for the Luftwaffe. Meanwhile, with the appearance of the small escort aircraft carrier, usually carrying FAA Hellcats, the MSFUs were abandoned with no tears being shed.

RCAF MSFU pilots included Vernon L. "Bill" Bowman, Felix Cryderman, Phillip E. Etienne, Tom Koch, Bruce Macpherson and Jack Sheppard. Koch's story is related in *Sixty Years: The RCAF and CF Air Command 1924-1984*. Little was known about Cryderman until research by Hugh Halliday in 2000 shed some light. Born in Sudbury in February 1917, as a young man he worked in prospecting and had some dealings in bush flying. He went overseas in 1939 with the Royal Canadian Ordnance Corps, but transferred to the RAF in August 1940, receiving a commission in April 1941. At first he flew with 411 squadron. From September 1941 to October 1942 he was in the MSFU, his first ship being SS *Empire Ocean*. So far nothing is known of this period of his career. Cryderman next flew with 222 Squadron (Spitfires), 1688 BDTF, then had a tour on 193 Squadron (Typhoons), finishing in February 1945. Eventually, he returned to Sudbury, but died in the crash of a Fairchild bushplane at Nakina, Ontario in June 1949.

October 1941 views taken by Bill Bowman of an MSFU "Hurricat" aboard *Empire Lawrence* on the North Atlantic. Note how the flaps are pre-set for launch.

An action shot that illustrates the MSFU story in a nutshell. (via Bill Stowe)

A rarely-seen photo of a CAM ship in convoy—*Empire Foam* ploughs through light seas, its MSFU Hurricane at the ready. The heavily-built launch rail stands out. (V.L. Bowman Col.)

A head-on photo from the rail. Sub Lieutenant Norman Towers (in RNVR uniform), posing with men from the ship's company, was the fighter director officer. The FDO's job being to vector his Hurricane by radar to any bogey. (V.L. Bowman Col.)

A close-up of the main Hurricat mount. Then, Bill Bowman in a May 1941 photo with 401 Squadron at Wellingore. He later joined the MSFU. (V.L. Bowman Col.)

Bill Bowman was a young man in Missouri when he decided to join the RCAF. Already a civilian flying instructor, he bypassed EFTS and SFTS, did some training at Central Flying School in Trenton, then received a commission. He was posted to Jarvis, moved to 118 Squadron at Rockcliffe, then got overseas to 401 Squadron. When word got around about some exciting new challenge late in 1940, Bowman volunteered—it was for the MSFU. After the usual training, he sailed from Liverpool for Halifax aboard Empire Lawrence. Next he made sailed to Gibraltar, and again to Halifax. These trips proved uneventful. Now he transferred to the USAAC to fly P-38s with the 474th Fighter Group in England. One day he fell victim to flak and ended up a POW for the last 9 months of the war. Postwar, Bowman had a 16-year career flying Constellations and Martins for TWA, then ran his own flying school for many years.

## Bruce Macpherson

MSFU pilot Bruce Macpherson was born in St. Thomas, Ontario on September 8, 1920. When he finished high school, he completed a year of pre-medical studies at the University of Toronto before joining the RCAF in 1941. This he did on the same day that his father, Dr. Arnold Macpherson, enlisted in the Canadian Army. Bruce's beginning steps in the RCAF included guard duty at Picton, ITS in Victoriaville, EFTS at Windsor Mills and SFTS at Summerside. He then sailed for the UK aboard the CP liner *Empress of Canada*.

Macpherson converted to the Spitfire at No.57 OTU near Hawarden. Early in 1942 he was posted to 412 Squadron at Digby. One day the CO, S/L John D. Morrison, was ticked off with Macpherson for having made a rough landing, and turfed him off the squadron (the CO soon was lost, disappearing on a sweep over France). Now Macpherson was in a quandary, but 412's MO, Bill Metzler had heard something about MSFUs and suggested the Macpherson put in for a posting there. Oblivious to what the MSFU was about, he volunteered and soon was at Speke, where the nature of this dangerous work became clear.

After some rudimentary training, Macpherson joined the ship's company of the 7457-ton merchant vessel *Empire Lawrence*, laden with Hurricanes for the Soviets. They sailed from Liverpool for Loch Ewe in the north of Scotland to join convoy PQ-16, bound for Murmansk. They set off westward, called in Iceland, then proceeded around its outer perimeter, sailed north close to the ice pack, then turned for Russia. Besides Macpherson, a South African MSFU pilot, Al Hay, also was aboard. On May 25, 1942 he was launched, shot down one intruder, harried another, then was rescued. Now *Empire Lawrence* had no Hurricat. On the 26th the Luftwaffe attacked while the convoy was off North Cape, Norway. Bruce Macpherson was on deck to witness the awful proceedings. He saw bombs falling towards his ship, felt it reel under the blows and

Sgt Bruce Macpherson as a new RCAF pilot at home in 1942 with his father, Dr. Arnold Macpherson, a major in the army, and his mother, Edna. (All, Bruce E. Macpherson Col.)

split asunder. As it broke up, Macpherson jumped overboard. *Empire Lawrence* soon disappeared, taking the captain, 11 crew and 3 RN gunners to their deaths. Macpherson, who was able to climb onto a Carley Float with some other survivors, was picked up by the corvette HMS *Hyderabad*. On January 1, 1943 he received an MiD related to this incident.

Finished with MSFU duties, Bruce Macpherson joined 93 Squadron as it worked up on Spitfire Vs at RAF Andreas on the Isle of Man. The squadron sailed for Gibraltar by convoy on October 20, from where it flew on to Maison Blanche then, on November 16 to its first forward base at Souk el Arba in the Tunisian desert. From here it engaged mainly in bomber escort work along with 72 and 111 Squadrons. During Macpherson's time, there seemed to be few skirmishes with the Luftwaffe, although one day, while flying as tail-end Charlie, Macpherson was shot down by an Me.109. He belly landed, but was soon picked up by a British army patrol. At their base he learned that a soldier, who had fallen asleep while on guard duty, was being court-marshalled. That night he was executed by firing squad. A few days later Macpherson was hit by return fire, while his section was attacking a Ju.88. Again he had to force land in the field, this time being returned to his unit by friendly Bedouins.

Bruce Macpherson and "Hutch" Hutchings in Malta, while bound for the UK after their North African tour.

From North Africa, Bruce Macpherson was posted home on a brief tour with No.1 (F) OTU in Bagotville. This took place over the winter of 1943-44. As usual there were several instructors there who had been on ops, including Bob Clasper and Paul Hurtibise of 93 Squadron, and Charlie Semple and Noel "Buzz" Ogilvie. Years later Macpherson expressed some doubt about the wisdom of sticking tour-expired fighter pilots into Bagotville as instructors, a task for which few had had an previous experience. Nonetheless, the system seemed to work.

In this period he married Pauline Richardson, a St. Thomas girl, who had been working as a message decoder at the trans-Atlantic flying boat base in Baltimore. Then he returned to the UK

Bruce and Pauline Macpherson in 1944—they were on Easter leave in Quebec City.

where, with the assistance of Jack Sheppard, whom he had met in MSFU days, he rejoined 412 Squadron. Now began another busy period, flying Spitfire Mk.IXs on the Continent. Macpherson was involved in the New Year's Day "festivities", 412 then being stationed at B.88 Heech, Holland. When the Germans struck Heech, 412 happened to have several Spitfires airborne. They met about 30 Fw.190s doing battle with some Tempests. F/Ls Doak and Macpherson each got a Jerry, but, as Macpherson recalled, the pilot of his seemed to bail out before being fired upon! On January 20, the tables were turned when, during operations around Arnhem, 412 lost aircraft piloted by F/Ls Macpherson (POW) and Richards (evaded), F/O W.J. Walkom (POW) and P/O Bruce S. McPhee (POW). In Macpherson's case, he ran out of fuel—he had to belly land for the third time. He immediately was taken prisoner, spending his first few hours playing chess with a German guard. His captivity (Stalag Luft 3) was relatively brief, the camp being liberated by the Americans near war's end. Macpherson reached England in a Dakota and soon was homeward bound on the *Queen Elizabeth*. Aboard ship he lost all his cash gambling, a typical story. Once home, he returned to medical school, graduated as an MD, then opened a family practice in Weston, near Toronto. There he served until retiring on January 1, 2000.

Bruce Macpherson (right) with the RAF 93 Squadron airmen who kept his Spitfire running smoothly.

# D-Day to War's End: 412 Squadron's March to Victory

| Date | Airfield |
| --- | --- |
| June 19, 1944 | B.4 Bény-sur-Mer, France |
| August 9 | B.18 Cristot |
| August 29 | B.28 Evreux |
| September 2 | B.26 Illiers l'Evêque |
| September 3 | B.44 Poix |
| September 6 | B.56 Evère, Belgium |
| September 21 | B.68 Le Culot |
| October 4 | B.84 Rips, Netherlands |
| October 15 | B.80 Volkel |
| December 6 | B.88 Heech |
| April 13, 1945 | B.108 Rheine, Germany |
| April 16 | B.116 Wunstorf |
| May 13 | B.152 Fassberg |
| July 6 | B.174 Uterson |
| March 21, 1946 | Squadron disbanded at B.174 |

Airing-out day for 412 at Bény-sur-Mer, Normandy. The squadron had set up here on D-Day + 13.

Bill Bellingham (in German head gear) poses at Bény-sur-Mer with a commandeered VW "staff car". Behind are the fancy quarters shared by 412 buddies Doak and Macpherson. Bellingham did not survive the war.

Some of the boys from 412: Lloyd A. Stewart, Bruce S. McPhee, W.A. "Bill" Aziz, J.A. "Jack" Swan, Bruce Macpherson, Bill Walkom and John Carr. On January 16, 1945 Stewart damaged an Me.262 jet. On April 28 he had to bail out, but was soon back to work. On April 30 he destroyed an Me.109. McPhee ended the war a POW. Aziz got an Fw.190 around Nijmegen on September 29, 1944.

A September 1990 trip took the Macphersons to England, where G/C Al Sheppard (RAF, ret'd) gave Bruce a flip in Spitfire PV202. Shown in 412 markings, it also had served on 33 and 412 Squadrons and in the postwar Irish Air Corps. It was credited with 3 e/a shot down.

The Macphersons at home in Toronto in January 2000. (Larry Milberry)

## From Alaska to Germany

As summer waned on Annette Island, Alaska some of the 118 Squadron pilots, who had been flying Kittyhawks for more than a year, were hoping to get overseas. Hugh Dickson of Truro, Nova Scotia was posted to 40-mm Hurricanes, but soon after joining 137 Squadron at Manston was lost on a shipping patrol around Ostende. Then, Allan Studholme got on Spitfires with 401 Squadron at Biggin Hill. On November 30 he went down in Holland to spend the rest of the war as a POW. A third pilot, P/O Bill Stowe, finally got his posting. On October 8, 1943 he sailed from New York aboard the *Queen Mary*. He spent from October 17 to December 6 at Bournemouth, then was posted to Spitfire OTU at Eshott. By this time he had 241:30 hours on

Spitfire II P8203/JZ-J was one of the aircraft flown by Bill Stowe at 57 OTU. The Mk.II could hit 370 mph with its 1050-hp Merlin XII. Built in 1941, P8203 also had served 266 and 123 Squadrons, was an ASR IIC on 277 Squadron, and ended with 5(P)AFU. A real survivor, P8203 was struck off charge on July 21, 1945.

Stowe's first operational trip was on April 18 in MB858. His log entry reads: "36 Bostons, bomb coastal guns." Now came many escorts for Bostons, Mitchells and Marauders, trips of 1:15 to 1:35 hours. For his first month Stowe flew 8 sweeps for 7:55 hours. Another sortie that appears often in his log is the "Noball"—escorting day bombers striking V-1 launch sites in coastal France. Other work included flying top cover for ASR Walruses or high speed launches. May 9's entry reads: "EB-F. Withdrawal cover, 8 Seafires, Pontiury Gael area, 1 Seafire buys it."

For May 14: "EB-G. Ranger. 1 staff car, 200 mi. over France. First time I fired." As D-Day approached, 41 Squadron got busier, the pilots sitting in readiness to intercept German photo recce aircraft. June 3-6 Bill Stowe flew 3 or 4 times a day. For D-Day he was not over the beaches. Instead, he noted: "Intercept a Mosquito". For June 7: "EB-B. Escort Typhoons, St. Peters Port, Guernsey. 1:00 hour. F/O Roby Robinson RAF hit by flack, bails out but is picked up dead. No Dinghy." As the pace towards victory quickened, Stowe's log would note many friends killed in action.

**Bill Stowe as a Harvard student at No.2 SFTS. (All, Bill Stowe Col.)**

Kittyhawks, so was a competent fighter pilot. On December 11 Stowe took a 30-minute familiarization flight in XO-H, a Mk.I "Spit". He flew the Mk.II (LV-D) on January 4, 1944 and finished at Eshott on February 16 with 60:25 hours on Spitfires. W/C Rook sent him on his way with an "Above average" grade. Stowe was posted to 41 Squadron at Tangmere (Spitfire Mk.XIIs). In 1999 he recalled what a thrill it was to fly the latest Spitfire, and what a high it was for a newly-arrived pilot to walk into the Officers' Mess to mingle at the bar with some of the great men of Fighter Command.

With a 1735-hp Griffon IV engine the Spitfire Mk.XII topped 390 mph in level flight. MB882/EB-B of 41 Squadron was built in November 1943 and served later at the Flight Leaders School.

**Peter Gibbs, Bill Stowe, mascot "Perkins" and EB-B at Lympne, an RAF station 45 miles southeast of London.**

In July, 41 Squadron was stationed at Lympne and busy on "Anti Diver" patrols—looking for V-1s. Although he chased the occasional V-1, it was tricky getting into position behind the speedy little "Doodle Bugs". One day as Stowe pursued one, a Tempest, then the RAF's fastest prop fighter, overtook him from behind to get the kill. By month's end Stowe's totals showed 13 sweeps, 9 recces, 6 ASRs and 1 Rhubarb for 49:50 hours on ops and 119:40 on the Spitfire XII. August saw No.41 on many strafing sorties across the Channel. On August 26, for example, Stowe noted going after 3 barges near Ypres. On a 1:20-hour Ranger on the 31st he shot up \a locomotive and some MT. On September 14 the squadron became part of 2 TAF and re-equipped with Mk. XIVs, a fast-climbing Spitfire. In mid-September, Stowe flew sorties around Arnhem, where Allied paratroops were having a rough time. The weather was duff for some days. Supply-dropping and glider-towing Dakotas, Halifaxes and Stirlings on "Skytrain" operations were getting clobbered by flak, as they came in under low ceilings. Stowe noted for September 19: "Anti-flak escort for Skytrain, return to Manston, weather u/s."

**Typical flak damage on Stowe's Spitfire XIV at Eindhoven—he was hit while strafing on the east side of the Rhine.**

**Bill Stowe's log entries for September 1944.**

By now Bomber Command was flying day raids deep into Germany, No.41 Squadron often escorting these. On October 6, Bill Stowe was on a 2-hour escort for Halifaxes attacking Sterkade. This was part of a 320-bomber operation to knock out oil refineries there and at Scholven. Three Halifaxes were lost at Sterkade. Stowe noted: "Land Antwerp, bombers get terrific flack from Ruhr." On the 14th he was on a similar do covering bombers to Duisberg. This was Op. Hurricane, intended to show the Germans what destruction the Allies could deliver—957 RAF/RCAF bombers dropped 4394 tons of bombs, but lost 14 aircraft. Several wings of Spitfires provided cover, but no Luftwaffe fighters met them. Meanwhile, the USAAF followed up with 1251 heavy bombers. That night Bomber Command sent 1005 bombers to Duisberg to continue that mayhem.

No.41 Squadron moved to the Continent on December 5, 1944, joining 125 Wing (41, 130 and 350 Squadrons), of 83 Group/2 TAF at B.64 (Diest). In 1999 Stowe recalled: "Our main role in at this stage was to fly armed reconnaissance well ahead of the army front lines. These trips usually were not planned in detail, but flights of 4, 8 or 12 pilots were given an area to patrol. Their first task was to seek and destroy enemy aircraft, then, enemy road, rail and barge transport. The altitude, usually 4000 to 8000 feet, and routes were at the flight leader's discretion."

On January 1, 1945 No.41, then at Y.32 Swartzburg/Ophoven, was shot up in the Luftwaffe surprise attack. Stowe was on a morning Battle of the Bulge patrol, but his section was not called back in time to help. All he noted for the day was, "Nothing around ... drome strafed." The Battle of the Bulge took place in this period. It cost 125 Wing 2 aircraft (S/L Phil Tripe, RCAF, CO of 130 Squadron, and F/O Peter Gibbs, RAF of 41 Squadron), shot down by US Army flak. Both men bailed out OK. On January 23 the squadron bagged a couple of long-nose Fw.190s around Munster. The ORB mentions Stowe regarding a clash with this type: "F/Lt Henry damaged one of the two he had engaged, and F/Lt Stowe, chasing another as far as Dortmund, damaged it with his .5s, having expended all his cannon on ground targets." On the 27th 41 Squadron moved to B.80 Volkel to join 124 Wing, flying top cover for its Tempests, keeping 109s and 190s off their tails.

Activity now was higher than ever. For March 2, after a patrol in the Velno-Munchen-Gladbach sector, Stowe recorded: "Danny Reid gets 1 jet Ar.234 destroyed." On March 17 No.41 moved to B.78 Eindhoven. It covered the Rhine Crossing around Wesel on the 24th. On April 5 Stowe was hit by flak, his prop and cowling being badly damaged. Five days later he noted: "Chase an Ar.234, no joy." Now he was posted to 130 Squadron as a flight commander. April 24 was especially interesting.

Patrolling north of Berlin, 41 Squadron spotted masses of aircraft; these proved to be Soviet Stormovik ground attackers and MiG fighters. The 125 Wing diarist later noted that this was on "a sweep in grand style", G/C Johnson leading 350 Squadron, W/C George Keefer—41 Squadron, S/L Woolley—130 Squadron. He concluded: "Having shown off our aircraft to the Russians, the rest of the day was spent in hard work, getting Huns while the stocks lasted." On his second trip that day Bill Stowe got an Me.109 and shared another with his wingman, WO Trevarrow (F/L S.R. Jeffreys, 125 Wing intelligence officer, attributed both to Stowe). Next day Stowe and several mates shared in an Me.262 caught landing at Lubeck (Jeffreys put it this way: "F/L Stowe and WO Ockenden shared in the probable destruction of an Me.262."). On April 25 No.130 Squadron, led by RCAF ace George Keefer, met 24 Fw.190s around Pritzwalk. In the fray two e/a went down, one to Keefer. Of this Jeffreys' noted of the enemy pilots, "very skilful and fought well." His reports are most lively, as typified by that for the afternoon of April 25, then for the 30th. This one paints a nice picture, where each man had his place:

[Stowe noted in his log: "Get one from astern and cause him to force land. Finish him off with Trev on ground. Open up on another with gyro sight at about 20° off, 500 yards and it explodes."] F/O Lord destroyed another... At 1045 hours a section of 350 Sqdn near Schwerin Lake sighted two gaggles of 30-plus Fw.190s at zero feet ... 6 e/a were destroyed and 1 damaged... And still they came ... 15 Fw.190s were seen ... 4 of these were duly laid on a plate and presented to the Wing... To vary the monotony F/Sgt Woodman destroyed a Siebel 204 ... it's all right, it wasn't ours!

F/L Jeffreys' report for May 2 begins by warning pilots to watch carefully before opening fire on any ground targets—at this stage they could be friendly. He reported on a great many vehicles between Crivitz and Schwerin:

On April 30, 1945 Bill Stowe and WO Trevarrow clobbered two long-nose Fw.190s. Gun camera film from Stowe is shown here. Then, his film of a typical flamer (near Ypres, August 31, 1944).

No.41 Squadron pilots Pete Hale, Vic Whale, Hugh Kelly, Wally Jallands, Mickey Moyle, Bill Stowe, Arnold Jolly and Junior Farfan. All were RAF except for Stowe (RCAF) and Moyle (RAAF).

*In the early afternoon a section of 41 Sqdn destroyed a Ju.188, shared by P/O Coleman and W/O Chalmers. They also damaged a Fw.190 and a Fiesler Storch on the ground on an unidentified a/c in the Printzwalk area. Once again 130 Sqdn came home with another scalp—this time a Siebel 204 destroyed by S/Ldr Woolley ... that completes the destruction of our 200th Hun since the Wing formed ... all this time Met and railway traffic was being clobbered in fine style... [30th] What a day! Victory rolls galore... It started when W/O Moyle, out on the first job of the day with a section of 41 Squadron, saw some 30 Fw.190s and Me.109s on a strip at the edge of a wood about 11 miles SSE of Schwerin. Later patrols kept a special eye on this area and at 0945 hours a section of 130 Squadron saw 9 Fw.190s ar zero feet at Banzkow. Other pilots attacked with the result that F/Lt Stowe destroyed one, caused another to crash land*

*It appeared to be jammed and unable to move either way... It was duly clobbered by a Section of 350 to the tune of 55 damaged... We suffered a loss during the afternoon when F/Lt Stowe got mixed up in the debris from ground targets just attacked and had to crash land west of Schwerin Lake either in our lines, or just beyond them [he had landed west of Lubeck]. He is believed to be safe and we look forward to seeing him in a few days. We had another victory at 1710 hrs when P/O Watkins, F/Lt Bengerter, F/Sgt King and F/O Van Eckh shared in the destruction of an Ar.234... an Me.262 in the same area only got away owing to superior speed... The Allied armies from the east and west continue to mop up what is left of Germany with great speed. A large-scale link up of the Anglo-American and Russian forces in the north is imminent.*

A battered Fw.190 seen by Bill Stowe at Twente, Holland. Wrecks were strewn across Europe by war's end. If found today, any one of them would be a priceless treasure and work would start to get it flying again!

The 124 Wing Me.108 in which Bill Stowe and Ron Collis flew to Tangmere.

A commandeered Siebel 104A. On June 18, 1945 Arnold Jolly and Bill Stowe flew this machine from Kastrup to Fassberg, where Bill had a posting to 401 Squadron.

An Arado Ar.234 jet bomber at Fassberg. A number of 234s were shot down by Canadian fighter pilots. S/L Arnold Jolly test flew this aircraft. Here, Bill Stowe is in the cockpit, Jolly in front. The rooftop feature is the periscope through which a pilot could watch his 6 o'clock position.

This Ju.188 also was at Kastrup. Fortunately, there were a few keen fellows to bring home photos of such interesting subjects. "We weren't allowed to take photos," is one flimsy excuse they didn't have to use when flipping through their albums years later.

**Garry Cooper, the Australian restorer of EB-E, while visiting Bill in Toronto in 1984. (Larry Milberry)**

As to Bill Stowe's force-landing, his log notes: "Picked up by recce patrol of 11th Armoured Division." He stayed two days with this small unit, overnighting in a barn with a crowd of German prisoners. Hereafter, he returned to 41 Squadron as a flight commander. On May 9 No.41 moved to Copenhagen—the boys were greeted as liberators. On June 18 Stowe was posted to 401 Squadron, prior to repatriation. He flew a bit there in the Spitfire Mk.XVI, then moved to 412. On August 11, 1945 he flew a Spitfire for the last time, a 20-minute hop from Luneberg to Lubeck. August 13-14 he and F/L Ron Collis of 41 Squadron took an Me.108 to Eindhoven, Knocke and Tangmere. Bill Stowe's war was over. He sailed for Halifax on the *Niew Amsterdam*. Like most he figured that his flying days were done. He enrolled in engineering at the University of Toronto. But that lacked enough challenge. In June 1948 he re-joined the RCAF as a member of 400 (Aux) Squadron at Downsview. On September 11 he flew the Vampire for the first time, taking up 17027. W/C Stowe, DFC, became commanding officer of 400 in 1952, remaining into 1954, when he left to devote more time to business.

In August 1981 Bill Stowe heard from an airline pilot in Australia, Garry Cooper, who was rebuilding Spitfire Mk.XIV RM797 EB-E of 41 Squadron. EB-E was an aircraft that Stowe had flown often—he considered it "his" Spitfire. EB-E had gone to Thailand in 1951, then languished for years. One day Cooper, from his perch atop an elephant, spotted it in a field! He arranged to acquire the hulk and transported it to Darwin. He was restoring EB-E when a cyclone roared through in 1974, severely damaging it. Cooper later shipped the wreck to Sydney. He now began tracing EB-E's history, contacting as many 41 Squadron survivors as possible, scouring the records, digging for photos, etc. For September 1944, for example, he determined that EB-E flew 6 ops: FSgt I.T. Stevenson—Fortress escort, F/L W.N. Stowe—sea patrol, F/O N.P. Gibbs—Skytrain, Stevenson again—Skytrain, and F/O J.F. Wilkinson. He unearthed 7 months of such records, the busiest being February 1945, when EB-E flew operationally 20 times. From what Cooper could determine, since joining 41 Squadron in September 1944 to EB-E's last days the following April, Bill Stowe flew EB-E 33 times; 24 other pilots flew it, but nobody more than 10 times. Through all his digging, Cooper did aviation history an extra service by reuniting squadron members. He located all of 41's surviving pilots in Australia. He found Jimmy Ware still residing at his 1944 address in New South Wales! Cooper persevered over the years, even locating the crash sites of some 41 Squadron Spitfires. In 1987 he found the spot where one pilot had crashed fatally near Munster. He met the landowner, who offered to drain his swamp so Cooper could comb the wreckage! By 2000 the project to get EB-E flying was on-going.

**Bill and his son Stephen, while Bill was gliding in California in 1999. (Larry Milberry)**

Once home, some aircrew left the war behind. Others had an on-going relationship with old squadron mates through organizations like the Canadian Fighter Pilots Association, or the Aircrew Association of Canada. At one reunion in the UK, Bill Stowe watched Spitfire II EB-Z of the Battle of Britain Memorial Flight lead a 41 Squadron Jaguar. In 1999 he attended a reunion of Luftwaffe pilots against whom he had fought. (Bill Stowe Col.)

## More Day Fighter Tales

A typical RCAF "Circus" (fighter-escorted, short-range bombing attack) operated on February 3, 1943 with 416 Squadron of the Kenley Wing and two squadrons from the Debden Wing escorting 12 Venturas. S/L F.H. Boulton of 416 was leading 9 Spitfire VBs. Crossing the French coast, Boulton observed: "The bombers ... flew through intense heavy flak and one was hit badly. I saw 3 of the crew bale out... The bombers were engaged by flak all the way to St. Omer." Some Fw.190s now came on the scene, Boulton engaging one, firing, seeing it smoke, then dive vertically. Boulton got behind 2 others. One accelerated, quickly leaving him in its wake: "From here on until I reached the French coast I continually kept breaking into attacking 190s. I finally ... made out near Mardyck with 3 aircraft in my section ... reaching sea level, I steered 300 degrees for home, notifying my No.2 with appropriate gestures that my R/T was u/s." In this same fray F/L R.A. Buckham shot down a 190 coming head-on; its pilot jumped. Heading out, Buckham spotted 2 Spits smoking, each with a 190 on its tail, but he was too far away to assist.

On March 8, 1944 the Kenley Wing was doing cover withdrawal for 60 USAAF B-17s bombing Rennes. They rendezvoused with the "Forts" over St. Lo at 1455. One Fort was smoking and descending, so Yellow Section sped to its side. When it reached the crippled plane, however, there already were 6 Spitfires on scene. Suddenly some Fw.190s appeared, F/L Godefroy (403) quickly shooting down one. F/O H.D. MacDonald (403) knocked the wing off another. F/L Magwood hammered a third, but nobody saw it crash.

An American in the RCAF, F/L A.F. Roscoe, DFC and Sgt B.R. Scaman of 165 Squadron (Spitfire VBs), had a curious experience on May 9. Flying from Dyce, they were vectored to a bogey hugging the deck. They flew full out, wanting to catch this fellow before he reached the coast. Control first had them go a short distance to sea, where Roscoe spotted the bogey about a mile inland on the deck and 13 miles from Aberdeen. He approached to find that a Ju.88 with wheels down, firing flares and waggling its wings. Roscoe took a position ahead, with Scaman behind. They led the Ju.88 into Dyce, landed and those aboard the Ju.88 surrendered. Another exciting "do" involved the RCAF Kenley and Hornchurch Wings on June 20. 1943. Kenley Wing's intricate report notes 3 Spitfires lost, 1 Fw.190 claimed:

*S/L H.C. Godefroy, DFC led the Wing and 403 (RCAF) Squadron, with 421 Squadron (RCAF) under command of S/L McNair, DFC. The role of the Wing was that of forward target support within the Abbeville-Amiens-Poix area, while 12 Bostons were bombing Poix aerodrome, closely supported by the Kenley Spitfire VB Wing ... The operation went on schedule, the Wing crossing out at Rye at 1245, where it joined up with the Hornchurch Wing. They began to climb at 1249, Kenley Wing crossing in at Quend Plage at 12,000 feet with Hornchurch above and to port. Thence, the Wing flew via Abbeville to Amiens, reaching 22,000 feet and gained 23,000 feet at 1316 hours. Appledore control then gave a vector of 010 degrees which was taken by the Wing, followed by Hornchurch until Aux-Le-Chateaux was reached at 24,000 feet.*

*Yellow Section 403 Squadron went down on some Fw.190s which were 1000 feet below and going in the opposite direction, but were not able to engage. 6 Fw.190s were seen by the squadron, then a further 12 were reported coming in behind ... at 24,000 feet. Unfortunately, these were first reported by a pilot of the Wing as Spitfires, and the Wing Leader ceased to turn to meet them. As they drew nearer they were then reported as Fw.190s and he again turned to meet them. 421 Squadron were able successfully to engage. S/L McNair (Black 1) fired at one head-on without seeing any results but, putting on some deflection, fired a 2-second burst at 350 yards and saw strikes all over the cockpit and fuselage. The e/a undercarriage came down, parts broke off and the aircraft ... went down. It is claimed as destroyed. Most of this was seen by S/L Godefroy ... F/O Ogilvie (Yellow 3) 403 Squadron saw an Fw.190 below him on which a bright orange flash occurred round the cockpit, and black smoke appeared from the underside of the fuselage. The e/a then went into a gentle dive. Similar results were described by S/L Godefroy ... who saw this aircraft attacked before the one attributed above to S/L McNair. In his opinion the aircraft could not have survived and it is therefore described as probably destroyed, although the pilot himself did not see the results of his attack as described in his personal combat report attached.*

*403 Squadron was now approaching Abbeville and 421 Squadron was seen to be momentarily on their port. The Wing Leader tried to re-form the two squadrons, but the cloud in this area at 18-20,000 feet prevented this ... Also, about 50 Fw.190s were now seen approaching Abbeville from the south at 27,000 feet. Hornchurch was not seen after the engagement began. The Wing Leader ordered 403 to dive out to the Somme Estuary, which they did, diving to 12,000 feet except for Blue Section which had become separated.*

*In the original engagement near Aux-Le-Chateaux, Blue 1, F/L [H.D.] MacDonald 403 Squadron, saw an Fw.190 on the port side and behind and, apparently, making for Blue 4 Sgt Windsor†. He ordered Blue Section to break quickly, but Blue 4 did not respond and was next seen ... with a thin stream of black smoke coming from his aircraft. During the R/T conversation ...(approximately 1320 hours) someone said, "I'm baling out. Cheerio." Sgt Windsor was the only one missing at this stage ... Hornchurch did not suffer casualties. A few minutes later, while Blue section was trying to re-form with the balance of 403 Squadron ... Blue 2 P/O Elliott‡ started to lose height. Blue 1 called to him to turn his oxygen up as Blue 2 had mentioned on the R/T that he was having trouble with his oxygen. Blue 2 half-rolled and dived away, followed by Blue 1 and Blue 3 from about 27,000 feet until Blue 2 was lost in the thin cloud at about 20,000 feet. The last time seen he was somewhere between Doullens and the coast at about 1325 hours. P/O Heeney Green 4 of 421 Squadron found that the oxygen tube of his mask came away from the Bakelite attachment to the mask when he was at 24,000 feet. He tried to re-attach the tube but ... lost consciousness, regaining it to find himself at 12,000 feet quite alone. He took refuge in some cloud ... he saw some aircraft which he attempted to formate on, until he realized they were Fw.190s. He quickly dived to ground level and, on his way to cross out south of Boulogne, he gave a short burst at goods wagons and a flak post...*

*P/O McWilliams*... made for the coast, but the section was intercepted by a gaggle of 20 Fw.190s. Blue 1 and 3 were flying line abreast giving each other cross cover, with Blue 1 on the port. Suddenly Blue 1 saw beside him on the port a 190 almost formating on him. This had not been reported by Blue 3, so Blue 1 called to Blue 3 to break. Instead Blue 3 did a gentle turn. Then one Fw.190 attacked Blue 3 and Blue 1 believes he saw strikes on Blue 3's wing. Blue 3 went into a spin with about 8 Fw.190s going down after him. At about 1333 hours Blue 1 called over the R/T "OK, McWilly, bale out while you've got a chance." This was the last seen of P/O McWilliams, near the coast south of Le Touquet. Blue 1 then climbed in tight turns to 38,000 feet at an ASI of 60 mph, closely followed by many Fw.190s which all fired at him but ineffectively. Having out-climbed them, he returned, crossing out over Hardelot.* († Sgt K.D. Windsor, who survived. ‡ John Charles Elliott, age 21 from Toronto, flying Spitfire BR637. * P/O Frank Cooper McWilliams, age 21 from New Westminster, BC, flying Spitfire LZ899. )

The watchword of the fighter pilot is "Check Six". In other words, keep watching behind (6 o'clock position) from where aces do their best work. Thus did fighters from the 1930s onward have a rear-view mirror in front of the pilot. Those who made a habit of ceaselessly glancing into it lived longer. On a sortie west of Falaise on June 30, 1944 F/L Dean H. Dover of 443 "Hornet" Squadron caught some Me.109s napping. The Hornets swarmed in, Dover zooming straight up on the tail of one: "I followed him upwards. He throttled right off and did a wing over. I did the same and followed him around." Now the Hun got a bit dozy, when he spotted a Spitfire chasing another 109. As he latched onto the Spit, Dover eased in and ended his career, later reporting: "The Hun ... had his finger well up and gave me plenty of time to shoot." These days the 109 pilot would be accused of "losing situational awareness". By getting fixated elsewhere and forgetting "his 6", he lost his life. While Dover's 109 went straight in, combat reports often describe a 109 or a 190 crashing in a shallow dive. Often it appeared that nobody was flying the plane—in many such cases, the pilot was dead, wounded or blacked out.

An operation of June 14, 1944 resulted from the Allies' determination to keep the Luftwaffe at bay in the critical time when the Normandy beachhead was building. The squadron involved was 443, then at Ford (the following day it crossed the Channel to B.3, Ste-Croix-sur-Mer. S/L Henry W. McLeod led 443 in escorting Lancasters to LeHavre. F/O R.A. Hodgins took off at 2250 hours—when, according to Double British Summer Time, it was still daylight—and climbed to 23,000 feet. Northeast of LeHavre his section spotted some aircraft, closed and recognized two Do.217s escorted by an Fw.190.

As 443 neared, a Lancaster appeared above the Dorniers, one of which fired upwards at it. Hodgins immediately attacked with a long burst, while both Dorniers fired on him: "I broke underneath and, as I pulled up ... the e/a's starboard motor ... burst into flame." This Dornier spun vertically, but Hodgins had other worries— an Fw.190 bounced him, perhaps spotting him in the light of the burning Dornier. Hodgins out-turned his attacker, spotted another Dornier, but realized that 4 other Spits were queued on it, so pulled off: "I gave chase to a further twin which turned out to be a Mosquito." The Spits were now ordered to withdraw, Hodgins landing at 2350. It must have been a hair-raising hour for the day fighter boys!

Hodgins debriefed, giving the details to his IO. While he had seen the Do.217 firing on the Lancaster, it wasn't clear that the Hun was using *Schrage Musik* ("jazz music") upward-firing cannons, a deadly weapon from which there was little salvation once it sighted in on a bomber. The unlikelihood of *Schrage Musik* is that the Dorniers were carrying underwing bombs. The fire directed at the Lancaster was probably a case of a sharp-eyed dorsal gunner grabbing an opportunity as his aircraft headed for its own target. (S/L McLeod of 443 also got a Do.217s this night, sending it into the water with a 6-second burst. He was killed in action around Nijmegen on September 27.)

S/L Jackson E. "Jack" Sheppard (5-0-0) served on 401 and 412 Squadrons. On November 26, 1943 he was escorting US Marauders, when his section peeled off to deal with Fw.190s rising from Achiet airfield near Cambrai. He harried one of these, making some strikes, but his quarry led him in a merry chase. He later reported:

"Hopping over trees and hedges, the pilot was taking such evasive action that he hit the ground 3 times with his propeller..." Sheppard finally got in his coup de grace, sending the "190" to its doom. He led 412 from April 12, 1944 to August 1, when he was shot down in Normandy, but evaded.

On October 3, 1943 FSgt Harlow W. "Bud" Bowker (411 Squadron) of Granby, Quebec had been shot down at sea, following Boston escort duty to Paris. Rescued, he soon was back in action, distinguishing himself with 430 Squadron at train-busting. In time he was posted to temporary test flying duties at 410 Repair and Salvage Unit, Tangmere. There he had some interesting action on May 22, 1944. Bowker was aloft in a newly-serviced Spitfire. He was to test its guns at a range near Selsey. Toodling along, but keeping his eyes peeled, he spotted two Fw.190s. The Germans were tight and low—when they climbed, Bowker said they did so to 30 feet! They didn't spot him, probably being too focused on not hitting the next fence. Bowker raked the aircraft on his left; the explosion disabled its neighbour. Bowker took no chances and blasted it with cannon. He rejoined 430 two kills richer. On July 2, 1944 he was busy on ops, dive bombing around Caen. That afternoon 430 shot down 4 German fighters, but they lost their ace, Bud Bowker.

W/C J.E. "Johnny" Johnson, who headed No.144 Airfield Wing in 1944-45, was legendary among his three RCAF squadrons. After the war he kept in touch with many Canadian friends, often flying across from England to visit. In June 1998 he attended the final annual convention of the Canadian Fighter Pilots Association, held in Ottawa. Johnson's men always viewed him as a fine leader (he flew with his wing on every occasion, amassing a formidable score), but he was implacable regarding weakness. One mistake and a pilot could expect a reprimand, or worse. In the air Johnson was a cool, ruthless warrior. Even so, he knew when to call it quits. After shooting down an Fw.190 near Villers-Bocage on June 16, 1944, he made himself scarce, later reporting: "I climbed into cloud as I was being engaged by intense and accurate light flak."

Johnson knew day by day what each man in the wing was accomplishing. On June 22, 1944, for example, there was a skirmish near Argentan between 144 Wing and 8 e/a. Six of the Jerries didn't go home. One-third kills on an Fw.190 went to FSgt R.A. McMillan, F/O W.W.L. Brown and F/O W.R. Chowen of 441. F/O J.W. Fleming (441) bagged a 190 ("I flew through the debris and upon returning to base found that a piece had been knocked off the propeller."). W/C Johnson, F/O J.T. Marriott, F/O W.R. Weeks and P/O F.B. Young of 442 each got a 109. Marriott's Hun bailed out. Later, Johnson reviewed the gun camera film of this turkey shoot. He noticed that S/L B.D. Russel also had made hits on the 109 later attacked by Marriott, so allocated each man a half-kill. Typical of W/C Johnson's terse combat reports is this (May 5, 1944):

*I was leading 144 Airfield on Ramrod 831, sweeping in the Lille area. When over Douai I saw 6 Fw.190s flying over the town at ground level. I detached 443 Squadron (S/L McLeod) to go down and search for these a/c. About 5 minutes afterwards I saw an Fw.190 flying west at 2000 feet. I closed from his port side and opened fire from 300 yards, closing from 20 degrees angle off to 5 degrees angle off. After 1-second burst e/a jettisoned hood and tank, and pieces were seen to fly off. I continued to fire and the pilot baled out at 400 feet and, unfortunately, landed safely by parachute. E/a crashed in a field 3 miles east of Douai. Claimed as destroyed. Rounds fired: 100 cannon, 480 mg.*

Any dog fight in a high performance fighter means plenty of "yanking and banking"—10 Gs can be pulled. This can be painful and disorienting, and cause loss of consciousness. High G is illustrated in a combat report from June 28, 1944. S/L H. J. Dowding of 442 Squadron was near Cabourg about 1700 hours, leading 6 Spitfire IXs. About 15 Fw.190s suddenly roared onto the scene and the classic fur-ball erupted, with fighters everywhere. Dowding latched onto one: "He half-rolled and went down doing aileron rolls. I followed, giving him the odd squirt, but didn't observe any results, as I was in a semi-blacked out condition most of the time. I broke off the attack at about 3-4000 feet and climbed up to the shambles again." G could have killed Dowding, as it had so many aircrew since WWI. The deadly effect of "G loss of consciousness" was not taken seriously in WWI and WWII. Today, it is a big topic wherever fighter pilots train. "G-LOC" is strongly suspected in the deaths of several Canadian CF-5 and CF-18 pilots since the 1970s.

On September 18, 1944 F/L Ron Lake was on a sweep in the Eindhoven-Aachen area at 12,000 feet when a dozen Me.109s appeared. He closed on one, firing. Its pilot weaved, allowing Lake to close, firing all the time. The e/a rolled straight in. On September 27 Lake shot down an Me.109 from a gaggle of 20: "There were strikes on the starboard wing and in the cockpit. As I broke away, his cockpit was in flames and the a/c dived into the ground at about a 70 degree angle." The same day P/O S. Bregman of 441 knocked down a 109. He peppered the target: "Pilot tried to bale out, but chute failed to open..." Lake caught two Me.262s at 9000 feet on September 30 east of Nijmegen. One disappeared in cloud. He fired on the other, observing a large piece falling off. The 262 pulled away trailing smoke, still much faster than Lake's Spitfire IXE. Lake reported: "At the beginning of the engagement the other e/a was in an excellent position to fire on me, but it did not do so."

# An Airman's Album

LAC Walter Edgar "Ted" Freeman, who was in 412 Squadron's servicing echelon from early days at Wellingore, saved many photos in a neatly-kept album. When he passed away in British Columbia in the 1990s, his album was saved from the trash bin by aviation fan Brian Musson. This selection is a small tribute to both 412 and Freeman. First is a postcard view of Wellingore, where 412 was stationed early in the war. Freeman's note on the back reads, "The Marquis of Granby and the Red Lion. I spent a lot of nights in them last winter with the boys."

An impromtu group shot at 412—14 erks with 2 pilots. Behind are P/O Stevenson, Doug Park, Bill Redie, Tommy Rusenstrom, Ted Freeman, Gordon Roach, FSgt G.S. Joe Gould (later KIA), Al Beam, Wilf Clark and Bruce Cox. In front are Frank Bauer, Johnny Quirk, Colin Wood, Percy Perry, Dusty Millar and Ron Penney.

FSgts Stewart William Pearce and George Stanley Gould in early 412 days. On April 10, 1942 Gould, age 21 from Toronto, died in a mid-air collision. On December 12 Pearce, age 23 from Toronto, was shot down over the Channel, while engaging Fw.190s.

Pilots of 412 "Falcon" Squadron at Heech, Holland in November 1944. Some were legendary figures in the day fighter world, as was Don Laubman (15-0-3). Back row: F/Os William H.L. Bellingham, George Harris, Louis Dunklemen, F/Ls F.T. "Freddie" Murray (5-1-2½), and James Basil "Joe" Doak, F/O W.J. "Bill" Walkom, F/O Vic Smith (on the cowling), F/L R.N. "Bob" Earle, F/O W.A. "Bill" Aziz, F/Ls Wilfred J. "Bill" Banks, DFC (9-3-1), Lloyd A. Stewart, Bill "Pappy" Barber and F.H. "Joe" Richards. Standing: F/L Bruce E. Macpherson, S/L Dave Boyd (the CO), P/O William Cowan, F/Ls Charlie Fox, DFC and Don Laubman, DFC and Bar, S/L Dean Dover, DFC and Bar, F/Ls P.M. "Phil" Charron, DFC and David R.C. Jamieson, DFC and Bar (8-0-1), F/Os McLeod and Bill Garson. September 26, 1944 was a red letter day for 412 Squadron, Banks, Bellingham, Charron and Laubman all registering Fw.190s or Me.109s shot down. Next day Fox and Jamieson each got 2 Fw.190s. F/O Aziz got one on the 29th. Bellingham, age 21 from Magog Quebec, and Charron, age 25 from Montreal, were lost on November 19, Charron while taking on a flock of Fw.190s. On November 26 Murray destroyed a Mustang in Luftwaffe markings. He got an Me.109 on December 5. On the same sortie Banks got 2 Me.109s and a probable. On January 1, 1945 he shot down a Ju.88—his 9th aerial victory. Richards destroyed an Fw.190 on December 14 (he was lost on January 20). Charlie Fox got a Ju.88 on December 29, while Murray tallied an Me.109. On January 14 Fox, who was on the last sortie of his tour, was shot down by flak, but survived. Stewart had occasion to bail out twice, first with 1435 Squadron over Sicily, when an Me.109 damaged his Spitfire, then with 412, when he had engine trouble. His Spitfire survived the latter event, flying into a tree where it hung up in one piece! On February 25, 1945 Fred Murray made a forced landing, while Cowan, age 25 from Lethbridge, had engine failure, bailed out, but struck the tailplane and was killed. On the New Year's Day raid, Doak, a 23-year old from Cowansville, Quebec, downed an Fw.190. Minutes later he was shot down and killed. Before going overseas Doak had been CO of 133 Squadron (Hurricanes) on Canada's West Coast. To get overseas he took a cut in rank, but felt that it was worth it to escape Tofino. Macpherson and Walkom went down on January 20, the first surviving, the second not. Earle, age 24 from Oliver, BC, was killed by flak near Haus, Germany on December 31, 1944, while attacking a rail car carrying a V-2.

Freddie Murray looking very operational. After the war the New Brunswicker married an American girl and settled in Erie, Pennsylvania, where he had a career in forest products.

Charlie Fox pokes his head through a gash in the wing of VF-F. He had been operating around Nijmegen from B.68 Le Culot, Belgium, when an Fw.190 shot him up. G/C G.R. McGregor of 126 (F) Wing recognized this good show by giving Fox a green endorsement. This had not been Fox's first close call. Flying a Harvard at Bagotville in June 1943, he was bounced by a Hurricane flown by P/O Sydney Smith Buckley of Ottawa. They collided—Buckley died, Fox bailed out. In 2000 Charlie Fox was active in the Canadian Harvard Aircraft Association, and involved revitalizing the BCATP airfield at Dunnville.

F/L E.C. "Lucky" Likeness of Ottawa and F/O Harlow W. "Bud" Bowker of Granby, Quebec were noted on 412 for aggressive ground attack work. Likeness (whose name had been Leichnitz) had joined the RCAF in 1937, serving first as a clerk. Remustered to pilot, he instructed at Dunneville. After shooting down an Fw.190 over France on May 10, 1944, he was hit, bailed out and became a POW. Bowker got an Fw.190 on October 3, 1943, while escorting Bostons near Paris. On November 8 he was wounded while strafing Me.109s at Lille. On July 2, 1944 he died on a dive-bombing sortie around Caen. (CF UK10737)

F/L Andrew Boyd Major Banks Ketterson of Montreal was 22 when shot down and killed near Ypres on March 4, 1944. He had been escorting American B-17s in Spitfire MJ306.

P/O Lloyd W. "Pip" Powell, DFC of 412 Squadron also was renowned for success against rail and MT targets. While on convoy patrol off the Scilly Islands on June 17, 1943 this 20-year old Edmontonian was shot down and killed by a 414 Squadron Mustang. On the same do F/O Frederick L. Vaupel of 414 Squadron (age 30 from Port Elgin, Ontario) was shot down by a Spitfire. One wonders, but somehow Spitfires and Mustangs got into a tragic balls-up. The previous summer F/O Vaupel's brother Raymond had died in a 148 Squadron Wellington.

Spitfire VB BL300 of 412 Squadron crashed at Wellingore on April 15, 1942.

Spitfire LFIXc MH883 of 412 was flown by George "Buzz" Beurling from December 1943 to February 1944. It later flew on 302 (Polish) Squadron and in 1947 went to Turkey. Charlie Fox, who was on his flight, recalled Beurling as a superb individual. He was very big on training, constantly urging his pilots to hone their skills in deflection shooting. "Needless to say, Buzz was deadly when it came to shooting pool in the mess", recalled Fox in 2000. "He never drank or smoked, but that didn't deter him when it came to having a good party. It was an honour to fly and fight with him."

LAC Ted Freeman (right) on his wedding day in England.

# Spitfire Gallery

Spitfire Mk.LFIXCs MK464/Y2-Y and MK777/Y2-Z of 442 "Caribou" Squadron roar off in Normandy. MK464 served after the war in France; MK777 in Belgium. The main task of the RCAF Spitfire wings after D-Day was armed recce—dive bombing and strafing. In this low-level environment, it was dangerous no matter what a fellow was flying. At first the region was rich in targets. July 19, for example, saw 127 Wing with claims of 60 MT flamers, 78 MT smokers, 84 other MT damaged and 6 tanks damaged. (CF PL33506)

Spitfire XIVE RN119/AE-J of 402 Squadron sitting on pierced steel planking in Holland. "PSP" let the Allies get right into air operations on the beachhead after D-Day. (CF PL42423)

Spitfire Mk.FIXC BS306/AE-A of 402 ready for a sortie from Kenley on November 24, 1942. AE-A first flew on August 24, 1942, going to 402 on the 26th. It later served with 416, then 421, 302 and 33 Squadrons. On May 22, 1944 it was lost in a mid-air collision with MJ309 at Mountfield, Sussex. (CF PL15055)

Armourers kept up with demands on any push. These were preparing 20-mm belts. In good weather Spitfires like these each would make several sorties a day. Besides armourers, the refuellers, mechanics, cooks, scheduellers and others would go full tilt from dawn to dusk. Spitfire Mk.V DB-S/EN579 of 411 Squadron appears in this view. It served postwar in Turkey. (IWN CH13319)

Spitfire XIV RM619/AP-D of 130 Squadron. In the spring of 1945 this RAF squadron was headed by a Canadian, S/L George C. Keefer, DSO and Bar, DFC and Bar. AP-D was shot down near Aachen on January 16, 1945, S/L Phil Tripe (RCAF) becoming a POW. (IWM CL1353)

Winter ops at Evère. AU-Y of 421 Squadron waits to start as a Hudson departs. The ground trades had an especially tough job when the weather was foul, but operations kept going. (IWM CL1779)

Mk.VIII AN-F/JF526 of 417 Squadron waiting to start at Canne, Italy on January 10, 1944. Beginning in June 1942 under S/L Paul Pitcher, 417 operated in Egypt, out across the Libyan and Tunisian deserts, on to Sicily, and through Italy, finishing at Treviso, north of Venice. There it disbanded on June 30, 1945 under S/L Dave Goldberg. (CF PL18758)

A water colour by Goranson showing 417 Squadron Spitfires at dispersal. (CF PL47612)

Spitfire PL775 was a Mk.XI on 541 Squadron, a PR unit that had its share of RCAF pilots. (IWM CH13497)

Two privately-owned Spitfires flew in Canada. CF-GMZ (ex-TZ138), sponsored by Imperial Oil, participated in the 1949 Tinnerman Race in the US. It did not return to Canada. Then, in the early 1960s "Red Indian" veteran, John N. Paterson, purchased this ex-Belgian Mk.LFIX. He flew it briefly from Fort William as CF-NUS (ex-NH188), then donated it to the National Aeronautical Collection for display in 421 colours. On June 15, while on a patrol led by F/L John McElroy, Paterson had shot down 2 Me.109s over Normandy. (NMST 10441)

# Spitfire Personalities

Canada's greatest ace of 1939-45 was George "Buzz" Beurling, DSO, DFC, DFM and Bar. Born in Montreal in 1921, he started flying at age 16. Anxious to get into the war, but turned down at an RCAF recruiting office, he sailed for the UK to join the RAF. September 1941 found him with 403 Squadron, but things began to perk only when he was posted to besieged Malta. Beginning in July he amassed a stunning record against the Germans and Italians. Next, he went home on a publicity tour. As a warrior and "lone wolf" he loathed this. Later he flew on 403 and 412, but the war was winding down. In the end Beurling tallied 31-0-9. Finding that peace was not his cup of tea, he signed on to fly for Israel. Taking off from Rome to fly there in a heavily laden Norseman, he crashed fatally on May 20, 1948. Various books have dealt with Beurling's exploits, by far the best being *Malta: The Spitfire Year, 1942*. (CF PL14940, NAC C85962)

G/C James E. "Johnny" Johnson (RAF), DSO and Bar, DFC and Bar was intimately associated with the RCAF. Having begun on 616 Squadron, he became OC at 610 in July 1942. March 1943 to April 1945 he headed 127 and 144 Wings (RCAF). An all-round good type, Johnson, as one of his men recalled in 1999, once led in singing the raunchy "North Atlantic Squadron" at a mess dinner in Celle. He kept in touch with his RCAF comrades into the 21st Century. (CF PL30459)

Percival S. "Stan" Turner (14-2-6) was in a class of his own. A CAN/RAF man, he fought first with 242 Squadron in the waning days of the Battle of France, scoring first on May 29, 1940—2 Me.109s destroyed. Next for 242 came the Battle of Britain. On September 15 alone it claimed 12 e/a, Turner getting an Me.109 and a Do.215. Subsequently, he was CO of 145 Squadron at Tangmere, 249 in Malta, and 134 and 417 in North Africa. All the time he steadily increased his score, but didn't always report a kill. W/C Turner finished the war in charge of 127 Wing. Afterwards, he held such RCAF posts as Air Attaché in Moscow. He retired in 1965. (CF PMR77-525)

Another key RCAF Malta pilot was P/O Claude Weaver of Tulsa, Oklahoma. Along with 30 other Spitfire pilots he reached Malta from the aircraft carrier *Eagle* on July 15, 1942. Two days later he made his first kill, an Me.109. He claimed 4 more on the 22$^{nd}$ and 23$^{rd}$ (Weaver has be called a "line shooter" regarding kills, e.g. no enemy unit reported losses for the time he was airborne on the 22$^{nd}$. The record is carefully documented by Shores, et al.) Flying a "Hurribomber" on July 27 (led by W/C Turner), Weaver attacked an airfield in Sicily. On August 27 he was credited with a Ju.88 kill. On September 9 he was shot down by a Macchi 202, belly-landed on a Sicilian beach and was captured. By this time he had 10½ confirmed kills. He escaped in September 1943 and returned to the fray, joining 403 Squadron by November. He shot down an Me.109 on December 31, an Fw.190 on January 21. On January 28, 1944 Weaver died in action around Amiens. (CF PL18481)

W/C J.F. "Stocky" Edwards, DFC and Bar was a superb pilot and leader (13-5-8). A Saskatchewan boy, he originally proved his mettle in North Africa, showing that the slower Kittyhawk could be deadly against the nimbler Me.109. In March 1944 Edwards became CO of 274 Squadron in Italy, but soon 274 was posted home on Tempests. Edwards finished his war on 127 Wing. Postwar he commanded 430 Squadron (Sabres) in France; he remained in uniform till 1971. (CF PL10237)

Among many RCAF men in the Mediterranean, Middle East and North Africa were these rough 'n ready-looking pilots from the North African desert in September 1942: P/O John E. Avise, Sgt W.J. Steele, FSgt H.M. Crompton, FSgt John R. Rebstock, F/O A.U. "Bert" Houle, P/O F.A. Wilson and FSgt William H. Stephenson, DFM. Avise (103 MU, from Mason City, Iowa) died ferrying a Hurricane in North Africa on October 17, 1942. Rebstock (213 Squadron, from Crystal Beach, Ontario) died in a Hurricane in Libya on November 19, 1942. Bert Houle (DFC and Bar, 11-1-7) later commanded 417, and was a leading figure in the postwar RCAF, e.g. with CEPE. Shot down on October 21, 1942, Stephenson became a POW. (CF PL10065)

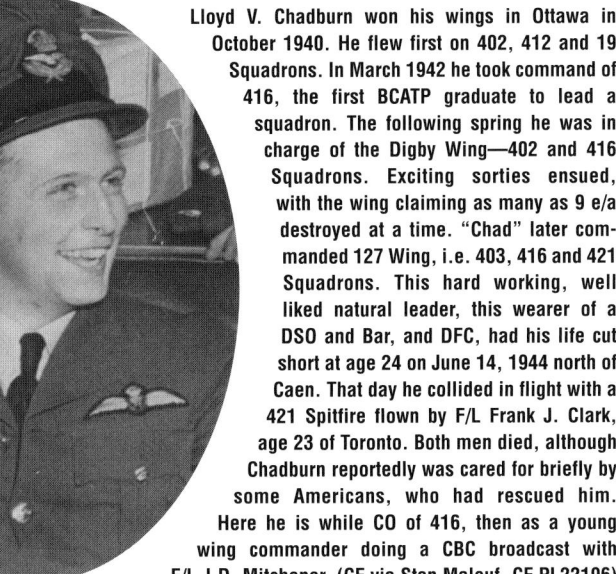

Lloyd V. Chadburn won his wings in Ottawa in October 1940. He flew first on 402, 412 and 19 Squadrons. In March 1942 he took command of 416, the first BCATP graduate to lead a squadron. The following spring he was in charge of the Digby Wing—402 and 416 Squadrons. Exciting sorties ensued, with the wing claiming as many as 9 e/a destroyed at a time. "Chad" later commanded 127 Wing, i.e. 403, 416 and 421 Squadrons. This hard working, well liked natural leader, this wearer of a DSO and Bar, and DFC, had his life cut short at age 24 on June 14, 1944 north of Caen. That day he collided in flight with a 421 Spitfire flown by F/L Frank J. Clark, age 23 of Toronto. Both men died, although Chadburn reportedly was cared for briefly by some Americans, who had rescued him. Here he is while CO of 416, then as a young wing commander doing a CBC broadcast with F/L J.D. Mitchener. (CF via Stan Malouf, CF PL22196)

Top RCAF Spitfire pilots at a Buckingham Palace investiture: W/C Robert W. "Buck" McNair, DSO, DFC and Bar; S/L Robert A. Buckham, DFC and Bar, US DFC; and W/C Hugh C. Godefroy, DSO, DFC and Bar. By war's end their scores were 16-5-14, 6½-0-3 and 7-0-3 for a total of 29½-5-20. Having begun on 411 Squadron, McNair fought in Malta with 249, then served on 411 and 403, and commanded 416 and 421. He was fished from the Channel (for second time) after bailing out on July 20, 1943. Finally, he commanded 126 Wing. He served postwar till his death in 1971. Buckham flew on 416 and 421, then commanded 403 Squadron and 127 Wing. He died in a flying accident in the Yukon 1947. Godefroy first flew on 401 in April 1941, led 403, was wing leader on 127 Wing and on A/V/M Broadhurst's staff at 83 Group. Like McNair, he was entitled to wear the Goldfish pin, having bailed out into the Channel. Returning home, Godefroy became a medical doctor. His autobiography *Lucky 13* is one of the best by an RCAF fighter pilot. Otherwise, the stories of such RCAF aces are best found in Hugh Halliday's *The Tumbling Sky*. (CF PL22285)

Inset, Hugh Godefroy at the 1998 CFPA barbecue held at Ian Ormston's farm near Kitchener, Ontario. (Larry Milberry)

Nova Scotians F/L Patrick T. O'Leary (Armdale), S/L Leslie S. Ford, DFC (Liverpool) and F/L G.U. Hill (Antigonish) during 402 Squadron days at Catterick in November 1942. On August 19 they were over Dieppe, where each scored, Ford getting 2 Fw.190s. O'Leary was lost off Dover on February 27, 1943. Ford would tally 6-0-2⅓ before dying off Holland on June 4, 1943. In March 1944 W/C Hill, DFC and 2 Bars (10-3-10), took over 441. He was forced down on April 25, evaded for a month, was captured, then liberated. Back home he studied medicine, became a renowned community leader, but died in a highway accident in November 1969. At his burial in Nova Scotia an Argus patrol bomber flew over to honour this great warrior and premier citizen. (CF PL15243)

Born in Claresholme, Alberta in 1919, F/O James E. Walker, DFC and 2 Bars began his career in 1941 flying Hurricanes with 81 Squadron in northern USSR. On September 12 he claimed his first victim, an Me.109. By November he was back in England, then he moved to North Africa. Early in 1943 took over 243 Squadron in Tunisia. After leave in Canada (where he married), he won command of 144 Wing in England. On April 26, 1944 he died after crashing an Auster near Tangmere. *Aces High* states Walker's score as 9-4-12, plus some shared victories. Here he is shown in North Africa astride a Matchless motorcycle. (NAC C37102)

F/L Richard J. "Dick" Audet, DFC and Bar (11½-0-1) of Coutts, Alberta joined the RCAF in August 1941. After EFTS in Quebec City and SFTS at Uplands he endured a long penance towing drogues. It was September 1944 before he was posted to 411 Squadron at Evère, near Brussels. There was lots of flying for the well-liked, sharp-minded Audet—52 sorties, but no chances to fire his guns. That changed on December 29 when, led by S/L E.G. "Irish" Ireland, 411 blasted out of Heech looking for trouble. It came in a gaggle of 109s and 190s. Within 2 minutes Audet shot down 5 to become an ace in a day! Other kills followed, e.g. on January 22 Audet destroyed an Me.262 on the ground, another in the air. On February 8 he had to jump from his flak-damaged Spitfire. Then, on March 9 the 22-year old was killed by flak, while strafing a rail target. (CF)

(Right) Beginning in October 1941 F/L Harry Deane MacDonald, DFC and Bar, of Toronto served on 401, 402 and 403, besides instructing at 53 OTU. After home leave he returned to operations in late October 1943. While ground strafing on February 11, 1943 he flew through a tree, but got home OK. Things were worse on June 20, when 403 was escorting Bostons over France. MacDonald and 3 mates became entangled with some Fw.190s. In the end, only he was left. On November 30, while on a Ramrod over Holland, he radioed about his engine acting up. S/L L.M. Cameron of 401 flew alongside, urged him to bail out and alerted ASR. Cameron could see the HSL speeding out from the English coast. But the Spitfire suddenly nosed straight into the Channel before MacDonald could escape. Here he sits in a Spitfire displaying the crest of his Toronto high school. (CF PL19227)

More RCAF aircraft fell to flak than to fighters. Happily, many pilots survived this dreaded enemy. Here F/L Robert D. "Dagwood" Phillip, DFC of 416 Squadron inspects his flak-damaged Spitfire after a January 9, 1945 sortie. (CF PL41349)

S/L R.I.A. "Rod" Smith, DFC and Bar of Regina started with 412 Squadron in June 1941, but didn't get any excitement until joining 126 on Malta in July 1942. There he teamed with his brother, Jerry, already a seasoned warrior. The Smiths shared in some kills, but Jerry was lost on August 10. Rod tallied 6 kills on Malta before being stricken by illness. He resumed flying at an OTU in the UK, then took home leave. In September 1944 he became CO of 401. His final score included 13 confirmed kills plus 1 shared. Postwar he practiced law and flew with the RCAF reserve. One good story is how Smith once sat the diminutive intelligence officer, S/L Hart Massey, on his knee and took him flying in a Spitfire! (CF PL29398)

P/O John Gillespie Magee of Washington, DC, who penned the famous poem "High Flight". A 412 "Falcon", he died on December 11, 1941 in a collision over England with an Oxford. (CF PMR82-238)

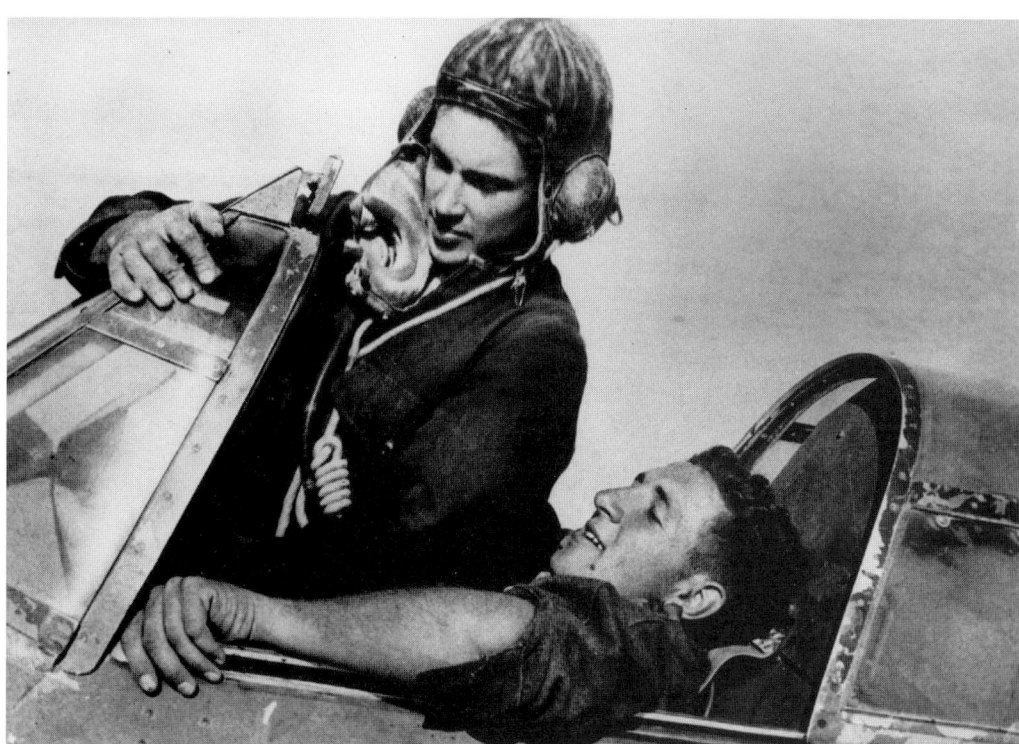

Born in 1921 of an American father and Canadian mother, George C. Keefer, DSO and Bar, DFC and Bar joined the RCAF in October 1940. In less than a year he was on Hurricanes with 274 in North Africa. There his most memorable sortie came on June 4, 1942. Spotting someone waving in the desert, he landed to meet a downed South African, Lt John Lane. The pair squeezed into the Hurricane (as demonstrated here, Keefer standing, Lane seated) and flew to safety. After 197 sorties on 274 Keefer joined 416 in April 1943, then commanded 412. One day he found himself in the Channel after bailing out. By good fortune a Walrus saved the day. Next, he was Wing Commander (Flying) on 126 Wing. On June 7, 1944 he bagged a Ju.88 and an Fw.190. After leave in Canada he headed 125 Wing (RAF). His final tally was 12-2-9. Postwar, Keefer worked at Canadair, then ran his own small business. (NAC C35623)

F/L Alan F. Aikman, DFC and Bar (8-1-4), while on Spitfires with 154 Squadron in North Africa. He already had fought over England and France, and was wingman to Paddy Finucane in July 1942, when the great ace was killed. Aikman's first North African victory was a Ju.88 in October 1942. Later he flew a tour on Dakotas with 436 Squadron in Burma. After the war he became a TCA/Air Canada captain. He died in St. Catharines, Ontario on March 21, 1991. (CF PL10283)

# Later Get Togethers

One of the best annual "dos" for the Canadian Fighter Pilots Association was at the farm of Ian Ormston (401, 411 Squadrons) near Kitchener, Ontario. In this 1999 view at Ian's, Spitfire pilots R.L. "Dick" Reeves (129, 403) and M.F. "Mike" Doyle (401), and Robert Bracken, author of *Spitfire: The Canadians*, look over some wartime photos. (All, Larry Milberry)

"Irish" Ireland (411, 442), Eric G. Smith (107, Mosquitos) and F.W. "Fred" Evans (421) at the 1998 CFPA convention in Ottawa. After the war Irish flew such types as the CF-100 and Voodoo; Eric flew Sabres and CF-100s; Fred, Sabres. Eric and Fred each did a Sabre tour in Korea.

Old friends from Spitfire days: J.D. "Danny" Browne (403, 421, 441), A/V/M Johnny Johnson and H.L. "Tex" Phillips (112, 412) reminisce at the 1998 CFPA reunion.

More of the 1998 CFPA reunion: Dean Kelly (249, 403, 412, 609), Hart R. Finley (1, 403, 416, 443), Joe Schultz (410, Mosquitos), Eric Smith (107), R.S. "Bob" Hyndmen (411) and Lloyd E. Hunt (93, 680, 443). These hard-fighting warriors did their share, wrecking plenty of the Third Reich's assets. After the war Schultz led in CF-100 and Voodoo operations, and finished as a leading man in flight safety. For the latter he was awarded the 1978 McKee Trans-Canada Trophy. Hyndmen, a Spitfire pilot turned official war artist, continued in art as a civilian. Dean Kelly led the way with Sabres and Voodoos, then finished his career spraying and fighting fires in clunky old Avengers. As the consummate aviator he enjoyed every minute of it.

Keith Scott (116, 403, 416), Chuck Darrow (416), C.T. "Cec" Brown (403) and Dick Reeves (129, 403) at a 1998 Remembrance Day lunch at the Toronto Cricket Club. After the war Keith had worked in PR at Ontario Hydro, Chuck in machine and auto parts, Cec in industrial relations at Ford, and Dick in construction supplies.

Douglas "Duke" Warren, DFC with J.R.D. "Joe" Schultz, DFC and Bar at the 1998 CFPA reunion in Ottawa. Duke and his identical twin, Bruce, flew Spitfires on 66 and 165 Squadrons, then stayed in the postwar RCAF. Duke flew Sabres in Korea. Bruce lost his life while test flying the CF-100 Mk.I. The story of the Warrens is found in *Royal Canadian Air Force at War 1939-1945* and *Fighter Pilot: A History and a Celebration*. Joe's story is summarized in our Night Fighter chapter.

Spitfire pilots A.R. "Andy" Mackenzie, I.F. "Hap" Kennedy and Larry Robillard at the Chateau Laurier in 1998. Andy flew Sabres in Korea, but was shot down and suffered a long imprisonment at the hands of his sadistic North Korean captors. Hap's book *Black Crosses off My Wingtip* is one of the outstanding RCAF WWII autobiographies.

## A Typhoon Pilot

The Hawker Typhoon was a very challenging fighter. Several books exist about it, chiefly *The Typhoon and Tempest Story* by Thomas and Shores, and *The Hawker Typhoon and Tempest* by Mason. The most valuable for RCAF purposes, however, is Hugh A. Halliday's *Typhoon and Tempest: The Canadian Story*. The Typhoon began with a 1937 concept to replace the Spitfire and Hurricane, themselves only in their infancy. The proposal was for a heavily-armed, 400-mph fighter. One concept, the Tornado, flew in October 1939 with a 1760-hp Roll-Royce Vulture. A version with the Napier engine flew in February 1940—this was the Typhoon. Both proceeded well, but production was delayed in favour of Hurricane and Spitfire deliveries. Meanwhile, there was an order for 500 Tornados and 250 Typhoons, but the former was cancelled due to poor Vulture performance. The first production Typhoon flew in May 1941; the type entered service in July 1942 with Nos.1 and 257 Squadrons.

The Typhoon soon proved no match for the nimble Me.109s and Fw.190s. Its forte became ground attack, squadrons specializing in delivering rockets or bombs. Occasionally, however, a Typhoon succeeded in air-to-air combat. Such an occasion occurred on May 18, 1944. No.183 Squadron was on a Day Ranger across the Channel in the Boissy-Beauvais area. Two Me.109s were encountered. F/L F.H. Pollock (Kitchener, Ontario) later reported: "In the dogfight that followed, I approached one Me.109G from 15 degrees till finally head-on ... I opened fire at about 200 yards, closing to about 30. I observed part of his starboard wing fall off before we broke. Following this, W/C [Bryan], fired at him and he exploded in mid air." On June 7 Pollock was shot down, but survived as a POW.

Many Canadians served on Typhoons; they suffered high losses. One who survived was Stafford Dean Marlatt. By war's end he had flown 2 tours on "Tiffies", plus 2 stints on Hurricanes, and logged 608:10 hours. "Staff" Marlatt was born in Oakville, Ontario on July 17, 1915. Following his studies at Appleby College, he started with the Imperial Bank in Toronto, but in October 1940 joined the RCAF. Training began on Fleet Finches at No.12 EFTS at Goderich, where he first flew on January 7, 1941 in Finch 4696 with instructor Len Fitton. EFTS finished on February 24. Years later he related how Fitton once had muffed a simulated forced landing demonstration. Misjudging, he got too low, their Finch ripping through a wire fence. Fitton told Marlatt that he was free to choose a new instructor, but the sprog, struck by Fitton's honesty, was delighted to stick with him.

Marlatt next went to No.1 SFTS at Camp Borden, where F/L Sharpe took him up on his first Harvard ride on April 1, 1941. Marlatt found the going rough in night flying and, reminiscing in 1995, credited his enlightened instructor, F/O Joseph W. Weis of Toronto, with getting him through. Rather than quitting on his student, Weis encouraged him. On May 15 Marlatt earned

SFTS graduates at Camp Borden with Wings freshly sewn onto their tunics. Stafford Marlatt is on the right. Beside him is Edwin Herbert "Eddie" Glazebrook of Valois, Quebec; then Percy Walker of Toronto. While Marlatt survived 2 Typhoon tours, his friends did not come home. Overseas, Walker joined 405 Squadron. On October 21, 1941 he was on night ops against Le Havre, France in Wellington Z8419. His plane and crew disappeared. On May 18, 1942 Glazebrook arrived on Malta, having taken off earlier that day from HMS *Eagle* with 10 other Spitfires. He soon was in the thick of it with 229 Squadron, fending off an Axis onslaught. On July 2 his Spitfire was damaged in combat, forcing him to belly land. On October 11 he shot down a Macchi 202. Next day he claimed a Ju.88, another as a probable, a third as damaged. On the 14th he damaged a Ju.88, then in a mêlée on the 15th, accidentally fired on W/C Thompson. Glazebrook's aim was good—Thompson crash-landed at Hal Far. Nonetheless, Glazebrook was awarded the DFC a few days later. On October 31, he departed Malta on a Liberator. Besides a crew of 6, there were 34 passengers—24 pilots (Buzz Beurling included) and 10 civilians. The Liberator crashed into the sea while overshooting at "Gib". Among the 14 who drowned was Eddie Glazebrook. (All, Marlatt Col.)

Going overseas aboard *Ausonia*. The ship sailed from Halifax, stopped in Iceland, then proceeded to Greenock.

his wings. Weis later went overseas. On the night of June 8, 1944 he was flying a 408 Squadron Lancaster on a raid to Achère, France. He and his crew were lost. From Borden, Marlatt went overseas, sailing on June 18, 1941 on the *Ausonia* from Halifax. It was slow going, sometimes no faster than 4 knots. The convoy put in to Iceland for several days. The RCAF contingent finally got away aboard a second vessel, reaching Greenock on July 12.

After Bournemouth, Marlatt was posted to No.61 OTU at Heston, his first flight being on July 30 in a Harvard. He was shocked at its rough condition, compared to Camp Borden's well-kept aircraft. After studying the cockpit for hours, manual in hand, and passing a cockpit blindfold test, Marlatt took up his first Spitfire (X4346) on August 6. Now came an intensive course, flying 2-3 times daily. He finished on August 31, ready for action with his 48:45 hours on Spits. He was posted to 247 Squadron at Predannack, Cornwall.

Marlatt, sure that he was going to a Spitfire squadron, was dismayed on reaching Predannack to find that 247 was a Hurricane squadron, but he adapted to squadron life. There were some daylight convoy patrols, but 247 mainly went on the prowl after dark. As there was limited ground radar control, and the Hurricane was unequipped for night fighting (other than having a coat of black paint), there were no successful intercepts. The Hurricanes often teamed with Douglas Turbinlites, equipped with searchlights in the nose. A Havoc would illuminate a target aircraft,

Pilots awaiting action on 247 Squadron during Hurricane days at Predannack near the southwest tip of England: Jimmy Greaves (RAAF); Allan "Gonfra" Burton (RAF) and "Wallers" Wallbank (RAF, with guitar). Behind is Frank Jones, an engine fitter. One day in the summer of 1943, Staff Marlatt was strapping into his Typhoon. Overhead, Burton was returning to land. Suddenly his Typhoon shed its tail. Burton didn't have a chance.

A pair of 247 Squadron night fighter Hurricanes.

When they were bombed out of their hotel in Exeter, 247 Squadron moved into the nearby St. George and Dragon Hotel. In 1998 Staff Marlatt recalled the bombing of Exeter and how one of 247's pilots saved some kegs of beer as the hotel went up in flames. These were set up nearby and served to people helping with the disaster.

This group constituted a 247 impromptu football team: P/O Ken Gear (RAF), Sgt Jimmy Greaves (RAF), Sgt Bruce McClellan (RAF), Sgt Staff Marlatt (RCAF), F/L Pete Pedigrew (RAAF), Sgt Allen Rene "Bob" White (RCAF), Sgt John H.A. "Jack" Ryan (RCAF), Sgt Russ Murray (RCAF), Sgt Forman (RCAF). The team was tossed together for a game with the army. Marlatt recalled: "Naturally, the army guys beat the hell out of us." White, from Amherst, Nova Scotia, was killed on night ops on June 27, 1942, having run out of fuel over France (Hurricane BN231). Ryan, a Toronto boy, was killed on May 25, 1942 while testing Hurricane BD951 at night near Exeter.

then a Hurricane would make a simulated attack. During Marlatt's time at this, there was little action. One night, however, a Hurricane shot up an RAF Halifax returning from ops. Other than this, Marlatt's main recollection was, "It was lots of fun".

On August 16, No.247 was detached to Middle Wallop to fly patrols pretending to be elements of other night Hurricane squadrons, detailed as "Hurribombers" supporting the ill-fated Dieppe assault. On September 21, No.247 moved to High Ercall, where it re-equipped with Typhoons. Marlatt flew his first on January 21, 1943. By this time his log showed 280 hours. The Typhoon took getting used to, especially its starting procedure, which could end with an engine fire. An erk was always standing by with an extinguisher. Then, takeoff had to be mastered. Going down the runway, a Tiffie swung hard to the right, so required left rudder to get airborne. Pilots also lived with the fear of engine failure. Their complex Napier Sabre, good only for 15-20 flying hours between overhauls, failed too often. When a pilot lost an engine, he made a quick exit, if not too low. Otherwise, he crash landed or ditched. Marlatt recalled that after each sortie on 247 Napier technicians inspected each engine. With time, reliability improved.

No.247 began on daylight ops, chiefly cross-channel sweeps, with the predictable casualties. Allan "Gonfra" Burton was one of the first. He died on May 18, 1943 in one of the infamous cases where a Tiffie lost its empennage. Staff Marlatt watched as Burton crashed, EJ977's tail fluttering down. This defect, the result of elevator imbalance, was partially rectified using doubler plates. On May 28, No.247 flew to Gravesend, the first of several moves that left members feeling like gypsies. By early July they were at 124 Airfield near New Romney to fly top cover for "Bombphoon" (fighter bombers) squadrons. There were ops to places like Beaumont, Courtrai and St. Omer. The latter, a key Luftwaffe station, was such a frequent target that the fellows referred to their trips there as "practice bomb runs".

On July 31, 19-year old FSgt Charles D. "Mac" MacIntosh from Kitchener, Ontario was sent into the sea by flak near Gravesline, France. Marlatt recalled Mac as a good sort with a talent for cartooning. On August 15, FSgt Eric Stuart "Mac" McCuaig and Alec Robertson (RAF) collided at New Romney. Both died. McCuaig, age 20, was from Bethany, Manitoba. Towards the end of this tour, Marlatt and Ken Gear were the only two ex-Hurricane pilots left on 247. Marlatt was pining to get home, something his CO, a Norwegian, helped with. His last flight was on August 24, 1943; he was posted on the 29th, when his log showed 397 hours. After a few week's leave he joined 133 Squadron, a Western Air Command Hurricane unit at Tofino. It had few pilots with overseas experience and did most of its flying in daylight and good weather. Some of the fellows considered this to be very operational. Marlatt was unimpressed by this or by

A scene during field exercises as 247 worked up to D-Day: F/O Alec Robertson, F/O Staff Marlatt, F/O Alan Burton, F/O Andrews (MO), F/O "Brains" Burns (IO), F/O Russ Murray, P/O R.S. Colquhoun and F/L Gerry Gray (later CO of 182 Squadron). Murray had done his EFTS at Mount Hope, near Hamilton, and SFTS at Dunnville, a few miles south of there. One day while stationed at Exeter, his engine failed on takeoff. He survived the ensuing crash, but, too severely injured to return to ops, returned to Canada to instruct. He stayed in the postwar RCAF, instructing at places like Centralia and Trenton, retired in 1967 as a wing commander, then worked in administration at the University of Western Ontario. One day Sgts Colquhoun, a Montrealer, and Marlatt were on leave in London. Marlatt chuckled when Colquhoun suggested that they stay at the posh Royal Automobile Club. Nonetheless, they called there and were welcomed—Marlatt was unaware that Colquhoun's father was a well-known club member. On January 29, 1944 Colquhoun was in with 3 other Typhoon pilots in shooting down an Fw.200.

FSgt Eric S. "Crusher" McCuaig of 247 Squadron. Staff Marlatt recalled him as a well-liked lad from Bethany, Manitoba. On August 15, 1943, while flying Typhoon JP505 of 247 Squadron, he collided with another Tiffie (JP487 flown by F/L A.W. Robertson) near Hawkinge, England. Both men died.

the fact that the Officers' Mess was dry. It didn't help when the CO and the "Adj" would come in and have the bartender "release" two beers for them. "I couldn't believe it," was Marlatt's only response. The main consolation was that each man had a monthly ration of a quart of liquor and 12 beers. At month's end one of Tofino's Bolingbroke would fly to Victoria and convert everyone's ration coupons into a hefty load of beverages. When the Bolingbroke returned, a great bash would ensue.

In April 1944 Marlatt was posted to Bagotville, this time preparing sprogs to be fighter pilots. A student couldn't have had a better instructor. There was a lot of fun in this isolated part of northern Quebec. One break was delivering Hurricanes for overhaul to Scoudouc, New Brunswick. Marlatt and his buddies would ferry via Mont Joli and Chatham, having the odd party along the way. One day instructors De Nancrede, MacLure, Marlatt and Westcott were routing via the US base at Presqu'Île, Maine. They envisioned a night or two of "cutting loose" but, following MacLure (leading in a Bolingbroke—none of the planes had radios), they got lost. Below was rugged Maine, whose forests loved gobbling up airplanes. They backtracked, praying to reach the St. Lawrence River and Mont Joli. They barely made it; De Nancrede ran out of gas seconds after touchdown. That evening the drinks were courtesy of MacLure.

Again keen for action, Staff Marlatt put in for overseas, his posting coming through in July 1944. By then he was a 470-hour pilot. In November he was at No.61 OTU at Rednal, hoping for a Spitfire squadron. Just before Christmas he went to Tangmere for a Typhoon refresher at 83 Group Support Unit. There he met W/C "Pissy" Passey, who one day asked to see his log. "Oh," he said, "I see that you have been flying Typhoons." The last thing Marlatt wanted was another Typhoon tour, especially since he had met some buddies in a bar in London, who told him, "Stay away from that Canadian Tiffie wing—they're dying like flies." Resigned to his fate, Marlatt asked for his old squadron, but the winco turned him down—new rules were that RCAF Tiffie pilots could only be posted to the RCAF. One consolation was that the winco invited Marlatt to his New Year's Eve bash.

Next morning at 0900 Marlatt was to go by Oxford across to Eindhoven, where No.143 (RCAF) Wing was stationed; but after partying all night, he missed his flight. He rolled out of bed at noon to hear that Eindhoven had been plastered by the Luftwaffe. He soon joined 439 Squadron there, making his first flight on January 4, 1945. Now came a steady string of ops, hitting trains, rail lines, bridges, road convoys, observation posts, gun emplacements, etc. The Typhoons carried 500- or 1000-pound bombs, but Marlatt's log shows a new development for February 8: "First show with cluster bombs." These were small devices, many packed inside a single large canister. When released, the canister split, releasing the bomblets, which descended individually. They exploded before penetrating the ground, so caused mayhem among troops. Marlatt related how trains were attacked 90 degrees to the tracks, making it more difficult for flak gunners to make a hit.

Since casualties were high, promotions came quickly at 439. On January 14 F/O J.A. "Joe" Côté was shot down, but survived. Two days later F/O J.D. "Jack" Sweeney went down after flak tore up his plane. He became a POW. F/O W.G. "Bill" Davis went down on March 30, 1945. He was on his second sortie of the day—earlier he had been hit by flak, but squeaked back. This time his section was after MT. As Davis picked up a truck in his sights, he would have seen it stop, then noticed a soldier jump from the cab and man the machine gun on the back deck. A quick duel ensued and the German won— Davis went in. Had he beat the German to the draw, Davis had his ticket home. Instead, he was killed on the 96th and last op of his tour. On April 2 F/O Donald G. Cleghorn was shot down near Hamburg. He was picked up by Germans who marched him into town to the jeers and threats of the locals, furious at how the RAF had flattened their city. Flanked by soldiers, lucky Cleghorn evaded the wrath of the citizenry.

About this time the Canadian wing was amazed at being sent squadron-by-squadron to a gunnery camp at RAF Warmwell. True, it was a respite from ops, but the Canucks got a chuckle that someone "upstairs" thought they still needed target practice! On one of the last wing efforts of the war 143 Wing mounted a 36-plane raid on May 2 on Lubecher Buch, where some German naval vessels were anchored. Some pilots were anxious about this business. They got in and out in a hurry, knowing that the war was ending. Who wanted to be the last man shot down? The raid resulted in 3 hits. "Great", someone cracked, "3 out of 72!"

One less German tank! The RCAF's "Bombphoons" could bring such results. The scene was at Vimoutier, France in August 1944. (NAC, Army Col. 39051)

Soon No.143 Wing was packing. Staff Marlatt, who finished as a flight commander with a DFC, flew to England in a Halifax, passed through PRC and sailed aboard *Empress of Britain*. To his rue, Marlatt was soon at his old desk in the Imperial Bank. Twitchy to get back flying, he became a partner in a small air service in Northern Ontario. That didn't last long, and with one final flight in a little Fox Moth bush plane on June 16, 1947, Marlatt turned elsewhere; he made a 33-year career with the Brewers Warehouse Co. in Toronto, retiring in 1980.

Staff Marlatt with his fiancée, Pat MacPherson. Pat was greeting Staff at Toronto's Union Station upon his return from overseas.

The type of target that Typhoons regularly were assigned after D-Day. Without such bridges the Germans could not readily withdraw, nor bring up fresh men, equipment and supplies. (Murray Castator Col.)

# Happy New Year!

Shortly after sunrise on New Year's Day 1945 the Luftwaffe blitzed Allied airfields in liberated Europe with operation *Bodenplatte*. Eindhoven lost many aircraft, some of which are shown here. But there were few casualties and aircraft losses quickly were made good. The Luftwaffe, however, sacrificed hundreds of irreplaceable aircraft and pilots. Bodenplatte proved to be a wasted effort. In men alone Germany sacrificed 151 aircrew killed and 63 POW. RAF/RCAF combat losses were 7 Typhoons, 7 Spitfires and 1 Tempest for 12 pilots killed.

The scene at Eindhoven during the attack. The best book about Bodenplatte is Norman Franks' superb *The Battle of the Airfields*. Among aircraft destroyed at Eindhoven were 25 Typhoons, 20 Spitfires, 3 Mustangs, 4 Ansons, 3 Austers, 3 Bostons, 1 Hudson and 1 B-17. Dozens of others were damaged, many later being declared unrepairable. In 1999, 438 Squadron fitter H.M. "Hal" Goode of Truro, Nova Scotia recalled how losses were made good within 48 hours. He was surprised to see that pilots stepping from replacement Typhoons at Eindhoven often were women of the Air Transport Auxiliary. (Bert Walsh Col.)

The 439 dispersal at Eindhoven showing Typhoon devastation. (Murray Castator Col.)

A Spitfire's death throes at Eindhoven. (RAF/Nesbitt Col.)

The pilots of 440 Squadron at air firing camp in Warmwell on May 6, 1945. In the back are Bob Gregory, Jake Muff, Don French, Ted "TR" Smith, R.H. "Tiny" Wilson, Pres Pearson, Bob Dean, Benny Dunn, Angus "The Beast" Scott, I.L. "Gunnar" Gunnarson, Paul Bissky and Neil M. Hughes. In the middle are R.S. "Bob" Gurd, "Holly" Hollingsworth, Jimmy Kerr, Don "Red" McMillan, H.K. "Obie" O'Brien, Lynn Roach, John Villiers, S/L Bob Coffey, Bill Clifford, Dave Leach, Robert M. Gray, Denning E. "Fats" Waller, Walter J. "Mac" McCarthy, H.T.C. "Cro" Taylor and Ken Smith. In front are Hal Keon, Donald V. "Dinny" Wright and two unknown instructors. The next day 440 returned to Celle in time to celebrate VE-Day. (Clifford Col.)

## Bill Clifford

William C. Clifford was born on February 23, 1923 and raised in St. Catharines, in Ontario's Niagara region. Having joined the RCAF in 1941, he began flying at No.12 EFTS in Goderich on September 2, 1942. He soloed on the 15th, graduated, then took advanced training at No.1 SFTS at Camp Borden. Soon after earning his wings he embarked in Halifax aboard *Louis Pasteur* (a famous liner from the St. Nazaire - Buenos Aires run of earlier times), and reached Liverpool on July 1, 1943. He flew his first Spitfire (TR-A) at 61 OTU on June 2, 1944. This was a dream come true—the young pilot had always savoured. But the relationship was brief—from Rednal Clifford was posted to No.3 TEU at Aston Down for Typhoon training (first flight, August 30), then joined 440 Squadron at Eindhoven. There he was pleased when assigned to "A" Flight, headed by Jim Beattie, also from St. Catharines. One of Clifford's first impressions was the rate at which men were lost on 440: "One of the first pilots to make me feel welcome and less apprehensive was F/O Ron Doidge, killed in action a few days later. Ron had survived a bale-out in the channel on D-Day, and a crash landing with 1000-pound bombs in mid-July. He kept coming back and was due to be tour-expired when it happened... The constant change of personnel went with the territory." In 1999 Bill Clifford recalled a few details of life at 440:

*On October 27 we moved our billets to a convent school in south central Eindhoven. The nuns graciously moved up to the second floor. We set up bunks in the classrooms, opened a bar in the reception area, and took over the kitchen and dining room. The courtyard was used for parking and the airmen dug pits for fires to heat drums of water for our daily ablutions. They'd just pour a 5-gallon jerry can of high octane petrol into the pit and throw in a match. Before long the water would be warm enough to wash and shave ... After a hectic day it was always nice to come home, get cleaned up, then go to the bar for a scotch. Our messing officer was a wheeler-dealer. He often had unexpected treats, which helped make up for the tinned stew or fried Spam—our usual fare. A couple of times we had raw oysters in the bar before dinner, and juicy steaks for the main course. We didn't know where he got these goodies and no one asked.*

When things were quiet (usually because of weather) the pilots amused themselves in discussion, reading mail and writing home, playing cards interminably and listening to records at their dispersal, pistol shooting (neighbourhood rats were their targets), taking flips in the squadron Auster, even learning to drive the Wing's 1500 CWT and 3-ton trucks. Some nights a few of the boys would drive into Brussels to party, yet be ready next morning for ops. As usual, there was some black marketeering— selling cheap booze at inflated prices to the local bistros (there was no compunction about this, since these same establishments recently had been doing business with the Germans). On the other hand 440 could show its heart. For Christmas 1944 it put on a St. Nicholas party for the local children: "I'll never forget it. In the mess tent, tables were set up with individual places for each kid. When they saw the food and Christmas treats laid out, they grew silent, just not believing it was for them. We had to get a Dutch adult to explain things, but even then they ate sparingly. Everyone left with a doggy bag—we could tell that there were hungry people at home." On Christmas Eve F/Os Cummings, Dunkeld and Harwood of 440 were killed in action; five others from 143 Wing also went down. Midnight mass was somber. Meanwhile, the Battle of the Bulge raged. On Christmas Day F/O Clifford flew on two shows: "There was fresh snow in the hills and this made it easier to track down and fire on anything moving... Harry Hardy had his controls shot away, bailed out and was back in time for dinner."

Clifford spent part of January 1 sheltering in a ditch, while the Luftwaffe plastered Eindhoven. From there he watched as a Spitfire shot down an Me.109. The pilot bailed out: "His chute opened and he floated to earth, only to be picked off by an over-zealous sniper... We pilots always dreaded the thought of that happening to us." Although 143 Wing lost most of its Typhoons, it was not long before it was back in business: "All of 'A' Flight climbed into a Dakota and headed for Tangmere to pick up new planes. In Chichester that night we blew the heads off a few mugs of beer. When the publican called, 'Time, gentlemen. Please,' he was ignored. A couple of guys produced some French cognac from their battle dress. Understanding the situation, the publican locked the door from the inside and we pressed on through the night. On January 3 we collected our factory-fresh Tiffies at West Hampnet and ferried them to Tangmere. There they were armed and serviced and had invasions stripes applied." Weather delayed departure till the 4th, then the boys had to stop overnight at Courtai due to weather.

Like a lot of Canadians in Holland, Bill Clifford was befriended by a local family. His was the Tebacs, whom he often visited. In his words, "This got me away from the tense situation that often existed in camp, and my possible over-indulgence at the bar, as well." In gratitude Clifford would scrounge something for the Tebacs from the 440 larder. The losses continued at 440. F/Os Gibbs, Passmore and Warrell died in February. F/O Johnny Flintoff was forced down, sheltered by a Dutch family, then liberated ("It was unusual to get a lost pilot back."). As to how many ops constituted a tour, always a topic of debate and disagreement, for 143 Wing Clifford explained: "That was up to the CO and the Winco. Some were credited in the 60s, others went into the high 90s. It depended on the individual's ability to handle the tension, and the availability of replacements."

On March 24 all of 2 TAF was on its toes for the Crossing of the Rhine. Clifford led 4 Tiffies covering Dakotas and Stirlings dropping paratroops and towing gliders. A flak emplacement harassed the gliders, so Clifford's flight was called in: "We were in touch with ground control, which had grid maps matching ours." The Tiffies silenced the flak. W/C Frank Grant later passed on thanks from the Army. The wing moved to Goch, Germany on April 12. On the 18th Clifford led again, this time striking the fortified town of Stukr. The bombing was accurate; again, congratulations were forthcoming. On April 20 Clifford logged his 92nd op, then led 440 to its new digs at Celle: "The advance party, ground crew and service personnel had been on the road from Osnabruk for 2 days. They were on hand to direct

**Walter McCarthy, Bill Clifford and Ted Smith at Niagara-on-the-Lake in May 1999. (Larry Milberry)**

us to our dispersals, then serviced and fuelled our aircraft, ready for business at first light next morning." Soon, however, the shooting ended and 440 went on leave. Clifford flew to Rednal to visit some buddies. On departure 4 days later he gave Rednal a taste of the Typhoon, dive-bombing the tower from 8000 feet. Now 440 returned to Celle, where a victory party was held: "Barrels of liberated German port wine were tapped and even mixed with scotch whiskey to chug-a-lug. Derelict German aircraft were shoved onto the field and torched." A few days later Clifford flew to Eindhoven, his Tiffy's ammunition compartments crammed with food for the Tebacs. After that reunion, it was back to Celle, where regular armed patrols were being flown. These reinforced the Allied victory, and let nearby Soviet forces know that they should stay in their own airspace. Clifford led such patrols, first going high in line abreast, then descending to tree-top level: "This was a lot of fun, but very hazardous... Perhaps our euphoria was stimulated by the very fact of our having survived the war."

Not every pilot could handle life on Typhoons. Clifford recalled one such: "He was a new arrival on 440 and having trouble coping. A really nice guy, he was terrified of the aircraft and wanted to be re-assigned. He felt badly about his apprehensions, but couldn't handle it." He soon was lost on ops. Another fellow was twitchy about flying and did not fit in socially. He ended as LMF. Clifford testified that his poor showing was due to a lack in flying experience and that, being the only 440 NCO, he felt out of place. The upshot was that the NCO was let off and sent home for discharge. Concluded Clifford: "I feel that, in this instance, a life probably was saved."

At the end of May 1945 Bill Clifford's tour ended. On June 1 he left Flensburg in an Auster flown by Ted Smith, heading to a nearby 'drome. There Clifford would catch a flight to the UK. Before long Smith pulled out a bottle—the hard-nosed fighter pilots polished it off. Another bottle was produced. A long time later Clifford recalled: "After Ted dropped me and my belongings on the tarmac, and disappeared in a cloud of dust, my next conscious moment was being billeted in Bournemouth. Some things are best forgotten." With mixed feelings Clifford boarded *Louis Pasteur* at Portsmouth. Still, it was wonderful to sail into the St. Lawrence: "When we passed a train going along the Quebec north shore, the engineer gave us the welcome whistle. Our captain acknowledged. That old familiar 'toot' made me realize that I was home and safe." The great liner docked in Quebec to a tumultuous welcome—cascading fireboats and military bands. Next it was on to Lachine by train, thence to Toronto:

*As we moved along, pulling in at every whistle stop, there were clusters of people watching intently for a familiar face to disembark. Many happy and vigorous reunions were witnessed, Some, however, were distressed to the point where they followed the departing train along the platform as far as it went. I started to think about the winter of 1944-45 and all those killed in action in 440 Squadron alone. I know that not all the next of kin had received positive confirmation about what had happened to their son, husband, father or brother.*

*I cried all the way to Toronto and vowed that on November 11 every year during the moment of silence, I would recite their names—Ron Doige (age 20), Buzz Harwood (21), Dunk Dunkeld (23), Duncan Cumming (22), Billy Gibbs (21), Deke Passmore (21), Bugs Byers (25), Frank Warrell (20), Frank Crowley (23), Percy Kearse (20), Al Sugden (21), John Cordick (25), Jack Reilly, Buck Jenvey (24), Bob Coffey (30). Chuck Gordon (20) completed his tour, but was killed in a flying accident in England in a Typhoon, shortly after leaving us.*

Bill Clifford now took a bus to Hamilton for a reunion with his mother, sisters and brothers. In St. Catharines his house was festooned with a big "Welcome home, son" sign and an RCAF flag: "I took them all down before going into the house. I don't know why. I suppose this was the initial attempt at putting the war behind me. I was discharged on September 28, 1945. I was 22 years old. My life was changed forever." Clifford now got into real estate, staying at it for 45 years.

## Buzza on Tiffies

John Buzza's first tour as a wireless operator is recounted in our Bomber Command chapter. But this airman was not content with one gung-ho tour. On leaving 106 Squadron in December 1942, he had one thing in mind—pilot's wings. In the interim he did some WOp instructing at 29 OTU at North Luffenham (February 6 to March 24, 1943 on the Wellington, Defiant and Lysander). When an opportunity came, he visited Canada House in London. There he found an encouraging air commodore, who suggested he apply to his station commander for remuster to pilot. The station commander, obdurate at first, released Buzza, perhaps after hearing from Canada House. Buzza now attended the Initial

F/Os Buzza and Mattock on the town in Ottawa in August 1943. Note that the would-be pilots still were wearing observer wings. (All, John Buzza Col.)

Training Wing in Torquay. The officers' course there included only 2 Canadians, himself and Al Mattock. All the students had seen action, so it was a great bunch. Buzza now thought, "At last I'm going to be what I've always wanted ... a Spitfire pilot." Anything short of fighters did not interest him. EFTS at Brough followed, with Buzza's first flight being on July 12, 1943 in Tiger Moth R4850. He soloed 5 days later. After 15 flights (11:50 flying hours) he travelled to the Personnel Receiving Unit at Warrington for a posting home. Once there he had some leave with his family, then took the train to Regina, reporting to No.15 EFTS, his first flight being on September 20, 1943 in Cornell 14487 with F/O Buckmaster. Buzza soloed on the 30th. The course was great, if a bit unusual - he and Mattock were the only officer students. The rest were sprog LACs. Buzza's last flight was on November 11. He had flown the Cornell for 67.70 hours. Now the system was determined to post him to bombers. He argued that he was too small to handle a big plane and won his case.

Buzza left Regina a happy boy on November 14, heading for SFTS on Harvards at Uplands. He began flying on December 2 with F/O Shukis, an American. They did 3 trips, then Buzza soloed next day. Log entries show the variety of exercises: formation, flapless landing, dual nav, back seat landing, precision landing, endurance, max rate turns, cross country, dive bombing. There also were the sequential tests: preliminary then final instrument tests, 50 hour progress test, preliminary then final night flying tests, wings nav test, wings clear hood test, etc. New friendships sprang up. F/O Eric G. Smith, who would distinguish himself on Mosquitos, was one of Buzza's instructors. They became lifelong friends. Come April 1944 Buzza finished SFTS. His assessment was typical—rarely did an examining officer praise a student. In Buzza's case, F/L Ross McAllister of No.2 SFTS gave him an average as pilot, and above average in bombing, but a "rough on controls" under the heading "Any points in flying or airmanship which should be watched."

Training continued—this is how the Allies ultimately would crush the enemy. Buzza packed for OTU at Bagotville. On May 23 he flew in Harvard FE503 with F/O Gerry De Nancrede, a Malta veteran. Other well-known names were on hand to guide the students, e.g. W/C N.H. Bretz, DFC (former CO of 402 Squadron, Kenley Wing, England), S/L A.U. Houle, DFC and Bar (former CO of 417 Squadron in Italy), S/L Gordon C. Semple, F/L Frederick J. Sherlock, DFC and F/L Milton E. Jowsey, DFC. Buzza flew the Hurricane on the 25th, then pushed through the course, making his last (and 86th) flight on August 10. This totalled 10:40 hours on Harvards, 74:10 on Hurricanes. He was done on September 23, 1944, but the training system still wouldn't give him up. He went to Greenwood, normally a Mosquito OTU, for 10 days of intensive training on Hurricanes under F/L Sherlock—doing battle formation, intercepts and low level flying.

Avid athlete Buzza joined in all sports activities. Here he is (farthest right, squatting) with the No.1 (F) OTU ball team at Bagotville over the summer of 1944.

Finally it was time for overseas—again. This followed leave and a tank identification course at Camp Borden. Once in England there was some waiting at Bournemouth, then news that he was going on Typhoons—disappointing for Buzza had his Spitfire fixation. Nonetheless, he turned up in February 1945 at No.56 OTU at Milfield, a station between Newcastle and Edinburgh. There he had his first sight of a Typhoon. Before flying the beast, there were lectures, at one of which on February 1 the instructor boasted, "You could crash this thing through a brick shithouse and come out the other side safe and sound." Shortly after these words the sprogs were standing outside when a Tiffie swished by, its propeller stopped dead. P/O N.L. Gordon set down, smashed through a stone fence, and died in the ensuing explosion. One of the student pilots wondered, "Do you think we should go back into class and ask that fellow about his theory?" (Gordon, age 20, from Canso, Nova Scotia, earlier had survived a Tiffie bail-out and a forced landing with 440.)

On February 4, 1945 Buzza strapped into a Typhoon and roared off for a 30-minute initial flight. He quickly was in the swing of things. Training was intensive, 65 Typhoon flights being made for 64:15 hours (there also was some Miles Master time). One day Buzza had the big Sabre engine in front of him quit. He managed to restart and flew home, where he reported the incident. Next day he noticed the paperwork for this machine with a note, "Ground checked, found serviceable." Buzza didn't like the sound of this, feeling that a test flight was in order. He mentioned this to the student about to take the bird

"Sure is much bigger than the Spit I dreamed of flying", is what Buzza scribbled on the back of this snap showing him and his Tiffie.

flying. Before long the student was reported missing. Happily he turned up OK, having force-landed, his engine dead. In Buzza's period No.56 OTU lost 6 Tiffies (3 fatal), at least 3 due to engine failure. But training costs from the beginning of the war were assumed to be high. On March 27, 1945 the winco at 56 OTU sent Buzza on his way with a "Proficient". At last he was a fighter pilot. Orders sent him to 438, where S/L J.R. Beirnes was the CO. On April 4 John Buzza and Al Mattock reached 438 at B.100 (Goch), arriving by Anson. They went straight to a debrief and listened as the pilots described how their 4-plane formation, which had been strafing MT east of Rheine, had been bounced by Me.109s. Lead and No.2 returned, but F/L Earl James McAlpine (age 23 from Walkerville, Ontario) and P/O William J. Kinsella (age 22 from Ottawa) died in this action.

**Repair and salvage work on Buzza's bent Tiffie after his prang at Hustedt.**

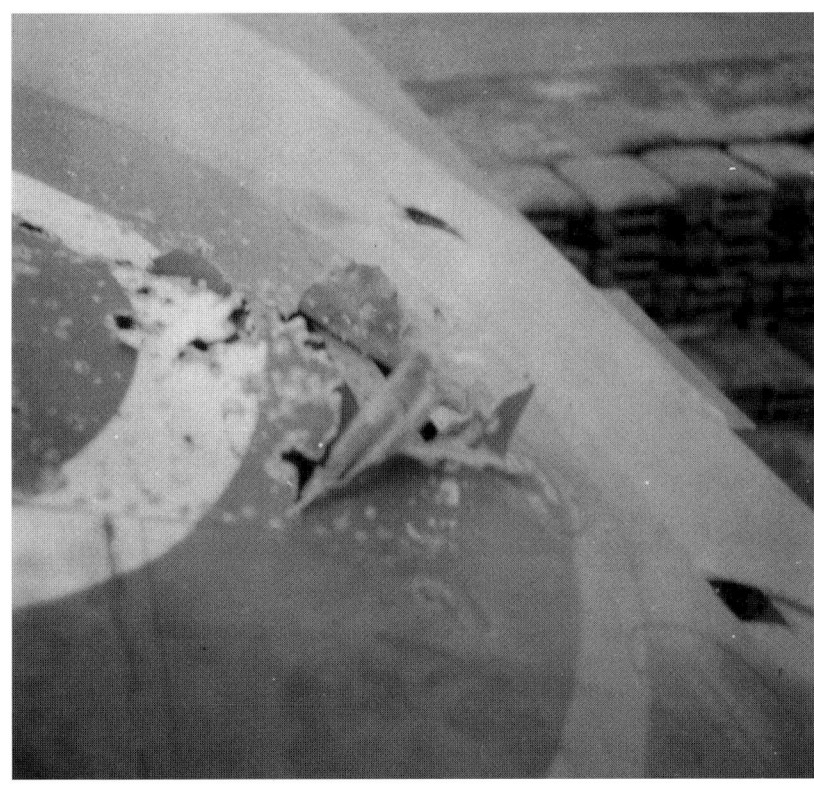
**Flak tore through Buzza's starboard wing one day, just missing the fuel tank and ammunition box.**

Late in the game as it was, Beirnes had work for his new pilots. In the 24 days from April 8 to his last on May 4, Buzza would complete 30 ops, plus 4 other Typhoon flights. As he recalled in 1999, most ops followed a pattern. He would fly No.2, sticking to Lead like glue. He would follow in as Lead dove and bombed, then would bomb himself, look for Lead, and get right back with him. He concluded that it was safest to stay to one side of Lead after the bomb run. As they climbed, any flak would be concentrated on Lead, but as gunners tended to under-lead a target, many No.2s were lost. Also, by keeping spaced, this would dilute the flak, since there were two clear targets for the gunners.

Buzza's first op involved dropping two 500-pounders on the town of Haselunne. On April 9 he bombed a target in some woods, then strafed horse-drawn MT. On the 12th 438 moved to B.110 Osnabruck. Next day he took severe flak, then got more of the same on the 14th, bombing a village where the Wermacht was dug in. This was one of many strong points which the British army had by-passed in its rush forward—it was left for Typhoons to clobber these. There was flak from the village; one shell exploded through Buzza's starboard wing, near the fuel tanks and ammunition boxes. Everything held together and Buzza landed at B.110. The senior pilot, whose name was stencilled on the new, 4-blade machine, was miffed at seeing how the sprog had brought it home.

The target on the 15th was a bridge. Returning, the Typhoons made their usual overhead break before going downwind in the circuit. For some reason, 24-year old F/O John G.S.J. Livingstone, a Montrealer, decided to do a victory roll. Such antics were discouraged and could bring a reprimand. For Livingstone there would be no reprimand—he lost control and crashed fatally. On the 16th F/O John K. Brown bailed out, perhaps having been damaged by a bursting bomb. On April 18 Buzza bombed an oil dump. After this trip he made a comment in his log, wondering if the target really hadn't been a decoy. On the 23rd he noted, "F/O Hartnett missing." Timothy Hartnett, a 23-year old New Brunswicker, was last seen entering cloud, so may have become disoriented and crashed. Two days later 438's target was rail lines around Kiel. There F/O Tom Jones went down in flames. Happily, he turned up, after making a last-second bailout.

On April 26 Buzza had a wing panel fly off, so had to return with two 1000-pounders. The same day he wrote, "Ted Brydon hit ground and exploded strafing." The 22-year old from Brampton, Ontario had struck a tree during his run. Much strafing was done in the following days, Buzza noting various "flamers". On a May 3 dive-bombing of a ship in Hohwacher Bucht harbour, his glycol tank was holed by his own bombs exploding. His engine overheated, then seized, as he overshot the strip due to a red flare. Quickly checking his options, he felt that he could make a downwind landing beside the strip. Airfield control agreed, so long as Buzza didn't block the runway. Within moments he was down, waiting for the rest of the wing to land. Thus did he survive his second prang.

On May 4, 1945 Buzza flew his last op, bombing a coastal vessel. Now the war ended and all that remained were some patrols from Flensburg (438's final home) and an airshow over Copenhagen. On June 1 S/L Beirnes was killed during an airshow mission; S/L Paul Bissky took over. On June 22 Buzza took up a Typhoon for the last time—a local Copenhagen flight.

W/C F.G. Grant of 143 Wing noted that he had completed 35:10 hours on ops (61:20 hours total with 438) and rated him "average" in all categories. Since joining the RCAF he had flown 456:50 hours. Now he and some buddies visited B.152 (Fassberg) for a few days. Some partying ensued, one result being that the fellows burned all the furniture they could find. Bad weather now engulfed Fassberg, making the boys realize what dummies they had been—there were no chairs, tables or beds left! Early in July, Buzza left for England on a Stirling. Soon he was on the waves again (for the 4th time in less than 3 years). His destination was No.4 Release Centre in Toronto. He became a civilian again on September 19, 1945, but on April 9, 1946 he re-enlisted.

**John Buzza at his home in Ottawa in 1999. (Larry Milberry)**

## Friendly Fire

Early in 1942 Gordon Kemp, who had grown up on Logan Avenue in east Toronto, tried enlisting in the RCAF, but was rejected—he had a nasal disorder. He had this surgically corrected and was accepted, first doing some high school courses at Central Technical School in Toronto. Then came No.1 MD, guard duty at Brantford, ITS in Toronto, 20 EFTS at Oshawa, 2 SFTS at Uplands, finally No.1 OTU at Bagotville. There the practice was to pair students with experienced pilots—Kemp's room mate was a pilot who had done a been with a Merchant Ship Fighter Unit. One of Kemp's fellow students was Perez Gomez, perhaps the only Mexican pilot in the RCAF.

**Proud LACs at No.20 EFTS in Oshawa—Gordon Kemp and Red Ellis. (Kemp Col.)**

Kemp went overseas aboard *Mauritania*, passed through Bournemouth and OTUs on Hurricanes, Spitfires and Typhoons, then was posted to 263 Squadron at Hurn. There he flew with such Canadians as Gus Fowler from Ferni, BC, Mac Hamilton of Sault Ste. Marie, Stanley Le Gear of Barrie, Ontario (killed in action on September 10, 1944), Pat McNenley of Massey, Ontario and Norm Woodward of Vancouver (killed in action on April 17, 1945). Kemp was with the squadron when it landed at B.3, Ste-Croix-sur-Mer on August 6, 1944. He flew his first operation on the 9th, then worked steadily to April 26, 1945, logging 110 sorties. He now was OTE (operational tour expired), pending a posting to Fighter Leader School.

The most memorable day in Gordon Kemp's tour was August 27, 1944, when he was on a wing sortie against an enemy flotilla. Upon reaching the target, the wing leader reported no anti-aircraft fire from the ships. Nonetheless, Control ordered the operation to proceed. Kemp rolled in with his lead and they fired at one of the ships. New to this work, Kemp didn't see where his rockets hit. The Tiffies then strafed the vessels. Great destruction was caused, before the do was called off. It immediately was clear that something very tragic had just happened—the ships were friendly. Naturally, nothing officially was reported about this, although the AOC of 84 Group visited B.3 to assure the 16 Typhoon pilots involved that they must not feel responsible—the Royal Navy accepted the blame. Somehow, its ships had been in an off-limits area. The cost to the Royal Navy was 117 killed, 153 wounded, HMS *Hussar* and *Britomart* (each 875-tons) sunk, and another ship disabled. It would be 50 years before word of this matter became public. By 1999 Gordon Kemp was the sole surviving 263 Squadron pilot involved on August 27, 1944. When he entered his logbook details for that day, he noted: "Went on a show after lunch—this time a shipping strike NW of Le Havre. 16 a/c, no flak (mainly because they were our own). Navy takes full responsibility but that doesn't help matters. 3 out of 6 sunk. Bad show." After a tour on Typhoons, Gordon Kemp of 263 Squadron was posted to Course 2 at the Fighter Leader School at Tangmere. His last flight was taking Tempest SN259 from Tangmere via Eindhoven to visit 263 friends at R.16 (Hildesheim, Germany) July 7 to 12, 1945. He finished the war with 630 hours, 186.5 on Typhoons.

In retrospect, Kemp enjoyed his tour, commenting in 1999: "I had no complaints about the Typhoon. You knew what it could do. If you treated it with respect, you usually had no trouble."

**F/Os John Spencer "Sandy" Colville and Gordon Kemp in front of the Lord Elgin Hotel in Ottawa on August 8, 1943, the day of their Wings Parade. They were posted to No.1 (F) OTU without leave, reporting in the next day. Colville died on August 18, 1944, hit by friendly fire while inadvertently strafing Canadian MT. The Colvilles of Bowmanville, Ontario suffered triple tragedy, first losing William in the crash of an 11 (BR) Squadron Hudson in Newfoundland; then Alexander, killed with his 408 Squadron crew over Germany; finally, Sandy. (All, Kemp Col.)**

This said, there was one dicey trip in MN974 at Tangmere. After takeoff Kemp had his throttle stick open. He consulted over the R/T with his CO, who suggested doing some simulated approaches on the cloud tops, coming in fast, then cutting his engine (and re-starting). After trying this, Kemp made a successful deadstick landing—a hot one at 160 mph. There was a suggestion that this merited a green endorsement, but higher-ups said that a pilot with Kemp's experience should be able to handle any emergency. Once home, Kemp returned to the Robert Simpson Co. in Toronto, then he and his wife Sally, a wartime WD (whom Gordon had met while both were at Simpson's in 1940) emigrated to the US. Kemp worked in mechanical drafting, eventually retiring in Florida.

**A flight of Hurricanes training near Bagotville. Gordon Kemp was flying BW878/6.**

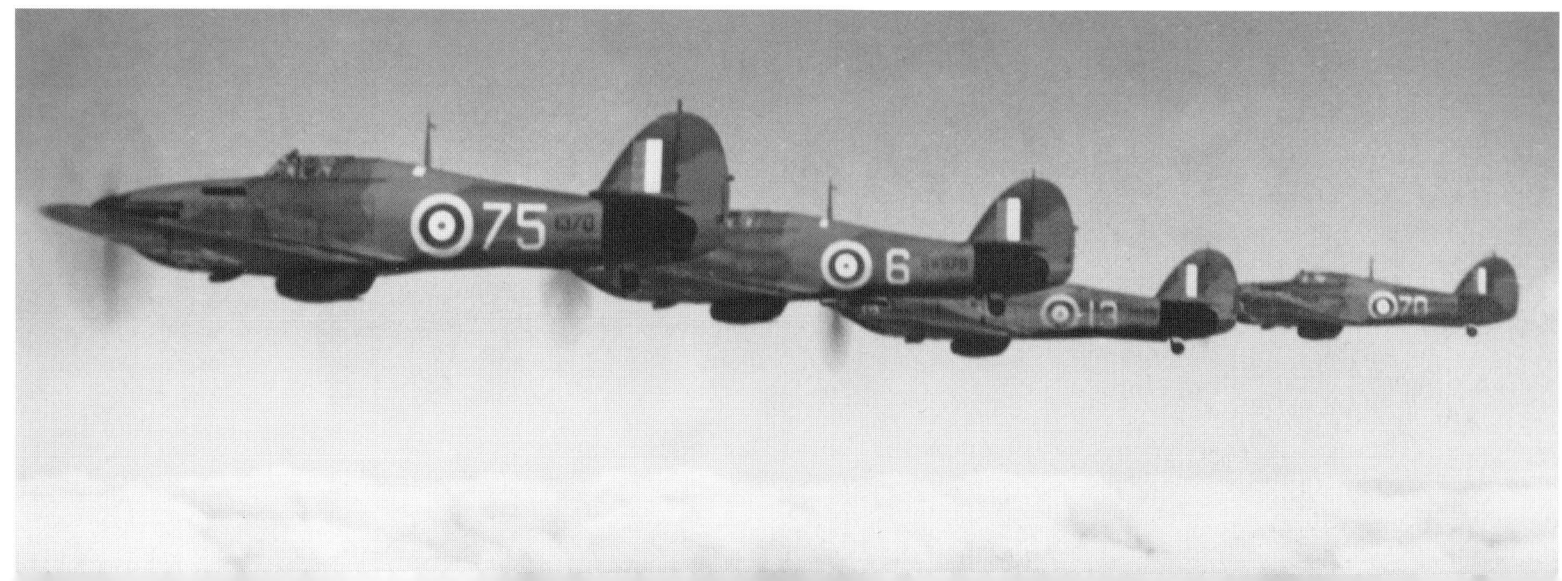

No.263 Typhoon pilots in Normandy in August 1944: First, Gord Kemp and "Mac" Hamilton, who survived their tours. Then, Frederick Stanley Le Gear and Norm Woodward (both RCAF), Ron Proctor, a Kenyan in the RAF, and Gus Fowler, RCAF. With his map tucked in one boot and revolver at the ready, Le Gear was ready for ops. Age 24 from Barrie, Ontario, he was lost on a shipping sweep off Holland on September 10, 1944. Stan's brother, Victor, survived a tour on 439 Squadron. Norman Paulle Courtney Woodward, age 23 from Vancouver, crashed into Harderwij harbour, Holland, while attacking a flak ship on April 17, 1945. By the turn of the century the passing of time had claimed many a Typhoon veteran. Mac Hamilton, for example, died in his native Sault St. Marie in June 1999.

A scene with 263 at B.89, Mill, Holland in early 1945: snoozing is Woodward, 4th from the left is Fowler. A card game appears to be in progress.

A dramatic view taken as 263 Squadron Typhoons launch RPs at fortified farm buildings in front of the Canadian 1st Army west of Calcar on February 22, 1945. The CO's rockets have found their mark. His wingman's (Gordon Kemp) are following.

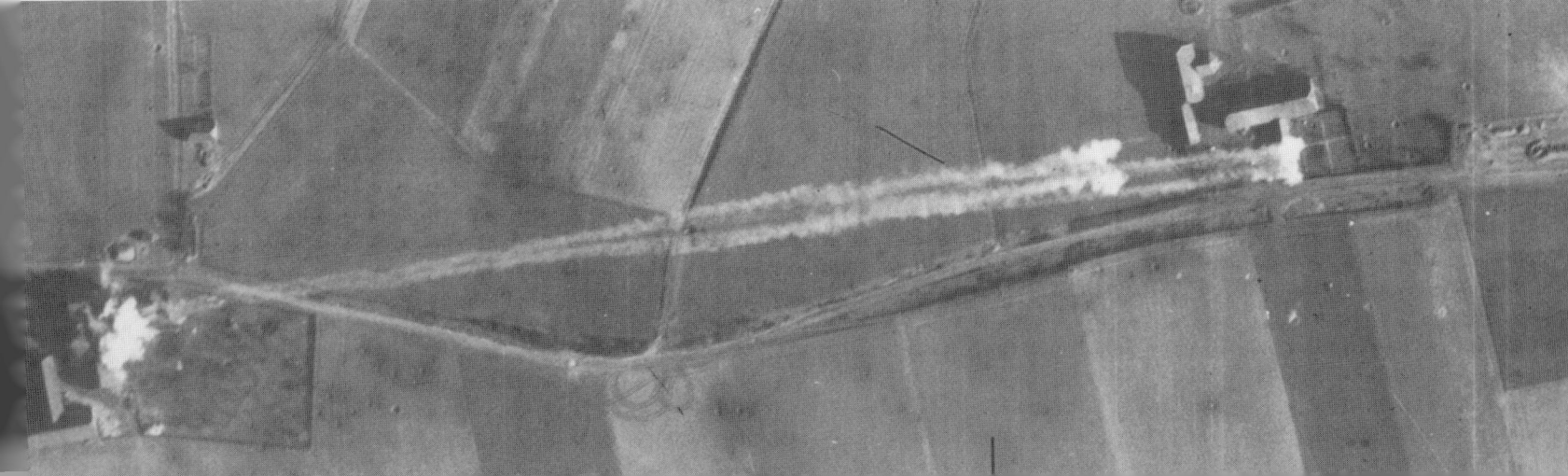

Gordon Kemp's favourite Typhoon at Antwerp in early 1945.

This should bring back a few memories: cartoonist Cairns' "Pay Day Pub Crawl—RCAF Overseas".

## Anecdotes from the Technical Experts

Keeping a wing with 75 to 80 Typhoons serviceable required a large support organization. With its pool of manpower, training system and foundation dating to pre-war days, the RCAF was well fixed for this. Typical of the hundreds of technical men at 143 Wing was Bertal Tolbert Walsh. Born on April 7, 1917, he had grown up on a farm at Bear Brook, east of Ottawa. After high school, in 1935 he followed some school chums and joined the RCAF. He was soon at Camp Borden in the Class of '37—206 students learning technical trades. He graduated as an airframe rigger with the rank of AC2 and was posted to No.7 (GP) Squadron at Rockcliffe. No.7 was mainly an experimental squadron. There were ski trials with the Delta, experimental engine fittings, trials with cockpit heating systems, propeller experiments, etc. As new types like the Anson or Oxford appeared, No.7's would evaluate them. For a young mechanic like Walsh this was a dream posting. In 1998 he recalled the day that F/L Gordon Truscott delivered the RCAF's first Grumman Goose, landing it on the Ottawa River, then taxiing up onto the tarmac—quite a spectacle for 1938.

With the war LAC Walsh transferred to No.11 (BR) Squadron, formed in October 1939 as the RCAF's first Hudson unit. Hudsons were delivered to Rockcliffe from the US, then No.11 ferried to Dartmouth. At this point Walsh's records went astray, so any promotion was delayed. It was December 1940 before he became a sergeant. Next he attended the RCAF engineering school in Montreal, graduated in late 1942, then did officer training. In April 1943 he joined No.123 Squadron at Debert doing army co-operation training with Harvards, Hurricanes and Lysanders. Walsh became the squadron's engineering officer.

By now some home squadrons had been re-numbered before embarkation overseas, e.g. No.123 formed the nucleus of 439 Squadron (Typhoons) in January 1944. In the summer of 1943, however, Bert Walsh reported to 150 Squadron, an Eastern Air Command "paper" squadron. He moved to Yarmouth and began accepting 13 Ventura patrol bombers. With little to do, he passed some of his time riding on Ventura acceptance flights. Into September he remained 150's sole member, although he heard that a S/L Mike Black would be arriving with 48 airmen. Nothing transpired, then word came that 150 would disband. Walsh was posted to what he called "useless duties" in EAC, e.g. helping S/L Shortreed of the Institute of Aviation Medicine develop an oxygen system for passengers on long Ferry Command flights. At last came news that he would be joining 439 Squadron's servicing echelon at Hurn, England. In May 1944 he sailed from Halifax aboard *Empress of Scotland*. The following photos from Bert Walsh's extensive album illustrate life on 143 Wing in wonderful detail.

## Memories of 143 Wing: Bert Walsh's Album

Bert Walsh astride a motorcycle in Normandy in August 1944. Beyond is Typhoon JR497/F3-F of 438 Squadron. F/O A.H. "Ab" Vickers was flying "F" on September 28, when he was hit by flak, bailed out and became a POW.

No.143 Wing airmen in a photo from Hurn before D-Day in 1944. Riggers, fitters, electricians, instrument techs and others, these men were the heart of the wing when it came to getting Typhoons on the line for operations every morning, then all through the day.

The Normandy sky full of Dakotas towing Waco gliders, as the invasion moves into high gear in June 1944.

A view of the B.9 dispersal taken from an Auster. Bert Walsh wrote on the back, "This has long since disappeared, so is not of use to enemy agents or censors."

No.143 Wing's technical men on the Normandy beachhead: Doug Seath, Don Brewster, Jeff Jeffries, Ray Collis, Pete Langille, Harry Thompson and Bert Walsh.

Ready for action—a 143 Wing Typhoon with 500-lb bombs.

Jeff Jeffries with the Wing's "liberated" Hudson Terraplane staff car. The right fender is marked "143 Wing RAF Kenley".

The first party in the officers mess at B.9 with G/C Paul Davoud going around the floor with a WAAF. Then, a general scene of merriment with officers, airmen and nurses.

The hangars at B.58 Melsbroek (Brussels) had been used by the Luftwaffe, then 143 Wing moved in.

Pete Langille takes some instruction on a Browning .50 machine gun.

WD's have some fun with Typhoon RB326/5V-V, thought to be at B.150, Celle in 1945.

Card playing out in the field at 143 Wing. Such games could involve sizeable amounts, so were not for the faint of heart. From this angle there's at least one $100 bill on the table.

An Ottawa *Citizen* clipping describing the action typical any day of the week at 143 Wing.

Typhoon pilots of 439 Squadron some time before August 12, 1944. Standing are W.D. "Don" Burton, DFC of Brantford, Ontario; Ivan W. Smith of Canmore, Alberta; John Stelter, R.H. "Bob" Laurence, DFC of Edson, Alberta; J.E. "Jimmy" Hogg, DFC of Ottawa; Charles L. Burgess, DFC of Fredericton; Ronald O. Moen of Saskatoon; J.O. "Tex" Gray, Pierre N. Bernhart of Vancouver; Bernard P. "Babe" Swingler of Port Arthur; Nick Rossenti, and Doc Moffat. In front are Pete Henderson, James A. "Jack" Brown of Berwyn, Alberta; Johnny Stitt, Ernie Allen of Edmonton, Raymond A. Brown, DFC of Toronto, S/L Hugh H. Norsworthy, DFC of Montreal (the CO), Maurice P. Laycock of Fairy Glen, Saskatchewan, Ken Fiset, DFC of Montreal, William Kenneth Scharff of Victoria and Mike Harrison (adjutant). Don Burton finished with 123 ops on 439, but previously had flown on 174 Squadron, where he began in November 1942. On August 8, 1944 Ivan Smith survived a crash after his engine quit, but on April 9, 1945 died in a Tempest at 56 OTU (RAF Brunton). Having enlisted in 1941, Laurence flew Hurricanes on 123 and 163 Squadrons in the HWE. On December 28, 1944 he was strafing trains, when his squadron was attacked by Fw.190s and Me.109s. After watching a Typhoon shot down, he tangled with one, then another e/a, claiming both without firing—he outflew each so successfully at low level that they crashed! In the New Year's Day brouhaha Laurence was airborne on a weather recce. Returning to Eindhoven, he and 3 mates mixed it up with the Huns. Laurence shot down one and got a probable. On March 23, 1945 S/L Hogg, then CO of 438, dived fatally into the sea off Portland during gunnery camp at Warmwell. Moen died along with Ernie Allen, shot down while bridge-busting on August 12, 1944. Pierre Bernhart later went to 56 OTU on Tempests, there to die in an accident on March 4, 1945. Babe Swingler survived a crash in B.9 in Normandy on June 25, but died on February 22, 1945. Jack Brown bailed out on November 6, 1944 and was captured. On September 26, 1944 Stitt, flying MN379, shot down a Ju.88 in the Nijmegen sector (MN379 was shot down on October 2, but F/O W.G. McBride force-landed it OK). On June 23, 1944 Ray Brown bailed out after his engine quit. He landed in the Channel, was rescued and resumed operations next day. Norsworthy force-landed in the English countryside on May 24, 1945 after an engine fire. On October 28, 1944 Laycock was shot down by flak and killed. Ken Scharff had suffered the same fate on August 19.

## Kingston Pilot, Two Local Men Escape Germans

WITH THE R.C.A.F. OVERSEAS, Nov. 8—(CP)—F.O. A. McBride of Kingston, Ont., couldn't help himself but he landed more than a little too close to the German lines for comfort.

When his Typhoon was hit by anti-aircraft fire as he dive-bombed Rhine railway lines, he had to set down in a field near Deurne, The Netherlands. Borrowing a bicycle from a friendly Dutch farmer he went to an American unit stationed nearby.

Then he learned he had landed only a few hundred feet from the German lines.

"When we got back," he said, "the Typhoon was blazing. The farmers told us 15 Germans surrounded it only a few minutes after I left and set it afire." He finally rode back to his own base from the American field with F.O. Bertal Walsh of 192 Waverley street, Ottawa, an engineering officer, and Flt. Sgt. Robert Malo of 327 Somerset street, Ottawa.

"We skedaddled out of there in a hurry," said Walsh. "It was too damn quiet and the forward troops were pretty jittery about going up with us."

An ESNA Show put on for the men at B.150, Hustedt, Germany on VE-Day.

After the ENSA Show there was great revelry at B.150. Here 143 Wing gets into the festivities, which include burning a few planes such as the long-nose JG26 Fw.190D9 in the second photo. There was hell to pay next day, since the Allied team recovering Luftwaffe planes for evaluation had wanted to save this one!

Artist Paul A. Goranson spent time with 143 Wing sketching, painting and coming up with several outstanding pieces that captured the atmosphere. Here the wing is shown disembarking in Normandy. Barrage balloons protect the beaches from the Luftwaffe. Next, "Typhoons over Lantheuil", a streetscape in the local village near B.9. The scene of devastation depicts Carpiquet in August 1944 with a 1000-plane raid under way on nearby Caen. The 440 Typhoon undergoing maintenance likely represents 419 RSU at work. The wild street scene shows the wing entering Brussels on September 6, 1944, led by Bertal Walsh in the jeep. Goranson had run ahead to make a quick sketch of this dramatic event. (CF)

## Other Technical Types

Leon Alfred Pedley, born on March 10, 1910, was another solid technical man on 143 Wing. He had come to Canada from the UK in 1926 under the British Immigration and Colonization Association, learned auto mechanics, then worked in garages in Kingston, Ontario. He enlisted in the RCAF in September 1939, going into aero engines. His first posting was to 118 Squadron (Goblins at Rockcliffe and Dartmouth, Kittyhawks in Alaska). In October 1943 No.118 re-formed in England as 438; the next month Pedley sailed aboard *Mauretania* to join it. From January to March 1944, No.438 flew Hurricanes at Ayr, where the weather couldn't have been worse, Pedley recalling how some RCAF blokes were caught stripping buildings of wood to burn in their stoves. "At Ayr a fellow guarded what little coal he had as if it were gold," he said.

Another scarcity was engine tools, so 438 purchased what ones it could on the black market in London, using a fund toward which pilots were expected to contribute. Soon 438 began re-equipping, flying its first Typhoon op from Hurn on March 20, 1944. Keeping the Typhoons serviceable was an endless challenge. Just before D-Day, for example, there were instructions to change all propellers from 3 to 4 blades. Pedley recalled how complete engine changes were done by the wing's heavy maintenance experts, No.419 Repair and Salvage Unit, always stationed nearby. Just before D-Day 143 Wing painted invasion stripes on its Tiffies, but it was too early. The stripes had to be scrubbed off, lest German photo recce notice the change. A day or two later the stripes were re-applied.

In 1998 Pedley recalled how 143 Wing had a great supply of vehicles, one for every job. Even so, there were special needs and captured German vehicles or commandeered civilian ones often were a solution. In the Falaise area Pedley liberated a small Renault van. One day engineer F/L Scruffy Langille came in with a big diesel generator on the back on a truck. Pedley got it going, and it soon was producing electricity for the station. Pedley also had a motorcycle with the requisite "Permission to drive an enemy vehicle" paperwork. On the Continent he recalled how 143 Wing would use 3-ton Bedfords to haul 12 500-lb bombs from the beachhead. For this the Canadian wing was reprimanded by the RAF, but the practice continued.

Operating from forward airfields and frequently relocating, 143 had to be innovative. Pedley described acquiring a quantity of ¾-inch bungee cord from a barrage balloon. This he wove across a frame made from scrap lumber, producing a comfortable bed. He and his mates concocted a water heater using a 45-gallon drum and a diesel oil drip to heat water for a makeshift shower. To Pedley the most ingenious scheme was building some huts at Eindhoven. This got the men out of their dank tents into cozy accommodations. The huts, built of scrap from a bombed-out hangar, were built to sit high and dry. They were arranged in a row, the "neighbourhood" being christened for Montreal's Ontario Street, famous for its whore houses. When 438 had to move across the field, all they had to do was fire up their ex-Luftwaffe half-track and tow the huts across. On the 1945 New Year's Day raid a number of huts were destroyed by the Luftwaffe.

A Typhoon pilot had his own two groundcrew; this little team worked smoothly. Many times, however, groundcrew would be saddened when a pilot failed to return from ops. Soon they were breaking in a replacement. Alf Pedley recalled W/C R.T.P. Davidson going missing in May 1944, then turning up in civvies in Amiens in September. Having been shot down, he had hooked up with the Maquis with whom he operated for some weeks. One night 438 Sgts Pedley and McKegney went into a local bistro. There they found some Maquis, who took them upstairs to show off some German prisoners. One of these, quoting the Geneva Convention (for which Germany had nothing but disdain), demanded that the Canadians take them to a proper military unit. The Canucks agreed to report the situation. Returning to camp in the dark, however, Pedley spotted someone in the shadows and ordered him to step forward. In the excitement, somehow Pedley's Sten gun went off, injuring McKegney. The half-corked 438 erks went to the medical tent, never getting around to mentioning the Germans.

At the end of the war Pedley was with 438 in Flensburg, waiting to return to the UK by sea, when a USAAF B-24 landed, en route to England. Pedley asked the captain about hitching a ride. This was OK, so long as he could get written permission from his CO, W/C Deane Nesbitt. The CO agreed, but the B-24 skipper also wondered if there were any souvenirs around. As if by magic Pedley came up with a sack of Luger pistols. This so impressed the Americans that they made him honourary captain, putting him in the co-pilot's seat. Once in London, however, Pedley was hauled in by two MPs. He was brought before a Customs official, who reprimanded him for sneaking back into the country. She searched his kit, but let him off the hook. Pedley never figured how the MPs knew where to find him in the hurly-burly of London.

Another memorable technical type on 143 Wing was F/L Cecil Garfield Langille, sometimes called "Pete" or "Scruffy" by his mates. Langille was born in Springfield, Nova Scotia on March 3, 1904. In 1927 he earned a geology

**Bert Walsh (right) with his old technical mate from 143 Wing days, Alf Pedley in Perth, Ontario, where they both were long retired by 2000. (Larry Milberry)**

degree at Acadia University, then studied at the University of Toronto, before returning to teach at Acadia. Langille was an entrepreneur—he ran a coal and building supplies business, and prospected. He enlisted in the RCAF at Halifax on April 2, 1940, attended No.1 MD in Toronto, then Technical Training School at St. Thomas. In October he was posted to Ottawa, working at No.2 SFTS and Test and Development Establishment. One day he survived the crash of T&D's ice research Hudson. On the recommendation of W/C T.R. Louden, OC of T&D, Langille was posted to the RCAF Aeronautical Engineering School in Montreal. In January 1944 he sailed to join 143 Wing. He went straight on course at Napier to learn the Sabre.

Once on duty Pete Langille excelled in technical affairs, being praised by his CO, G/C A.D. Nesbitt and immediate boss, S/L Don Brewster. While doing his job on the wing, Langille was not one to waste free time. One day he told Alf Pedley how, like many in the RCAF, he was cleaning up in the black market with such schemes as selling tires to Danish border guards. He also told of betting on the greyhounds, and buying shares in a Brussels casino. Trading in cigarettes was common. For a few packs a fellow could get almost anything from the Germans. The same went for common items like bars of soap—women would trade sex for soap, such were their privations. Pilots at 126 (RCAF) Wing, which finished the war at Uterson, outside Hamburg, had their own scheme, dealing in French wines and Champagne. They would stuff a Spitfire's ammunition boxes with dozens of bottles, then cross the Channel to stations like Tangmere. Much other loot traded hands. Some of the boys from 143 Wing recalled years later how some black market deals were made under cover of darkness in the cemetery at Flensburg, for it was strictly *verboten* for the Allies to fraternize openly with the populace.

G/C Paul Davoud of 143 Wing went for the exotic—he had an Me.108 communications aircraft crated and shipped to England for furtherance to Canada. The crates were labelled as Lancaster tail components. The sporty little plane got to England, but there it stopped, perhaps hauled in by Customs, perhaps left by Davoud, who had other things going on. He even had thoughts of restoring the Fw.190, sitting on its nose at Evère, but in very good condition. Many fellows had their pictures taken beside the spiffy fighter. One day there were orders from somewhere to move it, but as soon as it was jostled, it exploded, killing 8 and wounding others. The Fw.190 had been a booby trap all along.

At war's end the German commandant at Flensburg held a dinner, where he formally surrendered and handed over his base to G/Cs Paul Davoud and Deane Nesbitt. The same day a Soviet C-47 landed with some senior officers, also intent on accepting the German surrender. The Soviets were politely told that they were too late and should scram, which they did. On a somber note, while at Flensburg the men of 143 Wing visited the Belsen extermination camp. Later they organized a hunt, shot some deer, and delivered the meat for Belsen's survivors. In these days Germans still wore their uniforms; some even did guard duty. Sgt McKegney of 143 Wing, a prison guard in civvie life, liked to torment Germans on the street. He would select an officer in uniform, yell "Halt", then make a big deal out of frisking him. Many goofy stories were told of post-armistice goings-on. Bert Walsh recalled 143 Wing's RC padre, Father Michaud. One day he was de-lousing his uniform by soaking it in gasoline (while he smoked). As he was draping the uniform on a fence, it went up in flames! Another laugh was how Al Fitch of the Salvation Army was keeper of 143 Wing's stash of booze. Walsh experienced an interesting bit of black marketeering, when he traded Protestant padre Ashforth for a new jeep. All the padre wanted back was a bottle of scotch! Not quite so funny was how the Canucks occasionally played a trick on the locals. If a crowd was watching 143 Wing passing in convoy, a driver might backfire his engine. By this time the populace had been through a great deal, so people were edgy. At the sound of a backfire, some would duck for cover!

Another incident at Flensburg involved an Fw.200 Condor VIP transport that the RAF took as war booty. It was destined for Farnborough for flight testing, S/L J.C. "Big Joe" McCarthy, DSO, DFC being the pilot. McCarthy was an American in the RCAF, and a distinguished alumnus of 617 Squadron. One day in June he set off in the Condor for England, engineering men Jeff Jeffries, Bert Walsh and some others being aboard. Weather forced a turn-around, but they tried next day. Soon after leaving the coast, McCarthy lost an engine. Another quit, then, crossing the Thames Estuary, a third! The Condor squeaked into Farnborough, where the last engine died as McCarthy landed. What had led to such a near-disaster? The answer was simple—the RAF had topped up the Condor's engines with regular SAE oil, but they were designed for synthetic oil. RAF engineering had no idea about this, and the Germans weren't breathing a word!

Murray Castator of 143 Wing was born October 27, 1921. As a boy in Long Branch, Ontario he caddied at the Toronto Golf and Country Club, becoming an avid golfer himself. In 1938 he got into the Bryant Press in Toronto, where he began learning the printer's trade. He joined the RCAF on November 26, 1941, his first training being at Central Technical School in Toronto. Next he attended No.1 RCAF Technical Training School at St. Thomas, finishing as a rigger with special metal repair qualifications. Upon graduation he joined 118 Squadron, then flying Goblins. Through 1942-43 he was with 118 at Annette Island, then went overseas aboard *Mauritania*, sailing on November 2, 1943, disembarking at Liverpool on the 9th. Cpl Castator joined 438 Squadron at Ayr on January 8, 1944. From there 438 had several homes, including Hurn and Funtington. Along the way the "Wildcats" learned much of camping in the outdoors and convoying from place to place. The weather often was foul, so that was good training as well.

Castator brought home as many photos as possible. He took his own, swapped with buddies, and ordered batches of official PR shots. These he kept in a collection of albums and loose prints. Unlike many, he saved his negatives, so that in 2000 his collection was available for this project. While most who were overseas kept some kind of souvenir album or scrapbook, amazingly, some came home empty handed. It just wasn't their interest, although most later regretted this negligence.

G/C Deane Nesbitt, DFC (right) chats with Air Marshal G.O. Johnson, AOC RCAF Overseas HQ. S/L D.A. Brownlee looks on. As an original member of No.1 Nesbitt fought in the Battle of Britain. On August 28, 1940, when No.1 became the first RCAF squadron in combat, he shot down 2 e/a. He later commanded 111 Squadron (Kittyhawks) at home, 401 Squadron (Spitfires) in England, "Y" Wing (Kittyhawks) in Alaska then, in turn, 144 Wing (Spitfires) and 143 Wing (Typhoons) in the ETO. Postwar he returned to the family business in Montreal. (CF UK20819)

# Murray Castator's Album

The badge of 143 Airfield with its motto "Sua Cuique Pars". (Murray Castator Col. or, if known as RCAF photos, credits are noted.)

No.143 Wing comes ashore in Normandy in late June 1944. Then, going cross-country, and taking a lunch break. (2nd, CF UK20585, Murray Castator Col.)

Maintenence men of 438 Squadron at B.9 in Normandy. They would spend more than a year supporting their Typhoons as the Allies forged through France, Belgium, Holland and into northern Germany.

Canadian graves near Creully. After the war most fallen men were re-interred in war cemeteries. This grim but essential work was carried out by the Commonwealth War Graves Commission.

A view down Ontario Street. Then some of the erks pose for the PR photographer, reading the paper and *Esquire*, tinkering with a Luger pistol and writing home. Note the German machine pistol slung on the wall, and all the other interesting embellishments and accouterments. (CF PL33185)

Cpls "Von Castator" (left), Hanlon, Winters and Giles goofing around on Ontario St.

Murray with an armoured Luftwaffe tractor that came in handy at Eindhoven.

The RAF Regiment defended 2 TAF fields. Here is one of their 40-mm Bofors emplacements at Eindhoven. Note that weapons also included the .303 Bren gun, a good close-in AA weapon. During *Bodenplatte* the RAF Regiment shot down 45 Luftwaffe fighters. (CF PL43141)

Jim Hamilton and Murray Castator at Flensburg in August 1945. Beyond is Typhoon RB207/F3-T, which somehow survived the war.

Beautiful ruggedness—MN603 of 440 Squadron at Creully. It had seen action on D-Day.

No.439 Squadron at Flensburg. MN379/5V-E is nearest, RB441/5V-Z second. Note that Tiffies had 3- or 4-blade propellers.

RB205/FGG, the personal Typhoon of W/C Frank G. Grant, Wing Commander (Flying) at 143 Wing after October 1944. "FGG" was lost in *Bodenplatte*. The rows of 5-gallon jerricans in the foreground were everywhere around a 2 TAF base. Usually they came in on Dakotas.

Making do. RB207/F3-T of 438 taxis through puddles at Eindhoven, as some pilot practices dinghy drill. (CF PL42099)

A great operational scene as MP134 of 439 Squadron refuels at Eindhoven. On January 1, 1945 it was damaged in *Bodenplatte*. Then, on January 22 it exploded when hit by flak. Pilot R.G. Crosby was thrown clear, landed by parachute and evaded. The year before he had gone down in a 56 Squadron Tiffie, a story detailed by Hugh Halliday in *Typhoon and Tempest: The Canadian Story*.

Fitters at Eindhoven change an engine on one of their mighty Typhoons.

No.440 pilots F/O Robert John Reilly of Toronto and WO Johnnie Duncan of Paris, Ontario in a publicity shot at Eindhoven. On November 18, 1944 Reilly, flying MN475, was fatally hit by flak on a do against Hilfarth Bridge. On the 21st Duncan was hit, belly-landed, and became a POW. (CF PL33812)

Typhoons required an inordinate amount of technical attention. Here 440's MN716/I8-A "Diane" is serviced. Typhoon production totalled 3317, the majority being lost under fire or in accidents. On May 31, 1945 there were but 1149 remaining on strength. With one exception *all* were scrapped. (CF PL40739)

Sad times—the June 1, 1945 service at Flensburg for S/L Jack Beirnes, 438's CO. Padre S/L Handley Perkins officiated. G/C Nesbitt is nearest in the front rank. Burial was in the Aabenraa Cemetery in Jutland, Denmark. (CF PL44767)

Sports day at 143 Wing. (CF)

Murray's albums show various Luftwaffe aircraft. Included is this famous view of airmen around an Fw.190 at Melsbroek. In the end, this aircraft proved to be a booby trap. It exploded when moved, killing some US servicemen.

Flensburg was jammed with Luftwaffe aircraft in various states. Here are Fw.200 Condor and Ju.290 transports.

An airman looks over an Me.109 lying between a long-nose Fw.190 and a Ju.88.

Murray came home on the *Île de France*. He took this crowded deck scene as they neared Halifax, then snapped off a couple more frames as they docked. The scene would have been emotional—some aboard had not seen home for 6 years.

Murray Castator displays a battered but treasured piece of memorabilia—the kit bag that was his companion since joining up in November 1941. This was the bag 58 years later! (Larry Milberry)

## No.419 Repair and Salvage Unit: A Snapshot

Aircraft repair and salvage units were vital in daily operations. Should an aircraft be forced down, the nearest RSU would gather tools, a crane, flatbed truck, etc. and rush to the location. Should an aircraft be in ruins, it might be a matter of collecting explosives for disposal, salvaging a few components, and burying the main wreck. A complete aircraft could be disassembled and returned to base for cannibalization or repair. Naturally, there was never a shortage of work—after D-Day there were hundreds of fighters to salvage. No.419 (RCAF) RSU formed on January 24, 1944 in anticipation of D-Day. By April 1944 it numbered 305 men at Funtington. The unit diary for May 1, 1944 noted Typhoons JR315 and JR241 accepted for engine changes, along with Spitfires 633, 239 and 484 (numbers are not always precisely noted). On the 8th the diarist wrote: "Spitfires numbering MK149, MJ967, MK181 collided on the ground. No.149 Cat. E, 967 Cat. E and No.181 Cat. B. Spitfire 443 ripped underside of fuselage extensively."

June 6 brought jubilation at 419—this is what it had been awaiting: "All personnel excited this morning over radio news broadcasts with regard to the landing of Allied troops in France... air activity witnessed ... is something to go down in history." Business increased. For June 7 Typhoon EJ911 was at Tangmere with flak damage, MN345 was at Funtington with so much damage that 419 had to replace a starboard wing. Also: "Spitfire LFIX MJ887 taken on strength as possible Cat. B. Reported to have twisted airframe. Flown in from 127 Wing to Funtington site."

On June 19 No.419 RSU moved to a marshalling area near Fareham. There it waited until boarding ship on the 29th, sailing next morning. W/C W. "Bill" Skelding and his men were elated: "Dominion Day was celebrated by setting foot in France and by a fine display of fireworks provided by the enemy." On July 3 the unit began work: "Two Spitfires and one Typhoon accepted for repair." For the 15th: "Crash inspector and salvage crew were unable to pick up a damaged Typhoon due to proximity to enemy territory." This may have been the occasion when the Germans reached the Typhoon first and burned it. On July 22 there was some special activity: "Messerschmitt 109G14 picked up by No.5 salvage crew within a few hundred yards of enemy territory. This aircraft is in very good condition, having been brought down by a .50 bullet which pierced the oil tank. This job was packed up in record time of 2 hours and 20 minutes ... Hurricane IV taken on for repairs and Auster IV ... air tested and passed on to No.143 Wing." The Me.109 was shipped to the UK the next day. On July 26 another 419 crew was fired on by the Germans, so had to give up trying to salvage an Fw.190. Next day the same thing happened to a crew working on a Mustang. The statistics for 419 for July 1944 show what a vital unit it was:

No.419 RSU at work. First, recovering 403 Squadron Spitfire KH-T after a prang in Normandy. The pilot could have picked a better spot—he landed in a mine field! Even so, 419 got the plane apart, the heavy vehicles in, and the job cleaned up. At other times the task was trickier—German units could be at hand. Then, the gorgeous Me.109 recovered by 419 in July 1944 at Fontenay-le-Peniel. The crew included (from the left) LACs R.J. Allan and R. Watson, Sgt R.A. Edwards and LACs R. Williams and C. Dickenson. (CF PL31115, '30932)

| | |
|---|---|
| Aircraft accepted | 118 (12 types) |
| Aircraft repaired and returned to service | 69 |
| Aircraft shipped to UK for repair | 25 |
| Aircraft reduced to spares | 24 |
| Parachutes packed/serviced | 438 |
| Dinghies packed/serviced | 404 |
| Test parachute jumps | 20 |
| Aircraft components repaired | 155 |
| Aircraft guns repaired/serviced | 1029 |
| Engines repaired/serviced | 611 |
| Propellers repaired/serviced | 285 |
| M/T mileage | 69,985 |
| Petrol used | 8820 gal. |
| Correspondence and signals | 2612 |
| Mail handled | 20,000 |
| Meals served | 55,000 |
| Cigarette ration | 102,000 |
| Food drawn, prepared, consumed | 20.15 tons |
| On sick parade | 105 |
| Hospitalized | 32 |
| Evacuated | 8 |

September 23 was busy, 7 Typhoons and a Spitfire being salvaged and delivered for repair. Enemy fire fell on 419's camp at B.58 that night, destroying 2 aircraft and a truck. For the month 16 Typhoons, 12 Spitfires and a Hurricane were repaired; while the salvage section hauled in 15 Typhoons and 9 Spitfires. One each of B-26, P-47, B-24, Hurricane and Stirling were inspected; some help was rendered in their repair. By March 1945 No.419 was at Volkel, Holland. Even though war's end was near, lots of work remained—27 Tempests, and 11 Typhoons were repaired, and 32 aircraft salvaged. On May 18 members of 419 visited Belsen to distribute aid to the inmates: "Arrangements were made ... to have two Sergeant Majors and a Polish Jewess ... give talks to the unit personnel on conditions and atrocities committed by the Germans in various prison camps." Through June 1945 No.419 was at Fassberg and Lubeck. Business now was petering out with repairs only on 4 Tempests, 2 Typhoons and 1 Meteor (engine change); 6 aircraft were salvaged. On August 28, 1945 No.419 RSU disbanded under S/L Robert Cushley.

# Dave Davies' Album

Dave Davies, an LAC fitter on 439 Squadron, had grown up in Toronto. After finishing high school in 1939, he went to work in the food brokerage business. He joined the RCAF in 1942, trained as a rigger and completed a posting at No.6 SFTS before going overseas. In January 1944 he joined 439 at Ayr, where it was converting from Hurricanes to Typhoons, then served to the end at Flensburg in April 1945. Dave kept a good record as he went along, many of his pictures appearing in *Typhoon and Tempest: The Canadian Story*. Here is a new selection, starting with a snap of 439 erks looking good atop an early Typhoon at Ayr: (standing) fitter Cpl E.A. "Eddie" Zigayer from Montreal and photographer LAC Ian C. MacDonald; then LAC fitters Dave Davies and Jack Pinnell from Niagara Falls.

After months of training in the UK, 439 landed at B.9 Lantheuil in Normandy on June 27, 1944. Here LAC Mike Lipp gets used to life on the beachhead.

### No.143 Wing Statistics and Denouement: Nos.438, 439, 440 and 168 (RAF) Squadrons

| | |
|---|---:|
| Missions[†] | 1794 |
| Sorties[‡] | 12,805 |
| Ops hours | 12,881 |
| 1000-lb bombs | 7936 |
| 500-lb bombs | 9896 |
| 20 mm rounds expended | 1,078,000 |
| Rail cuts | 1210 |
| MT destroyed/damaged | 1577 |
| Tanks destroyed/damaged | 62 |
| Locomotives destroyed/damaged | 341 |
| Barges destroyed/damaged | 136 |
| Bridges destroyed | 16 |
| Locks destroyed/damaged | 2 |
| Pilots missing | 104 |
| Pilots returned from POW camps | 19 |
| Typhoons lost | 128 |
| Typhoons badly damaged | 148 |
| E/a destroyed, probable, damaged | 22 |

[†] This term is used by 143 Wing to describe an operation with one or more Typhoons.

[‡] This term is used by 143 Wing to describe operations by individual Typhoons, e.g., one mission of 36 Typhoons would total 36 sorties.

"Boche Basher" of 439 was the usual mount of F/O John Carr. Here it is at B.9 ready for a sortie.

Mid-day volley ball during a lull at B.9.

Really "out in the field". Wireless tech Cpl K. Ackland from Ottawa of 439 catches up on his reading, while the squadron sojourned at B.24 St. André in August 1944.

Dave Davies, H.S. "Stan" Webber from Toronto and Ted Wytsma from Red Deer, Alberta in an operational scene from Eindhoven. Dave was splicing cable.

LACs Bob Burton of Hamilton and Dave Davies at Eindhoven with F/O Jimmy Brown's Typhoon. Dave was painting the name "Peace River" on the kite. Then, F/O Brown from Peace River, Alberta by his plane.

Fitter Dave McCandless from Brampton, Ontario with his pilot, F/O William Adam "Nick" Gray of Toronto. Tiffie pilots wore heavy gauntlets to protect against fire in the cockpit. On September 26, 1944 Gray was on an Arnhem do, his section returning to Eindhoven late in the day. While preparing to land, he spotted an Fw.190, which he damaged with a quick squirt, then got down as fast as he could—with all its bomb craters Eindhoven was not the place for a night landing. Next day, Gray was hit by flak and went straight in to his death.

Dave Davies and armourer John Pesant from Montreal with a Hannoverean fraulein in April 1945—Canucks were not deterred by a strict injunction about fraternizing with German girls.

Some of those who kept the RCAF's Typhoons going—No.6439 Servicing Echelon at Celle on May 18, 1945. Behind are LACs G.L. "Len" Lemoins, Tom Fox, J.M. "Jim" Lapaire, K.G. "Ken" Burt, John F. Buckley, D.J. "Dave" Davies, George L. Browne, D.A. "Doug" McArthur, J.G. "Jerry" Hayes, W.M. "Bill" Brennan, "Stitch", O.A. "Al" Mitchell, E.H. "Harry" Duke, George Parry, K.C. "Ken" Dawson, E.E. "Johnny" Jonson, Bob Sault, Dave Ship, Stan Fisher, W.M. "Bill" Ross, A.R. Richardson, Cpl Richards, LACs J.C.A. "Shorty" Mayer, R.H. Pelton, Addley, unknown, J.J. "Joe" Deveau, Gerry Miller and Pat Fogerty. In the middle are Cpl C.G. Wigley, LACs M.P. "Mike" Zak, Norman V.R. Camp, Ian W. McCulloch, A.W. "Al" Chowne, Dave Epp, R.M. "Dick" Dopson, W.P. Neville, Roger Dauphinais, W.A. Marsh, S.A. "Stew" Bernier, J.E. "Jack" Shuttleworth, J.F.C. "Johnny" Pesant, J.L.D. "Leo" Brault, N.W. Mason, A.H. Gilpin, Haley, V.L. "Jet Job" Gray and J.E. Anderson. In front are LAC G.F. "Gar" Shields, Cpls H.S. "Stan" Webber, Eddie Zigayer, E.J. Donaldson, R.L. "Bob" Jones, Sgt S. Cowie, Cpl W.M. "Bill" Blackey, Sgt George Blyth, Cpl R.F. Ingram, F/O Jeffries (maintenance adj), F/L Walsh (squadron engineering officer), S/L Don Brewster (CTechO), F/L Langille (EO), FSgt Bob Malo (i/c 6439), Sgts Arnie Roode, Reg Cappleman, J.A. "Johnny" Gibb, Ross H. King, Cpls H.N.J. East, K. Ackland, P. Preziosi.

It took a small army to support a squadron of 18 Typhoons. On April 2, 1945, No.439 Squadron at Hustedt posted its servicing roster for gunnery camp at RAF Warmwell, near Bournemouth. More than 100 men from No.6439 Serving Echelon were listed with F/S R. Malo in charge (i/c) overall: fitters, 20 men, Sgt A.L. Roode i/c; fitters 20, Sgt R.H. King i/c; armourers 28 men, Sgt G. Blyth i/c; electricians 4, Cpl Wigley i/c; instrument 4, Cpl Ingram i/c; wireless 9, Sgt Gibb i/c; maintenance assistants 3, Cpl R.S. Brown i/c; photographers 2 and clerk 1. There also were 28 in repair and inspection and 12 in headquarters (e.g. drivers and cooks). Eight Dakotas (KN270 shown), escorted by Spitfires, completed the move. KN270 later served the RCAF/CF to 1970, then flew in the bush with Ontario Central Airlines as C-GCKE.

Eindhoven visitors. B-17 43-38018 on a pit stop—143 Wing helped it along. The same went for the B-24.

From B.150, Celle G/C A.D. Nesbitt, commanding officer of 143 Wing wrote "A Brief History of 143 Wing" for distribution to his men. His farewell message tells how it all came to an end for the RCAF's Typhoon specialists:

*On the 7th day of September 1945, No.143 (RCAF) Wing ceases to exist as a unit in the field. Before the disbandment takes place, therefore, I would give you this message of appreciation and farewell from the officers who have commanded this wing.*

*Through the trials of the formative period of the wing, through the pre-invasion fever, through the dust and mud of the beach-head and the wearying pursuit across France, Belgium and Holland, from the Rhine crossing to the final drive to the Baltic Sea, each of you has played his part with untiring energy and skill. In accomplishing the tremendous task set, you have given of your best, and as a result the achievements of the wing will live in history.*

*The deeds of the pilots in support of the Army and in the destruction of the enemy have been unsurpassed in warfare, and we pay our tribute to those members of the wing who have given their lives in achieving this great victory.*

*We have striven together to serve a common purpose toward a common end, and now that the goal is reached the time has come to say good-bye, and good luck to each one of you.*

# Typhoon Finale

Rarely does a decent new photo emerge of a Canadian Typhoon. Yet, in 1999 Murray Castator pulled this one from his files. MN553/5V-K of 439 Squadron had flown sorties against V-1 sites, dropping 1000-lb bombs. On August 12, 1944 flak claimed it around Le Pont de Vers, France. F/O E.J. Allen, age 23 from Edmonton, did not survive.

Views of MN740/MR-U of 245 Squadron. Such Canadians as Sam Bennett, G.L. Dakin, H.T. Mossip*, Arthur Miron*, John Thompson, Chester West and George Wharry served on 245 Squadron. The pilot of MR-U has made a smooth belly landing. Such prangs were common, as damaged Typhoons returned day by day. MR-U soon was repaired, finishing the war safely. *KIA (Murray Castator Col.)

Feel the rumble, smell the fumes as a 440 Typhoon gets ready for ops at Eindhoven. LACs Ivan Black and Del Christopherson assist P/O R.A. "Dick" Watson, who hailed from the tiny railroad town of Oba in Northern Ontario. On July 16, 1944 Dick's Typhoon exploded after being clobbered by flak. Hurled out, he parachuted into an artillery duel, was rescued by Allied soldiers, then put in charge of escorting German POWs to the rear. Later that day he reached B.4 from where someone flew him by Auster to 440 at B.5. He was flying the next day. After the war Dick ran a tourist operation in the North. He continued to fly into the 1990s. (CF PL40736)

Wireless technician Ed Adair (438 Squadron), fitter George Parry (439), rigger Fred "Bud" White (439) and rigger D.J. "Dave" Davies (439) attend the book launching for *Typhoon and Tempest: The Canadian Story* in Toronto on December 7, 1992. The solid interest of such veterans contributed greatly to RCAF heritage into the 21st Century. (Larry Milberry)

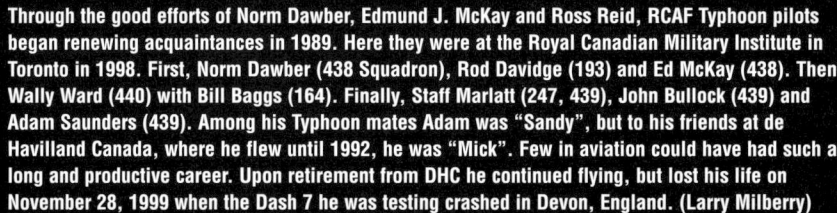

Through the good efforts of Norm Dawber, Edmund J. McKay and Ross Reid, RCAF Typhoon pilots began renewing acquaintances in 1989. Here they were at the Royal Canadian Military Institute in Toronto in 1998. First, Norm Dawber (438 Squadron), Rod Davidge (193) and Ed McKay (438). Then Wally Ward (440) with Bill Baggs (164). Finally, Staff Marlatt (247, 439), John Bullock (439) and Adam Saunders (439). Among his Typhoon mates Adam was "Sandy", but to his friends at de Havilland Canada, where he flew until 1992, he was "Mick". Few in aviation could have had such a long and productive career. Upon retirement from DHC he continued flying, but lost his life on November 28, 1999 when the Dash 7 he was testing crashed in Devon, England. (Larry Milberry)

Jack Cook (439), Graham Kennedy (137) and Clayton Leigh (182, 439) at a February 2000 get together at the Royal Canadian Military Institute in Toronto. RCAF war heroes Buzz Beurling and John Fauquier look on from their portraits. (Larry Milberry)

The monument at Noyers-Bocage commemorating 151 Typhoon pilots lost in the Normandy campaign of May-August 1944. In all 274 Typhoons were lost in this period, 91 in August alone. The idea for a memorial was suggested by Denis Sweeting, DFC (RAF) and supported all the way by the indomitable Jacques Brehin of *L'association pour le Souvenir des Ailes de la Victoire de Normandie*. Under Brehin the association had excavated Typhoon crash sites and arranged for pilots' remains to be properly interred. Charles le Baron constructed the monument using black granite from Zimbabwe and blue pearl granite from Brittany. The dedication on June 9, 1990 included the release of 151 doves of peace. The honour role, unveiled on June 8, 1994, includes the names of 41 Canadian pilots, 23 from 143 Wing. (via Bill Baggs and Ed McKay)

Typhoon pilots on May 19, 1999 at the Oban Inn in Niagara-on-the-Lake, where they were the guests of Sarah and Gary Burroughs. In front are Wally Ward (440 Squadron), Rod Davidge (193 Squadron), Ed McKay (438), John McCullough (1, 439), John Friedlander (181), Bob Hayes (174), Frank Johnson (174, over Bob's left shoulder), Stafford Marlatt (247, 439), John Thompson (245), Bill Clifford (440), Graham Kennedy (137), Norm Howe (175) and Jim Ruse (439). Behind are Clayton Leigh (182, 439), Bill Breck (198, 439), Walter McCarthy (438, 440), Russ Murray (247), George Lane (198), Crosier Taylor (440), Ted Smith (440), John Nixon (137), Murray Hallford (439), Bill Baggs (164) and John Bullock (439). Spitfire pilots had great respect for those who flew Typhoons—they knew what losses there had been. Al Bathurst (442) once teased about Ed organizing a Typhoon reunion—"You and which other pilot, Ed?" Ian Ormston (401, 411) once joked to Ed, "I can always tell a Typhoon pilot—he's the one who still has his hair's standing up!" (Larry Milberry)

TO THE GLORIOUS MEMORY OF THE 151 TYPHOON PILOTS WHO GAVE THEIR LIVES
DURING THE LIBERATION OF NORMANDY MAY - AUGUST 1944

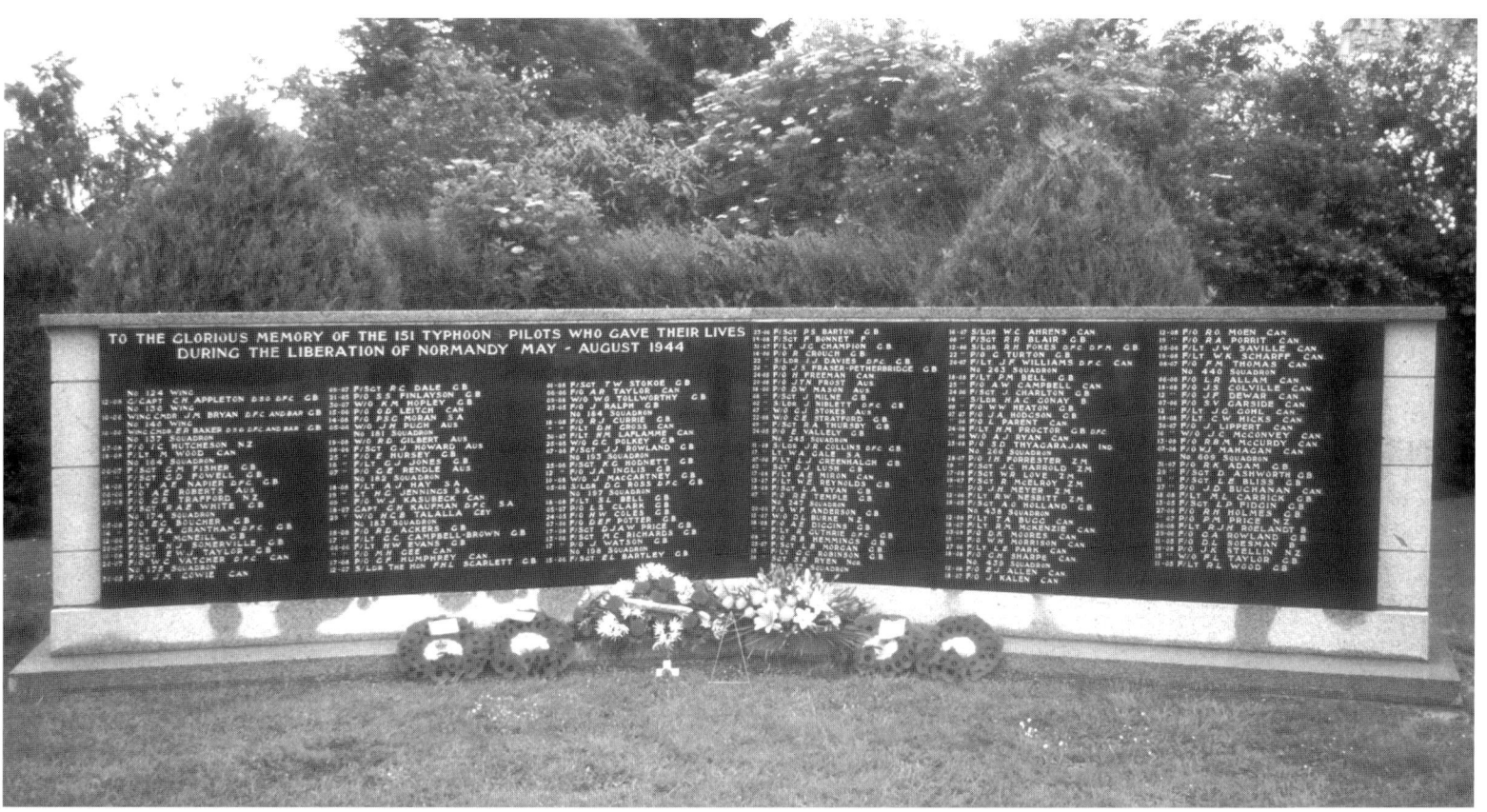

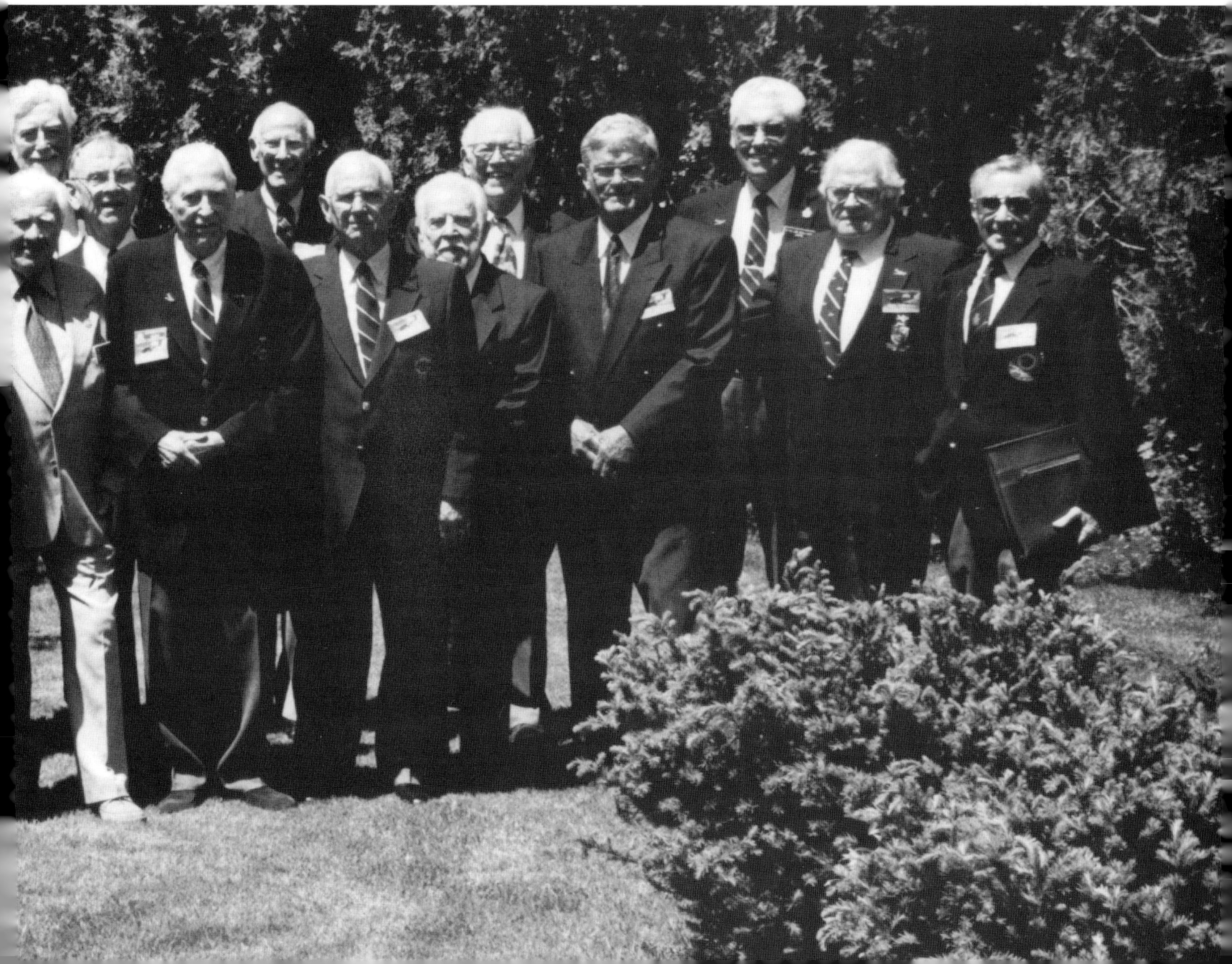

Jack Myles, Don Scott and Lawrence McMillan, Canadian Spitfire pilots with the PRU wing at RAF Benson in 1942. On the right a photo tech installs an aerial camera. Myles found that at 30,000 feet a PR Spitfire flew at about 360 mph, compared to 350 for a Mossie. (CF UK4755)

## Photo Reconnaissance

Knowledge of an enemy's plans and activities has always been coveted—the nature of his holdings (geography, weather, etc.), his emplacements, manpower and equipment, and the results of action taken against him. Only by diligent and sly means could such data be obtained, and an opponent would go to any length for it. After all, the least bit of news could turn the tide of war. The case of the Battle of Midway of June 1942 is a classic case. As to collecting information, nothing could contribute more than having a bird's eye view. In early times an observer might climb the highest tree or hill to get the big picture. Then came a great advance—the manned, lighter-than-air balloon, introduced in the late 1700s. In major wars that followed, balloons were part of an army's kit. With WWI, observation balloons flew in their thousands over the Western Front. Suspended from them in gondolas, skilled observers watched the enemy's every move. Accordingly, they then directed their own troop movements and artillery fire.

The airplane inevitably became the preferred means of aerial observation during WWI. While a pilot flew a prescribed route, his observer, with maps, perhaps a camera, maybe even a Morse transmitter or radio, would record as much battlefield information as possible. The mission done, they would turn for home. But the enemy was loath to let a reconnaissance crew escape, so kept his scouts on the prowl, ready to pounce. For this reason, the observer, kept his machine gun handy. If lucky, he also had a fighter escort. Such were the beginnings of aerial reconnaissance which, by 1939, had progressed considerably. Aircraft now were faster, flew further, carried greater payloads, and mounted improved cameras. Besides this, the art of interpreting aerial photos had advanced.

Jack Myles was born on January 13, 1923 in Saint John, New Brunswick, where his father was general manager at Eastern Bakeries Ltd. After high school Jack worked briefly with a plumbing company, then, in March 1941, joined the RCAF. He went to Toronto for Manning Depot, did guard duty at St. Thomas, Ontario, attended ITS at Victoriaville, Quebec, then was posted to No.21 EFTS (run by the Miramichi Flying Training School) at Chatham, NB. There he made his first flight in Finch 4440 on July 29, 1941, Mr. Chambers instructing.

Myles soloed on August 12 in Finch 4787; he made his last EFTS flight on September 23, finishing with 27:40 hours dual, 29:50 solo. He was posted to No.9 SFTS at Summerside, PEI for advance training on Harvards, making his first flight on September 29 in AH185 with P/O Calvert. The course totalled 94:20 hours. Myles received his Wings in December, leaving Summerside as a sergeant pilot. A note in his logbook stated, "No outstanding faults."

Before long Myles was aboard SS *Stratheden*, sailing for Great Britain. It was such a rough crossing that the grand piano in the ballroom was smashed and a lifeboat torn away. The latter was retrieved by a passing vessel, so there was concern that Stratheden was lost. Once in England, Myles, who had celebrated his 19th birthday at sea, spent a month in Bournemouth, another in Hastings. Finally his posting came—to No.17 AFU at Watton. There he trained on the Master from March 26 to April 23, 1942. May 3 to June 24 he was at Blackpool on a general reconnaissance course on the Botha. There S/L Donald M. Brass noted in his log, "Did consistently good work and showed natural keenness." (Brass was a prewar NZ/RAF type who had done a tour in Canada in the BCATP, then had a business in career in Canada after the war.)

Myles now went to No.8 OTU at Fraserburgh, Scotland, to convert to the Spitfire Mk.I (he flew aircraft L1001, R1001, R6968, R7020, R7146 and X4490). Finished on August 3, he advanced to "B" Flight on the Spitfire Mk.V. He flew another 27:30 hours before being posted on September 4 to No.541 (PRU) Squadron at RAF Benson. There, Myles hooked up with two other Canadians, Donald G. Scott and Lawrence "Mac" McMillan. Now came the point of all those intense courses—performing under fire.

At Benson, Myles flew the Spitfire Mk. V, IX and XI. These were stripped of weight, armament included, and fitted with cameras and extra fuel. There was a vertical camera for high altitude (its lens usually set to be out of focus below 20,000 feet). Camera settings were adjusted by a photo technician before takeoff, so could not be changed en route. Jack Myles noted in 1998: "The pilot controlled the time interval between exposures to provide a 60% overlap, so that every object on the ground appeared on two photographs. These could be manipulated under a stereoscope to provide a magnified, 3-dimensional image that provided much more detail than a single vertical print." Myles describes the daily routine of a PRU pilot at Benson:

*We required good light in order to take photos. Therefore, each day we had an early photo time and a late photo time. Unless we had an extra long sortie, we wouldn't start too early. We assembled at the Met office at 0800 and the senior Met officer briefed us on the weather for the whole of Europe. We then went to the intelligence office, gathering in front of a large map of Europe with coded pin flags identifying the first, second and third priority targets. The CO would indicate where he wanted sorties flown; the flight commander would assign them to pilots according to a roster system. An intelligence officer would brief each pilot, who would draw the required maps from stores, do his navigation, and complete his flight plan. The pilot then went to the dispersal to collect his flying gear, parachute, etc., proceeded to his assigned aircraft, and did his visual inspection. The ground crew helped him into his parachute and strapped him in.*

Myles flew his first operation on September 6, 1942, taking BR661 to Rouen and Le Havre. In 1998 he recalled:

*Since we flew alone, there was always a number of pilots at the dispersal, waiting their turn to fly. They always turned out to see off each flight, and welcome it home—great camaraderie! For my first flight there was a good turn out. I got buckled in, did my cockpit check and started up. Two ground crew draped themselves over the tail plane to hold it down, while I ran up the engine to test for the magneto drop. It sounded OK, but "Chiefy", our flight sergeant, signalled me to shut down. The ground crew made some adjustments. Then we started up again and did another test. Chiefy jumped up on the wing and said that he hadn't liked the sound of the engine, but it seemed to be OK now. However, if I had any doubts, I was to turn back. A nice start for a very nervous pilot's first Op!*

*I climbed out to the coast and crossed out at Beachy Head. It was a clear, sunny day. Looking back, I had my first view of the White Cliffs of Dover. The PR Spitfire had a plexiglass hood with a blister on each side containing a rear-view mirror. This gave a view of the blind spot behind the tail, but more importantly let the pilot check that he was not leaving a "con" trail that would give away his position.*

*When I was part way across the Channel on my way to Rouen and LeHavre, I checked the mirror and saw black smoke. "Engine trouble", I thought, and made a steep diving turn and headed home, relieved that I didn't have to proceed. Alas, when I looked again the "black smoke" turned out to be a white con trail, which only appeared black in the mirror. So I turned around again and completed my sortie.*

Other than the false start, Myles' first operation was routine. He returned with his photos after 2:10 hours. On a single operation a PRU aircraft could bring back photos of several targets. Important to Bomber Command was the damage done to a target a few hours earlier. Otherwise, a log such as Myles' shows trips to airfields, factories, rail yards, road junctions, bridges, canals, ports, radar facilities, and V-1 and V-2 sites. Photo interpretation—making sense of the photos of such potential targets, was an enormous task. In April 1998, speaking to 250 Wing, Air Force Association of Canada, Myles described this vital "PI" specialty:

*We took the photographs, but it was the interpreters who skillfully uncovered and reported the information they contained. Eventually, several hundred people were employed at this work at Medmenham on the Thames near Benson. Carefully screened, well-educated people from varying backgrounds, they worked in a number of sections. They were interactive, but each had its own specialty.*

*"Covers" was the name given to the batches of photos from a sortie. Covers from the latest sortie were passed from section to section and each took the information pertinent to them. One specialized in bomb damage and worked closely with Bomber Command. Another concentrated on shipping, and kept track of all major naval ships, but also merchant ships and even small coasters. Factories were the responsibility of another section. Photo interpreters could supply an amazing amount of detail—what was being produced, how much, even the destination of the products. Airfields and aircraft were a specialty. Peenemunde, rockets and jets were monitored by a special section. Another section continually monitored the Channel coast. When the invasion was mounted, they were able to provide up-to-date maps showing the location of every German gun emplacement, radar site, etc."*

On September 12 Myles flew a 2:40-hour recce to Cherbourg and Le Havre. For his first month he completed five ops for 19:10 hours. He soon learned the tricks of the trade, which were few and common-sensical. Pilots looked for the zone in the upper air where condensation trails would not form. Sticking in this air made it hard for the enemy to pick out an intruder. On the other hand, PR pilots kept a watch for "con" trails. These were avoided, for they could mean Luftwaffe fighters. PR pilots never followed the same course heading home as coming in to target, lest fighters be lurking.

On February 8, 1943 Myles logged his first and only trip at 42,000 feet (Spitfire XI EN151). He noted, "Chased home by e/a"—he was trailed for a bit by a Luftwaffe fighter, which he spotted when the sun glinted off its canopy. On February 20 Myles noted, "Hamburg, Wilhelmshaven. Took convoy in mouth of Elbe, U-boat yards, docks and old town of Hamburg. Docks and part of town at Wilhelmshaven."

A V-2 site in the Pas de Calais, the area of France closest to England. It was photographed on September 14, 1943 from a Spitfire flown by Jack Myles. Notice the willy-nilly effect of the bombing, most of which had been done in daylight. (All, Myles Col.)

This sortie also covered the island of Heligoland which had an airfield. On such a long trip (4:25 hours) Myles carried drop tanks. The trip over Wilhelmshaven likely was to record bomb damage of the previous night, when 338 RAF bombers struck the place (a day earlier 195 bombers largely had missed the city). For the night of the 19th/20th there was another major error, with no damage done to Wilhelmshaven. Yet a third raid (February 25-26) was equally a failure. This later was understood—the Pathfinders had been marking with out-of-date Wilhelmshaven maps!

**Damage recorded by Jack Myles at a German airfield in southern France on February 13, 1944. This bombing had been part of the Allied softening up of German bases in the months leading to D-Day.**

**An image from the Myles-Cawker sortie of March 8, 1944 to the Baltic Sea port of Gdynia. Note the capital ship alongside the centre dock. This is *Gneisenau*, which had been bombed at Kiel, then moved here, but never repaired. Gdynia was also home to Germany's first (but never commissioned) aircraft carrier, *Graf Zeppelin*. Also seen are drydocks, breakwalls, marine fabricating sheds, factories, roads, railroads, and residential areas. A photo interpreter would have gleaned much intelligence from this photo. If Bomber Command decided on a raid, it would have all the gen needed for planning. Except for a few craters the area appears mostly unscathed at this time, but this situation would not last. On December 18-19, 1944, for example, 236 Lancasters bombed Gdynia.**

Myles and Cawker made 10 flights, getting used to their new aircraft. The routine was different—the pilot operated the oblique camera, the navigator ran the others. The crew commenced ops on December 5, flying MM232 on a 4:15-hour trip to Annecy, France, site of a ball bearing factory (later hit by RAF bombers). Much interesting work followed, including on January 20, 1944—photographing an airfield near Toulouse from 26,500 feet. Over target they lost an engine and feathered the prop. They could not hold altitude till 12,500 feet. Now they set course across Spain for Gibraltar. Since they did not have maps for this region, Myles pulled his silk escape map, carried by every aircrew, from the lining of his trousers. "I had to practically undress to get at it," he recalled years later, "but with it we were able to get the outline of the coast and the approximate distance to Gib." He added that his one engine, "didn't even heat up." It was a fortunate trip—6:50 hours, 3:50 on one engine. Arriving at Gib, Myles had no radio, so fired a flare and landed. His duff engine was repaired,

A 2:40-hour trip in BS498 on May 13 focused on airfields. Myles noted targets at or near Beauvais, Bernay, Breux, Evreux, Le Havre and Paris. Photography was from 29,000 feet. On July 13, 1943 he was slightly damaged by flak in EN662. His last PR Spitfire flight was on August 20, 1943, a 3:05-hour recce to Emden in EN410. Myles had flown 44 ops and logged 420:45 hours, including 218:45 on Spitfires. Now he was posted to Mosquitos, starting with conversion flight at 410 Squadron at Coleby Grange. This began with some dual on Oxfords, then a Mossie flight with F/L March on August 27, 1943. He soloed two days later. Now Myles spent some time at 400 Squadron flying the Mustang, Auster I and Tiger Moth; but it seemed that he was spinning his wheels. Not about to be railroaded, he took an Auster to Benson and spoke with W/C Ring about joining 544 Squadron (PR Mossies). Ring agreed to bring on Myles and the navigator with whom he had crewed at 410 Squadron, F/O Hugh Roger Cawker (J11805).

Born in Winnipeg on June 24, 1922, Cawker had enlisted in May 1941, hoping to be a pilot. He began training in September at 19 EFTS, Virden, Manitoba. But after 13:55 hours, ceased training—his CFI, Andy Madore, considered Cawker lacking in "air sense". He remustered to air observer, training at No.10 AOS, in Chatham. There he passed with an overall 82%. A note from his instructor, F/L B.H. Wilson states, "Likable chap who should make a very good observer". His officer commanding noted, "A trifle cocky, nonetheless, a good worker." Bombing and gunnery at Mountainview, and an advanced nav course at Pennfield Ridge followed. From the latter he graduated with a commission. In June 1942 Cawker was posted to No.31 OTU (Mosquitos) at Debert, and had a brief posting to Ferry Command, being the navigator in July 1942 delivering a Hudson to the UK. Now he was with Jack Myles at Benson—they made their first Mossie flight on November 12, 1943 in MM246.

**An airfield near Gdynia photographed on March 8, 1944 by Myles and Cawker. Seen are runways, hangars, revetments, shops, barracks and several He.111 and He.177 bombers of KG1, a unit operating on the Eastern Front.**

giving him and Cawker time to relax briefly. On January 22 they made a test flight. Next day they returned to Benson in 4:25 hours, photographing the north coast of Spain en route.

March 5, 1944 saw Myles and Cawker over Peenemunde, the German V-weapon base. This day Myles, away for 5:25 hours, also covered airfields as distant as Rostock. On March 31 he was again over Peenemunde (33,000 feet), likely taking photos by which British intelligence could assess new activity there. For his 57th op Myles based from San Savero near Foggia, Italy, photographing around Belgrade. Departing one day, the escape hatch blew off, necessitating repairs. On April 8 he and Cawker took a slight flak hit on a 6-hour mission to Kassel. Another day they were despatched to Danzig and Gdynia. Along the way they diverted to the island of Bornholm to photograph what appeared to be a large supply vessel surrounded by many U-boats.

In the weeks before D-Day the crew photographed transportation targets (mainly rail). For June 1944 they logged 32:30 hours on six ops, then flew their last (Myles' 70th) on July 7— a 4:35-hour recce around Lyon and Montpellier in LR417. On July 16 his OC, S/L Steventon, wrote in Myles' log: "As a PR pilot, exceptional." All had not been pure duty at Benson. When not on duty, Myles and his squadron mates often went to London on leave, or visited pubs in closer towns such as Wallingford and Oxford. They were always coming up with new schemes for fun. Coming home from San Savero, for example, Myles thoughtfully brought a keg of Italian wine, adding a little extra cheer in the mess.

From Benson F/O Myles was posted to the Fleet Air Arm at RNAS Crail (HMS *Jackdaw*). There he joined 618 Squadron preparing for Op. Highball. Equipped with Mosquito VIs and XVIs, 618 was destined for the Pacific, where it would go after Japanese capital ships. PR Mossies would locate the targets, skip-bombing Mossies would go in, and the PRU boys would follow up with damage assessment. Myles now converted to the Barracuda torpedo bomber, which he enjoyed. There were practice landings on a runway painted like a carrier deck. Myles made 72 ADDLs (aerodrome dummy deck landings) on 16 Barracuda flights. Next he went aboard the aircraft carrier HMS *Rajah* in August, doing six takeoffs and landings, one ending in a deck prang. Now Myles rejoined 618 at Wick, where he again teamed with Cawker. They did more ADDLs, dicey activity, for the Mossie normally approached at about 120 mph. For ADDLs this was worked down to 100 mph, "just over the stalling speed", as Myles recalled: "There was little or no control once you came over the runway." On September 29 at Wick, Myles flew the Spitfire for the last time—a Mk.VIII on a local test. He now was required to make one Mosquito landing and takeoff aboard ship. These he did

Jack Myles qualified aboard HMS *Rajah* in August 1944. Even naval pilots could miss a landing, so one could sympathize if a landlubber from 618 Squadron, not used to pitching runways, wasn't perfect. Here F/L Clutterbuck of 618 waits in the cockpit of a Barracuda after a balls-up aboard *Rajah*. A note on the back of the photo reads: "Official photograph. Not to be communicated to the Press."

Jack Myles at home in Saint John in June 1998. (Larry Milberry)

with Mosquito "K" aboard HMS *Implacable* on October 10, 1944. He did this without his nav, then flew RN batsman Lt Hancock ashore as a passenger. In 1994 Myles noted, "I found Mosquito landings much easier than the single-engine Barracuda. Visibility was better, and the extra weight and power made it much steadier."

For the last day of 1944 Myles' log shows totals by type: Finch—57:30 hours, Harvard—95:00, Master—48:00, Botha—1:20, Spitfire—220:45, Tiger Moth—1:10, Oxford—6:55, Auster—12:10, Mosquito—250:30, Mustang—3:05, Barracuda—14:50, Anson—9:20. Of some 720 hours, 247:25 had been operational (115:05 on 44 Spitfire sorties, 132:20 on 26 Mossie sorties).

Myles and Cawker now sailed aboard the escort carrier HMS *Striker*. The Mosquitos, lacking folding wings, were inhibited and chained to the deck. The ships used the Suez Canal, forming into a convoy in Aden, then stopping in Ceylon for about two weeks. En route, Myles was quarantined with measles. The porthole in his cabin opened onto the elevator shaft, so he missed the sights along the way. In Ceylon 618 went ashore daily for meals in a fine mess. On December 24, 1944 they reached Melbourne, Australia. The aircraft were craned ashore and trucked to nearby Fisherman's Bend airfield.

Since it took about six weeks to get 618 ready, the crews had time to enjoy Melbourne. Myles and Cawker didn't make their first flight until January 23, 1945 (Mossie DZ524). Then 618 settled at Narromine, New South Wales, flying Mossies and Barracudas. Myles and Cawker made 14 flights in March, 14 in April. One was to Alice Springs in NS729, one sight there being a herd of wild camels. They pressed on to Darwin, visiting a PRU and staying several days. On April 25 they took part in an Anzac Day flypast of nine Mossies, visiting various towns around Narromine. They made their last flight on May 4, going to Sydney and back in 1:10 hours.

No. 618 seems to have been in never-never land in the Pacific. Some felt that the Americans viewed the Pacific War as their affair and that the RN and RAF weren't welcomed so near the end. Others mused that the Royal Navy was at fault. To be an effective force in skip bombing, the RAF would have needed three carriers, but would the RN commit these for an RAF show? In the end 618 did no Pacific operations. To Jack Myles this was a relief: "Fortunately the Yanks had already destroyed most of the Jap Navy. I certainly was not looking forward to flying off a carrier in rough weather with a deck littered with other Mosquitos!"

On VE-Day Myles and Cawker left Australia aboard *Nieu Amsterdam*. En route home they stopped in Durban, where Myles visited relatives. Then they sailed to Capetown and up the west coast of Africa. With the war over there was great revelry, including dancing under the stars on a fully lit deck. Once in England, Myles heard that he was to receive the DFC. This was presented by King George at Buckingham Palace on July 13, 1945. His citation referred to the emergency landing at Gibraltar. He also received the US Air Medal for doing five PRU ops for the USAAF. Its F-5 Lightnings did not have the range for these missions, so Benson obliged. MGen Eaker presented Myles' gong at USAAF HQ in Widewings, outside London.

In this period Jack Myles and Don Scott heard sad news of their old friend from PRU Spitfire days, F/L Lawrence McMillan, DFC, AM. On May 9, 1945, the day after the war in Europe ended, McMillan was sent in Spitfire PM142 to photograph German ships withdrawing from Danish waters. These had been attacked earlier that day by Russian aircraft. When McMillan came by, the Germans, twitchy about further attacks, shot him down. His body was never recovered. For breaking the cease-fire and knocking down McMillan, the German vessels were attacked by the RAF. McMillan may have been the last Canadian killed in the European theatre by hostile fire.

On August 5 Jack Myles sailed aboard the *Louis Pasteur*, which docked at Quebec on the 11th. He reached home in Saint John on the 14th and was discharged on the 26th. He began studying mining at Queen's University, but switched to architecture at the University of Manitoba. He married Ann Mott in April 1946. After graduation in 1950 he began a career with the Saint John firm of Mott, Myles and Chatwin, remaining in architecture until 1989. Thereafter, he enjoyed life at his home in Kennebacasis Park, outside Saint John, gardening, sailing, reading and giving the occasional historical talk about PRU days.

As to F/L Hugh Cawker, he proved a restless fellow. Upon leaving the RCAF in August 1945 he was awarded a War Service Gratuity of $1211.37, dispensed in 4 equal monthly payments. At first he hoped to study law at the University of Toronto. Early in 1951, however, he applied to the RAF. This seems to have come to naught and we next hear of him in the French Foreign Legion. He fought in Vietnam, lost an arm, and was awarded the French Military Medal and the Croix de Guerre with Palm. Eventually, he returned to Winnipeg. He died on October 13, 1966 at the Mica Creek Dam site on the Columbia River.

## Tactical Recce: Jack Seaman

John T. "Jack" Seaman (J5113) was born in Amherst, Nova Scotia on October 30, 1918. After high school he did a year at Mount Allison University, then worked as a CNR machinist in Moncton. He enjoyed this and the off-hours with his work mates. In 1940, however, he decided to try air force. He did his indoctrination at No.1 Manning Depot, guard duty at Camp Borden (where he got up in an Anson—his first flight), then ITS on Avenue Rd. in Toronto. Finally came flight school—No.4 EFTS at Windsor Mills, near Sherbrooke, Quebec. He flew first in Finch 4491 on November 9, 1940 with instructor Frank Cooke. Ten days later Mr. T. Marshall sent him solo.

After 49:10 hours in the Finch, Seaman began advanced training on Ansons at No.8 SFTS in Moncton, flying initially on December 23 with Sgt Giles in W1755. There were interesting characters at Moncton, including a student who was the son of actor Basil Rathbone, and Whitey Dahl, an American instructor who had flown in the Spanish Civil War. Rathbone had a special arrangement to live off-station and always seemed to be with some gorgeous babe. LAC James W.P. Skidmore from Pennsylvania was known as a card sharp, always eager to fish in a new sucker. LAC Hall from Georgia distinguished himself by beating up the Cape Tormentine ferry. Caught in the act, it cost him some unpleasant ground duties.

Students nearly always went their separate ways after earning their Wings. Jack Seaman, however, occasionally would hear news of his old buddies. There was Skidmore, age 23—on December 29, 1941 he was lost on No.10 (BR) Squadron (Digby No.744 failed to return from an anti-submarine patrol). Then there was Massey Williamson Beveridge, age 28 from Westmount, Quebec. His 409 Squadron Mosquito went down the night of August 7, 1944. His navigator, F/L J.W. Peacock, died, but Beveridge, by then OC 409 Squadron, escaped. On September 20, however, he also died on ops.

SFTS ended on April 3, 1941. Seaman's flying totalled 82:15 hours (day) and 10:00 (night). His posting after Wings parade was to Central Flying School at Trenton. There he qualified as an instructor and soon was teaching on Ansons at No.7 SFTS at McLeod, Alberta. Seaman's boss was W/C Manafrank Brown, a fellow with a gimpy leg. One day Seaman had a minor taxi mishap. Brown tore a strip off him, gave him a week of orderly officer duty and squelched his posting to Ventura OTU at Pennfield Ridge. Long afterwards Seaman was overseas. He and Jack Donovan had just landed their Mustangs at B.2 (in Normandy). Donovan kicked some dust and a winco, the same Brown, came hobbling out in a snit. He started to reprimand Donovan, threatening to ground him. But as soon as he recognized Seaman, he relented. Many times over the years Seaman would think about his air force career and about the "might have beens." What, for example, might have happened had he

RCAF squadrons overseas were introduced to the armed reconnaissance role on Lysanders, then Tomahawks. Luckily, they did not have to fly operations with these types, enjoying the luxury of Mustangs and Spitfires by that time. Here a 414 Tomahawk flies a training sortie from Croydon in 1942. (C.H. Stover Col.)

not been punished at McLeod—all too many crews had gone for the chop in the tricky Ventura.

Seaman finished at McLeod with a November 3, 1942 flight in Anson 7498. With 1201:45 hours he headed for OTU at Bagotville. There, as usual, there were some outstanding types. F/L Ian Ormston, back from Spitfire tours on 401 and 411 squadrons, was in charge of "D" Flight. F/L Reyno of the Battle of Britain also was instructing, as was a F/L Kent. He had had a scary tour—flying Wellingtons equipped with cutters to slice the cables anchoring barrage balloons. The idea was to fly straight at a balloon in an effort to cut its anchoring cable, that same cable designed to snatch an airplane and hurl it to the ground.

Of course, there had to be some sort of shenanigans at Bagotville. One day in January 1943 Jack Seaman and James Angus Francis "Frank" Halcro of Ottawa took Harvard FE395 to 19,000 feet. Halcro, not aware that Seaman had blacked out from hypoxia, pushed the nose down

A dramatic view from August 1943 of Mustang IA showing a typical offensive load of 500-lb bombs. There also were two .50 machine guns in the nose and two .303s and a .50 in each wing. Tactical recce pilots took photos of specific targets, but also bombed and strafed targets of opportunity. Their 1- to 2-hour sorties were always action packed. This print was stamped "Secret. Photographic Section, HQ Fighter Command, RAF". (RAF)

By war's end 414 had logged more than 19,000 hours, scored 29-1-11 in aircraft, and claimed 76 locomotives, 13 vessels, and many other targets; 19 pilots were lost on operations. Here is Mustang AM251 of 414 at Ashford, Kent on September 13, 1943. On June 19, 1944 this aircraft took F/L Roger Arthur Bromley, age 21 from St. Catharines, to his death. No one knows what befell him. Was it flak, an enemy fighter, mechanical trouble, weather? There were many variables in the dangerous world of low-level armed recce. (CF PL19833)

and built up speed. When he started pulling out, the Harvard groaned. By planting his feet on the instrument panel, he finally got level at 6000, but the plane was bent. Seaman's log reads: "Returned minus panels and wings very wrinkled." The lads cooked up a story to explain their stunt and somehow pulled the wool over their boss' eyes. The records show FE395 struck off strength some time later. Frank Halcro went on to fly with 139 Squadron. On September 12, 1944, then 23 years old, he went missing over Berlin in KB227, a Canadian-built Mossie.

Seaman finished OTU with 56 hours on the Hurricane. He was posted overseas, sailing in a convoy that was badly mauled by U-boats. After Bournemouth he went to 41 OTU, flying his first Mustang there on May 25, 1943. At the time he thought, "This really breaks my heart," for he had wanted Spitfires. He found the Mustang cockpit uncomfortable compared to a Hurricane, but loved the reassuring purr of the Mustang's Allison engine. After 45 hours on type, Seaman left in early July for No.4 Squadron at Gravesend, but transferred to 414 "Sarnia Imperial" Squadron in early October. His first sortie there was an armed photo recce on October 6. For each operation a pilot like Seaman would be briefed by an Army intelligence officer. They would study maps and discuss things such as the day's flak threat. But often there was little hard information. Even the target might be but vaguely described, whether a bridge, road junction, rail yard, tunnel or V-1 site. After a recce the pilot would be debriefed by an RCAF intelligence officer. Meanwhile, his film was rushed to Benson.

On units like 414, PR was different from the strategic recce of the lone, unarmed aircraft from Benson. Several Mustangs might go on the same op. On January 28, 1944, for example, Seaman was with F/L George W. Burroughs, F/L G. Wonnacott and F/O R.O. "Bob" Brown. This lasted 2:20 hours. The Canadians spotted two aircraft, thought to be Arado Ar.96s, which they shot down. Along came two Me.109s, which met the same fate. After the latter fight nothing was seen of Brown, a 28-year-old from Daysland, Alberta. What eventually came to light was that an Me.109 had shot him down near Chartres, France. In 1998 Jack Seaman remembered having been with Brown at Bagotville. They had sailed on the same ship, where Brown was checked out for anti-aircraft defence. When a Canso came into view one day, the ship's captain ordered him to shoot it down! Brown, knowing a Canso from any German type, held off. There were other losses on 414. On July 26, 1944 21-year old F/L Donald Cameron McLeod was shot down near Gisorp, France. The same day F/O John A. Levi had to bail out. He was captured and hospitalized by an enemy unit, eventually being liberated.

In the spring of 1944 No.414, operating from Tangmere, sometimes had a Spitfire escort. For March 30 Seaman noted that, after photographing the airfield at Abbeville, his escorts were nowhere to be seen! Another time he was to photograph a target moments before an RAF bombing raid, during the raid, and immediately afterwards! His job done, he pushed his nose down, got up to 450 mph, and beat his escorts home. In this period there was plenty of flying. For May 1944 Seaman logged 29:10 hours, 11:10 on ops. In August 414 converted to the Spitfire LF Mk.IX. In 1998 he recalled, "The Spitfire was the easiest thing to fly that you ever saw. We adapted naturally to it." One day 414 visited a neighbouring USAAF P-47 wing, hoping to scrounge some of the always-desirable American flying jackets. They succeeded at this, but an American colonel stopped to chat. "What do you fellas fly?" he asked. They answered "Spitfires", then the colonel jokingly cracked about how he used their roundels as targets. This was in jest, although there had been numerous cases of RAF fighters shot down by American P-47s and P-51s.

In the "who needs it" department, on June 20, 1944 Seaman had his canopy fly off, this making for a breezy trip back to B.3 in Normandy. Later that day he was with Wonnacott photographing a bridge. Coming home, Wonnacott led him on the deck through a Royal Navy balloon barrage! On June 25 Joe Rousselle taxied into Seaman's tail. On July 15 Seaman lost another canopy. Around Falaise on August 18 things got congested, Seaman noting in his log, "Very hazy. Main danger was collision with Tiffies pranging." On the 22nd he passed 100 hours on ops, and for the month logged 37:10.

For September 1, 1944 he wrote: "Saw 400+ MT and horse-drawn vehicles and guns" on a recce between Amiens and Abbeville. On the 12th Bob Cutting of 414 made an inadvertent wheels-up landing at B.56 (Evères). On November 11 F/L Jack Seaman finished his tour, flying MK289 from Tangmere to Eindhoven in 1:20 hours. At this stage his log showed 1651:25 hours, 167 on ops. He took up drogue-towing on Hurricanes at 577 Squadron at Castle Bromwich, near Birmingham, finishing this in July 1945.

One day Seaman and F/O Webb noticed an American Vengeance torpedo bomber on the ramp. They asked about it and figured that, since it looked like a big Harvard, it couldn't be too hard to fly. Someone said OK, and away they went for some good fun. Not long after, Seaman was homeward bound. His ship docked in Halifax, and he and some fellows went ashore to party. Seaman came back aboard ship for one of his worst nights of the war—he had contracted food poisoning! Once recovered, he headed back to the CNR shops to reclaim his old job; but he soon left again. He tried farming with his wife Audrey, and his brother. But civvie life wouldn't do. In April 1949 Seaman rejoined the RCAF in Moncton. His rank and classification? LAC machinist at St. Hubert. He had been told that there was no likelihood of a flying posting. Then Korea boiled over. Pilots were needed; but station commander G/C Archambeau was discouraging. He would berate his men, suggesting that some were deadwood. He even obliged them to prove that they weren't communists! When Seaman won his way back into flying, the negative old groupie told him, "All you're ever going to be is a flying instructor."

Members of 414 "Sarnia Imperials" Squadron on D-Day, when they were assigned to photo recce over the Normandy beaches. Behind and on the prop are F/L James A. MacKelvie, F/O B.B. Mossing, F/O Roger A. Bromley, F/L L.F. May, F/O R.C.J. Brown, F/L J.M. Robb, F/L J.L.A. "Joe" Rousell and F/O Harold D. Cougler (adj). Standing are F/O R.C. Ritchie, F/L G.W. Wonnacot, F/L George W. Burroughs, F/O Donald C. McLeod, F/L R.C. Hutchinson, S/L C.H. "Smokey" Stover, F/L J.C. O'Neill (MO), F/L David A. Bernhardt, F/L N.F. Rettie, F/O J.H. Donovan and F/O J.C. Younge. In front are F/O J.A. Levi, F/L Paul M. Brunelle, F/L K.A. Brown, F/O E. "Gus" Garry, F/L J.T. "Jack" Seaman and F/O W.G. Sherk. Over France on October 31, 1943 May and Brown shot down an ex-French Air Force Yale trainer, then a Ju.88. On January 28, 1944 Robert O. Brown, Burroughs and Wonnacot shot down 2 Ar.96s and 2 Me.109s, but Brown was lost to an Me.110. On June 7 Bromley and Mossing shot down a Ju.52 near Mortagne. Butcher and Brown were hit by flak, becoming POWs. Bromley and MacKelvie were KIA on June 18. On July 4 May, Rouselle and Younge fought off 12 German fighters, Younge destroying one. On July 26, 1944 McLeod died near Gisorp, France. Levi, also hit, bailed out, was captured, treated for burns and eventually liberated. (CF PL30031)

S/L C.H. "Smokey" Stover of 414 Squadron enlisted in the RCAF in March 1941. After EFTS at Windsor Mills and SFTS at St. Hubert he joined 414. He was in action at Dieppe on August 19, 1942. In a fight with an Fw.190 that day he clipped a telephone pole, losing a wingtip! On August 21, 1943 he shot down a Ju.88; but moments earlier had watched his wingman, Lou Theriault, slam into the ground. When 414 lost its CO (S/L H.P. Peters) on November 4, 1943, Stover replaced him. On June 23, 1944 he and F/L N.F. Rettie were mauled by 7 Fw.190s. Rettie got back to base (RAF Odiham), but Stover had to jump from low level. Postwar, he served on 420 Squadron (P-51Ds), while working for Shell Oil in aviation products. In the 1990s he was Honourary Colonel with 414 Squadron (T-33s) at CFB Comox. (CF PL10744)

On March, 2 1951 Jack Seaman signed out a Harvard in the same hangar at Trenton where he had worked in 1941. F/L John Cooper gave him his category test on May 7; Seaman was soon back instructing. S/L C.H. "Cam" Mussells worked it so that Seaman got back to the fighter world. In January 1953 he did a T-33 conversion at Chatham, then returned to Trenton. For April 1953 his log shows some good flying there: 20:30 hours on the T-bird, 4:20—Harvard, 0:40—P-51, 6:50—Expeditor, and 5:30—Lancaster. In August 1954 F/L Seaman was part of an RCAF recruiting push—Prairie Pacific, a flight of CF-100s, F-86s and T-33s visiting communities across Canada. Seaman, who had first flown a Sabre at Trenton on September 23, 1954, next began the Sabre OTU at Chatham, his first flight being on February 22, 1955 in a T-33 with F/O Larry Spurr. Following OTU, Seaman was posted to his old squadron, 414 at 4 Wing, West Germany. Besides the routine Sabre flying, Seaman did an exchange with the USAF at Landstuhl on the all-weather F-86D. Next he attended RCAF Staff College, but left the air force in 1964, his log showing 5284 flying hours. After taking his commercial licence at Central Airways in Toronto, he flew for Great Lakes Airlines in Sarnia. Later he flew DC-3s for Survair in Africa and the Arctic, then a Queen Air for Dow Chemicals of Sarnia. He finished as a check pilot for Transport Canada in Moncton, his final flight being on October 28, 1983. By then his log showed 11,550.8 hours.

Jack Seaman at his place in Moncton in June 1998. (Larry Milberry)

## Action on RAF Mustang Squadrons

F/L Robert L. Sutherland of Pictou, Nova Scotia was a typical RCAF fighter pilot in the RAF. He ended on No. 65 Squadron, flying Mustang IIIs. He excelled at this. On June 8, 1944 things were hot over Normandy, with every available Allied plane thrown into the invasion. No.65's record on the 8th brought it honour—shortly after sun rise the CO, S/L Westenra (RNZAF), F/L Milton (RAF) and F/L Sutherland (RCAF) each destroyed an Fw.190 around Gace and Dreux. The main task was to bomb stationary transport near Gace. As the Mustangs were so engaged, four enemy fighters appeared. Sutherland pursued them, then was joined by the CO, who later reported: "I saw F/L Sutherland fire at one e/a. which hit the ground and blew up, the pilot not getting out." The fight was hot, being all on the deck with speeds of 360 mph.

After 60 miles of chasing at full bore, the CO got his chance, clobbering an Fw.190. Milton, firing at another Hun, had his canopy suddenly shatter and was knocked stupid. Moments later he came to at 5000 feet, took inventory, then realized that he had had a bird strike. His mates confirmed that Milton's Hun had crashed. As to Sutherland, the story is best told in his own words.

Combat reports usually were written bare bones style. But there were some vivid reports, especially among night intruder pilots home after a 4- or 5-hour op. An adept artist can bring such a scene to life on canvas, so richly and sharply defined are the details. Sutherland's report is somewhere between the utilitarian and the detailed gem:

*I was flying No.4 in Presto Section. As I climbed after bombing convoy near Gace, I looked back to observe results of bombing, and saw an a/c below me crossing at right angles. It looked like an Fw.190, so I turned to get a better view, and then saw three more. Having identified them, I reported to Presto Leader and gave chase. After about three minutes I opened fire from about 800 yds. dead astern, 2-second burst. I closed to 600 yds. and fired another burst. There were strikes during second burst. The Hun took no evasive action, but continued east on the deck. The e/a then appeared to judder and I overtook fast, closing to about 100 yds. Just as I was about to fire again, the e/a dived straight into the ground, catching fire as it went down. I then followed the rest of the section, and saw Presto Leader firing at another 190. I saw the e/a pull up to about 800 ft. and the pilot bale out. At the same time to my left, I saw Presto 2 firing. After re-joining Presto Leader I saw a fire on the ground in the area of the last combat. The Hun I attacked had a yellow nose.*

Exactly 3 weeks after his episode around Gace, Sutherland was dive-bombing near Naigle, France in Mustang FZ173. While in a turn, his engine quit. He went into a spin and crashed fatally. Today he lies in the Canadian War Cemetery at Bretteville-sur-Laize, France.

Mustang IIIs of 19 Squadron in August 1944. Note the bubble canopy (i.e. the British-made "Malcolm hood") of this version. (IWM CH12876)

Another Canadian flying RAF Mustangs was Robert A. Haywood of Vancouver. On June 21, 1944 he was on ops with 19 Squadron. Around Conches his section was jumped by 15-20 Me.109s. They jettisoned their bombs to take on the opposition. The day went poorly for the Germans—four 109s lost, another damaged. Haywood got one, but was badly shot up. He was rushed to hospital upon landing back at base. Recuperated, he returned to squadron, but on March 13, 1945 died near Aberdeen after flying into wires while testing FB359.

## Canadians in India-Burma

While most attention is paid to RCAF fighter pilots in Europe, hundreds of others fought in North Africa, the Mediterranean, Far East, India-Burma and other distant areas. Their aircraft included the Beaufighter, Blenheim, Hurricane, Kittyhawk, Mohawk, Spitfire and Thunderbolt. Later, these men sometimes felt left out, since history had favoured Europe. Nonetheless, many historical articles cover the other theatres. Certain books deal extensively with the Canadians, *Mohawks over Burma* and *Bloody Shambles* being prominent. Hedley Everard's *A Mouse in My Pocket* recounts his time with 17 Squadron in Burma and China. Many other titles add further detail.

Donald Henry Crumb served in India-Burma. Crumb was born in Barrie, Ontario on May 19, 1921. His father, being with the Grand Trunk Railroad, sometimes had to move the family. So it was that the Crumbs came to 42 Wayland Avenue in the Main-Gerrard neighbourhood of east Toronto. There Donald attended Kimberly Public School and Malvern Collegiate. In August 1941 he joined the RCAF. Two friends from

AC2 Don Crumb on guard duty at Trenton in October 1941. This entailed working 2 hours on, 4 hours off, so there was little chance for a decent rest. Crumb concluded that the discipline and routine prepared him for later training. (All, Crumb Col.)

Wayland Ave. also joined—Larry Walker, who would go on Mosquitos, and Lou Wise, who would fly Kittyhawks in Alaska.

Crumb followed the familiar routine of manning depot (Toronto), guard duty (Trenton) and ITS (Toronto). Although guard duty usually was detested, he considered it useful in a beginning airman's career—it taught him much about air force life. He did, however, wonder when guarding the smashed airplanes being hauled in to No.6 Repair Depot. The sight of so many wrecks gave pause to a young AC2 thinking of being a pilot. At last Crumb was posted No.10 EFTS, managed under contract by the Hamilton Flying Training School Ltd. There on January 5, 1941 he flew for the first time, going up with Mr. McKay in Tiger Moth 5131 on skis (directional control was dicey and there were no brakes). On January 20 Crumb soloed—a 10-minute circuit in the same aircraft. On March 6 he flew his first solo cross-country, leaving Mount Hope in 5173 for London, then flying southward to St. Thomas

Sports-minded buddies at Dunnville. John Frederick Hart "Jackie" Williams (Toronto), Crawford Fisher (Toronto), Edward Gerrard Frezell (Toronto), J.L. "Gibbie" Gibson (Caledonia, Ontario) and Donald H. Crumb (Toronto). As it goes with many such photos, some didn't make it through. On November 30, 1943, while flying a Typhoon on a Ranger, Williams (257 Sqn) shot down an Fw.190. On December 4 he shared in a Do.217. But over the Argentan area on July 26, 1944 he reported being gravely wounded by flak. He did not get home and has no known grave. Fisher's tour was spent in India as a ferry pilot and he lived through the war. Frezell was not so lucky. On February 13, 1943, he died in a Wellington crash at 30 OTU in England. Gibson died in India while on 146 Squadron. On October 27, 1943, his Hurricane fell out of formation with a fuel problem and crashed.

several classmates from Dunnville boarded the *Cameronia* for a rough crossing to Greenock. Soon they were at Bournemouth where, to their surprise, some were sent to bomber OTUs (as single-engine pilots, all had expected fighters).

Crumb's posting was to No.5 (P) AFU at Ternhill, near Liverpool. There he flew first in Master 8573 on September 1, 1942. On the 15th he flew Hurricane L1848. Some flights are

Don Crumb took this snap of No.6 SFTS Harvards. Aircraft 3070 made it through the rigours of training, to serve after the war in the RCN. When disposed of in 1954 it became CF-MSZ.

Airmen always were on leave in London, where there were countless hangouts. A popular spot was the Canadian Legion Officers' Club near Victoria Station. Don Crumb saved this card as a memento of happy times.

and home. All went well and Crumb finished his course next day with a flight in 5893. Time in his log was 41:55 hours of dual, 34:59 solo, plus 12:30 in the Link trainer. Crumb would add many additional Link hours, something that, in 1999, he recalled as valuable, perhaps life-saving, training.

Now Don Crumb was posted to No.6 SFTS at Dunnville (Course 52), flying first on April 1, 1942 with FSgt Norman E. Kirk in Harvard 2773. He soloed on the 12th in 2810. Two or three Harvard or Yale flights per day now became the routine. Kirk, a 22-year old from Hamilton, would bring Crumb along, but that relationship ended sadly. On September 23, 1942 Kirk took off in Harvard 3180 with P/O James I. McIntyre, age 24 from Guelph. No sooner were they airborne than they dove into the Grand River. (The McIntyre family would suffer again—James' brother, Archibald, dying on April 21, 1944 in a Harvard near Aylmer, not far from Dunnville.)

Crumb enjoyed air force life. He chummed with several fellows who were keen on sports. On July 10, 1942 he flew Harvard FE335 to finish his course. He had added 149:20 hours of dual and solo time, day and night. His Wings Parade was on the 17th with Crumb's family and his sweetheart, Georgina Shorting, present. Following embarkation leave, Crumb and

listed in his log as "sodium" exercises. This indicated training with sodium-treated goggles, which made day seem almost like night, so that night flying training could be simulated without all the usual hazards. Crumb finished AFU with a flight in Hurricane N7642 on September 19, then was posted to No. 55 OTU at Annan, a damp and chilly Scottish 'drome. He began training on September 25, first in Master 8775, then Hurricane 3777.

Now came the busy schedule of OTU with formation flying, cine camera exercises, air firing, aerobatics, low level flying/map reading, night flying, etc. All this was squeezed in to about three months, December 8 being the date of Crumb's last trip. He finished with an "Average" assessment, having logged 91:15 hours of day flying. 5:30 at night, and 2:10 of dual. Rumour now was that the class could expect to go on Typhoons. There was surprise when some were posted to India. Details were fuzzy, but the sprogs, Don Crumb included, soon found themselves aboard HMTS *Dominion Monarch*, leaving on January 18, 1943. This would be a dreary ordeal, the ship

Don Crumb with Georgina (nicknamed "Jo"). Don was a new sergeant pilot about to go overseas. Jo wouldn't see him for more than 3 years.

sailing southward in the Atlantic one day, west, even north, on others (depending on U-Boat reports). Stops were made in Freetown (Sierre Leone) and Durban (South Africa) for water and supplies, the ship finally docking in Bombay on March 19.

Although the British needed reinforcements in India-Burma, where they had been suffering at the hands of superior forces, to Don Crumb and his flying mates there seemed to be little organization or purpose to the RAF in India. Pilots were shunted about on useless duties. Crumb started at the Aircrew Transit Pool at Poona (about 100 miles southeast of Bombay), doing Harvard and Hurricane practice flying. Next he was posted to 21 Ferry Control in Mauripur, Karachi. On August 9 he was ferrying Hurricane AG140, escorted by a Hudson. They got into weather and Crumb was obliged to set down en route. When the weather cleared, he departed, but got into trouble and flipped on takeoff. Once extricated, he was flown in a Tiger Moth to hospital in Jodpur. Then he spent a few weeks recuperating with a well-to-do Indian family, finally returning to the ATP, where he commenced flying in Harvard FE545 on November 22.

At last Crumb got a squadron—No.135 flying the Hurricane IIC at St. Thomas Mount, near Madras. There were several Canadians here, including Harry Blackburn (Dundas, Ontario) from Course 52 at Dunnville. The CO was a top-notch RAF type, S/L "Bunny" Stone, DFC, a veteran of the Battle of Britain and of fierce fighting in Burma. He led his pilots in intensive training, then 135 moved to Minneriya, Ceylon, where there was much training with the Royal Navy at nearby China Bay. On February 2, 1944 S/L Hawkins took over 135. For March, Crumb's log shows 16:20 hours on the Hurricane; for April—24:30.

May 18, 1944 was a red letter day, for F/O Reg Dunster (RCAF) delivered 135's first Thunderbolt I. On the 22nd Don Crumb flew the new type. Pilots loved the Thunderbolt from the start. Although they continued with Hurricanes for a few weeks, there was no comparing these fighters. Training slowed for about a month, due to a parts shortage, but 135 inevitably got into the shooting war. Meanwhile, the squadron lost one man. For his June 8 log entry Don Crumb noted, "F/L Budd Hart killed on height climb. Oxygen trouble." Eldon Budd Hart, a 22 year-old from Lewvan, Saskatchewan, had used another fellow's oxygen mask, but the fit was loose. It is likely that he blacked out before crashing in Thunderbolt HB977. This tragedy hit Don Crumb—the two had been best friends.

On June 6 No.135 took on some FAA Corsairs and Avengers in a practice dogfight. On June 26 Crumb was aloft in Hurricane "F" when he had a coolant (glycol) leak. His engine suddenly quit, but he had altitude. He turned toward base and glided to a safe landing, slithering around some barrage balloons at the last second. The Hurricane required an engine change, but had been saved. Besides flying, there wasn't a great deal to do at

Hurricane days on 135 Squadron, Madras: RCAF WO2 Byron "Buck" Beatty and F/Os John Holloway, Harry Blackburn, Don Crumb, Budd Hart (on wing) and Jim Slimon. From Lewvan, Saskatchewan, Hart (age 22) died on June 5, 1944 in a Thunderbolt crash on Ceylon.

The RCAF's all-Canadian officers baseball team in December 1943: Illman, Blackburn, Boyer, Anderson and Brown behind; Ivens, Crumb, Harris, Sunnucks and Walker (an American). Of the team, at least one later died—John A. Illman of Chatham, Ontario. On March 22, 1944 his 113 Squadron Hurricane ran out of fuel and crashed on a Japanese 'drome.

These 135 Squadron pilots have it made … they've just collected their monthly "26ers" of Canadian Club whiskey. Don Crumb is second from the left, Harry Blackburn is far right.

Don Crumb tees off at the Nuwara Eliya Golf Club (the Tea Planters' Hill Club). He noted on the back of this snap, "Don't laugh too long at this one, Georgina. The hankie was protection against the sun and the boots against leaches. You'd better hide this somewhere."

Tom Newton of Lawson, Saskatchewan soaks off some sweat in his galvanized tub at Jumchar.

Minneriya. One good place to relax, however, was the Tea Planters' Hill Club, where golf was a favourite pastime. It also helped that each pilot had a monthly ration of a "26er" of Canadian Club whiskey. Crumb liked to trade his to the Americans for a case of Pabst Blue Ribbon beer. Since he didn't smoke, he also traded his cigarette ration. Training continued, September 3 being typical—fighter affiliation. Crumb noted of this: "Curve of pursuit and range estimating, Liberator." On the 13th he remarked that a Canadian, F/O Fernand D. Jolicoeur of Ottawa, had died (in an American Catalina). This happened during fighter affiliation. As the Catalina evaded sharply, it flipped and went in. (Jolicoeur is buried in the National Cemetery at Fort Scott, Kansas, a rare case of a casualty apparently being repatriated from overseas for burial.)

On October 9 the squadron moved to Chittagong at the head of the Bay of Bengal. On the 16th Crumb noted, "Squadron's first operation this campaign—strafing". Henceforth, 135 was busy with ops almost daily—there would be no leave for months. On November 2 it moved again, this time to Cox's Bazaar on the coast about 75 miles south of Chittagong. From here 135 frequently escorted Liberators, or Dakotas dropping supplies to the army. On November 8 Crumb was on a search for a ditched Beaufighter. On the 16th he noted: "P/O Windle destroyed one Oscar II, Rangoon area." Three days later: "B-24s accurately bombed railway siding. Saw four Oscar IIs, damaged one." On many daily ops the Thunderbolts were heavily laden, departing with full internal fuel, two 500- or 1000-lb. bombs and ammunition for eight .50 cal. machine guns. Crumb logged 17 hours on ops in November.

**A Thunderbolt alights at a Burmese strip; then an overall scene as 3 "Jugs" beat up the flightline.** *AAF Manual No.50-6 Pilot Training Manual for the Thunderbolt* **notes of P-47 take-off performance: "Stay on the ground until reaching a speed of around 110 mph. Then lift the plane off the runway… Use rudder, not brake, to correct for torque on take-off…Be easy on the back pressure until you have at least 140 mph, then climb gently. The plane is sluggish before reaching its best climbing speed of 155-160 mph." For landing, the approach pattern speed was 150-200 mph, 125-130 on final. (IWM CF205, CF202)**

Ten 135 Squadron Thunderbolts return from operations. Then, Don Crumb's favourite Thunderbolt—"J-Georgina" at Jumchar. With 3 long-range drop tanks, it is configured for a 3 to 3½-hour sortie. Note its huge, 13-ft. propeller. Finally, P/O Bob Windle brought this severely damaged Jug home after colliding with Sgt Hargreaves.

The squadron moved to Jumchar (near Cox's Bazaar) on December 10. This 'drome had been a rice paddy. In the dry season such a strip was as hard and level as concrete. A week later Sgt Hargreaves and P/O Bob Windle collided. Hargreaves bailed out OK, Windle was able to land his damaged Thunderbolt. This was another example of the soundness of "The Jug"— the P-47's nickname. On December 24 No.135 did its first dive bombing, 12 aircraft accurately delivering 23 of 24 bombs. On January 2 the Japanese army was pummelled with 24 bombs (12 P-47s). Quite often the targets in this period were artillery positions. Railroads, river traffic and MT also were hit. If the Thunderbolts didn't get them by day, Beaufighters would—at night. Later in January came another squadron first: "Jap position Kangaw. Squadron's first napalm firing bombing." Napalming would become routine, although the tanks of this jellied gasoline had to be loaded at a US base—the RAF didn't have stock. A different sort of target was that of February 9: "Offensive recce... Jap cavalry ... 20-30 horses at Gelaung."

Visiting a US air base was not all work, for Don Crumb and his mates could enjoy an excellent meal in the US dining hall. Even ice cream was on the menu! A bonus if over-nighting was that US barracks had screened windows. This contrasted with conditions in the field, where RAF squadrons such as No.135 got by with few amenities, while putting up with oppressive heat and humidity, dust at other times, insects, scorpions, leeches and snakes, plus the real threat of malaria and dengue fever.

The day-to-day work continued. Crumb flew 28:05 hours on ops during March 1945. For April 8 he wrote: "Dive bomb and strafe No.1 dump, Prome. Good fires started! Light flak, a bit too close." This trip was 2:50 hours, shorter by a half an hour compared to most Thunderbolt ops. By comparison, one hour left a Hurricane or Typhoon getting thirsty for gas. On May 10 Don Crumb, by now a flight lieutenant, flew his last operation, flying "J- Georgina" from Akyab (where 135 had moved on April 24) on a 3:15-hour low-level strafing trip. His last flight came a few

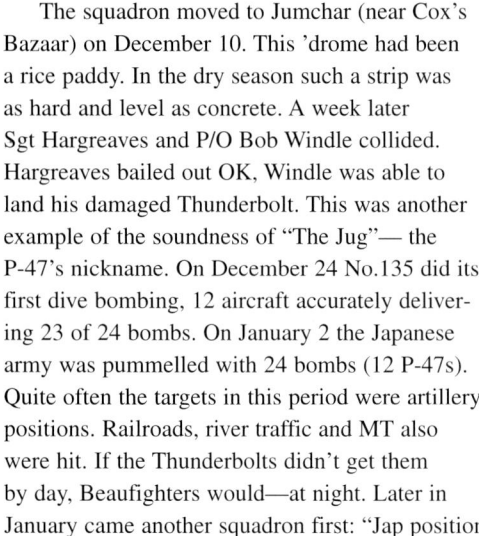

A recce photo (Akyab Island, November 21, 1944) of the kind used by 135 pilots in planning missions. Such photos enabled pilots to familiarize themselves with a target, know which buildings to hit, which to avoid, and where flak might be.

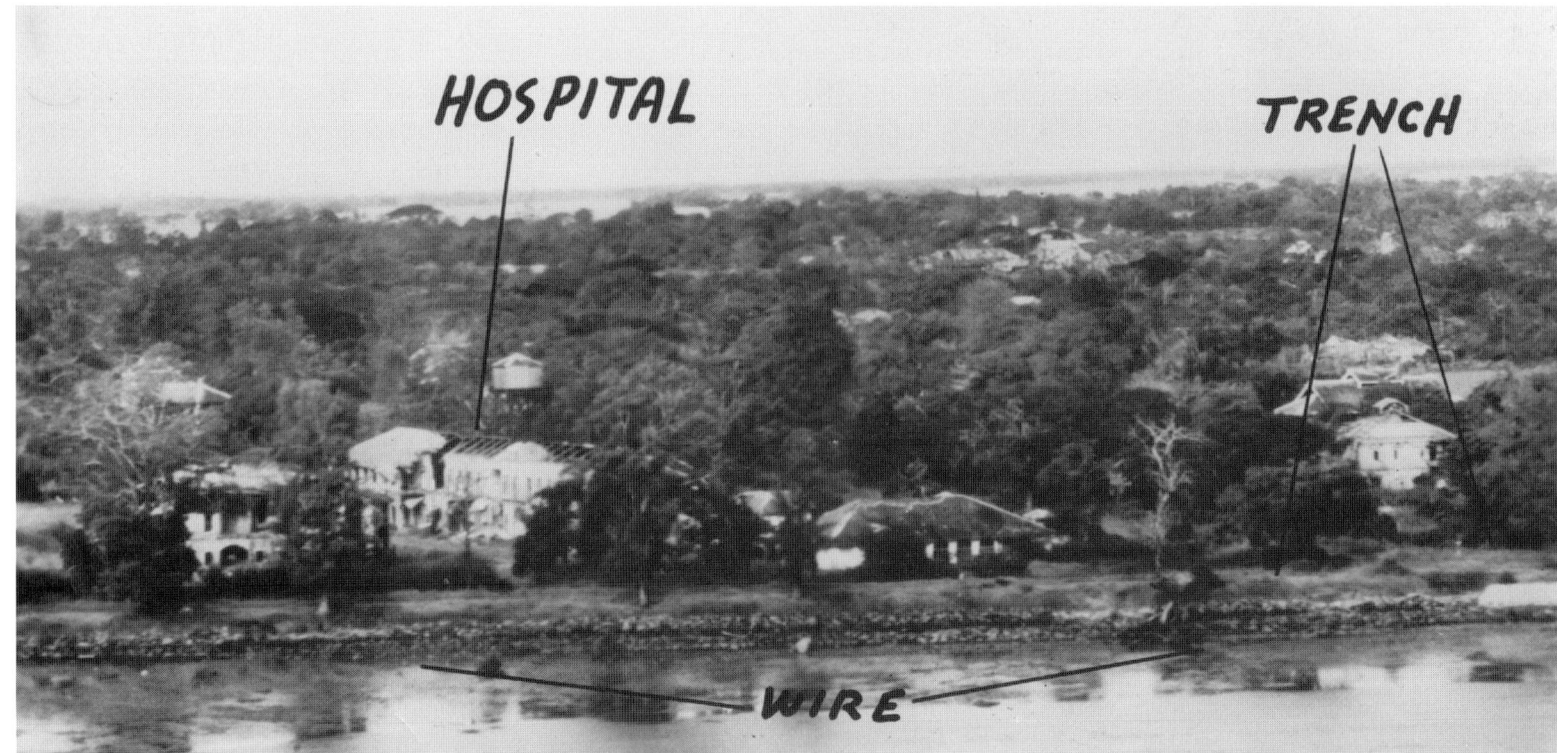

The Crumbs attended the 1992 aircrew reunion in Winnipeg. Here Don (right) stands with A/V/M Johnny Johnson and fellow 135 Squadron pilot, Jim Slimon. Behind is Bob Middlemiss, who had fought at Malta.

Don Crumb in November 1999. Note his special licence plate. (Larry Milberry)

days later—an air test in Thunderbolt KJ229. Now Crumb, like most weary, tour-expired pilots, was happy to keep his feet on the ground. On June 13 he took a 194 Squadron Dakota to Calcutta, from where an Anson flew him to Bombay. There he boarded HMTS *Canton* on July 12. The ship docked in Southampton on August 2. That day Crumb reported in at No.3 PRC at Bournemouth, moved along to R-Depot at Torquay, and boarded ship for home on August 21. Altogether, Don Crumb logged 676:15 hours, his operations totalling 132:50 on Thunderbolt, 75 on Hurricanes. Credited with two ops tours, he was awarded the Operations Wing and Bar

After battling three monsoon seasons, Don Crumb now was home, much to the joy of his family and Georgina, whom he married on December 8. Before long he joined Northern Electric. He and Georgina settled in Ajax, near Toronto, keeping busy with many activities, especially their growing family. Don's only flying was a few hours on Chipmunks as an RCAF reserve pilot. When Ottawa cut funds for that program in the early 1950s, he left flying for good. Meanwhile, he was much involved in community activities, such as being president of the Ajax Minor Hockey Association. When Ontario began offering the opportunity for drivers to acquire personalized licence plates for motor vehicles, Don quickly reserved two: "P47-JUG" and "BURMA". He wasn't going to forget that great adventure of his youth a half-century earlier.

## The Results

Whether CAN/RAF, on RCAF Squadrons or RCAF in the RAF, Canadian fighter pilots fared well during the war. Many became aces, i.e., having 5 or more confirmed kills. Those with at least 10 air-to-air victories are listed here, with the relevant wartime squadrons (data based on Hugh Halliday's *The Tumbling Sky*, but Shores' and Williams' *Aces High* also is indispensible in understanding the history of aces in the RCAF):

| Pilot | Score* | Squadrons |
|---|---|---|
| Bannock, Russell | 11-0-4 + 19 V-1s | 406, 418 |
| Barton, Robert A. | 13-4½-9 | CAN/RAF, 249 |
| Beurling, George F. "Buzz" | 31⅓-0-9 | 403, 249, 412 |
| Bing, L.P.S. "Pat"*** | 13-0-1 | 89 |
| Charles, Edward F.J. "Jack" | 15½-6½-5 | 54, 611 |
| Cleveland, Howard D. | 10-0-2 | 418 |
| Edwards, James F. "Stocky" | 13½-5-8 | 94, 260, 417, 92, 274 |
| Fairbanks, David C. | 14-1-4 + 2 V-1s | 501, 274, 3 |
| Fumerton, Robert C. "Moose" | 14-0-1 | 1, 406, 89 |
| Gordon, Donald C. | 11½-4-5 | 65, 274, 601, 417, 442, 402 |
| Gray, Ross G. | 10-0-12 + 2 V-1s | 418, 406 |
| Hill, George U. | 11⅘-3-11 | 421, 453, 403, 111, 441 |
| Houle, Albert U. "Bert" | 11-1-7 | 213, 417 |
| Huletsky, P.*** | 10½-½,-1 | 418 |
| Keefer, George C. | 13-2-7⅓ | 274, 416, 412, 125 |
| Kennedy, Irving F. | 12⁷⁄₁₂-1-0 | 263, 421, 249, 111, 93, 401 |
| Kent, John A. | 13-2-3 | CAN/RAF, 303, 92, Northolt Wing (Nos.303, 306, 308 Squadrons) |
| Kipp, Robert A. | 10½-½-3 | 418 |
| Klersy, William T. | 16½-0-3½ | 401 |
| Laubman, Donald C. | 15-0-3 | 412 |
| McElroy, John F. | 13½-2½-9 | 249, 421, 416 |
| MacKay, John | 10⁷⁄₁₀-0-6½ | 401 |
| MacKnight, William L. "Willie" | 18½-0-1 | 242 |
| McLeod, Henry W. "Wally" | 19-1-9¼ | 411, 603, 1435, 443 |
| McNair, Robert W. | 14-2-14 | 411, 403, 416, 421, 126 Wing |
| Mitchner, John D. | 10½-1⅚ | 402, 421 |
| Schwab, Lloyd G. | 11-1-0 | CAN/RAF, 112 |
| Smith, Roderick I.A. | 15⅕-½-1 | 412, 126 |
| Turnbull, John H. | 12½-0-0 | 125, 600 |
| Turner, Percival S. "Stan" | 14-2-6 | CAN/RAF, 242, 249, 134, 417, 127 Wing |
| Walker, James E. | 10-3-4 | 81, 243 |
| Weaver, Claude | 13½-1-0 | 185, 403 |
| Woodward, Vernon C. | 20-5-11 | 33, 213 |

* kills, probables, damaged  ** Fumerton's observer  *** Kipp's observer

Sopwith Dolphin fighters of No.1 Squadron, Canadian Air Force (England). No.1 flew some interesting types, including the Bristol F.2B and Fokker D.VII. Formed in November 1918, it disbanded the following January at Shoreham-by-Sea. (K.M. Molson Col.)

HS-2L G-CYAG was taken on strength in October 1920. A Trans-Canada Flight participant, it lasted only till September 1923, then was cannibalized and scrapped. (CF)

RAF Wapiti J9237 was on loan to the RCAF for ski trials from March 1930 to May 1932. Beginning in March 1936 the RCAF added 24 Wapitis of its own. At the same time as it was flying this clapped-out junker, advanced 1930s designs like the Douglas B.18, Dornier Do.17 and Vickers Wellington were in the air with other air forces. (Wm. Wheeler Col.)

The RCAF's largest aircraft to 1937 was its lone Ford Trimotor. It was used on forest dusting trials, to support the 1931 Trans Canada Air Pageant, and even operated on floats. It later was sold to Yukon Southern Air Transport. It was wrecked in 1939, run into on the ground at Vancouver by an RCAF Hurricane. Jack McNulty took this photo at Trenton in June 1934.

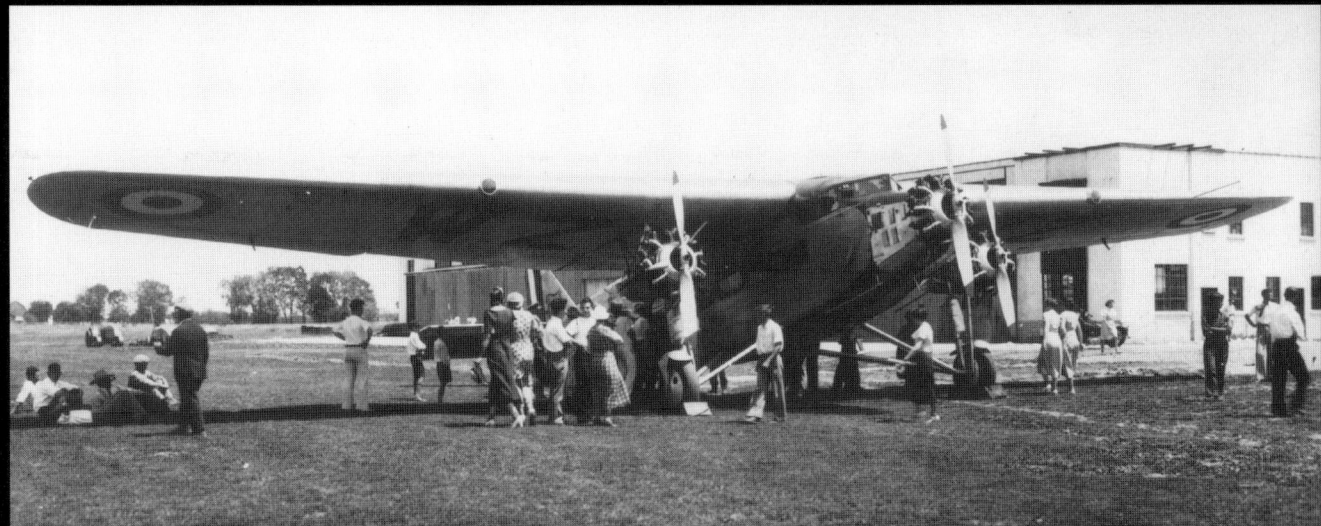

A rare gathering of RCAF bushplanes. Rusty Blakey, a young fellow starting out in aviation, took this photo on Ramsay Lake, Sudbury about 1935. Three RCAF Bellancas (nearest) and a Fairchild and Bellanca (farthest) had stopped in for fuel. Perhaps they were heading out on summer photo detachments, or were returning to Ottawa at season's end. An Austin Airways Waco is the other floatplane. Bellancas 604 and 605 are at the shore, with a crowd of locals gathered for the excitement. They were struck off strength in the summer of 1937, becoming CF-BFC and 'D with Canadian Airways, where they operated (and soon were wrecked) in Saskatchewan.

The Grumman Goose served well in the HWE as a versatile general purpose transport. Taken on strength in 1938, it was lost in Alaska on July 21, 1942. A classic story of survival and rescue resulted, as told in Air Transport in Canada. (David Thompson Col.)

One of the RCAF's Westland Wapiti bombers, a type that entered service in Canada in 1936. It's seen at Camp Borden in 1938, while with No.3 Squadron. Illustrating the RCAF's dire shortage of equipment in 1939, Wapitis conducted a few wartime operations, patrolling out to sea from a grass strip near Halifax. The fleet soon was relegated to ground training. Unfortunately, no RCAF prewar fighting aircraft were saved for museum purposes. (NAC PA92489)

Bolingbroke 9094 in a snapshot from Port Hardy. By the looks of the fellows' hair it was a breezy day! (Jack McNulty Col.)

Spitfire pilots Dagwood Phillip, Lloyd Chadburn and E.P. "Ted" Wood of 416 Squadron, while at Marthesham Heath in 1942. Chadburn led 416 on the Dieppe raid, at first driving off a gaggle of Fw.190s, then dispersing seven Ju.88s and hassling some Me.110s. For their efforts the "Oshawa" squadron tallied three confirmed kills. Chadburn later died in a mid-air collision. Wood later commanded 403 Squadron. Postwar he headed 413 and 421 Squadrons on Sabres. (Stanley Malouf)

Harvards from 6 SFTS, Aylmer, Ontario. Nearest is 3214, which remained on strength till 1960. (Alf Barton)

# Glossary

| | | | | | |
|---|---|---|---|---|---|
| A&AEE | Aeroplane and Armament Experimental Establishment | | the Channel. The aim was to tempt enemy fighters to take on the escorts and oblige the Luftwaffe to concentrate resources in specific areas. | GRS | general reconnaissance school |
| A/C | Air Commodore | | | GSU | group support unit |
| A/M | Air Marshal | | | HCU | heavy conversion flight |
| A/V/M | Air Vice Marshal | | | He. | Heinkel |
| ADDL | aerodrome dummy deck landing | | | HMCS | His Majesty's Canadian Ship |
| AFC | Air Force Cross (awarded to officers) | cluster bomb | a large container holding many small bomblets, which dispersed when the container split apart in the slipstream after being dropped | HMS | His Majesty's Ship |
| | | | | HMTS | His Majesty's Transport Ship |
| AFDU | Air Fighting Development Unit | | | hp | horse power |
| AFHQ | Air Force Head Quarters | CNE | Canadian National Exhibition | HQ | headquarters |
| AFM | Air Force Medal (awarded to NCOs and WOs) | CO | commanding officer | HSL | high speed launch |
| | | Col | colonel | HWE | Home War Establishment |
| AFU | advanced flying unit | COMM | communications squadron | IFF | identification friend or foe -- electronic transmitter to identify a friendly aircraft |
| AG | air gunner | CPA | Canadian Pacific Airlines | | |
| AI | airborne interception | CSO | chief section officer | | |
| ALG | advanced landing ground | CTechO | chief technical officer | IO | intelligence officer |
| AOS | air observer school | DC | depth charge | ITS | initial training school |
| Ar. | Arado | DFC | Distinguished Flying Cross (awarded to officers) | IWM | Imperial War Museum |
| ARP | air raid precautions | | | Ju. | Junkers |
| ASL | above sea level | DFM | Distinguished Flying Medal (awarded to NCOs and WOs) | LAC | leading aircraftman |
| ASO | assistant section officer | | | LAW | leading aircraftwoman |
| ASR | air sea rescue | DH | de Havilland | LCol | lieutenant colonel |
| ASV | anti-surface vessel radar | DHC | de Havilland Canada | Mailcan | Canada-overseas mail operations under 168 Squadron |
| ASW | anti-submarine warfare | Diver | code name for the V-1 | | |
| ATA | Air Transport Auxiliary | DND | Department of National Defence | Maj | major |
| ATC | air traffic control | "do" | air force slang -- any operation | MBE | Member of the Most Excellent Order of the British Empire |
| ATFERO | Atlantic Ferry Organization | Do. | Dornier | | |
| AW | airwomen | DOT | Department of Transport | MC | Military Cross |
| B&GS | bombing and gunnery school | DSC | Distinguished Service Cross | MD | manning depot |
| BAOR | British Army of Occupation of the Rhine | DSO | Distinguished Service Order | Me. | Messerschmitt |
| | | duff | air force slang -- poor, bad as in duff weather | MET | motorized enemy transport |
| BCATP | British Commonwealth Air Training Plan | | | Met | meteorological service |
| | | e/a | enemy aircraft | MiD | Mention in Despatches |
| BDTF | bomber defence training flight | EAC | Eastern Air Command | Mk. | mark of aircraft |
| BEM | British Empire Medal | EFTS | elementary flying training school | MO | medical officer |
| bogey | a target to be intercepted, identified and (possibly) attacked | ENSA | Entertainment National Service Association | MSFU | Merchant Ship Fighter Unit |
| | | | | MT | motor transport |
| BR | bomber reconnaissance | EPE | Experimental and Proving Establishment | MU | maintenance unit |
| Buzz Bomb | V-1 | | | NAC | National Archives of Canada |
| C. de G. | Croix de Guerre | ETPS | Empire Test Pilots School | NACA | National Advisory Committee for Aeronautics |
| CAC | Canadian Aviation Corps | F/C | flight commander | | |
| CAC | coast artillery co-operation | F/L | flight lieutenant | nav | navigator |
| CAF | Canadian Air Force (1919-23) | F/O | flying officer | NCO | non-commissioned officer |
| CAF | Canadian Armed Forces (1968-) | FAA | Fleet Air Arm | NF | night fighter |
| CAM | catapult-armed merchant ship | FB | fighter bomber | Noball | code name for a V-1 site |
| CAN/RAF | Canadian in the RAF | FB | flying boat | NPF | Non-Permanent Force |
| Capt | captain | FE | flight engineer | NWAC | Northwest Air Command |
| CAS | Chief of the Air Staff | FIDO | Fog Investigation and Dispersal Operation | NWSR | Northwest Staging Route |
| CB | Order of the Bath | | | OBE | Officer of the Most Excellent Order of the British Empire |
| CEF | Canadian Expeditionary Force | FSgt | flight sergeant (also FS) | | |
| CEPE | Central Experimental and Proving Establishment | Fw. | Focke-Wulf | ops | operations |
| | | G/C | group captain | ORB | operational record book |
| CFI | chief flying instructor | GC, GM | George Cross, George Medal -- decorations for heroic action off the field of battle. Available to military or civilians | OTE | operational tour expired |
| CGAO | Civil Government Air Operations | | | OTU | operational training unit |
| CGM | Conspicuous Gallantry Medal | | | P/O | pilot officer |
| Circus | Short-range, fighter-escorted day bombing ops against targets across | | | PFF | Pathfinder Force |
| | | | | PI | photo interpretation |
| | | GR | general reconnaissance | POW | prisoner of war |

| | | | | | |
|---|---|---|---|---|---|
| PPO | Provisional Pilot Officer | Rhubarb | small-scale fighter attacks of targets of opportunity | TTS | technical training school |
| PRC | personnel reception centre | | | u/s | unserviceable |
| PRU | personnel reception unit | RN | Royal Navy | USAAC | United States Army Air Corps |
| PRU | photo reconnaissance unit | RNAS | Royal Naval Air Service | USAAF | United States Army Air Force |
| PTU | port transit unit | RNZAF | Royal New Zealand Air Force | USAF | United States Air Force |
| R/T | radio telegraphy | Rodeo | fighter sweeps over enemy territory | USN | United States Navy |
| RAAF | Royal Australian Air Force | | | V-1 | Fieseler Fi.103 flying bomb |
| radar | radio detection and ranging | Rover | armed recce looking for any sort of target | V-2 | German A-4 artillery rocket |
| RAE | Royal Aeronautical Establishment | | | VC | Victoria Cross |
| RAF | Royal Air Force | RP | rocket projectile | VE-Day | Victory Europe Day, March 8, 1945 |
| Ramrod | similar to a Circus, but with the main objective to destroy a target | RSU | repair and salvage unit | | |
| | | s/e | single engine | VJ-Day | Victory Japan Day, August 14, 1945 |
| Ranger | Squadron- or wing-size intruder operations aimed at destroying ground targets and wearing down enemy fighter opposition. | S/F/L | sub flight lieutenant | | |
| | | S/L | squadron leader | W/C | wing commander |
| | | S/O | section officer | WAAF | Women's Auxiliary Air Force |
| | | SAAF | South African Air Force | WAC | Western Air Command |
| RCAF | Royal Canadian Air Force | SAC | Strategic Air Command | WD | Women's Division |
| RCN | Royal Canadian Navy | SFTS | service flying training school | WEE | Winter Experimental Establishment |
| RCNAS | Royal Canadian Naval Air Service | Sgt | sergeant | | |
| RCNVR | Royal Canadian Navy Volunteer Reserve | SOC | struck off charge | WETF | Winter Experimental and Training Flight |
| | | sweep | offensive fighter patrol to draw in enemy fighters | | |
| RD | repair depot | | | "winco" | wing commander |
| RDF | radio direction finding, i.e. radar | TAF | Tactical Air Force | WO | warrant officer |
| recce | reconnaissance | TCA | Trans-Canada Air Lines | WOp | wireless operator |
| RFC | Royal Flying Corps | TEU | Tactical Exercise Unit | WOpAG | wireless operator/air gunner |
| | | TOC | taken on charge | | |

**A formation of Ventura Mk.IIs of 34 OUT Pennfield Ridge over the coastline of New Brunswick.**

# Selected Bibliography

*405 Squadron History*, Craig Kelman & Associates, Winnipeg.

*416 Squadron History*, Hangar Bookshelf, 1984.

*434 Squadron History*, Hangar Bookshelf, 1984.

*A Battle for the Truth: Canadian Aircrews Sue the CBC over Death by Moonlight: Bomber Command*, The Bomber Harris Trust, Agincourt, Ontario, 1994.

*Action Stations*, Vols.1-10, Patrick Stephens Ltd., Wellingborough, England, 1979-87.

*Air Force Magazine*, Air Force Association of Canada, Ottawa.

*The RCAF Overseas: The First Four Years*, RCAF Historical Section, Oxford University Press, Toronto 1944.

*The RCAF Overseas: The Fifth Year*, RCAF Historical Section, Oxford University Press, Toronto 1945.

The RCAF Overseas: The Sixth Year, RCAF Historical Section, Oxford University Press, Toronto 1949.

Allison, Les and Hayward, Harry, *They Shall Grow Not Old: A Book of Remembrance*, Commonwealth Air Training Plan Museum, Brandon, Manitoba, 1991

Bailey, Frank W., "Who Was the Top Two-Seater Ace of World War I?", *Cross and Cockade* (American edition), Vol.24, No.1, Spring 1983.

Blatherwick, John, *Royal Canadian Air Force: Honours, Decorations, Medals 1920-1968*, FJB Air Publications, New Westminster, BC, 1991.

Boyer, Chaz, *The Short Sunderland*, Aston Publications, Bourne End, UK, 1989.

Brew, Alec, *The Defiant File*, Air Britain (Historians) Ltd., Tonbridge, England, 1996.

Brickhill, Paul, *The Dam Busters*, Evans Brothers Ltd., London, 1951.

Chorley, W.R., *Royal Air Force Bomber Command Losses of the Second World War*, Vols.1-6, Midland Counties Publications, Earl Shilton, England, 1997-98.

Christie, Carl A., "Alan Arnett McLeod", *Dictionary of Canadian Biography*, Vol.14, University of Toronto Press, Toronto, 1998.

Christie, Carl, *Ocean Bridge: The History of RAF Ferry Command*, University of Toronto, Toronto, 1995.

Cooke, Owen A., *The Canadian Military Experience 1867-1995: A Bibliography*, Directorate of History, Ottawa, 1997.

Cooke, Owen A., *Canadians in the British Flying Services: Enlistments, Manpower Losses, Honours and Awards, Operational Service; Analysis from the Computer Program*, Department of National Defence, Directorate of History and Heritage, Ottawa, February 1974.

Cosgrove, Ed, *Canada's Fighting Pilots*, Clarke, Irwin, 1965.

Cutlack, F.M., *Official History of Australia in the Great War of 1914-1918; Vol.VIII; Australian Flying Corps*, Angus and Robertson, Sydney, 1953.

Delve, Ken, *The Source Book of the RAF*, Airlife Publishing, Shrewsbury, England, 1994.

Dodds, Ronald, *Brave Young Wings*, Canada's Wings, Stittsville, Ontario, 1980.

Donald, David and Lake, Jon, *Encyclopedia of World Military Aircraft*, Aerospace Publishing, London, 1994.

Douglas, W.A.B., *The Creation of a National Air Force*, University of Toronto Press, Toronto,1986.

Dunmore, Spencer and Carter, William, PhD, *Reap the Whirlwind: The Untold Story of 6 Group, Canada's Bomber Force of World War II*, McClelland & Stewart, Toronto, 1991.

Duval, G.R., *British Flying-Boats and Amphibians 1909-1952*, Putnam, London, 1966.

Dye, Peter, "Nine Squadron RFC-RAF: An Analysis", *Cross and Cockade International Journal*, Vol.28, No.2 (Summer 1997), pp.73-109.

Ellis, Frank, *Canada's Flying Heritage*, University of Toronto Press, Toronto, 1954.

English, Allan D., *The Cream of the Crop: Canadian Aircrew, 1939-1945*, McGill-Queen's University Press, Montreal, 1996.

Everard, Hedley, *A Mouse in My Pocket*, Valley Floatplane Services, Picton, Ontario, 1988.

Fletcher, David C. and McPhail, Doug, *Harvard: The North American Trainers in Canada*, DCF Flying Books, San Josef, BC, 1990.

Fochuk, Stephen M., *Metal Canvas: Canadians and World War II Aircraft Nose Art*, Vanwell Publishing Ltd., St. Catharines, Ontario, 1999.

Foreman, John and Harvey, S.E., *The Messerschmitt Me.262 Combat Diary*, Air Research Publications, New Malden, England, 1990.

Franks, Norman L.R., *Battle of the Airfields: Operation Bodenplatte 1 January, 1945*, Grubb Street, London, England, 1994.

Franks, Norman, *Search, Find and Kill: The RAF's U-Boat Successes in World War Two*, Grub Street, London, 1995.

Franks, Norman L.R. and Bailey, Frank W., *Cross and Cockade (Great Britain) Journal*, Vol.4, No.1 (Spring 1973), pp.32-43.

Freeman, Roger A., *The Mighty Eighth: A History of the US 8th Army Air Force*, MacDonald and Co., London, 1970.

*From Boxkite to Jet: The Memoirs of an Aeronautical Engineer* (A/V/M E.W. Stedman's memoir), Mercury Series, Canadian War Museum, Paper No.1, Ottawa, 1972.

Golley, John, *Hurricanes over Murmansk*, Patrick Stevens, Wellingsborough, England, 1987.

Gomersall, Bryce B., *The Stirling File*, Air Britain (Historians) Ltd., Tonbridge, England, 1987.

Gowans, Bruce W., *Register of Canadian Civil Noorduyn Norseman*, Calgary, 1995.

Green, William, *Famous Bombers of the Second World War*, MacDonald, London, 1959.

Greenhous, Brereton, Harris, Stephen J., Johnston, William C. and Rawling, William G.P., *The Crucible of War*, University of Toronto Press, Toronto, 1994.

Gray, Peter and Thetford, Owen, *German Aircraft of the First World War*, Putnam, London, 1962.

Griffin, John A., *Canadian Military Aircraft: Serials and Photographs 1920-1968*, Canadian War Museum, Ottawa, 1969.

Halley, James J., *Royal Air Force Aircraft* (25 volumes), Air Britain (Historians) Ltd., Tonbridge, England, 1976-98.

Halliday, Hugh A., "Beamsville Story", *Journal of the Canadian Aviation Historical Society*, Vol.7, No.2 (Fall 1969).

Halliday, Hugh A., *242 Squadron, The Canadian Years*, Canada's Wings, Stittsville, Ontario, 1981.

Halliday, Hugh A., *The Tumbling Sky*, Canada's Wings, Stittsville, Ontario, 1978.

Halliday, Hugh A., *Typhoon and Tempest: The Canadian Story*, CANAV Books, Toronto, 1992.

Halliday, Hugh A., *Woody: A Fighter Pilot's Album*, CANAV Books, Toronto, 1987.

Hamlin, John F., *The Harvard File*, Air Britain (Historians) Ltd., Tonbridge, England, 1988.

Hatch, F.J., *Aerodrome of Democracy: Canada and the British Commonwealth Air Training Plan 1939-1945*, Directorate of History, Ottawa, 1983.

Hatch, F.J., *Air Training in Canada, 1914-1918*, Department of National Defence, Directorate of History and Heritage, Ottawa, circa 1973.

Hayward, Roger, *The Beaufort File*, Air Britain (Historians) Ltd., Tonbridge, England, 1990.

*History of 410 Squadron*, June 1941-June 1945, unpublished, circa 1961.

Jackson, A.J., *De Havilland Aircraft Since 1915*, Putnam, London, 1962.

Jefford, W/C C.G., MBE, *RAF Squadrons: A Comprehensive Record of the Movement and Equipment of All RAF Squadrons and Their Antecedents since 1912*, Airlife Publishing, Shrewsbury, England, 1988.

Johnson, Andrew, *Canadians in the British Flying Services: Statistical Report on the Computer Program*, Department of National Defence, Directorate of History and Heritage, Ottawa, August 1973.

*Journal*, Canadian Aviation Historical Society, Toronto, 1963-2000

Juptner, Joseph P., *U.S. Civil Aircraft Series, Vols.1-9*, Tab Aero, Blue Summit Ridge, Pennsylvania, 1994.

Kennedy, S/L I.F., *Black Crosses off My Wingtip*, General Store Publishing, Burnstown, Ontario, 1994.

Kostenuk, Sam and Griffin, John, *RCAF Squadrons and Aircraft 1924-1968*, Samuel Stevens Hakkert & Company, Toronto, 1977.

Leigh, G/C Z.L., *And I Shall Fly: The Flying Memoirs of Z. Lewis Leigh*, CANAV Books, Toronto, 1989.

Lewis, Peter, *The British Bomber Since 1914: Fifty Years of Design and Development*, Putnam, London, 1967.

Lewis, Peter, *The British Fighter Since 1912: Fifty Years of Design and Development*, Putnam, London, 1967.

Marion, Normand, *Camp Borden: Birthplace of the RCAF 1917-1999*, 16 Wing, Borden, Ontario, 1999.

Mason, Francis K., *The Avro Lancaster*, Aston Publications, Bourne End, England, 1989.

Mason, Francis K., *The Hawker Typhoon and Tempest*, Aston Publications, Bourne End, UK, 1988.

McVicar, Don, *North Atlantic Cat*, Airlife Publishing, Shrewsbury, England, 1983.

Merrick, K.A., *The Handley Page Halifax*, Aston Publications Ltd., Bourne End, England, 1980.

Middlebrook, Martin, *The Nuremberg Raid, 30-31 March 1944*, Penguin Books, Middlesex, England, 1986.

Middlebrook, Martin, *The Peenemünde Raid, 17-18 August 1943*, Penguin Books, Middlesex, England, 1988.

Middlebrook, Martin and Everitt, Chris, *The Bomber Command War Diaries: An Operational Reference Book 1939-1945* (revised edition), Midland Publishing Ltd., Leicester, England, 1996.

Milberry, Larry, *Aviation in Canada*, McGraw-Hill Ryerson, Toronto, 1979.

Milberry, Larry, *Sixty Years: The Royal Canadian Air Force and CF Air Command 1924-1984*, CANAV Books, Toronto, 1984.

Milberry, Larry and Halliday, Hugh A., *The Royal Canadian Air Force at War 1939-1945*, CANAV Books, Toronto, 1990.

Molson, Kenneth M., "The RFC/RAF (Canada): Its Squadrons and Their Markings", *Journal* of the Canadian Aviation Historical Society, Vol.22, No.3 (Fall 1984).

Molson, Kenneth M. and Shortt, A.J., *The Curtiss HS Flying Boats*, National Aviation Museum, Ottawa, 1995.

Molson, Kenneth M. and Taylor, H.A., *Canadian Aircraft since 1909*, Canada's Wings, Stittsville, Ontario, 1982.

Morgan, Eric B. and Shacklady, Edward, *Spitfire, The History*, Key Publishing, Stamford, England, 1987.

Moyes, Philip, *Bomber Squadrons of the RAF and Their Aircraft*, MacDonald and Co., London, 1964.

Moyle, Harry, *The Hampden File*, Air Britain (Historians) Ltd., Tonbridge, England, 1989.

Neate, Don and Davis, Mick, "35 Squadron RFC and RAF", *Cross and Cockade International Journal*, Vol.28, No.2 (Winter 1996), pp.194-202.

Ralph, Wayne, *Barker VC: William Barker, Canada's Most Decorated War Hero*, Doubleday Canada Ltd., Toronto, 1997.

Rawlings, John, *Fighter Squadrons of the RAF and Their Aircraft*, Macdonald and Co., London, 1969.

Roberts, R.N., *The Whitley File*, Air Britain (Historians) Ltd., Tonbridge, England, 1986.

Robertson, Bruce, *British Military Aircraft Serials 1878-1987*, Midland Counties Publications, Earl Shilton, UK, 1987.

Robertson, Bruce, *Lancaster: The Story of a Famous Bomber*, Harleyford Publications Ltd., Letchworth, England, 1964.

Saward, G/C Dudley, *The Story of Sir Arthur Harris: Bomber Harris*, Doubleday & Co., Garden City, NY, 1985.

Shail, Sidney, *The Battle File*, Air Britain (Historians) Ltd., Tonbridge, Engldand, 1997.

Shiras, George, "The Two Faces of Chivalry in the Air War", *Journal* of the Society of World War I Aero Historians, Vol.5, No.4 (Winter 1964).

Shores, Christopher, *2nd Tactical Air Force*, Osprey Publications Ltd., Reading, England, 1970.

Shores, Christopher & Williams, Clive, *Aces High: A Tribute to the Most Notable Fighter Pilots of the British and Commonwealth Forces in WWII*, Grub Street, London, 1994.

Silberbauer, Dick, "The Origins of South African Military Aviation, 1907-1919", *Cross and Cockade (Great Britain) Journal*, Vol.7, No.4 (Winter 1976), pp.174-180.

Smith, Blake W., *Warplanes to Alaska*, Hancock House, Surrey, BC, 1998.

Sturtivant, R.C., *The Anson File*, Air Britain (Historians) Ltd., Tonbridge, England, 1988.

Sturtivant, R.C. and Burrow, Mick, *Fleet Air Arm Aircraft 1939 to 1945*, Air Britain (Historians) Ltd., Tonbridge, England, 1997.

Sturtivant, R.C., Hamlin, John and Halley James J., *Royal Air Force Flying Training and Support Units*, Air Britain (Historians) Ltd., Tonbridge, England, 1997.

Sullivan, Alan, *Aviation in Canada, 1917-1918*, Rous and Rous, Toronto, 1919.

Swanborough, F. Gordon and Bowers, Peter M., *United States Military Aircraft since 1909*, Putnam, London, 1963.

Vann, R. and Bowyer, R.C., "15 Squadron RFC/RAF, 1915-1919", *Cross and Cockade (Great Britain) Journal*, Vol.4, No.2 (Summer 1973), pp.53-73.

Terraine, John, *The Right of the Line: The Royal Air Force in the European War 1939-1945*, Hodder and Stoughton Ltd., London, England, 1985.

Thetford, Owen, *Aircraft of the Royal Air Force since 1918*, Putnam, London, 1998.

Thetford, Owen, *British Naval Aircraft since 1912*, Putnam, London, 1982.

Thomas, Chris, *The Typhoon File*, Air Britain (Historians) Ltd., Tonbridge, England, 1981.

Thomas, Chris and Shores, Christopher, *The Typhoon and Tempest Story*, Arms and Armour Press, London, 1988.

Vincent, Carl, *Consolidated Liberator and Boeing Fortress*, Canada's Wings, Stittsville, Ontario, 1975.

Weicht, Christopher, *Jericho Beach and the West Coast Flying Boat Station*, MCW Enterprises, Chemainus, BC, 1997.

Wise, S.F., *Canadian Airmen and the First World War*, University of Toronto Press, Toronto, 1980.

Wohl, Robert, *A Passion for Wings: Aviation and the Western Imagination, 1908-1918*, Yale University Press, New Haven, 1994.

Yenne, Bill, *McDonnell Douglas: A Tale of Two Giants*, Crescent Books, New York, 1985.

# INDEX

*A Mouse in My Pocket* 270
Abbotsford, BC 129
Acadia University 202, 246
accidents 42, 54, 55, 58, 59, 67, 68, 84, 89, 94, 95, 104, 105, 130-133, 137-139, 146, 148, 151, 152, 159, 160, 166,179, 272
*Aces High* 224
aces (RCAF) 275
Ackland, Cpl K. 256, 257
Adair, Ed 259
Adams, G/C Oliver 177
ADDL 265
Addley, LAC (6439 SE) 257
Adrian, AC1 W.H. 147
Aerial Experiment Association 12
Aerial Survey Committee 61
aerial bombing 13, 21-28
aerial photography 12, 13, 60-72, 78-81, 83-85, 90, 179, 180
Aero Club of America 19, 32, 38, 66
*Aerodrome of Democracy* 116
Aeronautical Engineering Branch 51
Afelski, Frank 167
Aikman, F/L Alan F. 225
*Air Board,CAF, RCAF, 1919-1939* 11
Air Cadets 123
*Air Command* 11
Air Force Cross 17, 46, 96, 101, 182
air-sea rescue 56, 112, 149, 151, 160, 190, 207, 209, 215, 224, 225
Air Transport Auxiliary 231
*Air Transport in Canada* 62, 66, 75, 152
air mail 60
Aircraft Production Board 168
aircraft industry 168-177
aircraft names: "Boche Basher" 255, "Caledonia" 150, "Commando" 158, "Diane" 251, "Georgina" 274, "Gravel Gertie" 154, "Klondike" 189, "Le Tigre" 21, "Maggie" 153, "Marie" 145, "Northern Witch" 154, "Ruhr Express" 168, 175
aircraft skis 49, 74, 102, 117, 118, 177, 178-180
Aircrew Association of Canada 213
Airey, P/O 137
Airspeed Oxford 121, 134, 156
Ajax, Ontario 270
Akers, F/S/L 34
Aklavik, NWT 78
Akyab Island 274

Alaska 135-146
Albatros D-V 36
Alcock and Brown 100-103, 177
Alcock, F/S/L 32
Aldertin, Lt 26
Aldis Lamp 98
Aleutian Islands 134
Algonquin Park 107
Allan, LAC R.J. 254
Allen, F/O E.J 242, 258
Allerton Hall 164, 167
Allied Expeditionary Force 115
Alliford Bay 146
Allison engines 138
Alston, S.E. 158
Alzheimer's Disease 145
Amunsden, Capt Roland 92
Anderson, F/O C.M. 86, 101
Anderson, FSgt 70
Anderson, F/O D.R. 105
Anderson, Howard 120
Anderson, LAC J.E. 257
Anderson, F/S/L 36
Anderson, F/L N.R. 35, 59
Anderson, Lt P. 34
Andrews, F/O (247 Sqn) 229
Andrews, F/O Macdonald 188
Annan, Scotland 127
Annette Island, Alaska 136-146, 209
Annis, A/V/M C.L. 104, 106, 151
anti-aircraft systems 32
anti-submarine warfare 18, 19
Antwerp 210
Appleby College 227
Arado types:
    Ar.96 268, 269
    Ar.234 198, 210, 212
Archambeau, G/C 268
Arctic exploration 72-75
Armitage, LAW Doris E. 163
armourers 220
Armstrong Siddeley engines: 68, 82, 102, 119, 147, 178
Armstrong Whitworth types:
    F.K.8 18
    Atlas 97-100, 136
    Siskin 76, 97, 100-103, 109, 177
Arnhem 256
Arnold, Cpl 179
Aroostook, NB 197
Art Hope, Art 181
Arthur, F/O C.E.F. 109
artillery 13, 14, 15-18
Ashman, S/L Ralph 139, 150
*Asuma* (Japan) 93
ATFERO 150, 151, 156-159

Atkey, Alfred C. 34, 35
Atlantic Aircraft Corp. 73
Atlantic Canada Aviation Museum 54
Atlantic Ferry Organization 150, 151, 156-159
Atlas Steels Ltd. 174
Atomic bomb 178
atrocities 201-204
Audet, F/L Richard J. 224
Auster aircraft 224, 232, 233
Austin Airways 147
Austin, F/O C.C. 147
Australia 12, 161, 266
Aviation Detection Corps 149
*Aviation in Canada* 25
*Aviation in Canada 1917-1918* 46
Avise, P/O John E. 223
Avro Canada: Chinook engine 178
Avro types:
    504 trainer 48, 49, 55-58, 61, 62, 76, 77, 85-87, 95, 97, 109
    552 Viper 77, 85, 88
    Avian 95
    Anson 119, 120, 122, 132, 134, 149, 170, 179, 180, 182, 188
    Lancaster 150, 156-158, 168, 170, 175-177, 179, 180
    Manchester 175
    Tutor 111
Ayr, Scotland 194, 245
Aziz, F/O W.A. 208, 216
Azores 158
Babington, LCDR John 21
Baddeck No.1, No.2 12
Baggs, William M. 125-127, 200, 259, 260
Bailey, P/O Russel 185
Bailley, Frank 116
Bakewell, LAW Cherrie 165, 167
Baldwin, F.W. 12
Ball, S/O K.L. 161
balloons 12, 13, 33
Banks, F/Ls Wilfred J. 216
Bannock, W/C Russell 275
Banting, Dr. Frederick 178
Barber, F/L Bill 216
Barker Field, Toronto 116, 137
Barker, Chuck 123
*Barker, VC* 30
Barker, W/C William G. 29, 30, 55, 70
barnstorming 50
Barrie, Ontario 58, 270
Barton, Mr. 92

Barton, Robert A. 275
Barton, W/C R.A. 198
Baskerville, F/O P.G. 106
Bathurst, Al (442 Sqn) 260
Battle of Britain 110, 136, 142, 190-195, 222, 272
Battle of Britain Memorial Flight 213
Battle of France 190, 191
Battle of the Bulge 210, 232, 233
Bawtinheimer, Sgt Earl G. 152
Baxter, E.L. 137
Baxter, F/L F. 185
Baxter, P/O George 137, 144
Bayne, F/S/L 36
Beamish, F/L F.V. 103
Beamish, F/S/L 36
Beard, A/S/L Charles T. 12
Beardmore, F/O Eric W. 136, 192, 193
Beasley, F/O J.R. 200
Beattie, Bill 144
Beattie, Jim 232
Beattie, WOp Wilfred E. 58
Beatty, WO2 Byron 272
Beckett, Harry 128
Beech Aircraft
    Model 18 Expeditor 154, 178
    Queen Air 269
Beeman, Brig. W.G. 98
Beginnen, Eric 203
Beirnes, S/L J.R. 144, 234, 235, 251
Bell types:
    P-39 160
    P-63 160
Bell, Alexander Graham 12
Bell, Larry 178
Bell Telephone 202
Bella Bella, BC 146, 147
Bellanca Pacemaker 65, 78, 80, 90
Belleville, Ontario 125, 137
Bellingham, F/O William H.L. 208, 216
Belyea, Sgt Dorothy H. 163
Bender, Cadet C. 45
Bengerter, F/L 211
Bennet, P/O Sam 258
Bennett Lake, Yukon 160
Bennett, FSgt 70
Bennett, Donald C.T. 150, 156
Bennett, Lt Louis 38
Bennett, PM R.B. 104
Benoit, Capt P.S. 12
Berens River, Manitoba 89
Bermuda 158, 182
Bernhardt, F/L David A. 269
Bernhart, Pierre N. 242

283

Bernier, LAC S.A. 257
Bessonneau hangar 91, 94, 101
Beurling, F/L George 218, 222, 227, 259, 275
Beveridge, Massey W. 267
Bewsher, Capt Paul, DSC 21
Bickell, J.P. 150
Biehler, S/L John A. 112, 147
Biggar, Col O.M. 51
Billings, WO J. 148
Bing, L.P.S. 275
Birchall, Sgt K. 104, 106
Bishop's Field 171
Bishop, A/M W.A. 28-30, 104,124, 161, 193
Bissky, Paul 232, 235
*Black Crosses off My Wingtip* 190, 226
Black Market 246
Black, FSgt 70
Black, LAC Ivan 258
Black, S/L Mike 239
Blackburn Shark 112, 147, 149, 178
Blackburn, F/O Harry 272
Blackburn, P/O 106
Blackey, Cpl W.M. 257
Blackmore, Lt George J. 46
Blaiklock, F/O S.T. 192
Blanchard, F/L S.S. 81
Blatchford, W/C Howard P. 190
Blockley, Sgt H.T. W. 106
*Bloody Shambles* 191, 270
*Blue Skies* 190
Blyth, Sgt George 257
"Bobbie" 74, 75
Bodenplatte 145, 207, 216, 229, 231, 249
Boeing Canada 170
Boeing types:
    247D 5, 134
    B-17 151, 152, 155, 156, 158, 160, 257
    B-29 180
    Totem 106
Bohemier, WO J.E. 199
bomb types 27
Boomer, S/L Kenneth A. 136, 194-196
*Borden Flyer* 55
Borden, PM Sir Robert 39
Borland, F/O Alexander G. 200
Botwood, Nfld. 149-151
Boulton, P/O Bill 131
Boulton, S/L F.H. 214
boundary surveys 61
Bournemouth 126, 164, 185,197, 227, 262, 275
Bowers, Pete 103
Bowhill, Sir Frederick 150
Bowick, E.R. 181
Bowker, FSgt Harlow W. 215, 217
Bowman, Vernon L. 205
Bowyer, R.C. 18
Boyd, S/L Dave 216
Boyd, F/L W.G. 84
Boyle, F/O H.V. 198

Brachen, Robert 226
Brackley, H.G. 27
Bradford, FSgt A.J. 70, 106
Bradford, Robert W. 38
Bradshaw, G/C D.A.R. 130
Brady, F/L 200
Brass, S/L Donald M. 262
Brault, LAC J.L.D. 257
Breadner, A/V/M L.S. 33, 52, 55, 100, 150, 178
Brechnel, Sgt 201
Breck, Bill 260
Brehin, Jacques 260
Brennan, LAC W.M. 257
Breschfeld, Lt V. 24
Bretz, W/C N.H. 234
Brewster, S/L Don 240, 246, 257
Bridgman, F/O W.F. 201
Briese, F/O C.E. 104, 191, 192
Briggs, F/O 194
Briggs, S/L F.E.R. 179
Bristol types:
    engines 105, 179
    F.2B Fighter 22, 31, 34, 35, 61, 62
    Bolingbroke 121, 129, 130, 134, 139, 143, 146, 149, 152, 169, 178
British Commonwealth Air Training Plan 40, 112, 115-133, 161
British Commonwealth Air Training Plan schools/stations: 117,
    No.1 ANS 129
    No.1 (F) OTU 121, 126, 129, 197, 204, 207, 236, 267
    No.1 AOS 116, 120, 156
    No.1 B&GS 118, 132
    No.1 EFTS 137
    No.1 GRS 149, 163
    No.1 MD 137, 197, 236, 267
    No.1 SFTS 119, 124, 131, 227, 232
    No.1 TTS 246
    No.1 WS 119
    No.2 B&GS 123
    No.2 MD 181
    No.2 SFTS 137, 161, 197, 209, 234, 236
    No.2 WS 119, 122, 181
    No.3 B&GS 133, 179
    No.3 SFTS 6, 129
    No.3 WS 119
    No.4 EFTS 117, 131, 267
    No.4 WS 119
    No.5 ITS 125, 137
    No.5 MD
    No.5 OTU 129, 130;
    No.5 SFTS 188
    No.6 MD 162;
    No.6 SFTS 119, 122, 125, 126, 181, 200, 255, 271
    No.7 MD 162
    No.7 OTU 129
    No.7 SFTS 267
    No.8 SFTS 267
    No.9 AOS 156

No.9 SFTS 189, 262
No.10 AOS
No.10 EFTS 204, 271
No.11 EFTS 202
No.11 SFTS 120
No.12 EFTS 197, 200, 227, 232
No.13 SFTS
No.14 SFTS 163-166, 204
No.15 EFTS 234
No.15 SFTS 152
No.16 EFTS 131
No.17 SFTS 120, 181
No.18 EFTS 152
No.19 EFTS 117, 123, 264
No.20 EFTS 115, 125, 236
No.21 EFTS 189, 262
No.31 B&GS 128
No.31 GRS 149
No.31 OTU 129, 149, 264
No.34 OTU 133, 151, 152
No.36 SFTS 121
British Empire Medal 162, 163
Broadhurst, A/V/M 224
Brock, R.B. 46
Bromley, F/L Roger A. 268, 269
Brooker, Sgt Arthur C. 144
Brookes, A/V/M George E. 54, 62, 76, 83, 85, 100
Broome, Col E.L. 92
Brown, W/C Manafrank 267
Brown, Cpl 106
Brown, F/O E.P. 193, 194
Brown, LCpl F.S. 12
Brown, S/L J.D. 202-204, 226
Brown, James A. 242
Brown, Lt Jonathan M. 18
Brown, F/O John K. 235
Brown, F/L K.A. 269
Brown, W/C Manafrank 267
Brown, F/O Mark H. 191
Brown, F/O R.C.J. 269
Brown, F/O R.O. 268
Brown, Cpl R.S. 257
Brown, Raymond A. 242
Brown, A/C W.W. 80, 98
Brown, F/L W.W. (441 Sqn) 198, 215
Browne, S/L J.D. 202-204
Browne, LAC George L. 257
Browning machine gun 112
Brownlee, S/L D.A. 246
Bruce, Bo 128
Brunelle, F/L Paul M. 269
Brussels 244
Bruton, P/O 125
Bryans, F/O J.G. 80
Bryant Press 142
Brydon, F/O E.D. 235
Buchanan, Mrs. 203
Buckham, W/C Robert A. 214, 224
Buckley, LAC John F. 257
Buckley, P/O Sydney S. 217
Buckmaster, F/O 234
Buell, Elbert L. 128
Bullock, John 259, 260

Burcham, Cpl 162
Burgener, Lt Christian 25-27
Burgess, Charles L. 242
Burgess-Dunne aircraft 38, 39
Burke, F/O E.J. 71
Burmaster, Sgt 140
Burns, F/O "Brains" 229
Burns, Neil 197
Burpee, M.J. 62
Burroughs, F/L George W. 268, 269
Burroughs, Sarah and Gary 260
Burt, LAC K.G. 257
Burt, LCol Dave 199
Burton, Allan 227, 229
Burton, W.D. 242
Busby, F/S/L 33
Butts, F/L R.A. 151
Buzza, John 234, 235
Byers, F/L H.M. 233
Byrnes, Sgt 200
Cable, Sgt S.G. 106
Caen 244
Cagney, James 118
Cairns, F/O James R. 62, 67
Cairo, Egypt 158
Caldwell, Sgt C.S. 54, 87
Calgary 47, 53, 103, 106
Calvert, P/O 262
CAM ship 205-207
Cameron, D.R. 62
Cameron, Gord 201
Cameron, S/L L.M. 224
Camp Dundurn 100
Camp Niagara 97
Camp Petawawa 12, 61, 99
Camp Shilo 99, 100
Camp, LAC Norman V.R. 257
Campbell, A/V/M Hugh L. 203
Campbell, Sir Gerald 150
Canada House 234
Canada Wire and Cable 162
*Canada's Air Heritage* 126
*Canada's Fighting Airmen* 11, 29
*Canada's Fighting Pilots* 107
*Canada's Flying Heritage* 37
Canadair 225
Canadian Aerial Services Lyd. 62
Canadian Aerodrome Co. 12
Canadian Aeroplanes Ltd. 40, 42, 43
Canadian Air Board 28, 50, 51, 61, 71, 177
Canadian Air Force (1919-23) 17, 31, 35, 50
Canadian Air Force (England) 39
*Canadian Aircraft since 1909* 59
*Canadian Airmen and the First World War* 11, 20, 29
Canadian Airways (new) 153, 177
Canadian Airways (old) 28
Canadian Army Medical Corps 22, 46
Canadian Associated Aircraft 168
Canadian Aviation Corps 38, 39
Canadian Aviation Historical Society 11, 30, 38, 59, 66, 157

Canadian Car and Foundry 105, 113, 170-172, 184
Canadian Expeditionary Force 22, 28, 37, 38
Canadian Fighter Pilots Association 213, 226
Canadian Forces bases:
　Cold Lake 47, 199
Canadian Forces squadrons:
　410 47,
　414 269
Canadian Government Trans-Atlantic Air Service 175
Canadian Harvard Aircraft Association 217
*Canadian Historic Review* 29
Canadian Legion 271
Canadian National Exhibition 15, 103, 111
Canadian Pacific Airlines 5, 116, 150
Canadian Pacific Railway 156
Canadian Pratt & Whitney 168, 174
Canadian Transcontinental Railway 84
Canadian Vickers Co. 113, 134, 168, 170, 173
Canadian Vickers types:
　PBY 147
　Vancouver 69, 94, 106-108, 146, 147, 178
　Vanessa 60
　Varuna 64, 68, 88, 89, 94
　Vedette 50, 63, 65, 67-69, 78, 82, 88, 93-96, 98, 146, 147
　Velos 71
　Vigil 88
　Vista 88
Canadian War Museum 11, 19, 29, 103
Canadian Warplane Heritage 197
Canne, Italy 220
*Cannette* 137
Canol Pipeline 153
Cappleman, Sgt Reg 257
Capreol, Leigh 178
*Captains of the Clouds* 118
Carlisle, E.G. 158
Carlstrom, V. 40
Carmichael, H.J. 178
Carpenter, S/L 151
Carpiquet 244
Carr, F/O John (439 Sqn) 255
Carr, John (412 Sqn) 208
Carr-Harris, F/L B.G. 100 74, 75, 103
Carroll, John 62, 71, 80
Carscallen, F/O H.M. 106
Carswell, Doug 123
Carswell, F/O V.P. 123
Carter, F/L Albert W. 58, 78, 84, 87, 93
Carter, F/O Roy E. 204
Cartierville, Quebec 118, 169, 171

Castator, Murray 140, 142, 246-254, 258
Cauchon, N. 80
Cawker, F/O Hugh R. 264
Ceiffets, WO2 Dave 104
Central Airways 269
Central Canada Exhibition 103
Central Technical School 122, 236
Cessna T-50 Crane 6, 120, 179
Ceylon 266
Chaburn, W/C Lloyd V. 223
Chadderton, Clifford 29
Chadwick, Arnold J. 31-33
Chalmers, WO (41 Sqn) 211
Chambers, Mr. 262
Chanute Field 87
Charles, Edward F. J. 275
Charleson, J.C. 105
Charlottetown, PEI 149
Charron, Leo 66, 67
Charron, F/L P.M. 216
Chatham, Ontario 19
Chesterfield Inlet 787
Chinook engine 178
Chowne, LAC A.W.
Chowen, F/O W.R. 215
Christie, Carl A. 156
Christmas. F/O B.E. 192
Christopherson, LAC Del 258
Churchill River 62, 66
Churchill, Manitoba 71
Churchill, PM Winston 148, 158
Civil Aeronautics Administration 143
Civil Government Air Operations 51, 59-72, 97
Clark, F/L Frank J. 223
Clarke, Archibald 128
Clasper, Bob 207
Claxton, Capt W.G. 45
Clearwater, P/O 58
Cleghorn, F/O Donald G. 230
Clements, Cpl Dorothy 167
Clerget engine 48
Cleveland, Howard D. 275
Clifford, Bill 204, 232, 233, 260
Clutterbuck, F/L 266
Coates, F/O 125
Coffey, S/L R.E. 232, 233
Coghill, F/L F.S. 75, 97
cold weather trials/equipment 75, 78, 177-180
Coleman Lamp and Stove Co. 174
Coleman, P/O (41 Sqn) 211
Coleman, Sgt Ernest A. 188
Coleridge, Lt C.G. 45
Coles, Lt Eric M. 24
Collins, F/O Allan W. 200
Collis, Ray 240
Collis, F/L Ron 213
Collishaw, Raymond 10, 11, 29, 30
Colonna, Gerry 141
Colp, WO2 70
Colpitts, F/O 178, 179
Colquhoun, P/O R.S. 229
Colville brothers 236

Colville, F/O John S. 236
Comerford, L.P. 201
Commonwealth War Graves Commission 203, 247
Compact-O-Hangar 179
Compton, Lt Harry N. 37
Connell, Don 144
*Consolidated Liberator and Boeing Fortress* 129, 152
Consolidated types:
　B-24 Liberator 129, 130, 143, 148-150, 152, 158, 159, 180, 182, 227, 257
　Courier 95
　PBY Canso/Catalina 94, 143, 147, 148-150, 157, 170, 182
Controller of Civil Aviation 50, 51
convoys ON.202 148, ONS.18 148, PQ-16 206
Cook, Jack 259
Cooke, Prof. R.H. 61
Cookman, AC2 Murray 122
Coome, C.R. 125
Cooper, Col. 178
Cooper, Cpl 59
Cooper, Garry 213
Cooper, F/L John 269
Cooper, M.E. 94
Copeland, F/L J.C. 197-199
Copeland, June 199
Copenhagen 144
Corbett, F/O (1 OTU) 200
Corbett, F/O (1 Sqn) 193
Cordick, F/O J.M. 233
Corey, 2Lt Irving B. 17
Corness, Les 153
Cosgrove, Ed 107
Côté, F/O J.A. 229
Cougler, F/O Harold D. 269
Cousin, Sgt 70
Cowan, P/O William 216
Cowie, Sgt S. 257
Cowley, P/O Tom 92
Craig, George 180
Craig, J.D. 78
Cranwell, England 24
Creagen, Harry 11, 23, 32, 38
*Creation of a National Air Force* 60
Critchley, Brig. Arthur C. 28
Croden, Capt James E. 18
Croil, A/V/M G.M. 58, 76, 95, 97, 109, 161
Crompton, FSgt H.M. 223
Crosby, S/L R.G. 250
*Cross and Cockade* 11, 18, 28
Cross, Capt Alfred 17, 18
Crowley, F/O F.J. 233
Crumb, Donald H. 270-275
Crumb, Georgina 270, 275
Cryderman, Felix 205
Crysler, Lt C.A. 37
Cudemore, Capt C.W. 53
Cuffe, F/L A.L. 54, 55, 58
Culliton, F/O 179
Cummings, F/O D.H. 232, 233

Currie, Sir Arthur 39
Currie, F/L F.B. 146
Curtiss flying schools 19, 21, 32
Curtiss Museum 40
Curtiss types:
　"E" 40
　F.3 54
　HS-2L 50-52, 59-63, 67, 84, 85, 91, 92, 107
　JN-3 32,
　JN-4 41-47, 50, 54, 56-58
　P-40 (Kittyhawk, Tomahawk) 113, 129, 135-138, 267
　Helldiver 172
　Seamew 184
Curtiss, Glen 40
Cushley, S/L Robert 254
Customs patrols 92
Cutcliffe, LCol A.B. 163
Cutcliffe, Rosalie 163-167
Cutting, Bob 268
D-Day 219, 245, 254, 264, 269
Dagmar, John 128
Dahl, Whitey 267
Dalgleish, Jim 128
Dalzeil, W.J. 25
Dance, 2Lt C.C. 17
Darley, Cecil H. 21, 22, 33
Darrow, Chuck 201, 202
Dasey, FSgt 70
Dauphinais, LAC Roger 257
Davey, F/O 98
Davidge, Rod 259, 260
Davidson, R.D. 62, 66
Davidson, W/C R.T.P. 245
Davidson, Sgt T.M. 125
Davies, LAC David J. 255-258
Davis, LAC Kenneth F. 188
Davis, Sgt R.L. 104, 106
Davis, Sgt Ralph C. 104-106, 151, 152
Davis, F/O W.G. 229, 230
Davoud, G/C Paul Y. 241, 246
Dawber, Norm 259
Dawson, F/O E.D. 58
Dawson, LAC K.C. 257
Day's Business College 163
Day, J. 40
De Nancrede, C.S.G. 229, 234
De Pret, F/O 179
de Havilland (UK) types:
　D.H.4 17, 19, 20, 23, 24, 34, 48, 61, 62, 67, 75, 84, 86
　D.H.6 25
　D.H.9 17, 53, 61, 67
　D.H.60 Moth 69, 73, 89, 95
　D.H.82 Tiger Moth 128
　D.H.83 Fox Moth 151
　D.H.98 Mosquito 151, 158, 169, 179, 184, 264-266
　D.H.100 Vampire 178, 213
　Goblin engine 178
de Havilland Canada 134, 170, 259;
　D.H.82C Tiger Moth 115, 117, 131, 197, 271

de Niverville, A/V/M J.L.E.A. 108, 109
Dean, Bob 232
Dearaway, Cpl S.C. 70, 78, 101
Debden Wing 214
Debert, Nova Scotia 129
Deception Bay, Ungava 74
Defence Research Board 178
Delorme, E. 54
Denmark 144, 251
Department of Marine and Fisheries 73, 75
Department of Militia 12, 38, 46, 51, 97
Department of Munitions and Supply 168
Department of National Defence 168
Department of National Revenue 71
Department of Public Works 71, 80
Department of Railways and Canals 71, 73
Department of the Interior 62, 71, 78, 84, 91
Department of Transport 51, 149
depth charges 148
Deremo, 2Lt John C. 18
Derraugh, H.E. 128, 199
Deseronto, Ontario 45
Desloges, F/O 104, 193, 194
Detlor, Scotty 128
Deveau, LAC J.J. 257
Deville, E.G. 61
DFW C V 36
Diamond, S/L Gordon 136
Dickenson, LAC C. 254
Dickie, Alec 91
Dickson, Hugh 145, 209
Dieppe 224
Diller, Gordon L. 123
Diller, Cpl Harry S. 63, 70, 101, 123
Dilworth, Paul B. 178
Directorate of History 11, 29, 34, 91
Distinguished Flying Cross 15, 17, 18, 22, 24, 190, 191, 216, 227, 266
Distinguished Flying Medal 46
Distinguished Service Cross 22, 34, 129
Distinguished Service Order 28, 129, 223
Dixon, Mickey 167
Doak, F/L 207
Doak, F/L James Basil 216
Dobbin, E.C.W. 93
Dobson, Mr. 105
Dodds, Ron 11, 28
Dodds, Roy E. 17
Doidge, F/O Ron 232
Dominion Gasket173
Donaldson, Cpl E.J. 257
Donovan, F/O J.H. 267, 269
Dopson, LAC R.M. 257
Dornier types:
    Do.17 190, 191
    Do.215 191

Do.217 129, 202, 215, 271
Dorval 156-158, 181
Douglas types:
    A-20/Boston 156, 158-160
    B-18 150
    C-47 Dakota 143, 153, 154, 160, 257
    C-54 143
    Digby 106, 147-151, 179
    Dolphin 143
    MO-2B 60
    Turbinlite 227
Douglas, LAW 162
Douglas, W.A.B. 29
Douglas, Walter Norman 201
Dover, F/L Dean H. 214-216
Dow Chemicals 269
Dowding, S/L H.J. 215
Dowling, F/L I.M. 146
Dowsan, W/O S.S. 161
Doyle, M.F. 226
Dradon, Mr. 91
Drake, Clarence 128
Drew, George 11, 29
Dubuc, F/L Louis R. 151
Duggan, Dan 116
Duisberg 200, 210
Duke of Kent 150
Duke, LAC E.H. 257
Dunbar, Lt W. 47
Duncan, AC 101
Duncan, F/O C.J. 93
Duncan, WO J.L. 251
Duncan, J.R. 92
Duncan, Lt Richard 16
Dunkeld, F/O W.T. 232, 233
Dunklemen, F/O Louis 216
Dunn, Benny 232
Dunster, 2Lt 35
Dunster, F/O Reg 272
Duren, Germany 198
*Dust Clouds in the Middle East* 191
EAC Meteorological Flight 182
Eadie, Sgt R. 101
Eaker, MGen 266
Eardley, AC1 J.T. 67
Earle, F/L R.N. 216
East Grinstead 162
East, Cpl H.N.J. 257  K
Eastern Air Command 96, 106, 130, 134, 148-152, 182, 186, 239
Eastern Ontario Regiment 25
eclipse 62
Edmonton 121, 131, 150, 152, 153, 160, 187, 190
Edwards, A/M Harold 109, 150
Edwards, W/C J.F. 203, 223, 275
Edwards, Sgt R.A. 254
Edwards, F/O Robert L. 192
Edy, Alen L. 190
Eglin Field 180
Eisenhower, Gen. D.D. 115
Elliott, F/O (1 Sqn) 195
Elliott, P/O John C. 214
Ellis, F/S/L 33, 34
Ellis, Frank 37, 38

Ellis, LAC "Red" 236
Emery, F/O W.M. 71
ENSA 243
Enschede, Holland 204
Enstone, F/S/L 33, 34
entertainment and sports 140, 141, 229, 234, 238, 241-243, 255, 256
Epp, LAC Dave 257
Epp, P/O Jacob A. 185
Esquimalt, BC 97, 146
Estevan Point, BC 136
Etienne, Phillip E. 205
evaders 224
Evans, A/S/O Sylvia I. 161
Everard, Hedley 270
Ewart, P/O Fred J. 146
Ewart, Sgt 70
F-series flying boats 20, 43, 49, 53, 54, 84, 85, 91
Fairbanks, David C. 275
Fairchild Aircraft Co. 134, 170
Fairchild types:
    F.71 78, 80, 146
    FC-2/2W 60, 64, 65, 69, 71, 75, 78, 89, 94, 110, 177
    KR-21 70
    Cornell 6, 117, 178, 234
    Super 71 178
Fairchild, Sherman 71
Fairclough, Capt Arthur B. 37
Fairey types:
    Barracuda 265, 266
    Battle 112, 118, 132, 134, 166, 178, 179
    Swordfish 184
Fairfield, F/O C.E. 198
Fairview Cemetery 152
Falaise 198
Falkenberg, Capt C.F. 39
Fargo, ND 120
Faris, LAC Glen D. 188
Farrell, Lt C.M.G. 14, 146
Fauquier, G/C Johnny 164, 259
FBA flying boat 19
Federal Aircraft 170, 172
Fee, F/O C.J. 104
Ferrier, A/V/M Alan 71, 177
Ferry Command 137, 152, 156-159, 169, 179
Ffolliot, Lt C. 34
Fighter Leader School 236
*Fighter Pilot: A History and a Celebration* 226
fighter pilot scores (claims & kills) 34, 35
Finch, Teddy 128
Finley, Hart R. 226
Finucane, Paddy 225
Fiset, Col. E. 12
Fiset, Ken 242
Fisher, Crawford
Fisher, F/L R.F. 148
Fisher, LAC Stan 257
fisheries patrols 92, 93
Fitch, Al 246

Fitton, Len 227
flak 225, 230, 235, 242, 258
Flatt Davey, S/O Jean 161
Fleet Air Arm 265, 266, 272
Fleet Aircraft Co. 134, 170, 178
Fleet types:
    Fawn 111, 117, 134, 179
    Finch 6, 117,134, 262
    Fort 118
Fleming, Al (416 Sqn) 200
Fleming, Sgt Arthur 70, 75, 86, 94-96
Fleming, F/O J.W. 215
*Flight* 38
Flintoff, F/O John 233
Flood, F/O James M. 197, 198
Focke-Wulf types:
    Fw.190 190, 191, 198, 211, 212, 214,-217, 223-225, 227, 242, 243, 246, 252, 256, 269, 271
    Fw.200 205, 229, 246, 252
Fogerty, LAC Pat 257
Foggerty, Red 181
Fokker types:
    E-III Eindecker 36, 40
    D VII 15, 30, 36, 37
    Dr I 36, 37
    Universal 73-75
Ford Trimotor 103
Ford, S/L Leslie S. 224
forestry 83-85, 88, 89, 91,92
Forman, Sgt 228
Forrest, C.N. 179
Fort Fitzgerald, Alberta 78
Fort William, Ontario 113, 171, 172
Fort Worth 42
Foss, G/C R.H. 151
Foster, Lt George B. 14
Fowler, Gus 236, 237
Fox, F/L C.W. 198, 216-218
Fox, LAC Tom 257
Foynes 150
Franks, Norman 231
Franks, Dr. W.R. 178
Fraser, WO2 180
Fraser, Douglas 150
Fraser, LAC J.R. 83
Fraser, F/O T.G. 104
Fraser, Tom 128
Freeman, LAC Walter E. 216-218
Freestone, Harold 167
French, Don 232
French Foreign Legion 266
Frezell, Edward G. 271
Friedlander, John 260
friendly fire 236
Frye, Eric 66
Fullerton, F/L E.G. 92, 98, 104, 109
Fumerton, F/O 194, 195, 275
"G" force 215
G-suit 178
Gaines, Charles 197
Gallimore, J. 128
Gander 148-151, 156, 159
gardening 128

Garland, F/O M.L. 198
Garratt, F/L P.C. 58
Garry, F/O E. 269
Garson, F/O Bill 216
Gaspé 150
Gass, 2Lt C.G. 34, 35
Gayton, LAC G.F.G. 81
Gdynia 264, 265
Geale, C. 40
Gear, Sgt 70
Gear, P/O Ken 228
Geodetic Survey 71, 75, 78, 80
Gestapo 197, 204
Gibb, Sgt J.A. 257
Gibbons, Lt O.A.C. 54
Gibbs, F/O Peter 210, 213
Gibbs, F/O W.R. 233
Gibson, J.L. 271
Gilbert, F/O 203
Gilbraltar 264, 265
Giles, Cpl (143 Wing) 248
Gilmour, Sgt R.A.W. 83
Gilpin, LAC A.H. 257
Ginger Coote Airways 106
Glazebrook, Edwin E. 227
Gloster types:
    E.28 178
    Gamecock 100
Glover, Rex 123
Gobeil, F/O F.M. 94, 103
Godefroy, W/C H.C. 214, 224
Godfrey, A/V/M Albert E. 46, 60, 70, 92, 93, 95, 146
Goldberg, S/L Dave 220
Gomez, Perez 236
Gonyon, Harold M. 19, 20
Good, P/O Tom 133
Goode, H.M. 231
Gooderham, C. Grant 40
Gooding, Lt. 24
Goodwin, Sgt Frederick J. 151
Goose Bay 154, 158,159
Goranson, Paul A. 221, 244
Gordon, Donald C. 275
Gordon, W/C J.L. 73, 88, 92
Gordon, F/O K.A. 104
Gordon, Nelson L. 204, 233, 234
Gordon, F/O R.C. 98, 110
Gordon, W.I. 201
Gosport System 41
Gould, FSgt George S. 216
Govett, A/F/M 58
*Graf Zeppelin* 152
Graham, F/O 71
Graham, F/S/L C.W. 28
Graham, Sen. George 100
Grand, Pte Frederick W. 45
Grande Prairie, Alberta 89
Grandy, G/C Roy S. 54, 62, 71, 75, 78, 88, 89, 100, 105, 124
Grange, F/S/L Edward R. 28
Grant, W/C Frank G. 144, 233, 235, 250
Grant, Sgt J. 24
Grant, F/O P.J. 147
Gray, F/L Gerry 229

Gray, J.O. 242
Gray, Capt James 23, 24
Gray, Robert Hampton, VC 29
Gray, Robert M. 232
Gray, Ross G. 275
Gray, LAC V.L. 257
Gray, F/O William A. 256
Great Lakes Airlines 269
Greaves, Sgt Jimmy 227, 228
Green, LAC A.V. 101
Green, Bert 63
Green, S/L Charles D.B. 188
Green, S/L J.F. 148
Green, Dr. J.J. 179
Green, Cpl P.N. 101
Green, LAC S.N. 101
Greene, Sgt S.A. 70, 78
Greenhous, Brereton 29
Greenock, Scotland 137, 146, 197, 227
Greenway, Cpl 178
Gregor FBD-1 171
Gregor, Michael 171
Gregory, Bob 232
Griffin, F/L 180
Griffin, Frederick C. 62
Griffith, Lt Ewart T. 47
Group of Seven 42
Grove Cemetery 152
Grumman types:
    G-23 Goblin 113, 136, 171
    Goose 151, 178
guard duty 122, 125, 128, 129, 137, 236, 262, 271
Gubb, F/L 180
Gunn, W/C W.B. 203
Gunnarson, I.L. 232
Gurd, R.S. 232
Gurdon, 2Lt J.E. 34, 35
Guthrie, A/V/M Ken 110
Guy, William J. 151
Gwynne, LCol 184
Haddons, Eleanor 119
Haileybury, Ontario 84, 85, 107
Hain, Lt 21
Haines, C.W. 201
Halahan, Col. F.L. 20
Halahan, S/L 191
Halcrow, F/L A.F. 198
Halcrow, James A.F. 267
Hale, BGen E.A. 152
Haley, LAC (6439 SE) 257
Halifax, NS 51, 106, 182-185, 232, 253, 268
Hallford, Murray 260
Halliday, Hugh A. 127, 205, 224, 250, 275
Hamel, E. 116
Hamilton, Jim 249
Hamilton, Mac 236, 237
Hamilton, Ontario 46, 197
Hanbury, F/O 195
Hancock, Lt 266
Handley Page types:
    O/100 21, 22
    O/400 21, 25-27, 177

    V/1500 21, 28, 177
    Hampden 121, 129
    Harrow 179
*Hanging the Legend* 29
Hanlon, Cpl (143 Wing) 248
Hannah, Bud 128
Harbour Grace, Nfld. 177
Hardie, LAW "Mac" 167
Harding, F/L D.A. 95
Hardy, Harry 204, 233
Hargreaves, Sgt (135 Sqn) 274
Harling, F/L D.W.A. 200, 201
Harris, A/C/M 108
Harris, Sgt 190
Harris, F/O George 216
Harrison, Chuck 126
Harrison, Mike 242
Harrow, F/S/L 36
Hart, F/L Eldon 272
Hartnett, F/O Timothy 235
Hartsborn, AC2 Harry 122
Harvey, Cpl N.E. 83
Harwood, F/O C.F. 232, 233
Hassett, LAW E.M. 162
Hatch, Fred J. 40, 116
Hatton, F/O V.J. 109, 171
Havergal College 163
Hawker types:
    Hurricane 103-105, 110, 121, 126, 135, 137, 144, 146, 152, 171, 179, 180, 190-197, 200, 205, 206, 227, 228, 236, 267, 268, 271
    Osprey 106
    Tempest 210, 236, 242
    Tornado 227
    Typhoon 127, 198, 204, 227-261, 271
Hawkins, Mary 1163
Hawkins, S/L (135 Sqn) 272
Hawtrey, F/O R.C. 80
Hay, Al (SAAF) 206
Hay, Lt B.M. 12
Hay, D. 40
Hay, F/L M.M. 182
Hayes, Bob 260
Hayes, LAC J.G. 257
Haywood, Robert A. 270
Heakes, Lt F.V. 39, 83, 110
Heath, P/O James B. 188
Heeney, P/O 214
Heinkel:
    He.111 190-193, 265
    He.177 265
Hemming, LCol H.H. 203
Hems, S/L F. 172
Henderson, F.C. 20
Henderson, Pete 242
Henry, F/L 210
Hepburn, Donald S.R. 128
Hepburn, Reginald L.R. 128
Herbert, Sgt R.F. 106
Herriot, F/O J.K. 112, 147
Herron, William A. 151
Hettasche, Mr. 83
Hewson, F/L Henry W. 103

Hicken, Bert 181
Higgins, F/L 67
Higgins, F/L F.C. 62
"High Flight" 225
Hill, Brig. A.H. 100
Hill, Sgt Bill 131
Hill, G.M. 201
Hill, S/L G.U. 224, 275
Hill, W.D. 201
Hiltz, F/O Ab 152
*Hindenburg* 150
Hitchins, W/C Fred H. 11, 17, 34, 78
HMCS: *Armentieres* 97, *Aurora* 91, *Britomart* 236, *Dauntless* 98, *Fraser* 106, *Patriot* 91, *Patrician* 91, *St. Croix* 149
HMS *Adelaide* 93, *Apollo* 106, *Eagle* 223, *Franklin* 83, *Glorious* 191, *Hood* 93, 150, *Hussar* 236, *Hyderabad* 207, *Itchen* 149, *Jackdaw* 265, *Keppel* 149, *Rajah* 265, 266; *Renown* 148, *Rodney* 93, *Striker* 266; *York* 106
HMTS *Canton* 275, Dominion *Monarch* 271
Hoare, Brig. C.G. 41, 46
Hobbs, Maj Basil D. 53, 62, 66, 86, 91, 151
Hobbs, Jim 128
Hodges, F/S/L 33
Hodgins, F/O R.A. 215
Hogg, J.E. 242
Holland, Lt Herbert H. 16
Holland, Capt Hubert L. 62
Holland, Sgt John R. 45
Hollingsworth, "Holly" 232
Home War Establishment 134
Home-Hay, Capt 53
Hope, Bob 141
Hornchurch Wing 214
Hoseason, Cpl C.H.C. 83
Hotels, pubs and clubs: Anglo Swiss Hotel 185, Canadian Legion 271, Chateau Laurier 226, Lord Elgin 236, Marquis of Granby 216, Oban Inn 260, Red Lion 216, St. George and Dragon 228, The Orchard 193
Hotson, Fred W. 156-158
Houle, S/L A.U. 223, 234, 275
House, F/L 33
Howard, Dee 151
Howard, Sgt 70
Howe, Hon. Clarence D. 113, 168
Howe, Norm 260
Howell, F/L John 184
Howsam, Lt G.R. 39
Hoyt, S/L C. 184
Hudson Bay Railway 67, 71, 73
Hudson Strait Expedition 71, 73-75
Hudson Terraplane 240
Hughes, Lt D.J. 15
Hughes, Neil M. 232
Hughes, Sam 38

287

Hughes, W. 116
Huletsky, P. 275
Hull, F/O 93
Hull, F/L A.H. 97
Hulme, Jimmy 167
Humphreys, Lt C.J. 45
Hurtibise, Paul 207
Huska, K. 204
Huskisson, B.L. 34
Hutchings, "Hutch" 207
Hutchinson, F/L R.C. 269
Hutton, W/C 164
Huycke, Barbara 110
Hyde, F/O George G. 193-195
Hyndman, P/O Bob 137, 226
ice patrol 71
Immelmann, Max 31
Imperial Airways 150
Imperial Bank 227, 230
Imperial Conference 60
Imperial Defence College 96
Imperial Economic Conference 108, 109
Imperial Gift 48, 49, 56, 86
Imperial Munitions Board 40
Imperial Oil 92, 221
Ince, F/S/L Arthur 28, 40
India-Burma 270-275
Ingersoll, Ontario 204
Ingram, Cpl R.F. 257
Ingrams, F/L R.R. 148
Institute of Aviation Medicine 239
Interdepartmental Committee on Air Surveys 80
International Boundary Commission 61
Ireland, S/L E.G. 224, 226
Iroquois Falls, Ontario 202
Irvin Air Chute Ltd. 174
Irwin, G/C G.C. 163, 164
Irwin, Lt R.H. 12
Island Lake, Manitoba 65, 67
Israel 222
Iven, Herb 145
Iverson, Sgt 152
Jackson, LAC C 6
Jackson, A.J. 203
Jacobs engine 179
Jacobs, Mr. 204
James, A/V/M A.L. 68, 71
Jamieson, F/L David R.C. 216
Janney, Ernest L. 38, 39
Japan 93, 106, 113, 135, 136, 145
Jarred, F/L Arthur 137, 143, 144
Jean, L.J.R. 200, 201
Jeffreys, F/L S.R. 211
Jeffries, F/O Jeff 240, 246, 257
Jellison, S/L J.E. 80, 147
Jenkins, F.T. 180
Jenvey, David 204
Jenvey, F/O Donald E. 204, 233
Jenvey, Mrs. 204
*Jericho Beach and the West Coast Flying Boat Stations* 94
Jewell, FSgt 198
Jewett, Arthur 197-199

Johnson, LAC D.J. 188
Johnson, Frank 201, 260
Johnson, A/V/M George O. 39, 53, 97, 99, 110, 146, 161, 185, 246
Johnson, W/C J.E. 198, 215, 222, 226, 275
Johnson, F/O J.L. 63
Johnstone, F/O (1 Sqn) 195
Jolicoeur, F/O Fernand D. 273
Jolly, S/L Arnold 212
Jones, Frank 227
Jones, W/C L.L. 150, 190
Jones, W/C M. 182, 184
Jones, Cpl R.L. 257
Jones, F/O W.A. 100
Jonson, LAC E.E. 257
Jowsey, F/L Milton E. 234
Joy, D.G. 40, 172
Junkers types:
    J.I 15
    Ju.52 269
    Ju.88 129, 191, 193, 205, 207, 214, 216, 223, 225, 227, 242, 252, 269
    Ju.90 191
    Ju.188 212
    Ju.290 252
Juptner, Joseph P. 120
Kearse, Percy H. 204, 233
Keefer, W/C George 211, 225, 275
Keightley, Sgt W. 80
Kelly, F/O Charles E. 89
Kelly, Dean 226
Kemp, Gordon 236-238
Kemp, Sally 236
Kendall, F/O H.M. 161
Kenley Wing 214
Kennedy, Graham 260
Kennedy, S/L I.F. 226, 275
Kenny, Lt E.A. 39
Kenny, Samuel R. 151
Kenora, Ontario 53
Kent, F/L John A. 193, 275
Kenyon, P/O 106
Keon, Hal 232
Kerr, Jimmy 232
Kerwin, F/O John W. 105, 192, 194
Ketchikan, Alaska 136, 137, 140
Ketterson, F/L Andrew B.M. 217
Keystone Puffer 69, 87, 88
Kierstead, F/S/L R.M. 34
Kimball, F/O D.H. 198
Kimball, P/P/O G.F. 76
Kimberley Public School 270
King, FSgt (41 Sqn) 211
King, Al 158
King, Charles L. 17
King, Fred 171
King, PM Mackenzie 100
King, Sgt Ross H. 257
Kinnear, Sam 128
Kinsella, P/O William J. 234
Kipling, Rudyard 49
Kipp, Robert A. 275
Kirby, A.M. 22
Kirk, FSgt Norman E. 271

Kirkpatrick, C.J. 129
Kiska, Alaska 136
Klem, LAC 181
Klersy, William T. 275
Knapman, LAC T.S. 6
Koch, Lt Alfred 47
Koch, Tom 205
Kodak Co. 62
Kozoriz, George 133
Lac la Ronge, Sask. 89
Lac St. Jean, Quebec 61
Lachine, Quebec 125
Lake, F/L Ronald G. 197, 198, 215
Laking, John 141
Lambert, F.H. 61, 75
Lamorre, Harry 123
Lamoureux, Sgt 178
Lance, Sgt W.R. 151
Lane, George 260
Lane, Lt John 225
Langille, Pete 240, 241, 245, 246, 257
Lapaire, LAC J.M. 257
Larder Lake, Ontario 67
Larder, Clinton L. 151
Latta, Ruth 163
Laubman, F/L Don 179, 180, 216, 275
Lauder, Sir Harry 194
Laurence, R.H. 242
Laurentian Air Services 50, 93
Laurentide Air Service 107
Law, Lt 25, 26
Lawrence, F/L T.A. 73-75, 85
Lawrence, Ralph M. 128
Laycock, Maurice P. 242
Le Gear, Stanley 236, 237
Le Gear, Victor 237
Le Baron, Charles 260
Leach, Art 116
Leach, Dave 232
Leach, Lt John O. 46
Leckie, A/M Robert 51, 61, 91
Leclerc, F/L B. 202
LeGrave, Gerry 60
Leigh, Clayton 259, 260
Leigh-Mallory, G/C T. 110
Leitch, F/L A.A. 73-75, 97
Lemoins, LAC G.L. 257
Lend Lease 119
LeRoy, Robert 116
LeRoyer, Capt J.A. 58
Leslie, Lt H.T. 31
Leslie, Ralph 116
Lever Brothers 162
Levi, F/O John A. 268, 269
Lewis machine gun 21, 23, 24, 30, 31, 35, 37, 40, 107, 112, 133
Lewis, F/O 74, 75
Lewis, S/L A. 100, 106
Lewis, Peter 15
Lewis, FSgt R. 158
Leyland, Jack 201
Liberty engine 17, 61, 63, 64, 84
Likeness, F/L E.C. 217
Lincoln and Welland Regiment 151

Link Trainer 6, 123, 271
Link, Edwin A. 123, 125
Lion's Gate Bridge 94, 147
Lipp, LAC Mike 255
Little Grand Rapids, Manitoba 67, 89
Little, F/O 193
Livingstone, F/O John G.S.J. 235
Lochnan, F/O Peter W. 192-195
Lockheed types:
    Hudson 4, 113, 148-151, 156, 157, 159, 178, 179, 181, 220
    L.10A 152
    L.12 153
    Lodestar 143, 152-154
    P-38 Lightning 143, 160
    Ventura 133, 148-151, 156, 157, 190, 191, 197, 214, 267
Logan, S/L R.A. 54, 55, 71, 72
London, Ontario 62
Long Branch, Ontario 40
Lord, F/O (41 Sqn) 211
Long, Sgt (57 OTU) 197
Lord, Capt Frederick I. 37
Lord Beaverbrook 150
Lord Tweedsmuir 95
Lortie, Cpl 106
Lothian, George 148
Louden, W/C T.R. 179
*Lucky 13* 190, 224
Lumsden, Jack 125
Lymburner, Red 178
Lynch, Henry 128
Lynes, F/L J.E. 182
Macauly, Sgt 70
MacBrien, MGen J.H. 59, 60, 97
MacCarthey, F/O A.H. 126
MacCaul, P/O D.H. 97
Macchi 202 227
MacDonald Brothers 170
MacDonald, F/L E.G. 119
MacDonald, F/O H.D. 214, 224
MacDonald, LAC Ian C. 255
MacDonald, Ken 123
MacGregor, LAW Ruth 165
MacIntyre, S/L D.P. 129
MacKay, John 275
MacKellar, Doug 128
MacKelvie, F/L James A. 269
Mackenzie Air Service 139
MacKenzie, S/L A.R. 226
MacKnight, William L. 275
MacLachlan, E. 40, 91
MacLaren, Capt Donald R. 31, 39, 45
MacLaren, Sgt W.J.D. 6
MacLaurin, Clarence 40, 91, 92
MacLeod, Earl L. 91, 92, 93
MacPherson, Pat 230
Macpherson, Dr. Arnold 206
Macpherson, Bruce 206-208, 216
Macpherson, Edna 207
Macpherson, Pauline 207, 208
Madore, Garfield 144, 264

Magee, P/O John G. 225
Magor, F/S/L Norman A. 19
Magwood, F/L C. 214
Makas, Bill 181
Makepeace, Lt Reginald M. 31
Malo, FSgt R. 257
Malta 190, 207, 222, 225, 227
*Malta: The Spitfire Year 1942* 191, 222
Malton, Ontario 12, 116, 120, 168
Malvern Collegiate 270
Manning, W/C Ralph 11
Mannock, Mick 28
Markham, W/C Phillip 29
Marlatt, Stafford D. 227-230, 259, 260
Marriott, F/O J.T. 215
Marsden, F/O J.V. 202
Marsh, LAC W.A. 257
Marsh, Regie 117
Marshall, AC1 R. 62, 66
Marshall, George D. 45
Marshall, Lt H.A. 39
Marshall, T. 267
Martin B-26 143, 157, 168
Martin, Al 103
Martin, Bernard 103
Martin, F/L Bill 199
Martin, Hal G. 103
Martinside Elephant 35, 37
Mason, LAC N.W. 257
Mason-Apps, Gordon F. 61, 66
Massey, S/L Hart 225
Matthews, LAC Grant C. 188
Mattock, F/O Al 234
Maunsell, Maj G.S. 12
Maurice Farman 25, 26, 32
Mawdesley, S/L F.J. 79, 89, 95, 96, 106, 146
May, F/L L.F. 269
May, LAC Douglas G. 188
Mayer, LAC J.C.A. 257
McAllister, F/L Ross 234
McAlpine, F/O Earl J. 234
McArthur, LAC D.A. 257
McBride, LAW 162
McBride, F/O A. 242
McBride, F/O W.G. 242
McCabe, F/O Edward J. 199
McCall, Fred 47
McCallum, W.L. 201
McCandless, Dave 256
McCann, Pat 128
McCarthy, G/C 164
McCarthy, S/L J.C. 246
McCarthy, Walter J. 232, 233, 260
McCartney, Red 128
McClatchie, FSgt
McClellan, Sgt Bruce 228
McClung, F/O Vernon F. 199
McConkey, Lt T.W. 16
McCracken, Bob 201
McCrea, LAC G.W. 80
McCuaig, FSgt Eric S. 229
McCullagh, Lt 93
McCulloch, LAC Ian W. 257

McCulloch, Sgt J. 98
McCullough, John 260
McCurdy, John A.D. 12, 177
McDonald, LAC
McDonald, Alex 128
McDonald, Colin S. 78
McDonnel, S/L 125
McDonnell Douglas CF-18 Hornet 199
McDougall, G/C T.K. 203
McElroy, G.E.H. 29
McElroy, S/L John F. 200, 221, 275
McElroy, Lt Victor H. 15, 16
McElwee, Sgt W. Jr. 6
McEwen, A/V/M C.M. 39, 55, 76, 85, 107, 108, 161, 164
McFodgen, C. 67
McGill University 22
McGill, G/C F.S. 161
McGill, Lt William W. 47
McGillivary, Mr. 185
McGinnis, Lt J.A. 23
McGrandle, FSgt R. 100
McGregor, F/O Gordon R. 105, 191, 192, 217
McGregor, N.M. 201
McIntosh, J.A. 197, 198
McIntosh, LAC William A. 188
McIntosh, Wesley H. 152
McIntyre, F/O Archibald B. 271
McIntyre, P/O James I. 271
McKay, Edmund J. 127, 257, 260
McKay, Mr. 271
McKee J. Dalzell 60
McKee Trans-Canada Trophy 60, 226
McKeever, Capt A.E. 31, 39
McKegney, Sgt 245
McKell, S/L 150
McKinney, B. 116
McLean, N.B. 73
McLeod, F/O (416 Sqn) 216
McLeod, Lt A.A. 45
McLeod, F/L Donald C. 268, 269
McLeod, W/C Henry W. 126, 198, 215, 275
McLerie, Lt Allan G. 55
McManus, FSgt 178
McMillan, Don 232
McMillan, F/L Lawrence 262, 266
McMillan, FSgt R.A. 215
McMillan, F/O Stanley R. 146, 147
McNab, F/O E.A. 104, 105, 109, 100, 113, 136, 191-195
McNair, W/C Robert W. 214, 224, 275
McNaughton, MGen 97, 146
McNee, FSgt J.W. 94, 146, 147
McNeil, F/S/L Percy G. 28
McNenley, Pat 236
McNulty, Jack 103, 112, 117, 120
McPhee, F/O Bruce S. 207, 208
McQueen, LAC Donald G. 188
McWilliams, P/O Frank C. 214
Meere, Hermens 204
Menard, F/O J.J.M. 201

Mention in Despatches 16, 17, 28, 162, 202, 207
Merchant Ship Fighter Unit 137, 146, 205-207
Merritt, BC 53
Merritt, Edward M. 128
message snatching 99
Messerschmitt types:
  Me.108 212, 213
  Me.109 128, 190-193, 207, 211, 215, 223, 224, 227, 242, 252, 254, 268
  Me.110 192, 193; Me.262 198, 201, 208, 211, 215, 216, 224
meteorology 182
Metzler, Bill 206
Michalski, Sgt W.J. 104, 106
Michaud, Padre 246
Middlebrook, Martin 198
Mile 213, Nfld. 150
Miles types:
  Magister 194
  Master 127, 197, 262, 271
Military Cross 16, 22, 28, 31, 35, 37, 46
Military Medal 31
Millar, Sgt 70
Millar, F/O W.B.M. 193
Miller, LAC Gerry 257
Milligan, Sgt G.S. 191
Milliken, F/L 184
Milne, Cpl Alex J. 62, 66, 67
Miltemore, Earl 128
Milton, F/L (65 Sqn) 270
Ministry of Aircraft Production 156
Miron, F/O 197
Miron, F/L A.E. 258
Miscampbell, FSgt 178
Mission, BC 104
Mitchell, F/O (1 Sqn) 195
Mitchell, S/L J.D. 201, 223, 275
Mitchell, S/L J.F. 202
Mitchell, LAC O.A. 257
Moehne Dam 128
Moen, Ronald O. 242
Moffat, Doc 242
*Mohawks over Burma* 270
Moloney, F/O Peter J. 51, 52
Molson, S/L Hartland 135, 192
Molson, K.M. 59
Moncton, 268, 269
Monson, Cpl M. 106
Montreal 19-21, 31, 37, 45, 103, 108, 202, 217, 22
monuments and plaques 45, 188, 260, 261
Moore, Lt C.M. 37
Moore, Don 123
Moravian missions 83
More, Eric G. 128
Morfee, F/O Arthur L. 62, 71, 93, 94
Morgan, P/O James W.
Morrison, F/O (1 Sqn) 195
Morrison, S/L John D. 206

Morse, C.H. 84
Mossing, F/O B.B. 269
Mossip, F/L H.T. 258
Motor Transport Section 163-167
motorcycles 51, 224, 239
Mott, Guy E. 197
Mount Allison University 197, 267
Muegge, Herr 203
Muff, Jake 232
Mugge, Theodor 203
Mulhern, R. 116
Mullen, F/O Vernon W. 201
Mulligan, F/L 203
Mulock, Col Redford H. 28, 31
Mulvey, F/O J.B. 61
Munn, Mr. 92
Murphy, Ron 123
Murray, Const. 75
Murray, Don 158
Murray, F/L F.T. 216, 217
Murray, Sgt Russ 228, 229, 260
Mussells, S/L C.H. 269
Musson, Brian 216
Myles, Jack 262-266
Nain, Labrador 83
Nairn, G/C
Nanaimo, BC 10
Napier Sabre engine 227, 234
Napier, F/O 195
Narraway, A.M. 62
National Aeronautical Collection 22
National Archives of Canada 28, 37, 91
National Aviation Museum 15, 29, 30, 32, 45, 186
National Film Board 29
National Research Council 177-180
National Steel Car 112, 134, 157, 168, 170, 173, 174
NATO air training 40
Nault, F/L 200
Neal, F/O 195
Neal, J.F. 179
Nelles, Maj D.H. 62
Nesbitt, W/C A. Deane 136, 142, 144, 146, 192-194, 245, 246, 251, 257
Neville, LAC W.P. 257
*New Canadian Encyclopedia* 29
New Year's Day (1945) 145, 207, 216, 229, 231, 233, 242, 245, 249, 250
New Zealand 161
Newberry, F/C 33
Newbrigging, Cpl J.W. 83
Newfoundland Railway 184, 185
Newsome, Buck 136, 145
Newton, Tom 272
Niagara-on-the-Lake 260
Nicaragua 113
Nieuport types 28, 30, 32
Nijmegen 198
Nissen hut 167
Nixon, John 260
Noball 209
Nodwell, Capt R.J. 194

289

Noorduyn Aircraft 119, 170
Noorduyn Norseman 80, 82, 83, 129, 139, 146, 153, 169, 222
Norris, S/L R.W. 186, 192-194
Norsworthy, S/L Hugh H. 242
North American types:
    B-25 Mitchell 129, 130, 156-158,160, 180
    F-86 269
    Harvard 6, 118, 119, 124-126, 134, 138, 156, 164-166, 169, 182, 197, 200, 217, 262, 271, 272
    Mustang 160, 199, 267-270; Yale 119, 169, 179, 201, 269
North Bay, Ontario 53
North Sydney, NS 149
Northrop types:
    B-2 178
    Delta 80, 81, 83, 104-106, 166, 168, 177, 178
    Normad 118, 179
    XB-35 178
Northrop, John K. 178
Northwest Air Command 150
Northwest Staging Route 152, 154, 160
Northwest Territories 78, 81
Norway 190, 191, 206, 207
Norway House, Manitoba 67, 85, 89
Nottingham Island 73, 74
Noyer-Bocage 260
O'Brian, S/L G.S. 111
O'Brien, F/O J.T. 76
O'Brien, H.K. 232
O'Leary, F/L Patrick T. 224
O'Mara, Sgt Hank 200
O'Neill, F/L J.C. 269
Ocean Bridge 156
Ockenden, WO 211
Ogilvie, P/O Alfred K. 197
Ogilvie, Noel J., Sr. 78, 214
Ogilvie, Noel J., Jr. 207
Oldham, F/O Harold W. 151
Oldridge, AC1 78
Onderweegs, Mr. 204
Ontario Central Airlines 257
Ontario Provincial Air Service 38, 80, 107
Operations: Labrador 83, Noball 209, Prairie Pacific 269, Tiger Force 108,
Order of the British Empire 28, 45, 161, 162, 177
Ormshaw, Earnest 123
Ormston, Ian 260, 267
Orr, A/C Walter 152
Oshawa, Ontario 125
Ottawa 50, 53, 54, 103, 196, 226
Ottawa Car Co. 58, 95, 170
Ottawa *Citizen* 78
*Over the Front* 29
Owen, Lt E.R. 62
Owen, Lt. E.R. 180
Packard Merlin 175

Palmer, WO2 70
Palmer, P.E. 80
Pan American Airways 143
Pangborn, Clyde 157
Parkes, T. 40
Parkin, John H. 177
Parnall, F/L A.L. 129
Parrsboro, NS 177
Parry Sound, Ontario 38, 85
Parry, B.E. 201
Parry, LAC George 257, 259
Parsons, F/O L.M. 186
Parsons, WO 137
Partridge, S/L S.O. 179, 180
Pas de Calais 263
Passey, W/C 229
Passmore, Gerald L. 204, 233
Pate, Sgt W.G. 81
Paterson, John N. 221
Pathfinder Force 128
Patriarch, G/C Val 126, 200
Patterson and Hill 116
Patterson, Daphne 103
Patus, Sgt J.G.M. 200
Patuxent River 180
Peace Tower 46
Peacock, F/L J.W. 267
Pearce, FSgt Stewart W. 216
Pearkes, BGen C.R. 146
Pearson, Capt James W. 37
Pearson, Pres 232
Peberdy, Warner H. 40
Peck, Capt Brian 45
Pedigrew, F/L Pete 228
Pedley, FSgt Alf 140, 245
Peenemunde 265
Pelton, LAC R.H. 257
Pemberton, Capt F.D. 16
Penhold, Alberta 121
Pentland, Sgt Bob 137
Perkins, S/L H. 251
Perry, F/O W.R. 158
Pesant, LAC J.F.C. 256, 257
Peters, F.H. 68
Peters, S/L H.P. 269
Peterson, F/O Otto J. 192, 193
Petley, Bernie 136, 146
Petre, H.A. 12
Pfalz 17
Philadelphia 43
Phillip, F/L Robert D. 200, 225
Phillips, H.L. 226
photo/tactical recce 262-270
Picard, J. F.G.H. 200, 201
Picton, Ontario 25, 128, 129
pigeons 89, 91
Pilbrow, Pte S. 31
Pilcher, F/L Alan S. 151
Pinnell, LAC Jack 255
Pitcher, F/O Paul B. 192, 194 , 220
plaques and monuments 45, 188, 260, 261
Planche, LAC John P. 188
Pollock, F/L F.H. 227
Pollock, F/L W.R. 192, 194
Pook, P/O 180

Port Burwell, Quebec 73-75, 83
Port Nelson, Manitoba 71
Powell, G.J. 156, 158
Powell, P/O Lloyd W. 218
Power Jets Ltd. 178
Power, Hon. G.G. 150
POWs 145, 197, 200, 201, 204, 207, 209, 217, 223, 227, 229, 242, 251, 268, 269
Pratt & Whitney engines 60, 78, 177, 178
Presque Isle, Maine 133
Preston, Chris 201
Prestwick, Scotland 150-152, 155, 158, 193-195
Preziosi, Cpl P. 257
Price Brothers 50, 84
Price, A.K.
Prince Albert, Sask. 16
Prince Bernhard 150,152
Prince Rupert, BC 106, 136, 146
Princess Alice 161
Princess Patricia's Light Infantry 22
Princess Royal 161
Princeton University 61
Privy Council 59
Proctor, Ron 237
*Props on Her Sleeve* 163
Prouse, Margaret 167
Provisional Pilot Officer scheme 55, 95, 107-109
Ptolemy, William H. 44
Pubs, clubs and hotels: Anglo Swiss Hotel 185, Canadian Legion 271, Chateau Laurier 226, Lord Elgin 236, Marquis of Granby 216, Oban Inn 260, Red Lion 216, St. George and Dragon 228, The Orchard 193
Puffer, Al 128
Purdy, F/S/L Claude C. 20
Purdy, P/O Phillip H. 191
Pyrene Manufacturing Co. 174
Quebec 80, 233
Queen Victoria Hospital 162
Queen's University 128, 266
Quiddi Vidi Lake, Nfld. 153
Quigley, F/L H.S. 84
radio/wireless 99, 100, 118
Rainy Lake, Ontario 83
Ralph, Wayne 30
Ralston, J.L. 168
Rankin, Capt W.D. 192,194
Rannie, J.L. 75, 78
Rathwell, Sgt 70
Raymond, WO2 70
Raynes, F/L D.F. 147, 150
*RCAF Squadrons and Aircraft* 152
Read, F/S/L 19
Ready, Sgt J.M. 89
Rebstock, FSgt John R. 223
Red Cross 161, 163, 165, 203, 204
Redpath, F/C Ronald F. 36
Reeves, Harold 128
Reeves, R.L. 226
Regina 17, 234

Reid, Danny 210
Reid, Ross 145, 258
Reid, T.R. 71
Reilly, Maj B.R. 46
Reilly, F/O R.J. 233, 251
Reindeer Lake 62, 66
Renwick, F/L R.D. 182
Republic P-47 198, 200, 268, 272-275
Resseguier, W. 116
Rettie, F/L N.F. 269
Revelstoke, BC 53
Reyno, A/V/M E.M. 104, 126, 192, 193, 200, 267
Reynolds, Joseph B. 195
Ribble, Sgt 197
Richards, Cpl (6439 SE) 257
Richards, Cpl A. 101
Richards, F/L F.H. 216
Richards, J.C. 181
Richards, F/L W. 179
Richardson, LAC A.R. 257
Richardson, Pauline 207
Riddle, F/O W.I. 95, 106
Riess, Sgt Frederick 152
Riley, Lt George A. 47
Ritchie, F/O R.C. 269
Ritzel, F/O 152
Roach, Lynn 232
Robb, F/L J.M. 269
Roberge, Sgt 70
Robert Simpson Co. 236
Roberts, G. 158
Roberts, T. 158
Roberts, Wally 128
Robertson, F/O Alec 229
Robertson, Sgt G.D. 196
Robillard, Larry 226
Robinson, S/L E.L. 113
Robson, Cadet John 45
Roddie, W.G.D. 201
Rogers, William W. 28
Rolls-Royce engines: Eagle 21, 48, 59; Griffon 209; Merlin 112, 175, 209; Vulture 227
Rood, J.L. 148
Roode, Sgt A.L. 257
Rook, W/C 209
Roosevelt, President 156
Roscoe, F/L A.F. 214
Rose, Doug 121
Rosevear, Kenneth W. 128
Ross, F/O A.D. 59
Ross, P/O George H. 188
Ross, J.E. 75
Ross, LAC W.M. 257
Rossenti, Nick 242
Rosthern, Sask. 109
Roth, Sgt A.N. 106
Rough, Lt Herbert L. 24
Round, F/O S.A.R. 201
*Roundel* 11
Rouselle, F/L J.L.A. 268, 269
Royal Air Force/Royal Flying Corps squadrons:
    1 191, 227

3 15
4 14
5 16
8 18
9 14, 18
17 190, 270
18 34, 35
19 37, 223, 270
22 34
23 37
24 14
27 23, 24, 37
28 30
33 219
34 16, 62
35 18, 129
40 38
41 200, 209-213
46 31, 110
50 128
54 35
56 30
59 16
60 30, 35
65 270
66 226
72 207
79 38
81 39, 224
93 207
97 128
98 37
102 22, 23
106 16
107 226
111 191, 207
123 39, 209
126 225
130 210, 211, 220
132 128
135 272-275
137 209
139 268
140 187
145 222
146 145, 271
148 218
149 128
154 225
165 214, 226
174 242
177 128 201 44
183 227
193 205
203 15
207 121
208 33
214 25-27
222 205
226 133
229 227
235 205
242 222
243 224
245 258
249 191, 222, 224, 226

247 227, 228
257 190, 271
261 145
263 236-238
266 209
274 223, 225
277 209
302 219
303 193
315 190
350 210
541 128
457 190
461 148
541 221, 263, 264
544 151, 264
577 268
602 190
609 197, 226
610 222
613 190
615 193
616 222
617 128, 246
618 265, 266

Royal Air Force AFUs, OTUs, HCUs:
No.5 AFU 271
No.17 AFU
No.2 OTU 128
No.8 OTU 262
No.29 234
No.30 OTU 271
No.53 OTU 224
No.55 OTU 271
No.56 234, 242
No.57 OTU 197, 198, 204, 206, 209
No.59 OTU 204
No. 61 OTU 126, 127, 200, 202, 227, 229, 232

Royal Air Force misc. units:
No.3 EFTS 200
No.16 SFTS 202
No.21 Ferry Control 272
No.45 Group 157
No.83 GSU 200, 202, 229
No. 83 Group 210, 224
No.103 MU 223
No.124 Wing 210, 212
No.125 Wing 210, 211, 225
No.1688 Bomber Defence Training Flight 205
Aircrew Transit Pool 272
Armament School 96
Empire Test Pilots School 129
Royal Aeronautical Establishment 191
RAF Regiment 249
School of Army Co-operation 96
Staff College 96, 107, 108
Tactical Evaluation Unit 127, 198, 200, 204, 232
Women's Auxiliary Air Force 161

Royal Air Force stations/towns:
Annan 127, 271
Ashford 268
Aston Down 204, 232
B.2 Bazenville 267
B.3 Ste-Croix-sur-Mer 198, 215, 236, 268
B.4 Bény-sur-Mer 208
B.9 Lantheuil/Cruelly 240, 242, 244, 247, 249, 255
B.56 Evère 200, 201, 220, 224, 246, 268
B.58 Melsbroek 241, 252, 254
B.64 Diest 210
B.78 Eindhoven 145, 189, 210, 229, 231-233, 242, 245, 248-251, 256, 258
B.80 Volkel 210, 254
B.88 Heech 207, 216, 224
B.90 Petit Brogel 201
B.100 Goch 234
B.106 Twente 212
B.110 Osnabruck 201, 233, 235
B.114 Diepholz 202
B.118 Celle 233, 257
B.150 Hustedt 243, 257
B.152 Fassberg 254
B.154 Soltau 201, 202
B.158 Lubeck 254
B.166 Flensburg 233, 246, 250
Benson 187, 262-265
Biggin Hill 193, 209
Blackpool 262
Brunton 242
Castle Bromwich 268
Castleton 194
Colby Grange 264
Coltishall 187, 190
Copenhagen-Kastrup 212
Croydon 191
Digby 190
Drifield 195
Dunsfold 202
Dyce 214
Eschott 197, 198, 204, 209
Farnborough 187, 246
Fraserburgh 262
Hawarden 206
Hawkinge 199
Heston 227
High Wycombe 164
Honiley 127, 200
Hurn 127, 239, 245
Lympne 210
Manston 209, 210
Milfield 204, 234
Montford Bridge 127
North Luffenham 234
Northolt 191, 193
Odiham 269
Predannack 227
Rednal 126, 127, 200, 202, 229, 233
Skaebrae 198
Sumburgh 198

Tangmere 209, 222, 224, 229, 233, 236
Tealing 198
Ternhill 271
Torquay 234
Warmwell 230, 232, 242, 257
Warrington 234
Watton 262
Wellingore 206, 216, 218
West Malling 193
Wick 265
Y.32 Swartzburg/Ophoven 210

Royal Aircraft Factory types:
B.E.2 25, 32; B.E.12 10, 21
F.E.2b 26, 31
R.E.8 16, 18, 25, 26
SE.5 14, 30, 97
Royal Automobile Club 229
*Royal Canadian Air Force at War 1939-1945* 152, 204, 226
Royal Canadian Air Force squadrons:
1 (F) 104, 105, 110, 135, 191-195
2 (GP) 78
2 (AC) 99
3 (B) 100, 104-106, 112
4 (BR) 106, 107, 146, 147
5 (FB) 81
6 (BR) 106
7 (GP) 239
8 (GP) 80
10 (Aux) 111
10 (BR) 130, 147, 148, 150, 267
10 (FB) 105
11 (BR) 4, 148, 151, 239
12 (AC) 109; 12 (Comm) 152-154
14 143
15 (F) 109
18 136
110 (Aux) 111, 112, 191, 196
111 (Aux) 111
112 (Aux) 112, 191, 196
113 (BR) 113, 151, 152
115 (BR) 136, 139, 151
116 (BR) 150, 151
118 135-146, 209, 245
119 130, 152
122 151
123 239, 242
125 186
126 135
127 151
130 (F) 121
132 135
133 216
145 113, 151, 152
149 (BR) 151
150 239
161 152
163 242
164 152, 154
165 152, 154
166 152, 154

168 (HT) 152, 155
400 111, 112, 189, 196, 213, 226
401 144, 195, 206, 209, 212, 224, 225
402 112, 196, 219, 223, 224
403 214, 222, 223, 224, 226
405 128, 227
406 129
408 128, 129, 227
409 267
410 226, 264
411 196, 201, 215, 219, 224, 226
412 189, 206-208, 216-218, 222, 223, 225, 226
414 218, 267-269
416 200-202, 202, 214, 219, 223-225
417 219, 220, 223
418 196
419 129, 175
420 269
421 202-204, 214, 219-221, 224, 226
428 128
429 128
430 215, 223
435 154
436 154, 225
437 154
438 145; 439 229-231, 234, 235, 239, 245-253
439 237, 242, 255-258
440 204, 232-234, 258
441 128, 198, 199, 215, 224, 226
442 186, 199, 219, 226
443 214, 215, 226
Royal Canadian Air Force other:
 Air Historian 11
 Casualty Liaison Office 203
 Central Flying School 267
 Northern Ontario Mobile Unit 84, 85
 Overseas HQ (Lincolns Inn Fields) 162
 School of Army Co-operation 99
 Staff College 110, 269
 Technical Training School 164, 166
 Test and Development Flight 178
 Women's Division 161-167, 241
 "Y" Depot 126, 182
 "Y" Wing 146
 No.1 Canadian War Crimes Investigation Unit 203
 No.1 (P) Det 78, 80
 No.1 Photographic Establishment 90, 180
 No.1 Port Transit Unit 182-185
 No.2 (P) Det 80, 81
 No.3 PRC/PRU 126, 185, 202, 275
 No.4 CAC 149
 No.5 Canadian Hospital 193
 No.5 Equipment Depot 185
 No.6 Group 108, 161, 162, 164, 167
 No.6 (P) Det 78, 80, 109
 No.6 Repair Depot 271
 No.7 (P) Det 80
 No.7 (P) Wing 180
 No.9 (P) Det 80
 No.10 General Hospital 162
 No.12 (RD) 126
 No.22 (P) Wing 180
 No.126 Wing 217, 224, 225
 No.127 Wing 201, 219, 222, 223, 224, 254
 No.143 Wing 144, 229-261, 239
 No.144 Wing 144, 198, 215, 222, 224
 No.410 RRS 215
 No.419 RSU 244, 252
 No.6439 Serving Echelon 257
 Winter Experimental and Training Flight 179, 180, 187
 Winter Experimental Establishment 129
Royal Canadian Air Force stations/sub-stations:
 Aylmer 121, 163-166, 181, 271
 Bagotville 121, 126, 180, 197, 217, 229
 Bella Bella 106
 Botwood, 149-151
 Boundary Bay 129
 Calgary 181
 Camp Borden 6, 35, 38, 40, 48, 50, 52-58, 73, 76, 77, 86, 88, 95, 97, 101, 107-110, 117, 118, 124, 131, 227
 Chatham 269
 Cormorant Lake 59, 64, 72, 85, 88, 89
 Dartmouth 51, 52, 55, 68, 105, 136, 137, 147-152
 Debert 239, 264
 Dunnville 125, 271
 Gander 148-151, 159
 Gaspé 150
 Gimli 179, 180, 187
 Goose Bay 154, 158, 159
 Halifax 182-185
 High River 48, 58, 60, 67, 75, 84, 86, 87, 89, 91, 101, 105
 Jericho Beach 55, 60, 91-94, 97, 108
 Kapuskasing 179
 Lac du Bonnet 68, 88, 89
 Ladder Lake 72, 78, 88, 89
 Mont Joli 121, 149, 150, 182, 229
 Morley 51
 Mossbank 123
 Mount Hope 125
 Mountain View 123
 Ottawa/Rockcliffe 55, 67-70, 78, 79, 81, 99, 100, 105, 112, 114, 118, 152, 155, 180, 187, 239
 Patricia Bay 135
 Pennfield Ridge 133
 Roberval 64, 68, 84
 Sea Island 103, 104
 Shepherd 181
 Souris 181
 St. Hubert 105, 137, 156, 171
 St. Thomas 112, 164, 166
 Summerside 116
 Sydney/North Sydney 150
 Tofino 229
 Torbay 150, 186
 Trenton 6, 99, 111, 118, 122, 134, 269
 Victoria Beach 49, 59, 62, 66, 67, 85, 88
 Victoria Island 107
 Virden 117, 123, 181
 Watson Lake 181
 Windsor Mill 117, 131
 Winnipeg 55, 59, 62, 85
 Yarmouth 149, 150, 182
Royal Canadian Air Force vessels: *M.96 (Express)* 93, *M.291* 142; *Sekani* 136, *Takuli* 149
Royal Canadian Corps of Signals 73, 97
Royal Canadian Engineers 12, 15, 17, 60
Royal Canadian Military Institute 259
Royal Canadian Mounted Police 73
Royal Canadian Naval Air Service 39, 49, 50
Royal Canadian Navy 39, 149
Royal Canadian Navy Volunteer Reserve 152, 204
Royal Canadian Ordnance Corps 205
Royal Flying Corps (Canada) 37, 40-47
Royal Flying Corps (RFC) 10-47
Royal Flying Corps stations/facilities:
 Armour Heights 41, 44,
 Beamsville 41, 44, 45
 Camp Borden 40, 41
 Camp Mohawk 41
 Camp Rathburn 41
 Fort Worth 42, 45
 Hamilton 41
 Leaside 41, 44, 45
 Long Branch 41
 Toronto 41,
Royal Military College 95
Royal Naval Air Service 10, 20-22, 28, 32-34
Royal Naval Air Service squadrons/wings:
 1 28
 3 10, 28
 4 21, 32, 34
 5 32
 8 28
 10 28
 214 22
 216 19
 217 19
Royal Tours 104, 106, 110
Ruse, Jim 260
Russel, S/L D.B. 194, 215
Russell, Herbert J. 186
Russell, John B. 17
Russell, Neil G. 201
Rutledge, F/L H.H.C. 80
Rutledge, Lt W.L 39
Ryan, Sgt John H.A. 228
Ryan, Lt R.W. 39
Sadler, F/L J.A. 63, 93
Saint John, NB 262
Salt Spring Island, BC 130
Salter, Capt Ernest J. 35
Sampson, F/L Frank A. 110
Saskatoon 189
Sault St. Marie, Ontario 17, 53, 66
Sault, LAC Bob 257
Saundby, Capt R.H.M.S. 14
Saunders, Adam 259
Scaife, Lt C. Victor 47
Scaman, Sgt B.R. 214
Scharff, William Kenneth 242
*Scharnhorst* 191
Schiller, C.A. 116
Schneider, A.S. 168
Scholten, Peter 204
School of Military Aeronautics 34
School of Special Flying 54, 58
*Schrage Musik* 215
Schultz, Cpl J.F. 83
Schwab, Lloyd G. 275
Schwerin, Germany 211
Scothorn, Arthur 128
Scott, F/O Alexander G. 202-204
Scott, Angus 232
Scott, Arthur A. 202
Scott, Mrs. Clarissa 203
Scott, Donald G. 262
Scott, G/C J.S. 59, 60, 62, 67, 68, 73, 88, 89, 97, 100
Scott, Keith 201
Scratch, Sgt Don 130
Seaman, F/L John T. 267-269
search and rescue/air-sea rescue 56, 112, 149, 151, 160, 190, 207, 209, 215, 224, 225
Seath, Doug 240
Seath, F/O Larry 189
Sehl, 2Lt F.T.S. 15
Seiger, Henry 116
Selkirk, Manitoba 53
Semple, S/L G.C. 207, 234
Senneterre, Quebec 67, 78
SEPECAT Jaguar 213

Seymour, Capt Murton A. 46
Sgts Poulin, Sgt 70
Sharpe, F/L 227
Shearer, W/C Ambrose B. 51-53, 61, 97, 106
Sheppard, G/C Al 208
Sheppard, S/L J.E. 205, 207, 215
Sherk, F/O W.G. 269
Sherlock, F/L Frederick J. 234
Sherren, Percy C. 37
Sherwin-Williams 173
Shields, LAC G.F. 257
Shields, Capt W.E. 84, 86
Shilo, Manitoba 180
Ship, LAC Dave 257
Shipman, FSgt 190
Ships: *Aborjon* 184; *Andes* 184; *Argus Hill* 184; *Ausonia* 227; *Beaver* 184, *Olympic* 34; *Dutchess of Athol* 105; *Empire Kingsley* 186; *Empress of Asia* 93, 146; *Empress of Canada* 146, 206; *Empress of France* 60, 93; *Empress of Japan* 146; *Empress of Russia* 93; *Empress of Scotland (formerly Empress of Japan)* 183, 185, 239; *Empire Foam* 205; *Empire Ocean* 205; *Empire Queen* 205; *Empire Lawrence* 205, 206; *Erria* 184; *Fort Townsend* 184; *Île de France* 164, 184, 185, 252 *John A. McDougald* 149; *John Cabot* 185; *Kyle* 184,185; *Lady Nelson* 184; *Larch* 73, 75; *Louis Pasteur* 127, 157, 182-184, 201, 232, 233, 266; *Manchester Trader* 184; *Mauretania* 126, 137, 157, 164, 236; *Montcalm* 75; *Nieu Amsterdam* 183, 184, 213, 266; *Pacific Enterprise* 184; *Queen Elizabeth* 183, 207; *Queen Mary* 146,157,197, 209; *Riverview Park* 184; *Stanley* 73; *Stratheden* 262; *Strathnaver* 146, *Sydland* 184; *Uranienberg* 184; *Westmoor* 184; *William H. Gray* 184
Shiras, George 28, 29
Shook, F/C 33, 34
Short types:
    225 19
    S.23 150
    Sunderland 148
Shortreed, S/L 239
Shukis, F/O 234
Shuttleworth, LAC J.E. 257
Siberia 143
Siebel aircraft 211, 212
Siers, Tommy 177
Sikorsky XR-4 178
Silsby, Sgt 70
Silver Dart 12
Sim, F/L R.G. 198
Simard, Art 204

Simmons, George 160
Simpson, AC1 106
Simpson, Sgt Alex 133
Simpson, Bill (416 Sqn) 200
Sioux Lookout, Ontario 84, 106
*Sixty Years* 60, 152, 205
Skelding, W/C W. 254
Skidmore, LAC James W.P. 267
Slemon, A/M C.R. 80, 164
Slimon, F/O Jim 272, 275
Slocombe, George 125, 137
Small, Cyril H. 151
Small, S/L N.E. 113
Smallwood, George K.A. 128
Smith, F/S/L 33
Smith W/C Eric G. 164, 226, 234
Smith, Homer 40
Smith, Ivan W. 242
Smith, P/O Jerry 225
Smith, Ken 232
Smith, S/L R.I.A. 225, 275
Smith, R.M. 85
Smith, Capt Russell N. 47
Smith, Ted "TR" 232, 233, 260
Smith, F/O Vic 216
Smith-Barry, Maj R.R. 41
Smither, F/O R. 192
Smuck, Robert 116 Mitten, A 116
Snasdell-Taylor, P/O 180
Snider, Ted 123
Solandt, Dr. O.M. 178, 179
Sonnichsen, F/O A.K. 184
Soper, Sgt S.J. 191
Sopwith types:
    1½ Strutter 28
    Camel 10, 14, 15, 30, 33-35, 44, 97
    Dolphin 17, 37, 39
    Pup 28, 33
    Salamander 15
    Triplane 10, 28
Souris, Manitoba 120
South Africa 12, 161
Spanish Civil War 113
Sparks, Capt J.A. 17
Spartan Air Services 151, 180
Spicer, Jim 128
*Spitfire: The Canadians* 226
sports and entertainment 140, 141, 229, 234, 238, 241-243, 255, 256
Sprenger, F/O W.P. 192-194
Spurr, Larry 200-202, 269
Spurr, Nan 200
Squires, Hugh 123
St. Catharines, Ontario 232
St. Croix River 61
St. Donat, Quebec 65, 88, 148
St. Martin, P/O J.H. 58
St. Martin, Steve 47
St. Thomas, Ontario 17, 206
St. Vital, Manitoba 53
Stacey, Charles P. 11
Stalag Luft III 197, 207
Stapleford, F/L E.B. 146
Starkey, George 127

Starratt Airways 154
Stedman, A/V/M Ernest W. 21, 52, 59, 67, 68, 73, 88, 89, 177-180
Steele, Sgt W.J. 223
Steers, F. P. 78
Stelter, John 242
Stephenson, Sgt William H. 223
Sterkade 210
Stevenson, Lt Frederick J. 37, 38
Stevenson, FSgt I.T. 213
Stevenson, W/C J.G. 152
Stevenson, A/V/M L.F. 54, 106
Stevenson, P/O R.R. 148
Stewart, AC2 128
Stewart, Maj J.C. 97
Stewart, F/L Lloyd A. 208, 216
Stinson 105 156, Reliant 157
"Stitch" (6439 SE) 257
Stitt, Johnny 242
Stone, SD/L 272
Storey, Bert 123
Stover, S/L C.H. 269
Stowe, Stephen 213
Stowe, W/C W.N. 128, 136, 137, 144, 146, 209-213
Strategic Air Command 129
Straub, S.H. 201
Strickland, AC1 106
Strickland, Jack 181
Stubbs, Capt J.S. 17
Studholme, Allan 145, 146, 209
Sudbury, Ontario 205
Sueter, Cmdr Murray F. 21
Sueur, Cpl A.J. 62
Sugden, P/O A.W.E. 233
Sullivan, 2Lt Edward A. 46
Supermarine types:
    Spitfire 113, 127, 179, 180, 186, 187, 197-211, 212-227, 231, 234, 254, 262-264, 268
    Stranraer 107, 113, 146, 147, 149, 173, 178, 189
    Walrus 106, 209, 225
Survair 269
Sutherland, F/L Robert L. 270
Suttie, Sgt Grant C. 133
Sutton, A.D. 158
Swan, J.A. 208
Swanson, Carl R. 30
Sweeney, F/O J.D. 229
Sweeting, Denis 260
Sweetman, P/O Clare 152
Swift, Mel 181
Swingler, Bernard P. 242
Switzer, LAC Alexander F. 188
Sydney, NS 149
T. Eaton Co. 128
Tackaberry, G/C S/G. 102
Tailyour, S/L Keith 58
Takoradi 186
Tall, LAC 93
Tapley, F/O Rex 202
Tapping, P/O A. 54
Taverner, Cpl Ruth 162, 163
Taylor, LAC 181

Taylor, H.T. C. 232, 260
Taylor, W.R. 128
Taylor, W/O Winnifred May 162
Telfer, AC J. 101
Templeton, Lt William 91
Terroux, W/C S.A. 161
Terry, FSgt N.C. 74
test and development/experimental 88, 99, 102, 118, 149, 177-180, 239
Thatcher, W. 128
*The Aeroplane* 29, 45
*The Battle of the Airfields* 231
*The Bomber Command War Diaries* 198
*The Brave Young Wings* 11, 28
*The British Fighter since 1912* 15
*The Courage of the Early Morning* 29
"The Forest Watcher" 83
*The Hawker Typhoon and Tempest* 227
*The Kid Who Couldn't Miss* 29, 30
*The Memory of All That* 163
"The Night Bombers" 21
*The Sky Their Battlefield* 16
*The Tumbling Sky* 224, 275
*The Typhoon and Tempest* 227
*There Shall be Wings* 60
Thicket Portage, Manitoba 89
Thomas, FSgt R.I. 83, 106
Thomas, P/O W.E. 136
Thompson, Capt G.A. 53, 54
Thompson, Cpl H.I. 161
Thompson, Harry 240
Thompson, John 260
Thompson, Cpl L.S.
Thompson, Lyle 181
Thompson, Lt S.F. 35
Thornton, 2Lt 34
*Time Remembered* 163
Tinnerman Race 221
*Tirpitz* 129
Tofield, Alberta 144
Tomsett, F/L M. 151
Topographical Surveys Branch 60, 61, 68, 71
Toronto 18, 31, 34, 40, 45, 121, 137, 163, 216, 230, 255, 270, 275
Toronto Air Meet 12
Toronto *Evening Telegram* 136
Toronto *Globe* 14
Toronto *Star* 62
Towers, SLt Norman 206
Townley, Lt P.F. 39, 85
Trans-Canada Air Lines 148
Trans-Canada Air Pageant 94, 103
Trans-Canada Flight 53, 107
Trans-Canada Trophy 60, 226
Transport Canada 269
Trevana, F/O C.W. 192
Trevarrow, WO 211
Tripe, S/L Phil 210
Trochu, Alberta 134
Troup, C.R. 116
Trumley, Ken 127

Truscott, G/C Gordon 113, 178, 179, 239
Tudhope, F/L J.H. 58, 60, 75, 93, 97
Tupper, Kenneth F. 178
Turbo Research Ltd. 178
Turnbull, John H. 275
Turnbull, W.R. 177
Turner, F/L Freddie 146
Turner, Harold 128
Turner, W/C Percival S. 222, 223, 275
Turner, Sgt S.D. 83
Turner, Sir Richard 39
Tylee, A/C 53, 91
*Typhoon and Tempest: The Canadian Story* 127, 250, 255, 259
U-Boats 83, 149, 265:
    U-270 148
    U-341 148
    U-377 148
    U-402 148
    U-420 148
    U-422 148
    U-520 106, 147
    U-744 113
    U-658 113
Ucluelet, BC 147, 149
Umnak Island 143
Union Station (Toronto) 230
University of Manitoba 266
University of Toronto 23, 41, 46, 50, 179, 206, 213
University of Western Ontario 229
US Navy 143
USAAC 150
USAAF 151, 266
V-1 200, 209, 210, 263
V-2 263, 265
Vachon, Irenee 87
Valiquette, Mr. 74
Van Camp, P/O 106
Van Eckh, F/O 211
Van Nostrand, Cornelius I. 40
Van Vliet, P/O W.D. 93, 99
Vanbuskirk, LAW Lillias 165
Vancouver 53, 55, 152
Vanhee, S/L Archie 146, 147
Vankleek Hill, Ontario 189
Vann, R. 18
Varley, Fred 42
Vaupel, F/O Frederick L. 218
Vaupel, P/O Raymond 218
VE-Day 232
Verner, Lt J.F. 39
Vickers machine gun 15, 33, 103

Vickers Viking 59, 62, 66, 67, 85, 87, 92, 93, 97, 177
Vickers, F/O A.H. 239
Victoria Cross 29, 30, 45
Victoria Rifles of Canada 151
Victoria, BC 15, 24
Victory Aircraft 170, 175-177
Vidal Anson 179
Villiers, John 204, 232
Vimoutiers, France 230
Vimy Ridge 13
Vincent, Carl 129
Vincent, Lt 98
von Richthofen, Manfred 30, 36
Voss, Werner 29, 36
VW utility car 208
W. Hilchie, W. 116
Wabowden, Manitoba 67
Waco gliders 239
Waddington, Lt M.W. 31
Wakeham Bay, Quebec 73, 74
Wakeham, F/L G.G. 54, 55
Walker, F/O C.C. 60
Walker, David 161
Walker, W/C James E. 224, 275
Walker, W/O Kathleen 161
Walker, S/L Kelly 198
Walker, Larry 271
Walker, Percy 227
Walker, W. 116
Walker, A/S/O Wilhelmina 161
Walkom, F/O W.J. 207, 208, 216
Wallace, F/O (1 Sqn) 195
Wallace, F/O (115 Sqn) 146
Wallbank, Sgt 227
Waller, Denning E. 232
Walsh, Bertol T. 239-245, 257
Walton, Sgt K.W. 106
Wannamaker, Jack 123
War Service Gratuity 202, 203, 266
war art 220, 244
Ward, Frederick 128
Ward, Wally 259, 260
Ware, Jimmy 213
*Warplanes to Alaska* 152
Warrell, F/O J.L. 233
Warren, Ben W. 128
Warren, Douglas 226
Warwick, Doug 128
Wass, Gil 180
Watkins, P/O (41 Sqn) 211
Watkins, Lt L.P. 10, 14
Watson, LAC R. 254
Watson, P/O R.A. 258
Wattie, Charlie 128
Watts, Sgt J.H. 106
Waygood, Roy 128

weather service 57
Weaver, P/O Claude 223, 275
Weaver, P/O W.C. 61, 67
Webber, Cpl H.S. 256, 257
Weber, F/O R. 198
Weeks, Ernie 125
Weeks, F/O W.R. 215
Weese, Bob 123
Weicht, Chris 94
Weir, F/O (1 Sqn) 195
Weis, F/O Joseph W. 227
Wells, FSgt E.P.H. 97
Welsh, AC1 106
Wemp, Bert 136
West, WO C.E. 258
Westdale Collegiate 125
Westenra, S/L 270
Western Air Command 134
Western Canada Aviation Museum 50
Westlake, Jack 128
Westland types:
    Lysander 112, 173, 174, 178, 179
    Wapiti 100, 104, 105, 106, 112, 178
Wharry, F/O G.G. 258
Whealy, Arthur T. 10, 15
Wheeler, A.O. 75
"White House" (Rockcliffe) 180, 181
White, Sgt Allen Rene 228
White, Fred 259
White, Lt Harold A. 37
Whitehorse, Yukon 160
Whiteside, Arthur B. 22, 23
Whitford, Lt J 39
Whitman, Sgt William C. 152
Whitmore, Robert W. 151
Whitney House, Ontario 84, 85
Whittle, Frank 178
Wight, F/O J.R. 147
Wigley, Cpl C.G. 257
Wile, Floyd 128
Wilkins, G/C F.S. 130
Wilkinson, F/O J.F. 213
Williams, F/L 151
Williams, F/L Alvin T. 191
Williams, W/C D.J. 129
Williams, Gordon S. 93, 103
Williams, W/C J. Scott 35, 55, 58
Williams, John F.H. 271
Williams, K.J. 201
Williams, P/O Ken 200
Williams, LAC R. 254
Williams, T.F. 178
Wilson, S/L 106
Wilson, F/L B.H. 264

Wilson, Benjamin F. 100
Wilson, P/O F.A. 223
Wilson, J.A. 50, 59
Wilson, S/L Pete 136, 144, 145
Wilson, R.H. 232
Wiltshire, P/O H.D. 58
Windle, P/O Bob 273, 274
Windsor, F/L B.H. 54
Windsor, Sgt K.D. 214
*Wings* 29
Wings Parade 124-126, 200
Winnipeg 55, 66, 105, 137, 264, 266
Winny, FSgt W.J. 79, 106
Winship, FSgt 70
Winters, Cpl (143 Wing) 248
Wise, Lou 271
Wise, S.F. 11, 14, 20, 29, 46
Wiseman, S/L 180
Wolch, Sgt Theodore B. 188
Wonnacot, F/L G.W. 268, 269
Woodland Cemetery 152
Woodland, LAC Arthur 164, 167
Woodland, LAW Rosalie 163-167
Woodman, FSgt 211
Woods, Gerald 123
Woodstock, Ontario 18
Woodward, F/L Norm 236, 237
Woodward, Vernon C. 190, 275
*Woody: A Fighter Pilot's Album,* 190
Woolett, W.
Woolley, S/L 211
Wray, F/L Larry 178
Wreford, Lt W.J. 18
Wright brothers 66
Wright engines: 63, 73, 177, 178
Wright Field 178
Wright, Donald V.
Wright, F/O John W. 127
Wrong, F.H. 67
Wytsma, Ted 256
Yarmouth, NS 149, 150
Yellowknife, NWT 81
Yorkton, Sask. 120
Young, F/L 151
Young, P/O F.B. 215
Young, Sgt L.R. 6
Younge, F/O J.C. 269
Yuile, S/L A.M. 135, 136, 145, 146, 192-194
Zak, LAC M.P. 257
Zeppelins: 10, 14, 10, 13, 14, 20, 28, 66
Zigayer, Cpl E.A. 255, 257
Zobel, F/O 202

294

Another great BCATP class photo, this time with sprog LACs training at Oshawa in March 1943. (Bill Baggs Col.)

Mosquito Mk.XX KB187 was built in Toronto by de Havilland Canada in March 1944 and was in a batch for the USAAF -- the American serial appears above the Canadian one. (DHC, Joe Holliday)

A trio of prewar RCAF Tiger Moths formates over RCAF Station Sea Island, Vancouver in 1935. (Gordon S. Williams)